Handbook of Industrial Catalysts

FUNDAMENTAL AND APPLIED CATALYSIS

Series Editors: **M. V. Twigg**
Johnson Matthey
Catalytic Systems Division
Royston, Hertfordshire, United Kingdom

M. S. Spencer
Department of Chemistry
Cardiff University
Cardiff, United Kingdom

CATALYST CHARACTERIZATION: Physical Techniques for Solid Materials
Edited by Boris Imelik and Jacques C. Vedrine

CATALYTIC AMMONIA SYNTHESIS: Fundamentals and Practice
Edited by J. R. Jennings

CHEMICAL KINETICS AND CATALYSIS
R. A. van Santen and J. W. Niemantsverdriet

DYNAMIC PROCESSES ON SOLID SURFACES
Edited by Kenzi Tamaru

ELEMENTARY PHYSICOCHEMICAL PROCESSES ON SOLID
SURFACES
V. P. Zhdanov

HANDBOOK OF INDUSTRIAL CATALYSTS
Lawrie Lloyd

METAL-CATALYSED REACTIONS OF HYDROCARBONS
Geoffrey C. Bond

METAL–OXYGEN CLUSTERS: The Surface and Catalytic Properties of
Heteropoly Oxometalates
John B. Moffat

SELECTIVE OXIDATION BY HETEROGENEOUS CATALYSIS
Gabriele Centi, Fabrizio Cavani, and Ferrucio Trifirò

SURFACE CHEMISTRY AND CATALYSIS
Edited by Albert F. Carley, Philip R. Davies, Graham J. Hutchings,
and Michael S. Spencer

A Continuation Order Plan is available for this series. A continuation order will bring delivery of each new volume immediately upon publication. Volumes are billed only upon actual shipment. For further information please contact the publisher.

PREFACE TO THE SERIES

Catalysis is important academically and industrially. It plays an essential role in the manufacture of a wide range of products, from gasoline and plastics to fertilizers and herbicides, which would otherwise be unobtainable or prohibitively expensive. There are few chemical- or oil-based material items in modern society that do not depend in some way on a catalytic stage in their manufacture. Apart from manufacturing processes, catalysis is finding other important and ever increasing uses; for example, successful applications of catalysis in the control of pollution and its use in environmental control are certain to increase in the future.

The commercial importance of catalysis and the diverse intellectual challenges of catalytic phenomena have stimulated study by a broad spectrum of scientists, including chemists, physicists, chemical engineers, and material scientists. Increasing research activity over the years has brought deeper levels of understanding, and these have been associated with a continually growing amount of published material. As recently as sixty years ago, Rideal and Taylor could still treat the subject comprehensively in a single volume, but by the 1950s. Emmett required six volumes, and no conventional multivolume text could now cover the whole of catalysis in any depth. In view of this situation, we felt there was a need for a collection of monographs, each one of which would deal at an advanced level with a selected topic, so as to build a catalysis reference library. This is the aim of the present series, Fundamental and Applied Catalysis.

Some books in the series deal with particular techniques used in the study of catalysts and catalysis: these cover the scientific basis of the technique, details of its practical applications, and examples of its usefulness. An industrial process or a class of catalysts forms the basis of other books, with information on the fundamental science of the topic, the use of the process or catalysts, and engineering aspects. Single topics in catalysis are also treated in the series, with books giving the theory of the underlying science, and relating it to catalytic practice. We believe that this approach provides a collection that is of value to both academic and industrial workers. The series editors welcome comments on the series and suggestions of topics for future volumes.

Martyn Twigg
Michael Spencer

Lawrie Lloyd

Handbook of Industrial Catalysts

Lawrie Lloyd

 Springer

Lawrie Lloyd
Court Gardens 11
Bath
United Kingdom

ISSN 1574-0447
ISBN 978-0-387-24682-6   e-ISBN 978-0-387-49962-8
DOI 10.1007/978-0-387-49962-8
Springer New York Dordrecht Heidelberg London

Library of Congress Control Number: 2011931088

Printed on acid-free paper

Springer is part of Springer Science+Business Media (www.springer.com)

PREFACE

The use of catalysts in chemical and refining processes has increased rapidly since 1945, when oil began to replace coal as the most important industrial raw material. Even after working for more than 35 years with catalysts, I am still surprised to consider the present size of the catalyst business and to see how many specialist companies supply different operators. Now that each segment of the industry is so specialized no single organization is able to make all of the catalyst types that are required. The wide range of catalysts being used also means that it is difficult to keep pace with the details of every process involved. Unfortunately, there are few readily available comprehensive descriptions of individual industrial catalysts and how they are used. This is a pity, since catalysts play such an important part in everyday life.

Modern catalyst use was unimaginable a hundred years ago because catalysts were still chemical curiosities. The use of catalytic processes simply increased with the demand for new products and gradual improvements in engineering technology. Only now is it becoming true to say that catalyst design, which originally relied on luck and the experience of individuals, is becoming a more exact science. New construction materials have made plant operation more efficient and led to the development of better processes and catalysts. It is no coincidence that the two major wars of the twentieth century saw the rapid expansion of a more sophisticated chemical industry. Currently, some new catalysts are evolving from previous experience while others are being specifically designed to satisfy new consumer demands. This is demonstrated by the introduction of catalysts to reduce automobile exhaust emissions in response to environmental regulations. This has been one of the major catalyst growth areas of the past 20 years and the use of catalysts to control various industrial emissions is similarly important.

The demand for catalysts is still increasing particularly in the Far East, as expansion of the chemical and refining industries keeps pace with the increase in world population. As a consequence, the number of catalyst suppliers is still growing. All have the experience needed to produce large volumes of catalysts successfully and can give good advice on process operation, but different catalysts for the same applications are not always identical.

Ownership of key patents for catalysts and catalytic processes has led to licenses being offered by chemical and engineering companies. For this reason precise catalyst compositions are not often published, and while commercial products may seem to differ only in minor details, in a particularly efficient manufacturing process these can certainly improve performance. There are no catalyst recipe books, and details regarded as company secrets are hidden in the vague descriptions of a patent specification.

Competition among suppliers in a market where customers may only place large orders every few years has encouraged overcapacity in order to meet emergency requirements. At the same time, low selling prices and the high costs of introducing new products have reduced profitability. The recent spate of catalyst joint ventures reflects this.

Availability of reliable products must be guaranteed so that a customer's expensive plant will not have to close down or operate at a loss. Security of supply is clearly a major factor in catalyst selection. Indeed, for many years it was a strategic or political necessity as well as being of commercial importance. For instance, during the ColdWar era, most of Eastern Europe and China had to rely on their own domestic production capacity. At the same time, the big chemical companies in the United States and Europe, which had traditionally produced their own catalysts, began to buy the best available commercial products.

Since Sabatier published Catalysis in Organic Chemistry in 1918 many process reviews have been written on the industrial applications of catalysts and they provide a good deal of historical background. Lack of detail has meant, however, that catalyst compositions are not often included. In any case, earlier reviews are usually out of print and can only be found with difficulty from old library stock. Up-to-date information is badly needed.

Catalysts could, by definition, operate continuously, but those used industrially may lose activity very quickly. Some catalysts can then be regenerated at regular intervals by burning of carbon deposited during operation. Others have to be replaced following permanent poisoning by impurities present in the reacting gases. To avoid the necessity for parallel reactors or unscheduled interruptions to replace spent catalyst, efficient operating procedures have had to be devised for online regeneration or the removal of poisons from feedstock. The use of additional catalysts or absorbents to protect the actual process catalysts has become an important feature of operation. Catalysts are also deactivated by overheating. This sinters either the active catalyst or the support and occurs if the operating temperature is at the limit of catalyst stability, particularly in the presence of trace impurities in feedstock. Other problems can result from increasing pressure drop through the catalyst bed, if dust is entrained with process gas or if the catalyst itself slowly disintegrates.

It may therefore be necessary to replace catalysts many times during the life of plant equipment. Stability despite the presence of poisons becomes an important feature of the selection procedure to avoid unscheduled plant closures. Proper catalyst reduction may also be a critical step prior to operation to ensure optimum performance in the shortest possible time. This is not always easy and efforts have therefore been made to use prereduced catalysts and even to regenerate spent catalysts externally to restore as much of the original activity as possible. It should never be assumed that catalyst operation is straightforward. It is often a nightmare. And effort spent in solving problems or making improvements is time consuming. The provision of an efficient technical service has thus become an indispensable element of the catalyst business.

It is hoped that this extensive survey of industrial catalysis will stimulate a wider general interest in the subject.

The author thanks J.R. Jennings, M. S. Spencer, and M.V. Twigg for much help in bringing this book to publication.

Lawrence Lloyd
Bath, England

CONTENTS

Chapter 1

Industrial Catalysts

Chapter 2

The First Catalysts

Chapter 3

Hydrogenation Catalysts

Chapter 4

Oxidation Catalysts

Chapter 5

Catalytic Cracking Catalysts

Chapter 6

Refinery Catalysts

Chapter 7

Petrochemical Catalysts

Chapter 8

Olefin Polymerization Catalysts

Chapter 10

Ammonia and Methanol Synthesis

Chapter 11

Environmental Catalysts

1

INDUSTRIAL CATALYSTS

1.1. INTRODUCTION

The first industrial catalyst was probably the niter pot, which was used in the early sulfuric acid lead chamber process when it became known that oxides of nitrogen catalyzed the oxidation of sulfur dioxide. How was this important process—on which chemical development soon depended—discovered? Was it from the observation that cannons corroded or that condensation was acidic following the explosion of gunpowder? All the ingredients for chamber acid were there—sulfur, saltpeter, atmospheric air, and heat. Ostwald noted that "copious brown fumes" were evolved as gunpowder exploded, but did not make any comment on sulfur oxides.[1] Empirical observations, or inspired deductions, during the 1800s led to the introduction of several more important catalytic processes. The inevitable development of a chemical industry based on the use of catalysts followed from a mass of experimental observations, such as those shown in Table 1.1, accumulated after Berzelius[2] defined catalysts in 1835 (Figure 1.1).

Although the first catalyst was a gas, there are only a few homogeneous catalysts in use today. Most industrial catalysts are solids and operate heterogeneously in gas or liquid phase reactions.

Most of the basic ideas of industrial catalysis gradually evolved during the early period of development. The use of particular groups of metals for hydrogenation and oxidation reactions was investigated first in the laboratory and then industrially. Simple reactors with better control of operating conditions were

L. Lloyd, *Handbook of Industrial Catalysts*, Fundamental and Applied Catalysis,
DOI 10.1007/978-0-387-49962-8_1, © Springer Science+Business Media, LLC 2011

Figure 1.1. Portrait of Baron Berzelius.

introduced. The new processes accelerated the use of better steels and high-pressure technology, which, in turn, led to the development of further catalytic processes.

It was soon realized that in many reactions a support for the active metal not only made the catalyst more active and stable but also reduced the cost of the final product. Early supports for metal catalysts were natural or refractory materials such as asbestos, pumice, quartz, corundum, activated carbon, clays, firebrick, and kieselguhr. Even during the period from 1950 through the 1970s, graded river pebbles were often used as catalyst bed supports, and the original catalytic cracking catalysts from the 1930s were based on natural clays. Ammonia synthesis, one of the earliest large-scale industrial processes, still uses a granular catalyst made by the fusion of a pure natural magnetite.

The gradual evolution of more reliable supported metal catalysts required reproducible supports. Industrial processes for the production of pure alumina and silica were soon developed. This led naturally to the control and measurement of chemical and physical properties at all stages of catalyst production to ensure optimum surface area and pore structure. Controls were at first empirical, and quality depended on consistent production conditions. It was not until 1938 that techniques for measuring surface area and pore volume were introduced and modern methods of catalyst quality control and characterization began to evolve.[21]

TABLE 1.1. Some Examples of Catalysis before 1925.

Date	Process	Reference
1740	Sulfur dioxide oxidation in glass bell jars.	Ward[3]
1746	Sulfur dioxide oxidation in lead chamber.	Roebuck[4]
1788	Oxidation of ammonia to nitrogen oxides over manganese dioxide.	Milner[5]
1812	Hydrolysis of starch to glucose in acid solution.	Kirchoff[6]
1817	Ignition of combustible gases, such as coal gas, in air over hot platinum wire.	Davy[7]
1823	Absorption and combustion of ethanol to give acetic acid over spongy platinum—also combustion of hydrogen over spongy platinum (Dobereiner's Tinder Box).	Dobereiner[8]
1826	Reaction of hydrogen and chlorine over platinum.	Turner[9]
1831	Oxidation of sulfur dioxide with air over platinum.	Phillips[10]
1831	Platinum poisoned by hydrogen sulfide and carbon monoxide.	Henry[11]
1836	First definition of a catalyst.	Berzelius[2]
1839	Oxidation of ammonia to nitrogen oxides over platinum sponge at 300^0C.	Kuhlmann[12]
1860–1870	Oxidation of hydrochloric acid to chlorine over copper chloride.	Deacon[13]
1875	First sulfuric acid contact process plant making oleum from lead chamber acid.	Squire and Messel[14]
1876	Removal of sulfur and arsenic poisons from feed to Deacon process.	Hasenclever[15]
1888	Steam reforming of hydrocarbons over nickel oxide /pumice.	Mond and Langer[16]
1888	BASF operated first pyrites contact process plant using platinum catalyst.	
1889	First plant for partial oxidation of methanol to produce formaldehyde used platinized asbestos but changed to copper oxide.	Trillat[17]
1894	Sulfur recovery from reaction of hydrogen sulfide and sulfur dioxide over alumina catalyst.	Chance and Claus[18]
1898	BASF developed two bed contact process (first bed iron oxide/second bed platinum catalysts).	
1899	Hydrogenation of vegetable oils.	
1901–1904	Development of ammonia oxidation with platinum catalysts.	Oswald and Brauer[19]
1904	Haber starts work on ammonia synthesis osmium catalyst.	
1905	First nitric acid plant at Bochum (300 kg day^{-1}).	
1912	Polyvinyl chloride.	
1912	Patent for iron oxide/chromium oxide carbon monoxide conversion catalyst.	Wild
1913	First patent of methanol synthesis process.	Mittasch and Schneider
1914	First synthetic ammonia plant at Oppau.	Bosch and Mittasch[20]
1922	Fischer–Tropsch process.	Fischer and Tropsch[20]
1923	First synthetic methanol plant at Merseberg.	Pier and Winkler[20]

When using catalysts industrially it is important that the shape and size of the particles selected provide a proper balance between activity in the process and pressure drop through the reaction vessel. Thus, process design plays an important role in catalyst development. As catalysts are used and handled in increasingly large quantities, physical strength is one of the common factors in selecting any of the available shapes shown in Table 1.2.

Some catalysts can now be regarded as mini-reactors and are designed that way. For example, the auto exhaust catalyst is supported on a monolith small enough to fit underneath an automobile. On a molecular scale, metallocene compounds are single-site catalysts that are now being used to make poly-olefins more selectively. It is probably not necessary to emphasize that the industrial catalysts used in chemical and refining processes are not the same as the catalysts of theory. They all have well-defined features related to the basic demands of the process in order to achieve predictable and economic operation. These are shown in Table 1.3.

Catalyst manufacture is a specialized operation with producers working continuously to improve performance and quality. More than 90% of today's chemical and refining processes use catalysts. The world is dependent on cata-lysts for food, fuel, plastics, synthetic fibers, and many other everyday commodities, and there is no way that modern life would be the same if they were not available. Even so, operators have often ignored new refinements in various catalytic processes and have continued with a relatively inactive catalyst

TABLE 1.2. Common Catalyst Shapes and Sizes.

Shape	Process
Powder	Fluid catalytic cracking. Acrylonitrile production.
Granules	Ammonia syntheses.
Balls/spheres	Desulfurization. Hydrogenation. Catalytic reforming.
Pellets	Reactions in adiabatic beds.
Extrusions	Hydrodesulfurization. Sulfuric acid.
Rings	Hydrocarbon steam reforming. Butane oxidation.
Flakes	Vegetable oil hydrogenation.
Gauze	Nitric acid. Hydrogen cyanide.
Monoliths	Automobile exhaust purification.
Alloys	Hydrogenation.

TABLE 1.3. Essential Catalyst Properties.

Property	Effect
Activity	Rapid conversion of feed to required products at moderate operating conditions.
Selectivity	High proportion of required products compared with by-products.
Stability	Ability to resist thermal deactivation during operation.
Poison resistance	Ability to tolerate (absorb) trace impurities and maintain reasonable activity.
Strength	Physical ability to resist breakdown or excessive dust formation during handling and operation.

in an existing plant rather than spend money on new equipment. The advantages of a new process cannot be ignored indefinitely, but the fact remains that many plants will operate at equilibrium with an out-of-date catalyst, and this can still be cost effective relative to additional capital expenditure.

1.2. WHAT IS A CATALYST?

It is usual to define a catalyst as a substance that increases the rate of a chemical reaction but is not consumed in the process. This definition must be qualified because a catalyst cannot change the thermodynamic equilibrium of a reaction during operation. Rather, the role of a catalyst in industry is to accelerate the rate of reaction toward chemical equilibrium in processes to improve the process economics.

Industrial catalysts are often produced as oxides that may need activation before use. The principal catalytic components are intimately mixed with other components, usually by co-precipitation or impregnation from solution. The other components may act as promoters or supports. The role of the support may simply be to provide a porous framework on which the active materials are dispersed, but they also can have a key role in enhancing the lifetime of a catalyst, either by preventing loss of active surface area due to sintering or by absorbing traces of poisons. Many catalysts such as the platinum gauzes for nitric oxide production and Raney nickel, however, are already in the metallic form when supplied.

The metal oxide catalysts used for hydrogenation reactions are reduced to an active form of the metal before use. Apart from metallic platinum and silver, which are used to oxidize ammonia and methanol, respectively, oxidation catalysts are usually transition metal oxides. Acidic oxides, such as alumina, silica alumina, and zeolites are used in cracking, isomerization, and dehydrogenation reactions. These are only a few examples of the catalysts now being widely used. A more detailed list is given in Table 1.4.

1.2.1. Activity

Catalyst activity may be regarded practically as the rate at which the reaction proceeds on the catalyst volume charged to a reactor. The turnover number, or frequency, is the number of molecules of product produced by each active site per unit time under standard conditions. Because it is not practicable to calculate the active sites on a commercial catalyst, it is easier to measure the space-time yield (the quantity of product produced per unit volume of catalyst per unit time) during industrial operations. The space-time yield, or activity of a catalyst, determines the reactor size for a particular process. The ratio of volumetric gas flow per hour to catalyst volume under the design conditions chosen is known as the space velocity through the catalyst bed.

TABLE 1.4. Some Typical Catalysts Used in Industrial Processes.

Process		Catalyst
Hydrogenation	1	Nickel metal on a suitable support. (Precursor oxide reduced before use.)
	2	Raney nickel. (Aluminum extracted from nickel/aluminum alloy before use.)
	3	Precious metal deposited on a support such as carbon, silica, or alumina. (Usually no pretreatment required).
Dehydrogenation	1	Iron oxide promoted with chromium oxide and potassium carbonate.
	2	Chromium oxide on alumina support.
	3	Calcium nickel phosphate.
	4	Mixed copper oxide/zinc oxide. (Copper oxide reduced before use.)
Oxidation	1	Vanadium pentoxide and potassium sulfate supported on silica.
	2	Platinum/rhodium (10%) gauze.
	3	Iron oxide/molybdenum oxide.
	4	Silver supported on -alumina.
	5	Promoted bismuth molybdate.
Refining processes	1	Silica/alumina.
	2	Zeolites supported on matrix.
	3	Cobalt or nickel molybdates. (Oxides sulfided before use.)
	4	Platinum, often promoted with rhenium, iridium, or tin, on chlorinated alumina.
	5	Nickel or palladium supported on zeolite.
	6	Phosphoric acid supported on kieselguhr.
Ammonia and methanol production	1	Nickel supported on alumina or calcium aluminate rings.
	2	Magnetite promoted with chromia.
	3	Copper supported on alumina and zinc oxide.
	4	Nickel supported on alumina or special support. (All reduced before use.)
	5	Iron promoted with potash, alumina, calcium oxide.
	6	Copper supported on alumina and zinc oxide. (Precursor oxides reduced before use.)

1.2.2. Selectivity and Yield

The conversion of reactants to products and by-products during a chemical process is easily determined from the mass balance. The selectivity of a reaction is defined as the proportion of useful product obtained from the amount feedstock converted. Thus it is possible to obtain almost 100% selectivity and still have an uneconomic process if the conversion is very low. Many processes operate at less than 100% conversion to limit heat evolution, to achieve higher selectivity or because of thermodynamic limitations. In these instances, the unconverted feed must be recycled and conversion per pass can still be relatively low, but economic. The key parameter in these instances is the yield of the reaction, which is the conversion multiplied by selectivity.

An important advantage of using catalysts for any reaction is that the milder operating conditions give better selectivity. Low-selectivity catalysts are uneconomic, not only because feed is wasted and by-products have to be separated from products, but also because side reactions are often more exothermic and complicate reactor design. Although the formation of by-products must, in general, be prevented there are many examples of by-product sales becoming an important source of income. Often, when more efficient processes were developed, it was commercially attractive to introduce a process for making the by-product.

1.2.3. Stability

It is normal for catalysts to lose some activity and selectivity over their operational lifetime before finally needing replacement. However, certain aspects of maloperation can cause premature damage to a catalyst, leading to premature replacement. This is possible when:

- The catalyst is overheated and surface area decreases.
- A volatile component is lost at high operating temperature.
- Poisons in the feed deactivate the catalyst.
- The catalyst is overheated and active sites coalesce.

Catalysts often have short lives for any of these reasons and efforts must be made to obtain a more stable alternative or to prevent deactivation by modifying operation.

The performance of most catalysts can deteriorate during relatively short periods of maloperation, so the expected performance should be checked at regular intervals. By recording operating details such as feed and effluent composition and temperature profiles in the catalyst bed it is possible to assess abnormal operating features or feed purity. Appropriate adjustments can then be made. Some catalysts can be regenerated in situ while activity can be restored in others by identifying temporary poisons in the feed.

1.2.4. Strength

Plant operation will be adversely affected as pressure drop increases through the catalyst bed. This may be due to:

- Dust formed from catalyst disintegration during changing and operation.
- Entrained liquid cementing the catalyst.
- Entrained solids blocking the bed.
- Collapsed beds following mechanical damage to bed supports.
- Carbon formation from organic feedstocks.

These problems generally affect the temperature profiles in the bed and, possibly, the overall reaction. The catalyst must be strong enough to resist various forms of regeneration or reactivation.

1.3.   CATALYST PRODUCTION

Catalysts are produced in different ways depending on the chemical formulation and the severity of the chemical process in which they will be operated. The usual methods are listed in Table 1.5. Figures 1.2 and 1.3 show typical catalyst production facilities (See also Table 1.9 where some unit operations used in catalyst manufacturing are listed).

TABLE 1.5. Preparation and Application of Industrial Catalysts.

Preparation	Application	Catalyst
Impregnation	Suitable supports impregnated with soluble salts of the catalytic metals which are decomposed to oxides and reduced before use.	Catalysts containing small amounts of precious metals or easily impregnated amounts of base metals.
Precipitation	Carbonates/hydroxides of catalytic metals precipitated, decomposed and pelletted, extruded or granulated before reduction.	Catalysts containing high concentrations of base metals, which are required in a particular physical form.
Fusion	Metal oxides fused and chill cooled before crushing and sieving to required size.	Only used in relatively few catalysts.
Metals	Metal catalysts alloyed with a metal soluble in alkali. Catalyst active after removing soluble material. Alternatively, thin wires of catalytic metals can be woven into fine mesh and used directly. Metals occasionally used directly in granule form.	Only used in special applications.

Figure 1.2. Semi-technical unit for catalyst preparation, an intermediate stage in scale-up from laboratory to full-scale manufacture. Reprinted from *Catalyst Handbook*, 2nd ed., Ed. by M. V. Twigg, Wolfe Publishing, LTD., London, England, 1989, by kind permission of M. Twigg.

Figure 1.3. Typical production line for manufacture of catalysts by precipitation.

Many of the earliest catalysts were based on natural products or porous refractory materials that were available commercially, but improved catalysts were especially made as large-scale processes developed. Table 1.6 shows some of the materials that have been used as support.

The principal objective in catalyst production, however, has usually been to make a suitable catalyst for any large-scale process by the most reasonable economical procedure. Catalysts with a specific chemical composition have been established and appropriate standards for such physical properties as particle size and strength have been developed for most commercial processes. Target specifications provide for reasonably predictable operation at an acceptable pressure drop when used in standard plant equipment. After the initial period of deactivation, provided that the decline of catalyst performance is very slow and predictable, a process can be designed that is economic over the life span of the catalyst. A typical catalyst specification is shown in Table 1.7.

The most frequently used methods to produce catalysts are precipitation and impregnation. In both processes the catalyst precursors are usually converted to oxides by heating and the powders converted to solid granules or pellets. Catalysts often contain promoters that are added during the preparation stages. If

TABLE 1.6. Catalyst Supports.

Material	Function
	Essential properties of support:
Inert	Should not react with the active catalyst or take part in reaction.
Strong	Should not disintegrate during handling or use; low attrition loss.
Porous	Disperses the active catalyst to increase the activity and reduce the cost of expensive material.
	Early supports
Asbestos	Contact process.
Pumice	Deacon process, hydrogenation catalyst supports.
Kieselguhr (infusorial earth)	Fat hardening, hydrogenation catalyst support.
Bauxite/titanium dioxide	Dehydration reactions, catalyst support and cracking catalyst.
Carbon	Support for precious metals.
Metal salts, e.g., $MgSO_4$; $MgCl_2$	Contact process; olefin polymerization.
Quartz lumps	Used as an inert support for catalysts and also as a physical support at the bottom of a catalyst bed.
	Modern Supports
α-Alumina	Gamma and α-alumina used as a catalyst support in many reactions.
Silica	Produced as a pure catalyst support in a number of forms.
Activated carbon	Still used as a catalyst support in some processes.
Silica/alumina	Only used in special applications now that it has been replaced by zeolite in cracking reactions.
Cordierite	Used as preformed monolith in automobile exhaust treatment and other similar applications.

TABLE 1.7. Typical Catalyst Specifications.

Property	Test
Chemical analysis	Major components: wt% (either maximum/minimum level or acceptable range.)
	Impurities: wt% or ppm (maximum permitted level.)
	Loss on ignition: wt% (sample calcined to a loss-free basis at an appropriate temperature.)
Physical properties	Particle shape: pellets, granules, rings, extrusions.
	Particle size: length, diameter.
	Bulk density: $kg.liter^{-1}$; $lb.ft^{-3}$
	Crushing strength: lb or kg nominal.
Micromeritics	Surface area: $m^{-2}g^{-1}$
	Mercury density: $g.cm^{-3}$
	Helium density: $g.cm^{-3}$
	Mean pore radius: Å

Testing is usually done on a bulked representative sample from the daily production.

necessary, a support may be included with the solutions used during precipitation, as well as being impregnated directly with oxide precursors.

Many catalysts can be made by either of the two methods. Precipitation is usually chosen if a support is not porous enough for sufficient active metal to be loaded by impregnation. On the other hand, low concentrations of expensive precious metals are impregnated on to suitable supports particularly when they can be deposited on the surface of the support for greater efficiency. Conditions must be carefully controlled during catalyst production to give consistent quality. The checks carried out during all stages of production are listed in Table 1.8.

TABLE 1.8. Routine Testing During the Production of an Industrial Catalyst.

Test	Properties
Routine chemical analysis (products and raw materials)	Major metal components, metal impurities, well-mixed phases.
Crushing strength	Compression strength between faces (vertical) or across diameter (horizontal).
Attrition/tumbling loss	Dust formed during rotation in a tube under standard conditions.
Micromeritics	Surface area, helium and mercury density, pore volume, mean pore radius.
Particle size	Average length of pellets, average diameter of spheres, average length of extrusions.
Particle size distribution	Proportion of required, undersize, and oversize particles.
Particle porosity	Internal volume available during impregnation and operation.
Bulk density	Indication of pelletting or granulation efficiency.
Catalyst activity and stability	Determines preliminary reduction and operating efficiency.

Porosity and surface area are probably the most important properties of a catalyst as they control access to the active sites of the catalyst. Pore size and surface areas can be moderated in a number of ways to control selectivity and access of large molecules to the catalyst. It is interesting to remember that 500 g of catalyst with a surface area of 100 $m^2.g^{-1}$ has a total area of about 12 acres. A catalyst charge of 40 tonnes therefore has a total surface of almost 1 million acres. An amazing figure!

1.3.1. Precipitation

Aqueous solutions of the metal salts, usually nitrates or sulfates, are precipitated with an alkali such as sodium carbonate. Ammonium carbonate can be used to avoid residual sodium impurity, but it is relatively expensive. Occasionally precipitation conditions are controlled to form complex catalyst precursors and higher-activity products. If necessary, any support material can be added as a powder before the alkali is added.

When the precipitate has formed and settled, it is filtered and carefully washed before drying. Dried *mud* is then calcined at a temperature in the range $300^0 - 450^0C$ to decompose hydrates and carbonates. The calcined mud can then be grounded to powder and densified with a suitable lubricant before being pelletted. Many important practical details are involved in precipitating catalysts. The following are some points to remember:

- It is necessary to precipitate rapidly and maintain a uniform pH. This produces small active crystallites and well-mixed oxides in the finished catalyst. Under these conditions specific compounds form such as those described by Feitknecht, with the composition $(M^{2+})_6(M^{3+})_2 (OH)_{16} (CO_3)$ $4H_2O$. The Adkins catalyst, copper/ammonium chromate, is another example of applying a specific precipitation procedure.
- Slow precipitation of metals from solution is possible by the slow hydrolysis of urea or nitrite at about 90^0C. This procedure is quite slow and using it may be expensive for many catalysts.
- When preparing a nickel oxide/kieselguhr catalyst, reaction between the components leads to the formation of nickel silicates, which are difficult to reduce. This can affect operation and the catalyst may require prereduction before use. The addition of a promoter such as copper oxide allows reduction at a lower temperature.
- When calcining chromium catalysts in air any Cr^{3+} is oxidized to Cr^{6+} at temperatures exceeding 200^0C. Further calcination up to about 450^0C reduces the Cr^{6+} content to a practical level and avoids an exothermic reduction reaction in the plant.

1.3.2. Impregnation

Preformed, absorbent supports are uniformly saturated with a solution of the catalytic metals. Several impregnations may be needed to obtain the required metal loading. Supports must be strong enough to be immersed directly into the solution and any dust forming should not contaminate the catalyst surface. To avoid these difficulties, the supports can be sprayed with just enough solution to completely fill the pores. At this point the support suddenly appears to be wet and the procedure is known as incipient wetness. In some cases, when the support has been impregnated, the metals may be precipitated by immersion in a second solution.

After impregnation catalysts are carefully dried and most are calcined before use to decompose the metal salts to oxides. Care is necessary to avoid high concentrations of metals forming at the support surface as the solution evaporates. Most types of supports can be used to produce impregnated catalysts, although alumina or silica, in various forms, is usually chosen. The support should not, of course, react with the metal solution. For example, if the support is soluble in acid, there is a possibility of re-precipitation in an undesirable form.

1.3.3. Other Production Methods

Many other procedures have been used to produce catalysts:

- Platinum/rhodium alloy gauze is used as a catalyst in the selective oxidation of ammonia during nitric acid production and in the production of hydrogen cyanide. The wire in the gauze is only a few thousandths of an inch in diameter woven at 80 wires per inch. Several layers of gauze, up to about 8 ft in diameter, are used.
- Silver granules are used to oxidize methanol to formaldehyde. Raney nickel is produced by leaching aluminum from a nickel/aluminum alloy with alkali solution. Not all of the alumina is removed and the catalyst may be regenerated a number of times by alkali treatment.
- Mixtures of natural magnetite and various promoters are melted in a furnace at about 1600^0C and chill cooled by casting on a flat surface. The catalyst can be crushed and separated into appropriate size ranges before use in ammonia synthesis or Fischer–Tropsch processes.
- In large modern ammonia plants the synthesis catalyst is often used in a pre-reduced form. The catalyst is carefully reduced in a mixture of hydrogen and nitrogen and then re-oxidized with a mixture of oxygen and nitrogen. Less than 20% of the iron is re-oxidized, making plant start-up

much easier and ensuring that the catalyst is not pyrophoric before use. Some nickel oxide catalysts supported on silica are also pre-reduced.

1.4. CATALYST TESTING

Industrial catalysts must conform to a strict specification and physical and chemical properties are measured at all stages of production. The tests most often included in catalyst specifications are listed in Table 1.7.

1.4.1. Physical Tests

It is most important that the catalyst be strong enough to resist breakage and attrition. Fixed bed and tubular reactors are carefully filled with catalyst to ensure that the pellets or granules are not damaged and pack with a uniform density.

Strength is particularly important in processes in which catalysts are circulated continuously between the reactor and a regenerator. In the fluid catalytic cracking process significant daily additions of catalyst must be made to compensate for losses through attrition as well as catalyst deactivation.

1.4.2. Chemical Composition

Careful checks on the chemical composition of both raw materials and products are required to ensure that the specification is achieved. It is important to confirm that the desired chemical compounds are formed with the required crystalline form and particle size. It is very easy for the pH of the solution to change during precipitation reactions and influence the composition of the precipitate.

Elemental analysis of the bulk components in a catalyst is measured by X-ray fluorescence (XRF), which has now replaced traditional wet analysis. If required, electron probe analysis can also provide information on the distribution of elements in a catalyst particle. Bulk phases in a catalyst and crystallite size are determined by X-ray diffraction (XRD), as either a routine check or a diagnostic procedure to examine catalysts damaged during operation.

Thermogravimetric analysis (TGA) and differential thermal analysis (DTA) with either air or inert atmospheres show weight changes or heat evolution as catalyst samples are heated and intermediate chemical compounds decompose. The tests are therefore useful in providing information on temperature-related phase changes at different stages of catalyst preparation. Tests also show how discharged catalysts react during regeneration or oxidation. Temperature programmed reduction (TPR) in hydrogen is used to investigate similar changes during catalyst reduction over the appropriate temperature range. These procedures are widely used in catalyst development and routine testing.

Many other tests are used to measure the physical and chemical properties of industrial catalysts during development and routine examination. These are fully described in other publications but are summarized here in Tables 1.9 and 1.10.

1.4.3. Activity Testing

Activity measurements have been important ever since industrial catalysts were first introduced. BASF carried out some 20,000 tests during the production of a successful ammonia synthesis catalyst. Until the 1960s, however, when many new catalytic processes were developed, laboratory activity tests were fairly crude. It was only possible to screen different samples and obtain relative activities before new catalysts reached the semitechnical plant stage and commercial trials were considered possible. More recently, new testing equipment has been able to simulate plant conditions and provide accurate kinetic information for reliable design calculations.

TABLE 1.9. Physical Testing of Industrial Catalysts.

Test	Procedure
Chemical analysis	X-ray fluorescence (XRF) Samples emit secondary X-rays following bombardment with hard X-rays which allow complete elemental analysis.
Particle size	Simple measurement of solid particles or size grading of powders.
Crushing strength	Determined by compressing catalyst particles between anvils or by bulk crushing the catalyst in a standard container.
Bulk density	Packing weight per unit volume of loose or dense packed catalyst particles.
Attrition loss	Dust formed after the rotation of catalyst for a fixed period in a standard cylinder.
Crystalline phases	X-ray diffraction (XRD) identifies crystalline compounds by reference to standard tables. The proportion of each phase present in the sample can be calculated.
Crystallite size	Estimated from X-ray diffraction line broadening as crystallite size decreases.
Surface area/pore size	Calculated from the volume of a gas monolayer adsorbed by the catalyst and the known area covered by a gas molecule. Pore volume can be calculated from the helium density (helium fills the pores) and the mercury density (mercury does not fill the pores). The average pore radius assuming cylindrical pores is calculated as twice the pore volume divided by the surface area.

Thermogravimetric analysis (TGA)	Measures the weight loss as catalyst composition changes at increasing temperature with oxidizing or inert atmospheres.
Differential thermal analysis (DTA)	Measures the exo-or endothermal temperature changes taking place as catalysts are heated in oxidizing or inert atmospheres.
Temperature programmed reduction	Measures the reducibility of oxides in catalyst samples in reducing atmospheres.

There are several steps during the catalytic reaction before a gas molecule is available for reaction at the active surface. First, it must (a) pass through the gas film surrounding the catalyst particle, (b) diffuse through the catalyst pores and reach an active site, and (c) adsorb on the active site to react with adjacent molecules. Products then (d) desorb from the active site, (e) return through the catalyst pores to the catalyst surface, and (f) re-enter the gas phase.

This process is dynamic and may depend on the reacting molecules moving from site to site until reaction takes place and products can desorb. Under ideal conditions, there would be no film or pore diffusion limitations but, practically, these are often encountered in catalytic reactions, particularly when large rings or pellets operate at a relatively low linear velocity.

Simple screening tests can be developed for most reactions to compare the activity of different catalysts. It is important, however, to standardize operating conditions and to operate well away from equilibrium conversion, to obtain the most useful results. To avoid all diffusion limitations, catalyst samples are normally tested at a high linear velocity with small crushed particles. Test units operate with pure gas mixtures and the effects of typical poisons must be considered in separate tests. Until the 1960s, the screening tests operated at atmospheric pressure and compared the performance of new catalysts with an accepted standard at constant space velocity.

TABLE 1.10. Chemical and Structural Analyses of Industrial Catalysts.

Test	Result
Electron microscopy	Used to study the surface structure and composition of catalysts and has extended the use of optical microscopy in determining the characteristics of particles.
Scanning electron microscopy (SEM)	Electrons scan the surface of the sample and give a magnification of 20,000–50,000. It can focus on sizes down to 5 nm. This gives crystallite shape, size, and size distribution.
Transmission electron microscopy (TEM)	Higher magnification and resolution possible to give three-dimensional images of crystallites down to 0.5 nm and changes during operation.
Electron spectroscopy for chemical analysis (ESCA) better known as X-ray photoelectron spectroscopy (XPS)	Can analyze for all elements and atomic electron binding energies to give structural data and compound types in surface layers.

Auger spectroscopy (AES)	Elemental analysis of surface layers.
Secondary ion mass spectroscopy (SIMS)	Elemental analysis of surface layers.
High-resolution electron energy-loss spectroscopy (HREELS)	Types of chemical bond present.

Atmospheric pressure test units cannot give the information required for modern catalyst and process development and have been replaced with high pressure micro-reactors. To avoid diffusion limitations, micro-reactors operate at very low conversions, under isothermal conditions, using small quantities of crushed catalyst. Experimental catalysts are usually tested in the same process conditions over a range of gas rates to provide useful design information. The ratio of space velocities measured at the same conversion gives the relative activity of different catalysts and an indication of the catalyst volume needed to give the same performance. It is also possible, when using pulsed gas flow at a constant space velocity and process conditions, to measure the catalyst selectivity in a reaction over an appropriate temperature range.

Knowledge of reaction kinetics is required to enable engineers to design a catalytic reactor. To obtain this information it is usual to use full-size catalyst particles over the total range of plant operating conditions to determine an accurate rate of reaction. This became possible by the introduction of continuous stirred tank reactors. These give results directly for conditions that are almost uniform throughout the reactor with none of the gradients found in micro-reactors. Two popular types of high-pressure reactors are the Carberry reactor, in which the catalyst is held in a cross-shaped basket and rotated, and the Berti reactor, which recycles gas through a fixed bed of catalyst. Several catalytic reactors are listed in Table 1.11.

TABLE 1.11. Catalyst Activity.

Reactor type	Comments
Micro-reactor	Provides rapid automated screening of several catalysts. Requires small samples of small catalyst particle. Compares rate constants, i.e., activity directly. Allows initial life testing. Can be used for accelerated aging task. Eliminates unsuitable formulations.
Continuous stirred tank reactors	Use full size catalyst particles.
Carberry—rotating catalyst basket	Operates over full range of plant conditions.

Berti—internal gas recycle through stationary basket	Determines reaction rate and reaction kinetics directly. Advantages of stationary basket: • Well-defined gas flow • No catalyst breakage • Temperature measured directly • Smaller catalyst volume used • Space velocity close to full-scale operation
Also: Caldwell reactor—stationary catalyst basket.	
Robinson–Mahoney reactor—stationary radial flow basket.	

1.5. CATALYST OPERATION

In 1887 George Edward Davis gave a series of lectures at the Manchester Technical School describing the current technology in use in the chemical industry. By identifying combinations of the same basic operations, he found a novel approach to describe each process. This was the same concept that Arthur D. Little discussed in a 1915 report to the Massachusetts Institute of Technology, when he proposed the idea of unit processes. These suggestions led to a more systematic approach to the education of future chemical engineers and the design of chemical processes. Nowadays most chemical processes use combinations of catalysts in standard reactors that fit neatly into a range of different units.

1.5.1. Reactor Design

Chemical processes often combine several catalytic reactions and standard flow sheets have been evolved over the years by process operators and chemical contractors. Catalytic reactors are designed from knowledge of the reaction kinetics and the influence of operating conditions and feed gas impurities on catalyst performance. Catalyst volumes and operating conditions have therefore been optimized on the basis of experience and established process design, which means that design operation and catalyst life are generally reliable unless there are unexpected operating problems. Table 1.12 lists major catalytic processes and their industrial applications and indicates when they were introduced.

1.5.2. Catalytic Reactors

Reactors are always designed to ensure that the operating conditions do not damage the catalyst:

- For most processes this is possible with fixed adiabatic beds in which the temperature rise corresponds to the conversion. If the maximum temperature would damage the catalyst, several beds are used with interbed cooling using heat exchange or quench. The total catalyst volume is unchanged and is simply distributed among the beds to control the temperature rise and achieve the design conversion.
- With strongly exothermic or endothermic reactions tubular reactors are used to control selectivity or prevent catalyst deactivation. Tubes are either cooled or heated to maintain the catalyst temperature and approach to equilibrium.
- Poisons present in the feed to the reactor can be removed if necessary by installing guard beds. These can be either a specific absorbent or an additional volume of catalyst that is replaced when saturated.

TABLE 1.12. Some Important Catalytic Processes.

Process	Application	Year
Hydrogenation	Petrochemical syntheses.	1930–1945
	Fat hardening.	1902
	Refinery hydrotreating.	1950+
Oxidation	Sulfuric acid.	1900–1920
	Nitric acid.	1906
	Formaldehyde.	1920s
	Organic anhydrides, aldehydes, nitriles.	1950s
Cracking	FCC gasoline.	1940s
	Hydrocracking.	1960s
Reforming	Aromatics/gasoline.	1949
	Synthesis gas/hydrogen.	1920s
Polymerization	Polyolefins.	1950s
	Polygasoline/iso-octane.	1930s
Isomerization	Branched hydrocarbons.	1950s
Synthesis	Ammonia/methanol.	1915–1920s
	Fischer–Tropsch.	1923
Purification	Control of emissions from automobiles, power plants, organic pollution.	1970s

- In some reactions carbon deposition gradually deactivates the catalyst. If this happens an additional reactor may be installed in parallel that allows the on-line reactor to be isolated and regenerated as the plant operates continuously.
- In catalytic cracking units the catalyst is rapidly deactivated. A fluidized catalyst bed is used in which very small catalyst particles are fluidized in the flow of feed gas. Deactivated catalyst is circulated continuously through a regenerator and back to the reactor.

- If the temperature rise in an adiabatic reactor requires too many beds, the overall reaction can be made more selective and operation more economic by using a fluidized bed of catalyst at a uniform temperature.
- Continuous catalytic reformers circulate solid catalyst between the reactor and regenerator to achieve continuous operation.

The most widely used reactors are shown in Table 1.13, although batch reactors are not often used in full-scale operation.

1.5.3. Catalyst Operating Conditions

Catalyst life must be predictable to avoid unexpected shut down and lost production. It is necessary therefore to establish good operating procedures that achieve

TABLE 1.13. Reactors Used in Catalytic Processes.

Reactor	Catalyst loading	Temperature profile
Adiabatic beds	One or more packed beds in series or parallel—gas or gas/liquid feed.	Adiabatic temperature increase with interbed cooling to control selectivity.
Tube-cooled isothermal	Catalyst loaded into or around tubes.	Temperature rise controlled by suitable cooling medium; not perfectly isothermal.
Fluid beds	Very fine catalyst particles are fluidized in the reacting gas.	Used for exothermic reactions to give uniform bed temperature.
Batch/autoclave	Sufficient catalyst of appropriate size for batch of feed.	Heat of reaction controlled by heating or cooling coils.
Adiabatic gauzes	Layers of gauze supported on a large open mesh.	Usually a high adiabatic temperature rise through the gauzes.

the best life possible, so that all catalyst replacements can be arranged during scheduled maintenance periods. Normal deactivation caused by slow sintering of the catalyst, absorption of small quantities of poisons, or carbon deposition can be taken into account during plant design by installing guard vessels or using a known excess of catalyst.

Reliable commissioning is an important part of plant operation and is carefully planned with the assistance of engineering specialists and catalyst suppliers. Modern plants use hundreds of tonnes of different catalysts, which represent a significant proportion of the capital cost. The ordering, supply, storage, installation, and proper reduction to the active form take a considerable period and must be properly organized:

- Catalysts must be handled carefully during loading to prevent breakage and dust formation. They are loaded according to detailed specified instructions to give uniform packing.
- Start-up procedures involve drying and activating the catalyst, usually by reduction in controlled flows of hydrogen and inert gas at specified temperatures.
- Operation starts under controlled conditions to avoid hot spots in the catalyst bed or temperature run-away reactions.
- During operation, the conditions must be checked at regular intervals to maintain the design operating temperature and conversion to ensure the maximum catalyst life.
- The cost of closing down a large plant is so high that reliability of catalysts and process operations is essential.

Typical reduction and operating conditions for industrial catalysts are described in the appropriate chapters. Despite efforts to protect the catalyst from poisons or maloperation, it is still possible for problems to affect the catalyst. A few typical examples are shown in Table 1.14.

Steps must be taken to restore good performance following maloperation. Unfortunately this often involves shutting down the process or isolating the individual reactor to change the catalyst. To avoid loss of production it is possible to install a spare reactor for use in case of an emergency. For example, spare reactors are essential when removing acetylene from steam-cracker ethylene streams because polymers, known as *green oil*, saturate the catalyst pores. On the whole, however, once a process becomes established and *teething troubles* are sorted out most problems can be avoided or the process design modified.

1.6.   CONCLUSION

There was a remarkable interest in catalysts before 1900 considering the primitive state of industrial production at the time. Several catalytic processes that are still used today were being introduced on a relatively large scale. During 1900, for example, the worldwide production of sulfuric acid was about 4 million tonnes. Thereafter the introduction of catalytic processes played an increasing part in the expansion of chemical production. The pioneering work of Sabatier and Ipatieff demonstrated the potential of a wide range of catalytic reactions and the benefits of operation at high pressures. These ideas were gradually developed to cope with consumer demands.

The use of new catalyst-based technology, introduced by early chemical producers, was soon expanded by chemical engineering contractors, first in the

TABLE 1.14. Catalyst Deactivation or Maloperation.

Problem	Effect	Treatment
Dust	Blocks bed	Suck off catalyst and dust on top of bed or remove all catalyst and sieve.
Carbon deposit	Blocks catalyst pores	Regenerate by burning carbon in a stream of air either in the reactor or externally if the catalyst is not pyrophoric.
Compressor oil	Saturates the bed	Regenerate by burning oil in a stream of air preferably after removing the catalyst from the reactor.
Chemical poisons	React with and deactivate the catalyst	Poisoning is usually irreversible and catalyst is discarded; occasionally catalyst may be regenerated by suitable procedures.
Chemical effects	Loss of active component	The active catalyst may react with the support or a volatile impurity in the feed; the catalyst may also be volatile at high temperatures.

chemical and then in the refining industry. Since 1950 the petrochemical industry has introduced a wider range of very sophisticated new catalysts. After nearly 100 years of continuous development most chemical processes are now based on the use of catalysts.

REFERENCES

1. W. Ostwald, *Textbook of Inorganic Chemistry,* 1898.
2. J. J. Berzelius, *Jahres-Bericht* (1836); *Ann. Chim.* (1836).
3. F. S. Taylor, *A History of Industrial Chemistry,* Heinmann, London, 1957, p. 97.
4. W. Wyld, *The Manufacture of Acids and Alkalis,* Vol. 1, Ed. by Lunge, Gurney and Jackson, London, 1923, p. 4.
5. I. Milner, *Phil Trans Royal Soc.* **79** (1789) 300.
6. G. R. Kirchoff, *Schweigger's J* **4** (1812) 7.
7. H. Davy, *Communication to the Royal Society* (Jan 1817).
8. J. W. Dobereiner, *Schweigger's J.* **34** (1822) 91; **38** (1823) 321.
9. E. Turner, *Edinburgh Phil J.* **11** (1824) 99, 311.
10. P. Phillips, British Patent 6096 (1931).
11. W. Henry, *Phil Trans Royal Soc* **114** (1824) 266.
12. F. Kuhlmann, *Compt Rend.* **7** (1838) 1107; French Patent 11331-2 (1839).
13. H. Deacon, British Patent 1403 (1868).
14. W. S. Squire and Messel, British Patent 3278 (1875).
15. Hasenclever, *Berichte* **9** (1876) 1070; British Patent 3393 (1883).
16. L. Mond and C Langer, British Patent 12608 (1888).
17. A. Trillat, French Patent 199919 (1901); *Bull. Soc. Chem.* **27** (1902) 797; **29** (1903) 35.
18. R. F. Carpenter and S. E. Linder, *J. Soc. Chem. Ind.* **22** (1903) 457; **23** (1904) 577.
19. W. Ostwald and Brauer, *Chem Zeit* **27** (1903) 100.
20. A. Mittasch, Early Studies of Multicomponent Catalysts, in Advances in Catalysis, Vol. 2, Academic Press, New York, 1950, p. 81.
21. P. H. Emmett, Catalysis, Vol. 1, Reichold Publishing Corporation, New York, 1954, Ch 2.

2

THE FIRST CATALYSTS

Efforts to develop processes using catalysts were vital to the growth of the chemical industry. For many years, the *first catalysts* were most probably the result of trial and error and were based on the observations of scientists. When Berzelius defined catalysis, the examples he quoted did not include any industrial applications. For example, no mention was made of the lead chamber process or the Phillips patent proposing the use of a platinum catalyst for sulfuric acid production.

When the chemical industry began to expand, BASF improved the contact process and, following Haber's investigations, introduced ammonia synthesis, which provided a practical basis for catalyst design.

However, as Miles noted in his book on the contact process, published by Gurney and Jackson in 1925, the secrecy surrounding new processes slowed down the release of technical information. Miles tried to reverse a situation in which no process was completely described until it was out of date!

Many changes have now been made to the first catalysts since they were introduced, but they are still being used and, with more recent introductions, are indispensable in the modern chemical, refining and petrochemical industries.

2.1. SULFURIC ACID

Large-scale production of sulfuric acid began in about 1740 when Joshua Ward burned sulfur and niter in glass bell jars with a capacity as high as 66 gal. This procedure was improved in 1746 by Dr John Roebuck and Samuel Gardner at

L. Lloyd, *Handbook of Industrial Catalysts*, Fundamental and Applied Catalysis,
DOI 10.1007/978-0-387-49962-8_2, © Springer Science+Business Media, LLC 2011

their Birmingham, UK, *vitriol manufactory*, where they burned the sulfur and niter in *lead houses*. This was the critical step in producing tonnage quantities of sulfuric acid for the first time.

Later, Roebuck's factory in Glasgow incorporated a suggestion by J. A. Chaptal (Napoleon's Minister for Agriculture) that the sulfur and niter should be burned in an external furnace. While this meant that the sulfur dioxide and the nitrogen dioxide catalyst were passed into the lead chamber with a current of steam, it was some time before the process became really continuous. The size of the lead chambers increased from about 200 ft^3 in Roebuck's plants, producing about 25 lb of acid a day, to about 5000–10,000 ft^3 by 1820.

Large-scale production led to the price of acid falling from about £30 per ton in 1790–1800 to £3.5 per ton in 1820, when UK production of sulfuric acid was about 10,000 tons a year. Removal of the UK salt tax in 1825 reduced the price still further to £1.25 per ton.

2.1.1. The Lead Chamber Process

At first it was not known that niter, which was an essential part of the lead chamber process, acted as a catalyst. When Lavoisier showed that sulfuric acid contained only sulfur, oxygen, and hydrogen (1772–1777) it was realized that niter was not a component of chamber acid. Operators then assumed that it either made the sulfur flame hotter or supplied oxygen to the sulfurous acid.

At that time, acid was still being made in batches and no air was added to the lead house during reaction. By 1793 Clement and Desormes had suggested that the continuous addition of air would improve reaction, and in 1806 they defined the action of niter, which was clearly essential to the process:[1]

> " . . . nitric acid is only the instrument of the complete oxygenation of the sulfur: it is the base, nitric oxide, that takes the oxygen from the atmospheric air to offer it to the sulfurous acid in the state which suits it best . . . "

Clement and Desormes were also the first to observe the formation of *chamber crystals* that evolved nitric oxide and formed sulfuric acid when added to water. The nitric oxide was then available for recycling.

The basis of the lead chamber process was, therefore, to combine sulfur dioxide with the oxygen in air in the presence of a relatively small amount of niter. Some details of the process development are summarized in Table 2.1 and a typical lead chamber plant is shown in Figure 2.1. Despite this conclusion, many UK producers continued to operate the process in batches until as late as 1820. The fact that it was difficult to transport sulfuric acid meant that many small, *on-site* plants supplied users directly. This, of course, eliminated competition and delayed technical developments. Even after the eventual introduction of the contact process during the 1920s, the lead chamber process was still widely

TABLE 2.1. Development of the Lead Chamber Process.

Innovator	Procedure	Comment
1666: Fevre and Lemery	Sulfur burned with saltpeter (KNO_3)	
1740: Ward, Richmond, UK	Rows of 66-gal glass jars. One part KNO_3 and eight parts sulfur burned in a horizontal glass neck.	Sulfur trioxide dissolved in water in jar. Sulfur burned until acid strength high enough for use or concentration.
1749: Roebuck and Garbutt, Prestonpans, Scotland. Used lead chambers.	1-lb KNO_3 with 7-lb sulfur every 4 hours on iron trays.	Air replenished between batches of sulfur. Acid gravity usually 1.250 (33%) after six weeks! Concentrated to almost 50%. Yield about 110% based on sulfur. Several hundred chambers used at each site 70,000 ft³ (200 m³) at Prestonpans.
1793: Clement and Desormes	Continuous flow of air limited amount of KNO_3 required.	Confirmed air was main (90%) oxidant and KNO3 only an intermediate.
1807–1814: St Rollox, Scotland	Continuous sulfur burning—steam and air flowing through lead chambers.	
1827: Chaumy, Gay-Lussac tower	Nitrogen oxides absorbed at outlet of lead chambers to allow catalyst (oxides of nitrogen) recovery.	First use at Chaumy in 1842 and Glasgow in 1844. Little used until Glover tower became available.
1859: Glover tower	Nitrous oxides recovered from Gay-Lussac tower.	Slow acceptance. First use 1859 at Washington, Co Durham, UK.

used throughout the world until the 1950s—a typical example of industrial *catalytic* inertia.

Nitrogen oxides were lost to the atmosphere with the residual nitrogen during operation of lead chamber plants. This led Gay-Lussac to suggest in 1828 that effluent nitrogen be washed with chamber acid in a separate tower to dissolve the nitrogen oxides, which could then be recycled. Because it was difficult to liberate the nitrogen oxides without diluting the chamber acid, Gay-Lussac towers were not often used. Finally, in 1859, Glover passed the nitrous vitriol down through a second tower, where hot gas from the pyrites burner removed the nitrogen oxides, which were returned to the lead chambers.[2] This procedure was important as it also concentrated the acid. The combined Gay-Lussac and Glover towers were, therefore, the first catalyst recovery plants. A tower built in 1868 concentrated 73,000 tonnes of sulfuric acid to specific gravity 1.75 (80%

Figure 2.1. Lead chamber process for the manufacture of sulfuric acid. Reprinted with permission of the Science Museum, London.

H_2SO_4) from a total of 15,400 tonnes of pyrites. The tower cost £450 and annual repairs over a 6-year period cost only £11.[3] Even so, despite the bargain prices, Glover towers were slow to gain acceptance, and by 1890 were used by only about half of US acid plants. Until the lead chamber process was fully developed, concentrated sulfuric acid could only be produced by evaporation in glass or platinum vessels.

2.1.1.1. *Chemistry of the Lead Chamber Process*

Lunge and Berl proposed the following mechanism for the oxidation of sulfur dioxide:[4]

- Reaction with nitrous fumes:

$$SO_2 + NO_2 + H_2O \rightarrow HOSO_2 \cdot NO(OH) \tag{2.1}$$

- Oxidation of hydroxynitrosulfuric acid:

$$2\ HOSO_2 \cdot NO(OH) + O \rightarrow 2\ HOSO_2 \cdot NO_2 + H_2O \tag{2.2}$$

- Isomerization of nitrosulfuric acid:

$$2\ HOSO_2 \cdot NO_2 \rightarrow 2\ HOSO_2 \cdot ONO \tag{2.3}$$

- Nitrosulfuric acid then forms sulfuric acid with steam or sulfur dioxide:

$$HOSO_2 \cdot ONO + H_2O \rightarrow H_2SO_4 + HNO_2 \tag{2.4}$$

$$2\ HOSO_2 \cdot ONO + SO_2 + 2\ H_2O \rightarrow H_2SO_4 + 2\ HOSO_2 \cdot NO(OH) \tag{2.5}$$

Process efficiency depended on gas mixing in the lead chambers. Major improvements in operating the lead chamber process were the use of packed towers by Gaillard-Parrish and Peterson, the introduction of conical towers by Mills and Packard, and process designs introduced by Kachkaroff.[6] Chamber acid was dilute 65% sulfuric acid as produced but could be concentrated to 80% in the Glover tower.

2.1.1.2.　The Continuing Use of the Lead Chamber Process

Lead chamber plants were used for many years after the introduction of the contact process in which small companies made acid for their own operations. Table 2.2 shows that in both the United States and the United Kingdom production of chamber acid continued until well after the 1960s.

At least one lead chamber plant was still operating in the north of England in 1960. There were plumbers patching the lead chamber and carpenters regularly replacing the wooden ducting used to transfer acid from the chambers to the point of use.

TABLE 2.2. Gradual Introduction of Contact Process 1900–1975.

	United Kingdom		United States	
Year	Lead chamber	Contact	Lead chamber	Contact
1899	100%	—		New Jersey Zinc
1901	100%			General Chemical Co.
1920	More than 100 plants used	First contact process plant	Up to 80%	More than 20%
1929			About 65%	About 35%
1938	About 63%	About 37%	More than 90 plants remain	
1944	About 55%	About 45%		
1960	About 10%	About 90%	Few chamber plants remain	
1975	Small capacity lead chamber plants only. Contact acid > 96%.			

In fact new *lead chamber* plants were still being offered by a contractor in 1958.[6] A description of the *Kachkaroff process* shows how the old plants operated. The burner, which produced reaction gas, was followed by four large cylindrical reaction vessels in which sulfur dioxide reacted with oxygen, catalyzed by the circulation of *nitrous vitriol* and gaseous nitrogen oxides. Nitrous vitriol was a solution of 10–15 wt% nitrogen oxides in sulfuric acid. Nitrogen oxide losses were made up as required by a small ammonia oxidation unit.

More than 80% of the sulfur dioxide was oxidized in the first reaction vessel. Reaction was rapid and the surface of the liquid was violently agitated. A consequence of this mixing was that as sulfuric acid was produced the gaseous nitrogen oxide catalyst dissolved to form more nitrous vitriol.

The circulation of nitrous vitriol through the final three reaction vessels was regulated to balance the conversion of the remaining sulfur dioxide. Inlet temperature to each reaction vessel was controlled by cooling the circulating acid solution. A volume of the nitrous vitriol equal to the sulfuric acid being produced was removed from the circulating liquid when the reaction was completed and pumped to a denitration tower linked to an acid concentration tower. During denitration of the nitrous vitriol in the first tower, the sulfuric acid was diluted from 80–85% to 60–68%. Cool acid was then concentrated by passing it down the second, concentration, tower, which also cooled the hot gases leaving the burner. Cooled burner gas then passed through the reaction vessels to continue the cycle.

Overall loss of nitrogen oxide catalyst was equivalent to about 0.15 tons of nitric acid per ton of sulfuric acid produced, which was lower than in the conventional, less sophisticated lead chamber process plants. Sulfuric acid could be obtained as either 77–82% or 60–80% solutions. An advantage of the more modern process was that the low-temperature operation allowed the use of PVC as piping, tower cladding, and storage tank linings.

2.1.1.3. *Raw Material for Sulfuric Acid Production*

During the nineteenth century, the main source of sulfuric acid was Sicilian sulfur and most US plants continued to import sulfur from Sicily until the 1890s. However, in 1838 the king of Sicily gave an export monopoly to a French company, which increased the price from £5 to £14 per ton, and most European companies considered it necessary to find an alternative raw material. Iron pyrites was known to burn forming sulfur dioxide and, despite the presence of arsenic impurities, was used from about 1825. Spain became the major supplier of pyrites in Europe and sulfuric acid production was often associated with copper smelting. By 1860, most European plants were using pyrites, although, later, as spent oxide (sulfide iron oxide) started becoming available from gas works, it, too, was used as a source of sulfur.

TABLE 2.3. Development of the Contact Process.

Innovator	Comment
1831: Peregrine Phillips	British patent 6096 (1881) described process of forming sulfur trioxide with a platinum catalyst. Used stoichiometric volumes of sulfur dioxide and oxygen. Inspired further research.
1837: Clement	Writing to Schneider felt that the contact process would be widely used within 10 years.
1844: Schneider	Demonstrated that a pumice catalyst could produce sulfuric acid without lead chambers. He did not claim use of platinum but this is probable [*Dingl Polyt J* **56**, 395 (1847); **69**, 354 (1860)].
1846: Jullion	British patent 11425 (1846) claimed a platinum catalyst supported on asbestos for the first time. Catalyst also used for a range of other reactions.
1852: Wohler and Mahla	Found that chromium and copper oxides oxidized sulfur dioxide. Copper metal inactive—the first comment on oxidation with oxide catalysts. Showed iron and copper were reduced and oxidized during reaction. Findings later applied to Mannheim process [*Ann. Chim. Pharm.* **81**, 255 (1852)].
1850s: Deacon	Patented use of copper sulfate in process and first to observe that reaction rate faster with an excess of oxygen.
1853: Robb	British patent 731788 (1853) protected the use of pyrites cinders as catalyst.
1853: Hunt	British patent 1919 (1853) protected the use of silica as a catalyst support.

Note: These processes could have reduced costs of niter and lead used in the lead chamber process while improving production rate. Development was slow owing to a lack of technical experience and innovation. Demand for acid, in particular, was still small.

Herman Frasch developed a process to recover cheap natural sulfur in 1891 that was used first in Louisiana and then in Texas, and eventually became the major source of supply, particularly in the United States. The recovery of refinery sulfur by the modified Claus–Chance process now provides a further enormous supply of pure sulfur throughout the world, and this has largely replaced the Frasch process.

2.1.2. Contact Process Development

The contact process was patented as early as 1831 by Peregrine Phillips, the son of a vinegar maker in Bristol.[7] His process involved a heated porcelain tube containing finely divided platinum or wire to convert a stoichiometric mixture of sulfur dioxide and air to sulfur trioxide. The process was not commercial at that time for several reasons, including the engineering problems associated with circulating hot corrosive gases, the availability of acid resistant materials, and poisons in the sulphur dioxide. Table 2.3 shows the main developments leading the granting of Phillips's patent in 1853 and demonstrate the interest in a new *contact* process.

By 1875 Squire and Messel had patented a form of the contact process that used dilute chamber acid as the raw material to avoid any problems with catalyst poisons.[8] Chamber acid was decomposed by heating and the poison-free sulfur dioxide thus produced was oxidized in air using a platinized pumice catalyst. Sulfur trioxide could then be used to produce oleum, which was needed for the synthesis of alizarin by the newly developing dye industry. The same procedure is still being used to recover sulfur from sulfuric acid wastes. At about the same time Clemens Winkler set up a contact process plant in his factory at Freiberg using more or less the same process but did not apply for patent cover immediately.[9] All of the processes up to the 1890s used the same exact ratio of sulfur dioxide and oxygen to provide sulfur trioxide. Efforts were made by Schroder and Hannich to use higher pressures.[10] Messel also suggested burning sulfur in pure oxygen to avoid dilution of the reaction mixture with nitrogen.[11] A typical modern sulfuric acid plant is shown in Figure 2.2 and the most important developments are described in Table 2.4.

Figure 2.2. Modern sulfuric acid plant using the contact process.

TABLE 2.4. Introduction of the Contact Process for Oleum.

Innovator	Comment
1875: Clemens and Winkler *Dingl. Polyt. J.* **218**, 128 (1875); **223**, 409 (1877).	Described experiments to produce oleum using 8.5% platinum on asbestos with pure oxygen (73.3% conversion) or air (47.4% conversion). Used stoichiometric ratio of SO_2/O_2. Pure sulfur dioxide from decomposing sulfuric acid. Sulfur trioxide absorbed in water to form oleum. Their results were not thermodynamically possible—Ostwald later claimed it delayed developments [*Z. Electrochem.* **8**, 154 (1902)].
1875: Squire (and Messel)	British Patent 3278 (1875) resulted from high oleum price. Used a platinum catalyst supported on pumice with a stoichiometric mixture of SO_2/O_2 made from decomposing H_2SO_4 in a platinum still (70% recovery of SO_3). Plant at Silvertown produced three tons of SO_3 per week. Patent mentioned that this avoided catalyst deactivation with dust and, probably arsenic although *poisons* were not recognized.
1875–1880: Jacob	Operated a contact process oleum plant in Germany at first from decomposed chamber acid but later from sulfur burning (43% free SO_3). Jacob sold his plant to Meister, Lucius, and Bruning at Hoechst, who still made oleum in 1925.
1879: Thann Chemical Works, Alsace	Acquired an improved oleum process design from Squire. Burned Sicilian sulfur and washed gas at 4 atm pressure. Mixed SO_2 with stoichiometric volume of air and formed SO_3 using platinized asbestos. Output 1.5 tons of SO_3 per day and dissolved in concentrated H_2SO_4.
1880s: BASF	Began to use the same process as Thann, producing such large volumes that the oleum price fell. Production increased from 18,500 tons during 1880 to 116,000 tons by 1900.

Further development of the contact process did not rely on a better catalyst but depended on better methods to remove poisons and clean the gases produced by roasting pyrites, which, by then, had replaced sulfur as the preferred source of sulfur dioxide. In attempting to overcome the difficulty, the Mannheim process used a bed of relatively inactive iron oxide to guard the main bed of a platinum catalyst. New Jersey Zinc and the General Chemical Company in the United States built plants of this kind in 1899 and 1901, respectively.

A 1901 lecture by Rudolph Knietsch[12] described the work carried out by BASF during the period 1880–1900.[13] As might be expected, the early process developments he described were mainly empirical. They concerned washing of pyrite gas, determination of the most efficient sulfur dioxide/oxygen ratios with excess oxygen for use in the feed gas, and absorption of sulfur trioxide in 98% acid to produce sulfuric acid. This information had been confidential until the paper was published. Perhaps the most important detail, apart from the use of excess oxygen, was the cooling of the gas during reaction in tube-cooled reactors to improve conversion, which was not part of earlier processes.

In a further significant advance, De Haen first demonstrated the use of vanadium pentoxide catalysts in 1900, following a suggestion by R. Meyers in Germany in 1898.[14] There was little further progress at that time because the original vanadium pentoxide catalysts were relatively unstable and much less active than the platinum catalysts then available.

The contact process was developed as a matter of urgency during World War I because the effective nitration of toluene required the catalytic use of concentrated sulfuric acid to generate the active species, NO_2^+. Nitration of toluene, of course, yields the military explosive, TNT. The increased demand for platinum could not be met economically, so that from 1914 on vanadium catalysts had to be introduced rapidly to expand sulfuric acid production. About one-third of German sulfuric acid at that time came from the contact process, but vanadium catalysts were not used extensively in other parts of the world until the mid-1920s.

Porous supports were used to make commercial platinum sulfuric acid catalysts. These included asbestos, kieselguhr, and silica gel. Rather surprisingly, water-soluble carriers such as magnesium sulfate were also successful and made platinum recovery more convenient. A flow sheet of a typical modern susphuric acid plant is shown in Figure 2.3. Table 2.5 describes four typical industrial catalysts.

TABLE 2.5. Contact Process Catalysts Containing Platinum.

Producer	Comment
1898: BASF[15]	Platinized asbestos produced by impregnating asbestos with platinic chloride solution followed by reduction with formaldehyde. Operated up to 10–12 years in several 10–20 cm layers. Contained 8–10% platinum. Tubular reactors were still designed for vanadium catalysts until 1950s. Agreement with Grillo up to 1898.
1902: Grillo Company	Grillo Company recovered sulfur dioxide from zinc blende smelters. Pre-1898 tightly packed tubes, using up to 15 layers of 8–10% platinum on asbestos converted 25% sulfur dioxide and 75% air. From 1898 used calcined magnesium sulfate sprayed with platinic chloride to give 0.1–0.3% platinum[16] on finished catalyst. Feed gas contained less sulfur dioxide.[17]
1998–1999: Mannheim Tenterloff Process	First bed loaded with burnt pyrites containing copper. Second bed, with up to 200 tubes—about 12 cm diameter, full of asbestos sponge soaked with platinic chloride solution reduced with formaldehyde.
Davison Chemical Company	Used silica gel impregnated with ammonium chloroplatinate to give 0.1% platinum. Claimed to use less platinum than other catalysts and to resist arsenic poisoning.

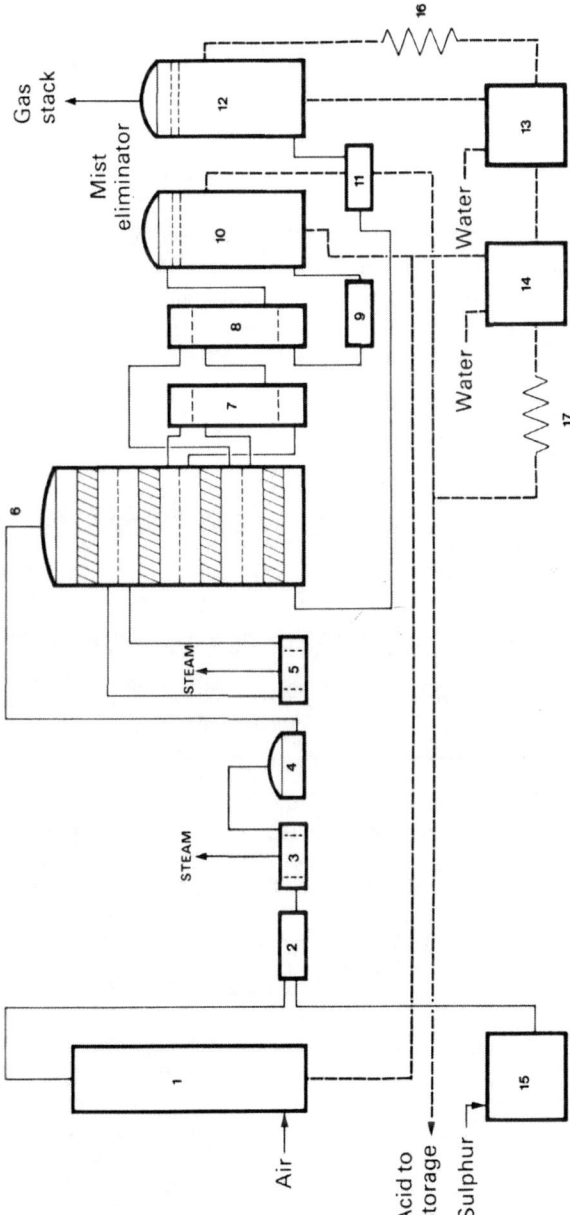

Figure 2.3. Schematic flow diagram for a typical sulfur-burning double-absorption sulfuric acid plant. 1. Drying tower; 2. sulfur burner; 3. waste heat boiler; 4. hot gas filter; 5. waste heat boiler; 6. four-pass converter; 7. hot interpass heat exchanger; 8. cold interpass heat exchanger; 9. secondary economizer; 10. interpass absorption tower; 11.economizer superheater; 12. final absorption tower; 13. final absorption tower circulating tank; 14. drier and interpass absorption tower circulating tank; 15 sulfur melter; 16. acid cooler; 17. acid cooler. Reprinted from *Catalyst Handbook*, 2nd ed., Ed. by M. V. Twigg, Wolfe Publishing, LTD., London, England, 1989, by kind permission of M. Twigg.

The performance of platinum catalysts depended strongly on the form of the support, which influenced crystallite size. Operating life depended on the poisons present in the sulfur dioxide and was originally about two years. With process improvements to remove poisons, the catalyst life eventually increased to about 10 years. Most reactors contained tubes that were cooled by exchange with cold inlet gas. The Grillo reactor contained trays with a form of heat exchange.

Vanadium soon replaced platinum as the most economic catalyst for the contact process. The *strike* temperature at which reaction began was higher but, when necessary, a *striker* layer of platinum catalyst was used until better process designs were available. The big advantages of vanadium pentoxide were that it was both cheaper and less affected by common poisons such as halogens, phosphorus, arsenic, selenium, tellurium, and mercury. Platinum catalysts were probably replaced by vanadium sometime before 1930. Table 2.6 lists several of the early vanadium catalysts.

Sulfuric acid catalysts containing vanadium pentoxide are characterized by complicated recipes and manufacturing procedures.

Following the introduction of vanadium pentoxide by De Haen it was discovered empirically by Slama and Wolf that alkalis improved catalyst activity and these have been used in all the catalysts produced ever since. They were specified as important for stability by both BASF and the General Chemical Company. Vanadium pentoxide catalysts had been used for more than 20 years before Frazer and Kirkpatrick[22] showed that the addition of alkali led to the for-

TABLE 2.6. Contact Process Catalysts Containing Vanadium Pentoxide.

Producer	Comment
1920: Slama-Wolf[18]	Used by BASF from 1920 and in the United States by General Chemical Co. from 1927. Made by mixing ammonium metavanadate and potash with kieselguhr (50:56:316). Dried, granulated and calcined at 480^0C in air and sulfur dioxide.
1932: Seldon Corporation[19]	Ammonium metavanadate mixed with potash and potassium aluminate combined with a gel formed by sprinkling kieselguhr with potasssium silicate (*zeolite*). Dried, pelletted (4–6 mm cylinders), calcined in air and SO_2 to fuse V_2O_5. Used in tubes cooled with feed gas.
1933: Monsanto[20]	Silica gel with ammonium metavanadate and potassium hydroxide. The first of many catalysts made by Monsanto and used worldwide.
1932: General Chemical Co.[21]	Developed by Joseph. A mixture of caustic soda, potash, and vanadium pentoxide added to wet mix of fine kieselguhr, potassium sulfate, and *tragacanth* gum. Dilute sulfuric acid added to neutralize alkalis. Mixture evaporated before granulation and extrusion. Calcined 600^0C.

mation of an active liquid catalyst melt held in the pores of the support. This consisted of liquid potassium pyrosulfate with dissolved vanadium pentoxide. Catalyst activity is about the same for vanadium pentoxide contents in the range 2–10%, although higher levels give longer lives. Catalyst performance depends on the porosity and stability of the support, which controls and stabilizes the liquid melt to determine the life of the catalyst. Vanadium catalysts resist the effects of most poisons, although vanadium may be volatile in the presence of halogens.

The main cause of operating problems is related to the deposition of entrained dust on the relatively wide but thin layers of catalyst in the reactor. Dust, which is usually carried into a reactor with feed gases, can also form by catalyst disintegration and leads to an increase in pressure drop. Feed gas from the roasting of pyrite, anhydrite, or smelter gases is more likely to cause dust problems, but *pure* sulfur can also contain 0.5% ash, which should be reduced to less than 0.002% by filtering the liquid sulfur or using gas filters before the reactors.

If pressure drop through the catalyst beds (or *passes* as they are normally called) increases then the catalyst can be removed during a normal shut-down period and sieved before being examined and replaced for future use. Catalysts in the form of rings or other shapes have now been introduced in order to minimize pressure drop problems. A normal average life of a catalyst in sulfuric acid plants is usually more than 10 years.

2.1.3. Modern Sulfuric Acid Processes

Modern vanadium pentoxide catalysts have been developed on a more scientific basis than those discovered empirically during the 1920s. Following Slama and Wolf's use of alkali by to improve catalyst activity, Frazer and Kirkpatrick and then Kiyoura,[22] realized that the catalyst was molten and filled the support pores during operation. Extensive investigation then led to an understanding of the ideal catalyst structure[23] and the reaction mechanism.

It is obvious now that the pores should be large enough to hold the melt as a thin film without being completely filled. Furthermore, since this is an equilibrium reaction that is adversely affected by higher temperatures, the solid catalyst should melt at a sufficiently low temperature to alleviate this equilibrium limitation. This was just about possible although the silica gel supports, prepared with an appropriate pore size and volume to increase low-temperature activity, tended to sinter at the operating temperature required in the first bed of a multi bed reactor.[24] It was often beneficial to use kieselguhr supports to increase operating stability, so some operators used a stable catalyst in the first bed with more active formulations in the remaining beds.

The catalyst melt contained vanadium compounds dissolved in a mixture of alkali pyrosulfates,[25] which melt at a lower temperature as the atomic number of the alkali metal increases. Potassium sulfate containing a small proportion of sodium sulfate was usually chosen, because it was cheaper than the higher-atomic-weight alkali metals such as cesium. Satisfactory operation was therefore possible at a time when environmental controls were minimal and the price of sulfuric acid catalysts extremely low.

During the experimental work following the discovery that the sulfuric acid catalyst was actually a liquid held in the pores of the silica support, several observations were significant in understanding the way in which sulfur dioxide was oxidized:

- The vanadium pentoxide dissolved in melted alkali pyrosulfate.
- Activity and long life were associated with the reduction of pentavalent vanadium when alkali sulfates were used in catalyst production.
- Sulfate or pyrosulfate ions were not found in the melted catalyst and the degree of sulfation was normally in excess of that required to form pyrosulfate.

Fresh catalyst contains vanadium pentoxide, potassium/sodium sulfate, and silica in the ratios required for high activity and stability. During stabilization with sulfur dioxide it is likely that the vanadium pentoxide first dissolves to form pyrosulfate, which then reacts to give a mixture of polymeric ions. The melting point of the catalyst has been found to depend on the alkali metal/vanadium pentoxide ratio, up to about four, as well as the atomic weight of the alkali metal. The liquid state allows rapid formation of the polymeric ions, which are the active catalyst.[26] According to Boreskov, they correspond to specific compounds with a general composition $V_2O_5 \cdot nK_2O \cdot mSO_3$, where $n = 2$, 3, or 4 and m is approximately $2n$. Crystals with these compositions had been isolated from melted catalyst. At high sulfur dioxide concentration, melts also contain $K_2O \cdot V_2O_4 \cdot 3SO_3$. Polymers formed by direct absorption of sulfur trioxide or copolymerization of existing ions often have molecular weights that exceed 1000.

The redox mechanism of sulfur dioxide oxidation was first explained by Mars and Maesson,[27] who proposed that the oxidation to sulfur trioxide led to the reduction of pentavalent vanadium in the polymeric ions. The resulting tetravalent vanadium was then reoxidized by adsorption of molecular oxygen. On occasions when a catalyst is partially deactivated by process gas containing a high sulfur dioxide concentration and low oxygen concentration, it is possible to regenerate the catalyst simply by heating in air at about 450^0–500^0C.

2.1.3.1. Catalyst Preparation

Catalyst is prepared by mixing a silica sol made from potassium silicate with vanadyl sulfate or ammonia metavanadate and precipitating with ammonia. The silica used can be either fresh or a solid such as kieselguhr. A sol mixed with

TABLE 2.7. Chemical Composition and Physical Properties of Modern Sulfuric Acid Catalyst.

Composition (wt%)	
V_2O_5	6–8
K_2O	8–10
Na_2O	1–2
SO_3	20–30
SiO_2	55–65
Physical properties	
Bulk density	0.4–0.6 kg.liter-1
Attrition loss	< 10%
Dimensions	6-mm diameter extrusions
Surface area	2–5 m^2 g^{-1}
Pore volume	0.5–0.6 ml g^{-1}

kieselguhr can react to give the *zeolite* support described by early catalyst producers. No filtering or washing is required and after drying the powder can be formed into shapes by the usual methods. Finished catalyst has the composition and physical properties shown in Table 2.7.

A list of some US patents describing catalyst preparation between 1935 and 1981 was given by Donovan, in Leach: *Applied Industrial Catalysts*, Vol. 2, Ch. 7, Academic Press, 1983.

Before use the catalyst is generally pretreated in a stream of air containing a low concentration of sulfur dioxide. This sulfates the vanadium compounds and avoids an undesirable exothermic reaction when operation begins. However, final sulfation in the reactor does assist in start-up by increasing catalyst temperature faster than using heated feed gas and heat exchange between the beds.

2.1.3.2. *Sulfuric Acid Plant Design*

Conventional contact process sulfuric acid plants operate with four adiabatic catalyst beds, or passes. Heat of reaction is removed after each bed by heat exchange to generate steam or by quenching with cold air.

Up to the 1960s sulfur trioxide was recovered by a single absorption stage at the outlet of the fourth bed. A typical plant design limited the catalyst volume used in the first bed by the minimum inlet temperature of up to 420^{0}C, at which the catalyst was active, and the approach to the equilibrium conversion of sulfur dioxide to sulfur trioxide at about 600^{0}C. By coincidence, the maximum reasonable operating temperature to avoid deactivation in the first bed of catalyst was about 600^0–650^{0}C. Catalyst volumes in the final three beds were also designed to maximize conversion within the same limits of inlet temperature and equilibrium. This made it difficult to achieve more than 98.5% conversion consistently in four beds, even with large volumes of catalyst, as shown in Table 2.8B.

Better conversion was achieved when double absorption plants were introduced on a large scale during the 1960s.[28] As shown in Table 2.8B, the first three beds were operated in the same way as in the conventional plants but when gases left the converter to be cooled they also passed through an intermediate sulfur trioxide absorber. On re-entering to the final bed, a better equilibrium conversion, in some cases up to 99.8%, was achieved as a result of the lower sulfur trioxide content. Double absorption is now widely used despite the extra cost of equipment. The process was even more useful as all of the new plants being built in the United States needed to use double absorption or stack scrubbing systems to comply with strict environmental regulations from 1966 introduced.

2.1.3.3. *Cesium-Promoted Catalysts*

The use of catalysts in which some of the potassium is replaced by cesium provided the more active catalyst anticipated from earlier development work.[29] A striking temperature as low as 320^0C was reported in a full-scale four-bed plant, and operation was possible at a stable bed-1 inlet temperature of 370^0C.

TABLE 2.8. Operation of Single Absorption and Double Absorption Sulfur-Burning Sulfuric Acid Plants.

	Bed 1	Bed 2	Bed 3	Bed 4
A: Four pass *single-absorption* converter. Production 250 tonnes.day^{-1}. Heat exchange cooling between beds 1–2, 2–3 and 3–4 with sulfur trioxide absorption after bed 4. Feed gas 9% sulfur dioxide and 10% oxygen. Catalyst loading 190 liters of catalyst per tonne of acid per day.				
Catalyst volume (m^3)	10	11	22	23
Inlet temperature (^0C)	420	445	435	428
Outlet temperature (^0C)	600	500	450	430
% Conversion SO_2 to SO_3	65–70	88–90	96	98.5 (SO_3 Abs.)
B: 1000 tonnes.day^{-1} *double-absorption* converter. Heat exchange cooling between beds 1–2 and 2–3 with sulfur trioxide absorption after bed 3 and bed 4. Feed gas 10.5% sulfur dioxide and 10.5% oxygen. Catalyst loading 165 liters of catalyst per tonne of acid per day.				
Catalyst volume (m^3)	30	42	42	50
Inlet temperature (^0C)	414	440	440	425
Outlet temperature (^0C)	613	500	495	440
% conversion SO_2 to SO_3	65–70	85–88	96 (SO_3 Abs.)	99.8 (SO_3 Abs.)

Operation of the cesium catalyst at a much lower inlet temperature than the potassium-promoted catalyst achieved a sulfur dioxide conversion in the range 99.2–99.6%. This was comparable to a double-absorption plant but with a lower capital cost apart from increased heat exchange capacity and a slightly more expensive catalyst. It allows producers to use existing four-bed single-absorption units and meet environmental demands without the capital expense of a new plant.

2.1.3.4.   *Sulfuric Acid Plant Operation*

Sulfuric acid plants are designed with optimized catalyst volumes and bed inlet temperatures to give a reasonable approach to equilibrium in each bed to achieve the maximum possible conversion of sulfur dioxide to sulfur trioxide. As shown by the examples in Table 2.8, this results in a significantly smaller volume in bed 1 than the remaining beds. The total catalyst volume used normally corresponds to a *loading* of 180–220 liters of catalyst per tonne of sulfuric acid produced per day although many plants use more, depending on conditions and the source of the sulfur dioxide. Lower volumes of catalyst are normally used in double-absorption units.

2.1.3.5.   *Improved Catalyst Shapes*

The main problem in operating sulfuric acid plants using an extruded catalyst is usually increasing pressure drop through the reactor, which can result from dust or other impurities in process gas, which generally deposit at the top of the first bed. The catalyst must, therefore, be removed at intervals, screened to remove the accumulated dust, and replaced. On average the first bed has to be sieved at intervals of 1 to 3 years and the remaining beds at longer intervals. Catalyst life usually exceeds 10 years.

   Modern catalysts are now supplied in a variety of shapes, all with the same composition. These allow longer continuous operation, at a lower pressure drop, by distributing the dust to prevent the formation of a crust. Shapes are available as rings of various diameters, often with fluted surfaces (ribs) and simple fluted extrudates.[30] Use of any shaped catalyst can also offer more than 30% reduction in pressure drop and, often, increased activity to allow more operating flexibility. The best combination of shapes to be used in particular plants is recommended by catalyst suppliers.

2.2. THE DEACON PROCESS

Scheele's discovery of chlorine in 1774 was soon followed by its use to bleach cotton and linen. The Deacon process, which made use of one of the first industrial catalysts to be especially designed rather than discovered empirically,[31] was used to produce chlorine from about 1870, until it was superseded by the electrolysis of brine.

2.2.1. The Process

In the Deacon Process, chlorine was produced from hydrochloric acid, the low-value by-product of the Le Blanc process, by catalytic oxidation with air. Deacon used a copper chloride catalyst that could combine with oxygen and hydrochloric acid and form chlorine through an oxidation/reduction cycle.

2.2.2. Operation

The overall reaction can be summarized as follows:

$$4 \ HCl + O_2 \xrightarrow{CuCl2} 2 \ Cl_2 + 2 \ H_2O \qquad (2.5)$$

Hurter, in 1883, suggested that the reaction mechanism involved three stages:[32]

- Thermal decomposition of cupric chloride by heating at 500^0C in a stream of 40% hydrochloric acid and air:

$$2 \ CuCl_2 \rightarrow Cu_2Cl_2 + Cl_2 \qquad (2.6)$$

- Oxidation of the cuprous chloride by air:

$$2 \ Cu_2Cl_2 + O_2 \rightarrow 2 \ CuO \cdot CuCl_2 \qquad (2.7)$$

- Hydrolysis of the cupric oxychloride with hydrochloric acid:

$$CuO \cdot CuCl_2 + 2 \ HCl \rightarrow 2 \ CuCl_2 + H_2O \qquad (2.8)$$

The overall reaction is exothermic and controlled by equilibrium.

It was found that despite lower equilibrium yields, the optimum reaction rate was achieved in the temperature range 400^0–450^0C, giving only about 70% conversion of hydrochloric acid to chlorine.[33] Water formed during the reaction had to be removed from the gas leaving the first reactor and a second reactor included to increase conversion up to about 85%.

Operation at high temperatures led to catalyst problems because copper chlorides are volatile and chlorine corroded the equipment. It was reported in 1921 that the cost of copper lost per ton of *bleach* produced was one shilling![34]

At the same time, the low melting point of copper chloride meant that the catalyst operated as a liquid in the pores of the *baked clay* support. The process could not be used successfully on a large scale until the sulfur and arsenic impurities in the hydrochloric acid gas were removed by scrubbing with hot sulfuric acid, which is an early example of gas purification to remove catalyst poisons.[35]

2.2.3. Catalyst Preparation

Deacon catalyst was prepared by impregnating a suitable porous and heat-resistant solid—firebrick and pumice could be used—with an aqueous solution of copper chloride. The final catalyst contained about 10 wt% of copper chloride.

2.2.4. Development

Although the Deacon process was only used for about 40 years, it is still of interest for two reasons: as an example of a catalyst selected by logical rather than empirical procedures and as an illustration of the need to remove poisons from process gases. Derivatives of the Deacon catalyst are still used in the production of ethylene dichloride, by the oxychlorination of ethylene.

2.3. CLAUS SULFUR RECOVERY PROCESS

Sulfur recovery from sour natural gas or refinery off-gas streams is not only essential to avoid air pollution but also has become the main source of elemental sulphur, the key raw material for sulfuric acid production.

Claus sulfur recovery plants were reintroduced in about 1950, when a shortage of Frasch sulfur was anticipated. Subsequently, capacity was increased further with the need to process crude oils with high sulfur content. By the 1990s some two-thirds of US refineries, representing almost 90% of the crude oil treated, had acid gas treatment facilities to recover more than 60% of the total sulfur in the crude. This proportion will increase even further to meet new Environmental Protection Agency sulfur emission regulations, and virtually all refinery projects include new sulfur units or plans to expand existing facilities. The additional sulfur produced will result in the closure of Frasch sulfur capacity.

World production of Claus sulfur was about 22 million tonnes during 1998 (production capacity 45 million tonnes)—an important contribution from such a relatively uncomplicated catalytic process.

2.3.1. The Claus Process

The Claus process to recover and recycle sulfur in the Le Blanc process, based on the procedure suggested by C. F. Claus in 1883,[36] was introduced in 1887 by A. M. Chance. *Alkali waste* containing calcium sulfide was suspended in water, and hydrogen sulfide was generated by pumping carbon dioxide through the slurry:

$$CaS + CO_2 + H_2O \rightarrow CaCO_3 + H_2S \qquad (2.9)$$

Sulfur could then be recovered by passing the hydrogen sulfide, in a stream of air, through a kiln containing an iron catalyst.

The modern two-step process converts hydrogen sulfide mixed with a stoichiometric volume of air to sulfur. In theory, one-third of the hydrogen sulfide is oxidized to sulfur dioxide in a carefully designed furnace, while the remaining hydrogen sulfide reacts with the sulfur dioxide to produce sulfur in two or more reactors containing a suitable catalyst.

$$2 H_2S + 3 O_2 \rightarrow 2 H_2O + 2 SO_2 \qquad (2.10)$$

$$SO_2 + 2 H_2S \rightarrow 2 H_2O + 3 S \qquad (2.11)$$

In practice, up to 70% of the reaction can take place in the furnace before the gas is passed directly to the reactors.[37] During the 1950s Claus plants operated at 90–95% conversion and tail gas containing the residual sulfur compounds was passed into the refinery fuel gas system.[38]

Complete conversion of hydrogen sulfide cannot be achieved in Claus plants because the reaction is limited by equilibrium, and unavoidable side reactions in the furnace lead to the formation of carbon disulfide and carbon oxysulfide, which are difficult to remove. Recent environmental legislation has required that the overall conversion of hydrogen sulfide should now exceed 99%. To achieve this requirement, new processes have been developed which can be added on to the tail of existing Claus plants, to meet the new target.

2.3.2. Claus Plant Operation

Sulphur often occurs in crude petroleum as a complex mixture of organosulphur compounds, such as sulfides, thiophenes and benzthiophenes. These are fairly intractable compounds and need to be converter to hydrogen sulfide prior to separation using the Claus Process. This is usually achieved by hydrogenolysis of the sulphur derivative over cobalt/molybdenum or nickel molybdenum catalysts. Hydrogen sulfide is then separated from hydrocarbons by absorption in diethanolamine solution, but is usually contaminated with carbon dioxide.

Acid gas from refinery streams contains 70–90% hydrogen sulfide, whereas acid gas recovered from natural gas is often more diluted. The hydrogen sulfide content of feed gas to the furnace has a significant effect on both plant and catalyst operation.

With hydrogen sulfide concentrations greater than 60% the flame temperature is stable and all the acid gas and air pass directly to the furnace. With concentrations less than 60% it may be necessary to preheat the gas mixture or even to split the stream so that 37% of the hydrogen sulfide burns in the furnace and the remainder goes directly to the first catalytic reactor.

During combustion, some 60–70% of the hydrogen sulfide is converted directly to sulfur. Flame temperature depends on the hydrogen sulfide content of the feed gas and can reach almost $1300^\circ C$ with more than 90% hydrogen sulfide.

Careful furnace and burner design is essential to maximize sulfur formation and to ensure complete combustion of hydrocarbon impurities that would otherwise damage the catalyst. Residual oxygen in gas from the furnace can also poison the catalyst. A significant excess of air must be avoided. At the temperature of the flame, nitrogen and oxygen will combine to form a small amount of nitric oxide. This will catalyse the oxidation of sulphur dioxide to the trioxide, which will form sulfates on the catalyst and lead to deactivation. Side reactions affecting the process also take place in the furnace. For example:

- Carbon disulfide forms by reaction of sulfur with hydrocarbons:

$$CH_4 + 2\,S \rightarrow CS_2 + 2\,H_2 \qquad (2.12)$$

$$CH_4 + 4\,S \rightarrow CS_2 + 2H_2S \qquad (2.13)$$

  Both reactions are rapid and the carbon disulfide concentration at equilibrium reaches a maximum of almost 20 ppm at $950^\circ C$. Carbon disulfide formation becomes negligible as the furnace temperature approaches $1300^\circ C$.

- Hydrogen sulfide cracks to form hydrogen, which produces carbon monoxide by the reverse water gas shift reaction with the carbon dioxide. Carbon oxysulfide is formed by reaction of the carbon monoxide with sulphur:

$$H_2S \rightarrow H_2 + S \qquad (2.14)$$

$$CO_2 + H_2 \rightarrow CO + H_2O \qquad (2.15)$$

$$CO + S \rightarrow COS \qquad (2.16)$$

- Carbon oxysulfide and carbon disulfide also form by reaction of hydrogen sulfide with carbon dioxide:

$$H_2S + CO_2 \rightarrow COS + H_2O \qquad (2.17)$$

$$2\ H_2S + CO_2 \rightarrow CS_2 + 2\ H_2O \qquad (2.18)$$

The concentration of carbon oxysulfide reaches a maximum of 10–15 ppm at about 1100^0C but then declines as the flame temperature reaches about 1300^0C. Because the flame temperature is proportional to the hydrogen sulfide content of the acid gas, both carbon disulfide and carbon oxysulfide concentrations are more significant when treating gases containing lower concentrations of hydrogen sulfide, particularly in the range 50–75%.

A typical modern Claus sulphur recovery plant uses several reactors to achieve the equilibrium conversion of hydrogen sulfide. The complex gas mixture from the furnace is cooled to condense sulfur and then reheated before it enters the first catalyst reactor. There are generally three catalytic reactors in series containing a catalyst in series, with coolers at each reactor outlet to condense sulfur as it forms. Typical operating conditions are shown in Table 2.9. Inlet and outlet temperatures in each reactor are controlled at levels high enough to prevent condensation of sulfur on the catalyst.

Owing to the equilibrium limitation, it is not possible to achieve 100% conversion simply by adding more reactors and catalyst to the plant. Overall con-

TABLE 2.9. Operation of Claus Sulfur Recovery Plant.

Capacity	35 tonnes day^{-1} sulfur		
Flow rate	1100–1400 Nm^3 h^{-1} acid gas H_2S content 92% in feed; 7.5% entering reactor 1		
Fuel gas composition	H_2S: ~ 0.35% SO_2: ~0.16% COS/CS_2: ~ 20 ppm		
	Reactor 1	Reactor 2	Reactor 3
Catalyst volume (m^3)	4.6	4.6	4. 6
Inlet temperature (^0C)	240	215	200
Outlet temperature (^0C)	320	235	204
H2S in outlet gas (%)	2	1	0.3
Overall conversion %		98.0–98.4	
Equilibrium conversion (%)		98.8	

Note: The Reactor 1 conditions are a compromise between the need for a temperature as high as 350^0C for a maximum conversion of COS/CS_2 and a lower temperature to achieve better approach to equilibrium for H_2S/SO_2. The reaction rate in Reactor 2 is low owing to the lower H_2S/SO_2 concentration. Only a small amount of remaining H_2S/SO_2 is converted in Reactor 3.

version increases gradually within the following ranges as the number of reactors is increased: two reactors 94–96%; three reactors 96–98%; four reactors 97–98.5%.

A number of tail gas treatments have been developed to increase sulfur recovery efficiency to more than 99%:[37]

- *Cold bed absorption* involves adding two additional Claus reactors in parallel, operating below the dew point of sulphur at 120^0–140^0C. The equilibrium conversion to sulphur is favoured at the lower temperature, but sulphur condenses in the pores of the catalyst, leading to temporary deactivation. Thus, when one bed is saturated, flow is switched to the parallel bed while the first is regenerated at 300^0C. Up to 99% conversion can be achieved.
- All residual sulfur compounds in tail gas are hydrogenated in a bed of cobalt/molybdate/alumina catalyst, operated at 300^0C, after the addition of a suitable volume of hydrogen to the Claus reactor tail gas. The hydrogen sulfide formed can then be recycled to the Claus process furnace. This treatment improves conversion up to 99.9%.
- Alternatively, after hydrogenation of sulfur compounds, the residual hydrogen sulfide can be selectively oxidized to sulfur. A suitable catalyst is 5–6% ferric oxide supported on low-surface-area, high-pore-volume silica. During reaction the iron oxide is converted to ferrous sulfate. Conversions of up to 99.9% are possible, but if the hydrogenation step is omitted, only the residual hydrogen sulfide is oxidized and overall sulphur recovers depends on the volume of residual carbon oxysulfide and carbon disulfide in tail gas.

2.3.3. Claus Process Catalysts

The Claus plants built by A. M. Chance in 1887 used firebrick impregnated with an iron salt as the catalyst. Claus' original idea probably originated from the use in the United Kingdom of bog iron ore, a form of hydrated iron oxide, to remove hydrogen sulfide from town gas. During use, the bog iron ore was converted to a mixture of iron sulfides and free sulfur. Additional free sulfur could then be formed by exposing a partly spent oxide absorbent to air in a series of regenerations. After several *revivifications*, spent oxide contained up to 50 wt% sulfur, which blocked gas flow through the bed. The mass could then be used as a source of sulfur in sulfuric acid production:

$$Fe_2O_3 \cdot H_2O + 3\ H_2S \rightarrow Fe_2S_3 + 4\ H_2O \tag{2.19}$$

$$Fe_2S_3 + 1.5\ O_2 \rightarrow 3\ S + Fe_2O_3 \tag{2.20}$$

Overall this corresponded to the Claus reaction:

$$3 H_2S + 1.5 O_2 \rightarrow 3 S + 3 H_2O \qquad (2.21)$$

Claus felt that the cyclic process could be simplified if the hydrogen sulfide were converted to sulfur by an iron catalyst in a kiln. It was later found that dried Weldon process mud or bauxite would operate as a catalyst at a lower temperature than ferric oxide. This not only extended the life of the kiln but also increased sulfur yield.[39] Problems with blocked beds were overcome as technology evolved and proper reactors containing solid catalyst particles were developed. Thus, the modern Claus sulfur recovery process originated from the statutory obligation to remove sulfur from town gas in Victorian gas works.

Different catalysts were used when the Claus process was reintroduced in refineries in 1940–1950. Bauxite, for example, which was already available in refineries to hydrodesulfurize straight-run naphthas, is a variable mixture of gibbsite and boehmite with iron and silica impurities. When calcined to activate the alumina, it is converted to a catalyst with about 1–12% ferric oxide supported on γ-alumina. Bauxite catalysts were successfully used in the Claus process, giving a sulfur conversion greater than 90%.[38]

Eventually, sulfur recovery was introduced on a larger scale, particularly in Canada, and higher conversion was required to limit sulfur emission.[40] Pure activated alumina catalysts were then introduced in the form of strong spheres that improved gas flow and reduced pressure drop. Alumina catalysts are still the most widely used and give excellent results under normal conditions. However, more stable and active catalysts are needed in some plants,[41] where they have been shown to operate more successfully in the presence of residual oxygen and to be particularly active for the conversion of carbon oxysulfide and carbon disulfide.[42]

The formation of carbon oxysulfide and carbon disulfide in the furnace leads to problems when high overall conversion is required. Alumina catalysts are not sufficiently active to convert carbon disulfide and oxysulfide in the first *reactor* unless a high temperature is reached at the bottom of the bed. When this is not possible, the bottom third of the bed can be loaded with either an iron-promoted alumina or a newer titania catalyst. Cobalt/molybdate/alumina catalysts were also tested in early attempts to hydrogenate the impurities, but it was found that conditions in the first reactor favored sulfur dioxide hydrogenation instead. All of the catalyst types used in Claus sulfur recovery plants are described in Tables 2.10 and 2.11.

2.3.4. Catalyst Operation

Typical Claus plant operating conditions are shown in Table 2.9. Temperature in the first reactor is a compromise between the need to remove any carbon oxy-

TABLE 2.10. Catalysts Used for Claus Sulfur Recovery.

	Standard alumina	Oxygen-sulfation guard
Al_2O_3 (wt%)	93–95	~91
Na_2O (wt%)	0.3–0.35	Traces
SiO_2 (wt%)	0.02–0.03	0.4
Fe_2O_3(wt%)	0.02	8
Loss on ignition (wt%)	4.5–6.5	Loss free basis
Bulk density (kg liter^{-1})	0.6–0.7	Less than 1
Surface area (m^2 g^{-1})	340–380	225–275
Pore volume (ml g^{-1})	0.5–0.6	0.5
X-ray diffraction	γ-Al_2O_3	γ-Al_2O_3
	Standard titania	
TiO_2 (wt%)	85–95	
NiO (wt%)	0–6	
Bulk density (kg liter^{-1})	0.9	
Surface area (m^2 g^{-1})	100–140	
Pore volume (ml g^{-1})	0.3	

sulfide and carbon disulfide that may be present and achieve maximum hydrogen sulfide conversion. High temperatures favor the removal of carbon sulfides and lower temperatures favor sulfur formation.

There are several typical catalyst operating problems. The most common is the deposition of elemental sulfur in the catalyst pores at low temperature. Alumina catalysts are soon saturated with sulfur if the operating temperature is less than 270°C. The *macro* pore volume of catalysts should, therefore, be high, and have the smallest particle size possible, consistent with a reasonable pressure drop at maximum space velocity. This increases the rate of diffusion in and around the catalyst particles. Operating temperature in the first reactor should also be high enough to increase the rate of reaction and avoid sulfur deposition.

TABLE 2.11. Catalysts Used for Claus Plant Tail Gas Treatment.

Tail gas hydrogenation		Tail gas incineration	
CoO (wt%)	3.0–4.0	Fe_2O_3 (wt%)	5–6
MoO_3 (wt%)	12.0–14.0	SiO_2 (wt%)	Balance
SiO_2 (wt%)	< 1		
SO_4 (wt%)	< 2		
Al_2O_3 (wt%)	Balance		
Loss on ignition (%)	1.5		
Surface area	250 m^2 g	Surface area	40–45 m^2 g^{-1}
Pore volume	0.6 ml g^{-1}	Pore volume	0.8 ml g^{-1}
Form	Extrusions 3 × 6 mm	Form	Granules
Bulk density	0.6 kg liter^{-1}	Bulk density	< 1.0 kg liter^{-1}

Coke is deposited in the first reactor, particularly when the flame temperature is low, if hydrocarbons are not completely oxidized in the furnace. This can be avoided with proper furnace design. Hydrothermal sintering of the catalyst is

During normal operation the first reactor catalyst will not normally contain more than about 5–10 wt% of sulfur.

possible during start-up or shut-down if the temperature exceeds $500^{\circ}C$ in the presence of water, and this will reduce activity. If sulfur trioxide forms in the furnace from oxidation of sulfur dioxide, it will react with the catalyst to form aluminum sulfate, which also reduces catalyst activity.

The catalyst can also be sulfated at lower temperature by a complex series of reactions with sulfur dioxide, and the catalyst can contain up to 3% of combined sulfur under normal operating conditions. It has been suggested that sulfur dioxide is strongly chemisorbed by surface hydroxyl groups to give a sulfite intermediate. This reacts with sulfur vapor to give a thiosulfate intermediate that reacts, in turn, with a neighboring hydroxyl to form sulfate. This does not necessarily deactivate the catalyst.

The presence of even a few hundred parts per million of oxygen, however, can cause immediate catalyst deactivation. Sulfate is produced on the active sites,[42] by interaction with sulfur dioxide resulting in an affect similar to that of sulfur trioxide. The effect of sulfation is more severe in the second or third beds, which operate at a lower temperature.

Oxygen *poisoning* can be reduced to a certain extent by placing a layer of alumina containing iron or nickel oxide above the catalyst in the first reactor. This converted oxygen to water but at high oxygen levels ferrous sulfate was formed. A further benefit of these guard catalysts was that a higher proportion of any carbon disulfide and carbon oxysulfide in the gas could be converted.

Sulfation was more effectively controlled by the use of titania catalysts, which were not affected by oxygen concentrations of several thousand parts per million. This was partly because the thiosulfate intermediate on titania is unstable above $100^{\circ}C$, and because surface sulfates on titania are more easily reduced with hydrogen sulfide.[43] This means that a titania surface is free from sulfate, whereas sulfate blocks an alumina surface. Further advantages of using titania are that it can operate at a higher space velocity than alumina and convert a greater proportion of any carbon disulfide and carbon oxysulfide present.

2.4. AMMONIA SYNTHESIS

Ammonia, or alkaline air, was isolated by Priestley in 1724, who found that it could be decomposed by electric sparks to give an increased volume of an inflammable gas. Later, it was shown that the decomposition product was a mixture of hydrogen and nitrogen and that the reaction was reversible because 100% decomposition was not achieved at the elevated temperature required.[44]

Ammonia could be formed when a mixture of nitrogen and hydrogen was exposed to electric sparks. Ramsay and Young also found that traces of ammonia formed when hydrogen and nitrogen were passed over a heated platinum/titania catalyst on a porous support. Hlavati and the Christiania Minekompanie in Norway both produced some ammonia using a supported titanium catalyst, with or without platinum, and disclosed the use of supported catalysts containing the oxides of antimony and bismuth, and alkali or alkaline earths containing small amounts of platinum.[45] Dufresne (alias Charles Tellier) demonstrated the production of ammonia in a cyclic process by heating spongy *titaniferrous* iron alternately with nitrogen and hydrogen and suggested operation at 10 atm pressure.[46] A similar cyclic process, devised by De Motay, reacted redhot titanium nitrides alternately with hydrogen and nitrogen.[47]

Le Chatelier began to work on the high-pressure formation of ammonia in 1901, but discontinued his experiments following a serious explosion.[48] By 1900 it was thought that it should be possible to synthesize ammonia from its elements, but it was not yet known whether a suitable industrial process using catalysts could be developed.

2.4.1. Sir William Crookes

Both Liebig, in the book he published in 1840,[49] and Lawes, in his work at Rothamsted in 1845,[49] recognised that nitrogeneous substances such as ammonia or nitrate were essential for healthy plant growth. Lawes, in particular, stressed that additional nitrogen in the form of mineral fertilisers would be required. The *nitrogen* problem had become widely recognised as a serious issue towards the end of the nineteenth century, since by 1890, it was clear that the available quantities of sodium nitrate from Chile, Chile saltpeter, would not be sufficient to meet the anticipated future demand. It was equally clear that other sources of supply would soon be required.

It is therefore not surprising that Sir William Crookes chose this subject for his presidential address to the British Association for the Advancement of Science in Bristol in 1898.[50] He had already demonstrated his *flame of burning nitrogen* in 1892 by combining the nitrogen and oxygen in air to form nitrogen oxides at high temperatures. He appealed to chemists for other, more economic and practicable methods to fix atmospheric nitrogen to supply the fertilizers needed to produce feed for a growing world population.

The direct production of nitric oxide from air at high temperatures in an electric arc by the Birkeland and Eyde or Cyanamide processes was feasible, but could only be used in locations with abundant and cheap hydroelectric power. This clearly was not the long term answer, and a series of significant advances initiated by Ostwald at Leipzig, following discussions with William Pfeffer, soon followed. Ostwald, himself, worked on the catalytic synthesis of ammonia, and its oxidation to nitric acid.

TABLE 2.12. Rates of Growth in World Population and Ammonia Production.

	World population (billions)	Synthetic ammonia capacity (million tones year^{-1})
1804	1	—
1927	2	~ 1
1960	3	1.6
1974	4	80
1987	5	145
1999	6	175

Note: Only approximately 70–80% of ammonia used as fertilizer.

Since the introduction of the more economic Haber process in 1913, it has been interesting to compare the worldwide increase of ammonia production with the increase in world population. Details are shown in Figure 2.4 and in Table 2.12.

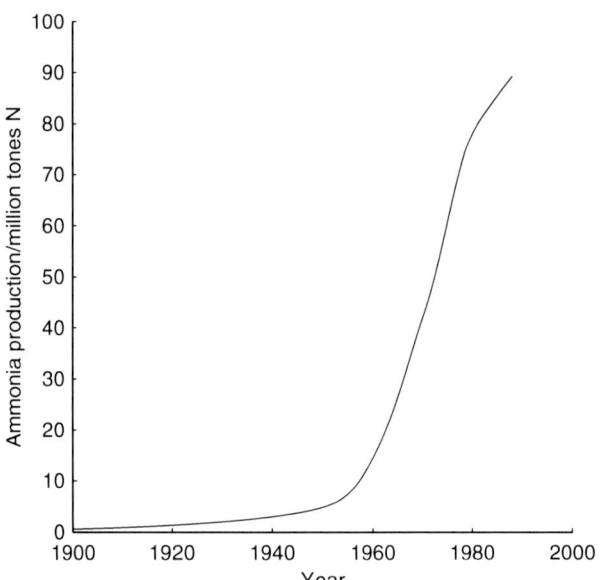

Figure 2.4. The growth of world ammonia production. Reprinted from *Catalyst Handbook*, 2nd ed., Ed. by M. V. Twigg, Wolfe Publishing, LTD., London, England, 1989, by kind permission of M. Twigg.

2.4.2. Development of the Ammonia Synthesis Process

Despite all the preliminary investigations, the production of ammonia using catalysts on an industrial scale was not possible until 1913 at Oppau, because it was not until then that the reaction was properly understood, a practical catalyst had been developed and suitable equipment available. The ammonia process resulted from the theoretical and experimental work of Ostwald, Nernst, and, particularly, Haber, who, with a number of associates, began a systematic investigation of the reaction over a wide range of temperatures and pressures to determine the equilibrium constants. Haber also measured the activity of a number of different catalysts.[51]

Ammonia synthesis has been described as the first example where the knowledge of thermodynamics led research to the most practicable industrial process.[52] We now know from the thermodynamics of the reaction that the formation of ammonia is favoured by low temperatures and high pressures. It was thus possible to devise the conditions required for economic operation at low equilibrium conversion and then to develop a catalyst and the high-pressure equipment that was not available at the time.

Ostwald produced ammonia in laboratory experiments at atmospheric pressure using an iron wire catalyst and claimed that he obtained a relatively high yield.[53] He had, however, nitrided the iron during its pretreatment with ammonia and withdrew his patent application. From about 1904, Haber and a group of coworkers, funded by the Margulies brothers from Vienna, began to investigate the equilibrium conversions in the ammonia synthesis reaction using an iron catalyst at the Technical University of Karlsruhe. Although the conversion at atmospheric pressure was too low for an industrial process, it was known that conversion could be increased at higher pressure. Nernst, who used theoretical calculations to query some of Haber's early experimental results, experimented with Jost at pressures up to 75 atm.[54] He used catalysts including iron and manganese, but felt that the process would still not be commercially attractive because the conversion to ammonia was less than 1%. Haber, however, was more optimistic due to his experience with osmium and uranium carbide catalysts. He realised that despite the low equilibrium constant, the process could work at high pressures, and he achieved up to 6% ammonia in the gas stream in experiments at 200 bar. This suggested that a process could be feasible provided that the synthesis gas was recycled continuously in a loop and that the product ammonia was removed from the synthesis gas after each cycle through the catalyst. The patents, which were issued in 1908 and 1909, had many of the features of the modern process, including the recirculation of synthesis gas and the use of heat exchange between the gases leaving and entering the reactor.[55]

2.4.3. Commercial Application of Ammonia Synthesis Catalysts

The commercial process was developed after 1910 when Haber began his collaboration with BASF. Carl Bosch, who was in charge of the development, began to look for an efficient, cheaper catalyst. Osmium could be operated successfully at 550^0–600^0C and 175–200 atm giving an ammonia conversion up to 6%. It was, however, expensive, poisonous, unstable in air, and, more important, almost unobtainable. These were not the qualities required for an industrial catalyst. Furthermore, the iron reactor then available was found to suffer from hydrogen embrittlement under operating conditions and could explode. Uranium, the other active catalyst favored by Haber, was also expensive and, unfortunately, was rapidly poisoned by traces of water and oxygen in the synthesis gas.

One of the first innovations made by Bosch was the introduction of a comprehensive program of catalyst testing using thirty specially designed laboratory units. These are described as using only 2g of catalyst—a tremendous achievement in those days. Alvin Mittasch was in charge of the testing program.[56]

It was thought that iron would be the best catalyst, despite its relatively poor activity in earlier investigations. In one of the fortunate coincidences that are typical of industrial developments, a particular kind of magnetite from Sweden that Mittasch found in his laboratory was used in the tests. It gave excellent results and, even now, is used for industrial catalyst production. It will continue to be so until a better catalyst is discovered or the particular deposit in Sweden is exhausted.[56]

An intensive investigation of catalyst promoters was then undertaken, and by 1910 an alumina-promoted iron catalyst was produced that had the same activity as the previously favored osmium and uranium types. This was followed in 1911 by an alumina/potash-promoted iron catalyst that was more stable.[57] Finally, a few years later, calcium oxide was discovered to be a third promoter.

During tests full-scale operating procedures were worked out and catalyst poisons, including sulfur compounds, chlorides, phosphates, arsenic, and relatively common oxygen compounds such as water and carbon monoxide, were identified. By 1922, when several full-scale ammonia plants were operating, a total of about 20,000 tests had been completed![58]

Despite the novelty of the new process, a small pilot plant was rapidly constructed in 1909 so that metallurgical and operating problems could be investigated. The first full-scale, 30-tons.day^{-1}, ammonia plant was then built at Oppau in 1912 and was operating by 1913. By 1916 production had been increased to 250 tonnes.day^{-1} and a further plant was operating at Leuna with a capacity of 36,000 tonnes.year^{-1}, which had increased to 240,000 tonnes.year^{-1} by 1918. An early ammonia synthesis converter is shown in Figure 2.5.

2.4.4. The Haber–Bosch Synthesis Reactor

The development of a high-pressure synthesis reactor was difficult because the carbon steel available for fabricating the shell quickly burst as a result of decarbonization (hydrogen embrittlement) as hydrogen diffused through the steel and removed carbon to form methane. Problems with decarbonization were overcome by the use of a soft iron lining in the carbon steel shell. The outer shell was also drilled so that any hydrogen passing through the liner could escape.

The ammonia synthesis reaction is exothermic, but as a result of the low conversion heat loss from the small early reactor exceeded the heat of reaction.

Figure 2.5. An early ammonia synthesis converter. Reprinted with permission from the Imperial Chemical Industries PLC.

Therefore, additional heat had to be supplied to the catalyst bed and this was done in a number of ways:

- In early units the reactor was heated externally, by gas burners, although this had the disadvantage of further weakening the shell.
- A special air burner at the top of the catalyst bed could increase the gas temperature and was used until 1922, despite the poisoning effect of water on the catalyst.
- Larger reactors only used a gas heater at start-up with reverse gas flow to avoid catalyst poisoning.
- From 1920 better steel alloys that resisted embrittlement became available.

By 1925, new reactors had been developed by using improved chromium/vanadium steel alloys and internal heat exchangers. The outer shell was protected from overheating by passing the cold synthesis gas down the annular space between the shell and the catalyst basket as it entered the vessel. A typical plant was operated with the catalyst temperature in the range 500^0–650^0C and at a higher pressure, up to 300–350 atm, which allowed higher conversion and easier ammonia removal by water scrubbing. While these conditions should give a theoretical conversion in the range 8–11%, the actual conversion was only 7–9%.

A reactor, producing 20 tons.day^{-1} of ammonia weighed about 70 tonnes, was 12 m long, and held a basket of 80 cm internal diameter. It took about 3 days to change a deactivated catalyst and restart operation. In those days, in order to increase production, typical ammonia plants operated with several small reactors rather than a single large one.

2.4.5. Conclusion

The production of ammonia during the early 1900s stimulated the increasing use of industrial catalysts. Development of the synthesis catalyst set a pattern for all other catalysts subsequently used in chemical and refining processes.

Theoretical and experimental effort had shown that the process was feasible. This was followed by the development of practical equipment and full-scale operation. A relatively cheap and reliable catalyst was thoroughly tested and produced economically in what were then large volumes. Finally, both the process and catalyst were gradually improved as the scale of operation expanded.

The pioneering work of Haber, Bosch, and Mittasch led to a process which has survived in more or less the same form as it is used today. Their achievement led to the introduction of chemical engineering, high pressure technology and consolidated the ideas of unit processes. New materials were developed for use with hydrogen at high pressures.

From 1940, when synthesis gas first was produced from natural gas rather than coal, single-stream ammonia plants were developed and the process was subject to an ongoing series of improvements. Improved catalysts based on the same natural magnetite were made as the internal structure of magnetite and the function of the promoters could be investigated with modern analytical procedures. Catalyst life with purer synthesis gas can now exceed 15 years.

Although better reactor designs were introduced, the use of almost 200 tonnes of catalyst in a single vessel led to problems with packing, activation, and pressure drop. Furthermore, spent catalyst is very pyrophoric and large volumes of spent catalyst are difficult to deal with. Catalyst reduction could last for almost a week, so the first modern catalyst innovation was prereduction and stabilisation of the catalyst before it was loaded into the converter. This made plant start-up more efficient. Attempts to provide a more uniform, pelletted catalyst were not successful and crushed granules are still used.

Since the early 1980s there have been several catalyst developments, including the use of cobalt oxide with magnetite to increase activity. The most significant, however, is the successful use of a ruthenium catalyst supported on a special carbon and promoted with cesium and barium. Although still expensive, cost and availability should not restrict the use of ruthenium in the way that osmium was excluded by Bosch, provided that the metal is recycled.

2.5. COAL HYDROGENATION

2.5.1. The Bergius Process

Bergius began his experiments on the high-pressure hydrogenation of coal using small autoclave reactors as early as 1911. His aim was to increase the yield of liquid products from coal carbonization and, like Ipatieff, he worked at the time that BASF was developing its new high-pressure ammonia process. By 1921 he had built a small, continuous, semitechnical unit at Reinau/Mannheim with horizontal stirred reactors and was obtaining encouraging results. These units operated until 1927.[59]

In the plant, a slurry of coal and heavy oil was hydrogenated using about 4 wt% luxmasse as the catalyst. Luxmasse, which is rich in iron with some titania, is the residue from bauxite after alumina extraction. Hydrogenation conditions were in the range 450^0–480^0C and 100–150 atm of hydrogen, yielding 40–50 wt% of liquid hydrocarbons, depending on the type of feed used. Residual solids and heavy oils could be recycled.

Bergius certainly recognised the relationship between his work and the catalytic hydrogenation of heavy crude oil fractions, relative to the newly introduced thermal cracking.[60] Thermal cracking of crude oil fractions was first used in refineries around 1911–1912 to increase the yield of gasoline. By about 1924 the thermal cracking process was an essential part of refinery operation, particularly in the United States, where more than 25 million motor vehicles were registered. The attraction of a more efficient catalytic process was obvious and led big oil companies such as Royal Dutch Shell, who apparently funded some of Bergius' work, and Standard Oil, who later worked with I. G. Farben, to take a keen interest at a time when crude oil reserves were thought to be declining.

2.5.2. Commercial Development by I. G. Farben

In 1920 after the war, demand for synthetic ammonia had fallen and Bosch felt that the high-pressure ammonia plant at Leuna might be converted to hydrogenate coal. Experimental work began at Oppau, and the Bergius patent rights were acquired in 1925, at the time that I. G. Farben was formed. Soon afterward I. G. Farben decided to convert the plant at Leuna to produce 100,000 tons.year^{-1} of oil products. Operation started in June 1927, but it was some time before the technical problems were sorted out and the plant could be operated successfully. Costs were therefore extremely high and operation was stopped. The plant was restarted in 1931, when plans were made to treble capacity and to build three more plants in other parts of Germany in 1935.[61]

    I. G. Farben needed to develop efficient, sulfur-resistant catalysts and to improve the process. Two converters were operated in series. Light oils formed in the first converter were removed by distillation before further hydrogenation of the residue took place in the second converter.

2.5.3. Cooperation between I. G. Farben and Standard Oil

I. G. Farben and Standard Oil began to talk about coal hydrogenation in 1925, and in 1927 they signed an agreement to cooperate in the research and development of oil hydrogenation. At that time, Standard Oil decided to build two gas oil hydrogenation units, each with a production capacity of 40,000 tons.year^{-1} of petrol, solvents, lube oil, and kerosene at Baytown, New Jersey, and Baton Rouge, Louisiana.[62] Hydrogen for these plants was to be made in the first commercial hydrocarbon steam reformers using a process and catalyst developed by I. G. Farben. Standard Oil planned to use the low-molecular-weight waste gases from the hydrogenation process as the hydrocarbon feed to the steam reformers. Standard Oil acquired the world rights to oil hydrogenation in 1928.

2.5.4. Commercial Developments by ICI

There was, of course, a worldwide interest in producing gasoline by coal hydrogenation, and those companies which developed ammonia processes were able to establish production facilities. In 1927, ICI acquired the patent rights of the British Bergius syndicate and started to work independently on the coal hydrogenation process. To suit local conditions it was decided to modify operation and produce gasoline from bituminous coal. Coal was chosen because tar, the preferred feed, was not readily available in the quantities needed. A large pilot plant was built by 1929 and was in operation until 1931, by which time it had been established that at least 60 wt% of gasoline could be produced from coal. A full-scale plant was then designed to start operating in 1935 to produce a nominal 100,000 tons.year^{-1} of gasoline.[63]

2.5.5. International Cooperation

In 1931 the four major companies interested in the hydrogenation process— I. G. Farben, Standard Oil (New Jersey), Royal Dutch Shell, and ICI—became associated in the International Hydrogenation Patents Company to pool their patent rights and exchange technical information.[63]

2.5.6. Coal Hydrogenation Processes

Coal hydrogenation processes were being developed at a time when there was no known theory of catalysis. High-pressure equipment was not generally available and was, therefore, very expensive. For special applications, potential operators had to design reactors and valves themselves. For these reasons progress in developing coal hydrogenation in the 1920s was fairly slow. However, because of the continuing fears that crude oil supplies would decline, work on the project went ahead and a range of new catalysts was developed. The most active chemical companies in Europe were I. G. Farben in Germany and ICI in the United Kingdom, and they were also working on a wide range of other catalytic processes at the time. Similarly, the international oil companies Standard Oil of New Jersey and Royal Dutch Shell were also introducing new catalytic processes for use in refineries.

These activities made significant contributions to both sides during World War II for the production of aviation gasoline. Subsequently rapid developments led to many other chemical and refinery processes based on catalysts. These are listed in Table 2.13.

Table 2.14 shows the total production of oil products in Germany and aviation gasoline in the United Kingdom by catalytic hydrogenation of coal or creosote from 1935 to 1946.

It was found during pilot plant testing that maximum yields of liquid hydrocarbons could only be obtained if the coal, or later tar and creosote, was partly

TABLE 2.13. Processes Developed from Coal Hydrogenation

Direct	Indirect
Coal hydrogenation	Hydrocarbon steam reforming
Tar/creosote hydrogenation	Liquid/gas phase reactors
	Catalyst sulfiding
	Hydrodenitrogenation
Gas oil hydrogenation	Hydroforming (catalytic reforming)
	Hydrotreating
	Hydrocracking

hydrogenated as a slurry in one reactor and the conversion completed by vapor phase hydrogenation in a second reactor. Operating conditions and the catalysts used depended on the feed to the process and whether the product required was a mixture of oil products or simply gasoline.

2.5.6.1. *The I. G. Farben Process*

The I. G. Farben process was used to produce mixed oils.[64] Originally the coal, slurried with heavy oil and a molybdenum catalyst, was hydrogenated at 400^0C and 200 atm. Light and middle oils were then separated. The residue was again hydrogenated at 450^0–470^0C and 200 atm, this time with a cobalt sulfide catalyst, to produce more light and middle oils. Residual heavy oil was recycled as a slurry with more coal. About 75% of the coal was converted into useful light and middle oils. Eventually I. G. Farben used sulfided iron catalysts at 700 atm to achieve higher conversion.

TABLE 2.14. Production of Oil and Gasoline from Coal, 1935–1946.

Year	Feed		
ICI (high-octane gasoline)[a]	Coal	Creosote	Gas oil
1935–1939: total tons	170,000	320,000	—
1940–1946: total tons	—	630,000	~ 1.2 million
Germany (liquid oil products)	Coal	Tar/pitch	Brown coal
By 1945: tons.year^{-1}	740,000	910,000	~ 2 million

Note: Operation was at different pressures owing to the different catalysts used by ICI.
[a]Iso-octane production by ICI from waste C4 gases was about 10,000 tons.year^{-1} from 1941 and 60,000 tons.year^{-1} from 1943.

2.5.6.2. The ICI Process

ICI produced gasoline.[65] Coal was slurried with heavy oil (bp > 400°C) and 2% of an iron oxide catalyst (later an improved stannous oxalate catalyst was used) and hydrogenated at 420°C and 250 atm. During operation in the tall, narrow reactor the slurry was agitated by the upward flow of hydrogen. The ratio of hydrogen to slurry was about 1000, with a residence time of up to 2 h.

The liquid product was separated into gasoline (< 200°C), middle oil (200°–300°C), and heavy oil (> 300°C) fractions. Heavy oil was recycled with more coal to the first reactor for further hydrogenation. Middle oil was then hydrogenated in a second, vapor phase reactor at 480°C and 250 atm with 1 m^3 of hydrogen per kg of oil at a residence time of 3 min. The catalyst was pelletted tungsten sulfide. Gasoline was again separated and the residual middle oil recycled. The yield from both stages exceeded 60 kg of gasoline per 100 kg of coal. ICI operated two stages at 240–260 atm: stage 1 to convert either coal, a heavy oil slurry, or heavy creosote to middle oil that was then cracked in stage 2 to produce gasoline. Details are shown in Table 2.15.

TABLE 2.15. Coal and Creosote Hydrogenation Process and Catalysts.

Conditions	Reactor details	
First stage (liquid phase)	47% coal/heavy oil slurry	Creosote > 375°C
Catalyst/ton feed	Tin oxalate (0.02%) NH$_4$Cl (0.2%)	Tin oxalate (0.01%) CCl$_4$ (0.04%) iodine (0.02%)
Temperature (°C)	465	445–475
Product:		
heavy oil (%)	6	—
middle oil (%)	43 (to second stage)[a]	70 (to second stage)
gasoline (%)	11	11
C$_1$–C$_4$ (%)	20	17

	Early operation with one reactor	Later operation with two reactors		
		No. 1: 1939–1945	No. 2: 1939–1942	No. 2: 1942–1945
Feed ton^{-1} catalyst	1.1	1.2	1.3	1.0
Temperature (°C)	420	385	400	370
Conversion (%)	54	Removes nitrogen	65	55
Yield (%)	92	Fed to No. 2	88	85–75
Octane	68	—	75	85–75

Note: One of the first catalysts was sulfided ZnO/MgO/MoO$_3$, which gave low yields. Replaced with WS$_2$, which produced low-octane gasoline. Two vapor phase reactors in series used WS$_2$ in the first, with either WS$_2$ on Terrana clay (fuller's earth) or 10% FeF$_3$ on kieselguhr in the second to produce high-octane gasoline.
[a]Second stage = vapor phase.

2.5.7. Catalysts for Coal Hydrogenation

Originally Bergius felt that coal hydrogenation could not be catalyzed because the large quantities of sulfur present would poison the catalysts. He added luxmasse simply to absorb sulfur from the products although, coincidentally, the combination of iron oxide with titania and alumina was an excellent choice of catalyst. Since his first tests, however, the industrial use of the process has depended on catalysts that were developed more or less empirically. It was soon realized that the processes involved in hydrogenating coal were more complex than the simple reactions described by Sabatier and Ipatieff. Different catalysts such as iron oxide or iron sulfide, probably with traces of other metal oxides, were required. These catalysts could be used in the presence of sulfur and were, in fact, even more active when sulfided.[66] Several studies reported that iron, nickel, cobalt, tin, zinc, and copper chlorides were effective catalysts and claimed that ammonium molybdate was particularly active.

An early I. G. Farben patent[64] used a molybdenum catalyst in the first stage of the hydrogenation, probably based on a 31 ZnO:15 MgO:54 MoO_3 mixture, and a cobalt sulfide catalyst in the second stage. At about the same time ICI used an iron oxide or tin-plated iron catalyst in the first, liquid phase reactor.[65] ICI subsequently used stannous oxalate in their first reactor with the addition of ammonium chloride to neutralize the alkaline ash and maintain catalyst activity. Alkalinity was a common problem with all coal feeds and, eventually, I. G. Farben plants opted to increase operating pressure to 700 atm, which allowed them to use a simple iron catalyst which was more resistant to alkali. Later they also used sulfuric acid to neutralize the alkalinity. Until 1935 the preferred vapor phase hydrogenation catalysts used in the second reactor seemed to be the zinc oxide/magnesia/molybdena catalyst. This could only operate at high temperatures and gave relatively low liquid yields with correspondingly high levels of gas formation.

By 1930 I. G. Farben had introduced a new tungsten sulfide catalyst that was extremely active in both the cracking and hydrogenation stages of the process and produced high yields with all feeds. A disadvantage was that often sulfur had to be added to resulfide the catalyst. Despite the higher proportion of gasoline produced compared with early catalysts, the octane number (68–70) was low, because the use of tungsten sulfide results in a decreased aromatic content. When higher-octane gasolines were required, the pure tungsten sulfide catalyst was modified to 10% tungsten sulfide supported on activated montmorillonite (Terrana clay). The new catalyst was just as active but produced gasoline with an increased octane number.

It was found, however, that the catalyst was poisoned by feeds containing more than 5 ppm of nitrogen. This meant that it could only be used directly with crude oil fractions and not with coal or coal tars. Nitrogen poisoning could be avoided by partial hydrogenation of the feed over tungsten sulfide at a low temperature before the final hydrogenation step with supported 10% tungsten sulfide. This produced some low-aromatic gasoline in the first reactor, which was removed by distillation before the final hydrogenation in the second reactor which produced high-octane gasoline. In order to achieve a somewhat higher octane number, at the expense of yield, in the 1940s ICI began to use an iron fluoride catalyst supported on hydrogen fluoride– activated superfiltrol.

2.5.8. Creosote and Other Feeds

The ICI plant was converted to use creosote feed during the war years. This was originally for safety reasons but later coal became too expensive and was in short supply. Creosote could be easily hydrogenated and gave reliable operation in a plant producing more than 100,000 tonnes.year^{-1} of 100-octane aviation gasoline (Figure 2.6).

Gas oil was another readily available alternative to coal and required only half of the amount of hydrogen needed compared with creosote. It was used in a second hydrogenation plant operated by Shell, ICI, and Trinidad Leaseholds, Ltd. (Trimpell, Ltd) to produce more than 300,000 tonnes.year^{-1} of 100-octane gasoline. Butane, a by-product from both plants, was converted by the new UOP

Figure 2.6. ICI creosote hydrogenation plant. Reprinted with permission from Johnson Matthey.

process to iso-octane for use in gasoline (Figure 2.7). In May 1940 the 10,000-tonnes.year^{-1} iso-octane plant at Billingham was the first in the world to use the process.

While coal hydrogenation was not a commercial success it was certainly a wartime strategic necessity for Germany and the United Kingdom. It is probable that the introduction of catalytic refinery processes would have been delayed if Standard Oil, Shell, and ICI had not used the coal hydrogenation catalysts for gas oil hydrogenation and developed the other *refinery-type* processes mentioned previously in Table 2.13.

Figure 2.7. First UOP iso-octane plant at ICI. Reprinted with permission from Johnson Matthey.

2.6. THE FISCHER–TROPSCH PROCESS

The possibility of producing fuels from carbon monoxide and hydrogen (synthesis gas) was investigated as far back as 1913 using alkali-activated cobalt and osmium oxide catalysts supported on asbestos. A mixture of alcohols, aldehydes, ketones, and fatty acids with some aliphatic hydrocarbons was produced at 300^0–400^0C and 100–200 atm. The work was, of course, also stimulated by the development of high-pressure equipment and the production of synthesis gas by BASF.[67] From 1923 Franz Fischer and Hans Tropsch continued work on what they called the *Synthol* process at the Kaiser Wilhelm Institute at Mulheim-Ruhr. Using alkalized iron turnings as a catalyst, they were able to produce *Synthin*, an oily liquid at 400^0–450^0C and 100–150 atm. At high pressures the oil produced by the Synthol catalyst consisted mainly of oxygenated compounds but further experiments at about 7 atm did provide a mixture of olefins and paraffins. Nickel and cobalt catalysts also produced methane and higher hydrocarbons at atmospheric pressure and 200^0–250^0C, but the catalysts were quickly deactivated under these conditions. As all the tests with nickel and cobalt catalysts were carried out at atmospheric pressure, on the assumption that this gave the most desirable product, iron catalysts, which operated more effectively at higher pressure, were excluded.

Full reviews of the early work are given by Robert Anderson[68] and H. H. Storch.[69] The synthesis of hydrocarbons from hydrogen and carbon monoxide *synthesis gas* by Fischer, Tropsch, and their associates became known as the Fischer–Tropsch process. This has been used and developed extensively since 1955 by Sasol, in South Africa, where it is still known as the Synthol process.

Fischer and Tropsch went on to test a number of catalysts based on nickel and cobalt supported on thoria and kieselguhr, which were considered more promising than iron. Their experimental work is a classic example of catalyst and process development that has probably since been followed by many other investigators.

Catalyst preparation evolved through the use of the mixed metal oxides, the decomposition of mixed metal nitrates, and the precipitation of carbonates and hydroxides from solutions of metal salts. Because powders were difficult to handle, granules and pellets were produced using binders and tested in a variety of shapes and sizes.

The first promising catalyst was introduced by 1931 and contained a high proportion of nickel oxide supported on a mixture of thoria and kieselguhr. The convention widely used at the time was to describe composition as 100 parts nickel, 18 parts thoria, 100 parts kieselguhr. Catalysts made with cobalt rather than nickel were more effective but could not be considered commercially at that time because cobalt was not available in sufficiently large quantities. The same problem had, of course, faced Haber and Bosch in the replacement of osmium by iron oxide for the ammonia synthesis catalyst.

A disadvantage with the nickel catalyst, which was active but less selective than cobalt in the reaction, is its short operating life. One reason for this may have been the need to activate the catalyst by reduction at 450^0C before use. The reason for this short lifetime is that nickel silicates are formed during catalyst preparation by reaction of the nickel compound with kieselguhr. These can only be reduced at the relatively high temperature of 450^0C. Nickel crystallites are very prone to sintering at this temperature and the catalysts are somewhat deactivated even before they are charged to the reactor. Silicate formation during catalyst preparation has been a continuing problem with nickel/kieselguhr hydrogenation catalysts. A small amount of copper was added to allow catalyst reduction at a temperature closer to the operating level of 178^0C. Unfortunately, the copper also sintered and catalyst activity was actually decreased.

Subsequent work included attempts to use less of the more expensive materials in the catalyst recipe and to replace them with cheaper *diluents*. This was done empirically adding manganese oxide, magnesia, and more kieselguhr. Although it was a novel approach at the time, it is now a routine procedure in cutting the cost of a catalyst!

Ruhrchemie built a large pilot plant in 1934 and tested a 100Ni:25MnO: 10Al2O3:100 kieselguhr formulation. This approach demonstrated another important stage in catalyst development, namely that small-scale atmospheric pressure tests do not highlight full-scale operating problems. In this case the problem was the need for efficient heat removal from exothermic reactions. The pilot plant also confirmed that although nickel was not very selective, deactivation could be partly restored by regeneration in air and rereduction in hydrogen at 400^0C. A further problem was an unacceptable loss of nickel during operation, presumably as a result of nickel carbonyl formation.

Roelen[70] took over the responsibility for catalyst development in 1934 and began further testing of the previously successful cobalt catalyst used by Fischer. He decreased the quantity of expensive materials and also investigated the effects of copper on reduction temperature. The catalyst developed by Roelen, 100Co:5ThO2:8 MgO:200 kieselguhr, was used in the four large plants built by Ruhrchemie in 1936 and operated during the war. By the end of the war nine plants were using the catalyst with a capacity of about 700,000 tonnes of hydrocarbons per year.

The introduction of magnesia to what seems an already complicated mixture is interesting mainly because it was also included in other nickel catalysts such as the raschig-ring catalysts for steam reforming. It is now realized that the molecular dimensions of magnesia are similar to those of cobalt and nickel oxides, and that magnesium can replace cobalt and nickel in solid solution within a crystalline lattice. This can make catalyst reduction easier and result in the formation of smaller, more stable metal crystallites.

Research on precipitated iron catalysts continued while the first commercial plants were being built as part of the program to find a cheaper catalyst. Results were not encouraging until the operating pressure was raised to 15 atm.[71] This increased both the hydrocarbon yield and the catalyst life. Several catalysts, including a typical fused ammonia synthesis catalyst, were compared in 1936 in a government competition at Schwarzheide. Different promoters were tested, and potash was found to increase activity and selectivity. Later, in 1940 Pichler and Buffeg tested supported ruthenium catalysts at pressures up to 1000 atm and obtained high-molecular-weight waxes.[72]

2.6.1. Postwar Development of the Synthol Process by Sasol

There was no further full-scale operation of the wartime plants after 1946. However, experimental work continued in the United States and the United Kingdom as well as in Germany because of possible future oil shortages. The lack of oil in South Africa had led Anglovaal to take out a license for the construction of a Fischer– Tropsch plant in 1935,[73] but owing to various delays the plant was not actually built until 1955, when Sasol, which was owned by the South African Government, went ahead with the project. The Sasol 1 plant, at Sasolburg, produced up to 200,000 tonnes.year^{-1} of hydrocarbons using two processes. The first, licensed by Lurgi and Ruhrchemie, used fixed bed tube-cooled ARGE reactors, containing a precipitated iron/copper/silica catalyst to provide heavy liquid hydrocarbons and waxes. The second, licensed by Kellogg, operated with a circulating fluid bed of crushed fused iron catalyst and made hydrocarbon gases and gasoline.[74]

The fixed bed process worked well. It was not easy, however, to circulate the fluidized dense iron catalyst through the reactor and back through the separator without further development. Good operation was eventually made possible and the process was successful in the Sasol *Synthol* reactor. Typical product distributions for the two processes are shown in Table 2.16.

Early catalysts used in the Synthol process were produced as follows:[74]

a)   Ruhrchemie catalyst for the in fixed bed tubular reactor:
   - Solutions of cobalt, thorium, and magnesium nitrates were added to a sodium carbonate solution up to pH 7.
   - Kieselguhr was added and the slurry filtered, dried, washed, and calcined.

b)   Alternative recipe used by Sasol:
   - Precipitation of basic carbonates and hydroxides from copper and ferric nitrate solutions with sodium carbonate solution.
   - Filter and wash precipitate before slurrying the solid with potassium silicate solution to give 25 g SiO_2 per 100 g iron.
   - Wash with dilute nitric acid to remove any excess potassium leaving 5 g K_2O per 100 g iron.
   - Filter and extrude partly dried filter cake. Dry to less than 10% water.
   - Add other promoters.

TABLE 2.16. Product Distribution from the Fisher–Tropsch Process.

Product	Fixed bed	Fluid bed
C_1–C_4	13	43
C_5–C_{11} (gasoline)	18 (ON35)	40 (ON65)
C_{12}–C_{18} (diesel)	14 (CN75)	7 (CN55)
C_{19}–C_{23} (jet fuel)	7	—
C_{24}–C_{35} (medium wax)	20	> C_{19} 4
> C_{35} (hard wax)	25	—
Water soluble neutral	3	5
Acid	0.2	1

Note: Light gas + LPG reformed to hydrogen and cracked to ethylene. α-Olefins separated for use in polyethylene or detergents. Alcohols and ketones extracted.

c) Fused magnetite catalyst used in transported fluid bed reactors:
- Fuse magnetite, such as mill-scale, with potash, alumina, silica, and other promoters at 1500^{0}C.
- Chill cast molten mass and mill to required size grade.
- Oxides such as Al_2O_3, MgO, TiO_2, and CrO_3 were claimed to form solid solutions whereas K_2O and SiO_2 remained on crystal boundaries. The composition was not published.
- Catalyst must be reduced before use.

Following the successful development of the Synthol process, Sasol went ahead and built two larger plants at Secunda, Sasol 2 and Sasol 3, which were based on coal. A further plant using the Synthol reactor but with natural gas as feed was built at Mossul Bay by Mossgas.

The atmospheric pressure adiabatic reactors used during the war by Ruhr-chemie were not very efficient, and although research on reactor design continued, nothing could be done at the time to make any improvements. Metal plates, 7 mm apart, were stacked vertically in the reactor. During operation, the catalyst between the plates was cooled by water running through horizontal tubes passing through the plates. This arrangement was inefficient at the low gas hourly space velocity through the catalyst.[75]

At the time Ruhrchemie made some minor improvements by using concentric tubes in the medium-pressure reactors, with catalyst in the annular space and cooling water flowing around the tube and through the inner space. This was still inefficient at low gas space velocity. The ARGE reactors used by Sasol in 1955 were conventional boiling water tubular reactors with gas recycle to limit heat evolution. A typical wartime reactor contained 1250 tubes, whereas the early Sasol reactor used more than 2000 tubes.

Present-day reactor design is much more efficient and, as a result, catalyst operation has been improved. The nonadiabatic reactor recently developed by UOP is similar to the original Ruhrchemie design, with the catalyst packed between the vertical plates of a heat exchanger.[75] This allows the feed and coolant to flow in any direction to control the temperature more efficiently in both exothermic and endothermic reactions. Pressure drop through the catalyst bed can also be selected. The advantages are a better approach to reaction equilibrium at high rates with better selectivity and less recycle. Nonadiabatic reactors provide a good example of how interesting ideas in catalytic processes could not be developed even 50 years ago because the technology was not available.

Fluid bed operation proved to be difficult in the original trials held by Standard Oil and Hydrocarbon Research Incorporated in the United States.[77] The main problems were obtaining a uniform density throughout the bed and proper gas mixing. The circulating fluid bed used by Kellogg and developed at Sasol eventually operated very well. Operating conditions for both Sasol processes are summarized below:[74]

a) Fixed bed tubular reactor:
- Reactor with more than 2000 tubes, each 12 m long, containing 20 liters of catalyst per tube.
- Operating pressure 27 atm, temperature 220–250^{0}C. Fresh feed (1.8 H$_2$: 1 CO) at space velocity 500 h^{-1}.
- Initial conversion 40% per pass.
- Catalyst life less than 1 year, depending on operating severity (e.g., to produce high-molecular-weight wax.) The catalyst sinters in the presence of water. Prereduction provides preshrinkage of the catalyst. Preshrinkage is also possible by heating the catalyst in liquid wax.
b) Transported fluid bed:
- Reactor 46 m high (original design 2.3 m internal diameter).
- Operating pressure 22 atm; reactor outlet temperature about 340^{0}C. Fresh feed 100,000 m^3 h^{-1} (6 H$_2$: 1 CO).
- Initial conversion 85% per pass.
- Catalyst life was originally 40 days but now increased.
- Reduction of catalyst in hydrogen before use at high linear velocity.
- Catalyst is carbided during reaction with some reoxidation by water produced during the reaction. Potash promotes carburization and retards oxidation. Alumina increases surface area and activity.
- Catalyst can be added or removed during operation to improve operation.

Slurry bed reactors using heavy oil to support the catalyst have been tested and can operate over a wider range of operating conditions and feed gas compositions than the fluid beds.[77] Sasol has now developed an improved Sasol advanced Synthol (SAS) reactor to produce high-grade distillate. The Sasol slurry phase distillate process (SSDP) has been tested in a demonstration plant at Sasol 1 since 1993. The SAS reactor is said to use an iron-based catalyst similar to the one used in its original plants, whereas the SSDP process uses a cobalt catalyst.

The Sasol processes are of interest now that the conversion of synthesis gas-to-liquid products is being developed by a number of companies. The gas-to-liquid (GTL) process is of particular interest in the production of sulfur-free distillates.

2.6.2. The Importance of Gas-to-Liquids as Gasoline Prices Increase

The major developments of gas-to-liquids (GTL) technology have arisen due to the availability of cheap by-product natural gas or associated gases in remote areas. The capital expense of new plants can be offset against future increases in crude oil prices. An advantage of the GTL products is the low content of sulfur, metals, and other impurities. GTL plants that are currently operating on a large scale; the Mossgas process using natural gas, was reported to be using the largest steam-reforming train in the world.

Since the end of the 1990s several companies have been developing processes to supply liquid hydrocarbons. Synthesis gas is produced and treated in variations of the Fischer–Tropsch process. High-boiling C_{10}–C_{40} liquid and wax products can be converted to sulfur-free, low-boiling products in the C_{10}–C_{20} range by hydrocracking. Since 1955 only the Sasol processes have been tested extensively.

Sasol produces synthesis gas from coal by partial oxidation or from natural gas by steam reforming. The first version of the Synthol process was upgraded to use the advanced Synthol reactor. Both Synthol processes use fluid catalyst beds. A new SSDP has now been introduced and should soon be operating. Distillates and waxes can be produced (25 atm; 240^0C).

Shell operated its middle-distillate synthesis process (SMDS) in Malaysia from 1993 to 1997, but closed the facility in 1997 while better Fischer–Tropsch catalysts were being developed for future operation. Synthesis gas produced by the Shell partial oxidation process was converted to distillates and waxes in a tubular high-pressure Fischer–Tropsch reactor. A slurry phase reactor using an improved catalyst is being developed (40–46 atm; 120–130^0C).

Exxon developed an advanced gas conversion process (AGC-21) that produces synthesis gas by combined partial oxidation and steam reforming in a fluidized bed. A multiphase, slurry Fischer–Tropsch reactor has also been developed. Syntroleum uses an autothermal air/natural gas reformer to produce synthesis gas. A fluidized bed Fischer–Tropsch reactor has been developed (20–35 atm; 190–230^0C).

The catalysts likely to be used in those processes are listed in Table 2.17. It is clear that although many innovations will be included, the catalysts will, as far as possible, be well-tested types already operating in other processes.[78] So far only Shell, which is using a cobalt/metallocene catalyst, seems to have developed something new. Following recovery of the low-boiling, liquid products, the

TABLE 2.17. Catalysts Used in Gas-to-Liquids Production

	Synthesis gas	FT conversion
Sasol	Conventional steam-reforming catalysts.	1. ARGE process. Iron catalyst. 2. Advanced Synthol. Iron catalyst. 3. SSPD process. Cobalt or iron catalyst.
Shell	No catalysts in partial oxidation except purification of feed.	Proprietary metallocene catalyst to provide better product range.
Exxon	Conventional reforming catalysts.	Proprietary cobalt-based catalyst.
Syntroleum	Autothermal reforming using a ceramic membrane (as catalyst?)	Proprietary cobalt-based catalyst.

high-molecularweight hydrocarbons and waxes will be upgraded in hydrocracking units. Gas-toliquid process plants will, of course, be expensive and in order to be profitable will have to be built on a large scale. However, the recent surges in the price of crude oil suggest that more of these processes will become necessary in the future.

REFERENCES

1. Clement and D´esormes, *Ann. Chim.* **59** (1806)329.
2. W. Wyld, *Manufacture of Sulfuric Acid (Chamber Process) Series on Manufacture of Acids and Alkalis*, Vol. 2, Ed. by Lange, Gurney and Jackson, London, 1924, p. 124.
3. ICI, *Ancestors of History*, Kynoch Press, 1950, p. 99.
4. G. Lunge and E. Berl, *Angew Chem* **19** (1906) 807, 857, 881; **20** (1970) 1713; *Trans. Am. Inst. Chem. Eng.* **31** (1935) 193.
5. A. M. Fairlie, *Trans. Am. Inst. Chem. Eng.* **33** (1937) 563.
6. Kachkharoff Process, *Sulfuric Acid Plants*, Simon Carves, Stockport, UK, 1958.
7. Phillips, British Patent 6096 (1831).
8. W. S. Squire and Messel, British Patent 3278 (1875).
9. C. A. Winkler, German Patent 4566 (1878).
10. Schroder and Hannich, British Patent 9188 (1887).
11. Messel, British Patent 186 (1878).
12. R. Knietsch, *Berichte* **34** (1902) 4009, 4086; *J. Soc. Chem. Ind.* **21** (1902) 172.
13. BASF, German Patents 113933 (1898); 15947-50 (1901).
14. De Haen, German Patent 128616 (1900); US Patent 687834 (1901).
15. A. M. Fairlie, *Sulfuric Acid Manufacture*, Rheinhold, New York, 1936.
16. Grillo Co, British Patent 251158 (1898).
17. Grillo Process, *J. Soc. Chem. Ind.* **22** (1903) 348.
18. F. Slama and H. Wolf, German Patent 291792 (1921); US Patent 1371004, (1921).
19. Seldon Corp, British Patent 170022 (1920).
20. Monsanto, US Patents 1933067 (1933); 1933091 (1933); A. O. Jaeger, *J. Ind. Eng.* **21** (1929) 627.
21. Gen Chem Co, US Patents 1371004 (1921); 1887978 (1932).
22. J. H. Frazer and W. J. Kirkpatrick, *JACS* **62** (1940) 165; Kiyoura, *Kagaku Kogyo* **10** (1940) 126.
23. P. Davies, British Patent 895624 (1962); US Patent 3186794 (1965).

24. H. Livbjerg and J. Villadsen, *Chem. Eng. Sci.* **27** (1972) 21.
25. G. H. Tandy, *J App Chem* **6**, 68 (1956); H. F. Topsoe and A. Nielsen, *Trans. Dan. Acad. Tech. Sci.* **1**(1948) 3, 18.
26. G. K. Borescov, *Dokl. Acad. Nauk. USSR* **171** (1966) 648.
27. P. Mars and J. G. H. Maessen, *Proc 3rd Int Conf Catalysis* **1** (1965) 266; *J. Cat.* **10** (1968).
28. Gen Chem Co, US Patent 1789460 (1931); Bayer, *Kirk Othmer* **27** (1983) 191.
29. *Topsoe Topics,* July, 1992.
30. *Topsoe Topics* April, 1979; *Catalysts and Chemicals Europe: Sulfur,* Nov–Dec, 1984.
31. H. Deacon, British Patent 1403 (1868).
32. F. J. Hurter, *J. Soc. Chem. Ind.* **2** (1883) 106.
33. K. von Falkenstein, *Zeit. Phys. Chem.* **65** (1909) 371.
34. E. B. Maxted, *Catalysis and its Industrial Applications*, Churchill, London, 1933, p. 360.
35. Hasenclever, *Berichte* **2** (1876) 1070; British Patent 3393 (1883).
36. Claus Process, *J. Soc. Chem. Ind.* **7** (1888) 162.
37. *KTI Newsletter,* Spring (1995).
38. Nelson, Vlcek, and Graff, *Oil Gas J.*, July 6 (1953) 64.
39. Carpenter and Linder, *J. Soc. Chem. Ind.* **22** (1903) 457; **23** (1904) 577.
40. Coward and Barron, *Oil Gas J.* August 29 (1983) 54.
41. Grancher, *Hydrocarbon Processing*, July (1978) 155; *Int. Sulfur Symp.*, Calgary, Al, October 23 (1977); Nongayrede, *Oil Gas J.*, August 10 (1987) 65.
42. Dupin and Voirin *Hydrocarbon Processing*, November (1982) 189.
43. J. W. Geus, *Environmental Catalysts*, ACS Symposium Series 552, Ed. by J. N. Amor, 1993.
44. Deville, *Comp Rend* **60**, 31171 (1865); Donkin, *Proc. Roy. Soc.* **21** (1873) 281.
45. Hlavati, Aust. Pat. 45/2938 (1895); Christiania Minekompanie, French Pattent 225183 (1896).
46. Du Fresne, German Patent 17070 (1881); French Patent 138472 (1881).
47. Du Motay, French Patent, 923346 (1871).
48. Le Chatelier, French Patent 313950 (1901).
49. J. von Liebig, *Organic Chemistry in its Application to Agriculture and Physiology*, (1840); J B Lawes, *Address to Agriculturists of Great Britain* (1845).
50. W. Crookes, *Address to British Association for the Advancement of Science*, Bristol, 1898.
51. F. Haber with van Oordt, Le Rossignol, Greenwood, et al., *Berichte* **40**, (1907) 2144; *Z. Anorg. Chem.* **43** (1904) 111; **44** (1905) 341; **47** (1908) 42; *Z. Electrochem.* **14** (1908) 181, 513.
52. F. Sherwood Taylor, *A History of Industrial Chemistry*, Heinemann, London, 1957, p. 429.
53. W. Ostwald, *Lebenslinien* **2** (1926–1927) 279.
54. W. Nernst, *Z. Electrochem* **16** (1910) 96; Jost, *Z. Electrochem.* **13** (1907) 521; **14** (1908) 181, 373; *Z. Anorg. Chem.* **57** (1908) 414.
55. German Patents 235421 (1908); 252275 (1909); 223408 (1910).
56. A. Mittasch, *Advances in Catalysis*, Vol. 2, Academic, New York, 1950, p. 81.
57. German Patents 249447; 254437; 258146; 262823 (1910).
58. M. Appl, *Nitrogen (l00th issue)*, March/April (1976) 47–58.
59. F. Bergius, *Z. Angew. Chem.* **34** (1921) 341; **37** (1924) 400; *Engineering* **120** (1925) 675; *Proc. World Petroleum Conf.*, Vol. ii, 1935, p. 282.
60. US Patents 1251954, 1342790; British Patents 18232 (1914); 5021 (1915); 148436 (1920); German Patents 303272; 303332; 307671.
61. M. Pier, *Petrol Times* **33,** April 6, 13, 20; May 4 (1935); *J. Soc. Chem. Ind.* **54** (1935) 284.
62. N. Gard, Thirty years of steam reforming—A review of ICI developments and experience, *Nitrogen* No. 39, January/February, 1966.
63. K. Gordon (a) Oil From Coal, *Trans. Inst. Mining. Eng.*, Vol. 82, Part 4, 1931, pp. 348–363; (b) Development of Coal Hydrogenation by ICI, *J. Inst. Fuel*, December, 1935; December, 1946.
64. I. G. Farben, British Patent 320473 (1929).
65. ICI, British Patents 335215; 338576; 363445; 392459 (1929).
66. Bataafshe Petroleum Maatschappij, British Patent 348243.
67. BASF, German Patent 293787 (1913).

68. R. Anderson, Catalysis, Ed. by Emmett, Vol. 4, Reinhold, New York, 1956, Ch. 2–4.
69. H. H. Storch, *Advances in Catalysis*, Vol. 1, Wiley, New York, 1948.
70. C. C. Hall, S. R. Craxford, and D. Gall, *Final BIOS Report 447*, Items 22 and 30.
71. H. Pichler, *Advances in Catalysts*, Vol. 4, Wiley, New York, 1952.
72. H. Pichler and H. Buffleg, *Brennstoff-Chem* **21** (1940) 257, 273, 285.
73. A. H. Stander, *Financial Times Symposium*, London, September, 1975.
74. M. Dry, The Fischer–Tropsch Synthesis (a) *Catalysis, Science and Technology*, Ed. by Anderson and Boudart, Vol. 1, Springer-Verlag, Berlin, 1981; (b) *Applied Industrial Catalysis*, Ed. by Leach, Vol. 2, Academic, New York, 1983.
75. H. H. Storch, N. Golombik, and R. B. Anderson, *Fischer–Tropsch and Related Syntheses*, Wiley, New York, 1951.
76. S. T. Arakawa, R. C. Mulvaney, D. E. Felch, J. A. Petri, K. Vandenbussche and H. W. Dandekar, *Hydrocarbon Processing*, March (1998) 93.
77. C. C. Hall, N. Gall, and S. L. Smith, *J. Inst. Petrol.* **38** (1952) 845.
78. *Oil Gas J.*, January 31 (2000) 74.

3

HYDROGENATION CATALYSTS

3.1. THE DEVELOPMENT OF HYDROGENATION CATALYSTS

Prior to 1900, catalytic hydrogenation was not really seen as general reaction type, and the only examples known were a few specific, seemingly unrelated, reactions of hydrogen with both organic and inorganic compounds. These reactions took place over a reactive surface, which we now know to comprise a catalytic surface. A selection of these reactions is shown in Table 3.1 It was not until the early twentieth century that the foundations of catalysis were established, initially by the work of Sabatier and Senderens.

3.1.1. Sabatier and Senderens

Modern catalysis began with the systematic hydrogenation of reactive organic compounds by Professor Paul Sabatier and Abbe Jean-Baptiste Senderens in 1897.[1] They generally use nickel oxide catalysts to effect addition of hydrogen to unsaturated hydrocarbons or to the functional groups of other organic compounds. Reactions were generally carried out in the vapour phase and the effects of poisons or *anticatalysts* noted. This early study continued until about 1920 and the co-workers who were also involved are listed in Table 3.2

Sabatier was first attracted to the use of nickel as a catalyst when he saw details of the newly introduced Mond process, in which nickel metal was purified by the formation and decomposition of nickel carbonyl.[2] The fact that nickel combined with gaseous carbon monoxide suggested that other unsaturated molecules might react in a similar way. Sabatier later described the methanation

L. Lloyd, *Handbook of Industrial Catalysts*, Fundamental and Applied Catalysis,
DOI 10.1007/978-0-387-49962-8_3, © Springer Science+Business Media, LLC 2011

TABLE 3.1. Early Hydrogenation Reactions.

Author	Reaction	Reference
Dobereiner	Confirmed Davy's observation of ignition of combustible gas with air using spongy platinum. Also combined oxygen and hydrogen.	*Schweiger's J.* **34**, 91 (1822); **35**, 321 (1823)
E. Turner	Produced hydrogen chloride from hydrogen and chlorine over platinum.	*Ed. Phil. J.* **11**, 99, 311 (1824)
F. Kuhlmann	Hydrogenated nitric oxide to ammonia with platinum sponge.	*Comp. Rend.* **7**, 1107 (1838)
B. Corenwinder	Reacted iodine with hydrogen using platinum sponge.	*Ann. Chim. Phys.* **34**(3), 77 (1852)
H. Debus	Formed methylamine by reducing hydrogen cyanide with hydrogen using platinum black. Also reduced ethyl nitrite to ethyl alcohol and ammonia.	*Annalen.* **128**, 200 (1863)
M. Saytzeff and H. Kolbe	Reduced nitrobenzene to aniline with platinum black.	*J. Probt. Chem.* **4**(2), 418 (1871)
De Wilde	Hydrogenated acetylene and ethylene at 20^0C using platinum black. Regenerated catalyst to recover activity.	*Berichte* **7**, 352 (1874)
W. Karo	Hydrogenated acetylene selectively to ethylene over palladium, using base metal to improve selectivity.	German Patent 253160 (1912)
G. Lunge and J. Akunov	Reduced benzene with palladium or platinum black to cyclohexane. ($20–100^0C$).	*Z. Anorg. Chem.* **24**, 191 (1900)

reaction, in which he converted carbon monoxide to methane with hydrogen using a nickel metal catalyst.[3]

The series of reactions studied by Sabatier led him to formulate new ideas on the mechanism of catalytic reactions. He also reached a number of conclusions that were useful in the development of catalytic reactors and industrial processes. Some of these conclusions are listed in Table 3.3.

TABLE 3.2. Co-Workers with Paul Sabatier, 1897–1919.

Co-worker	Date	Number of papers published
Jean Baptiste Senderens	1897–1905	94
Alphonse Mailhe	1906–1919	100
Marcel Murat	1912–1914	19
Leo Espril	1914	12
Georges Gaudion	1918–1919	11

TABLE 3.3. Sabatier's Work and its Application to Industrial Catalysts.

Process change	Possible improvements
Use of support	Metals in a finely divided state.
	Increased activity by increasing surface area.
	Use of less metal reduces cost.
	Regular shape decreases pressure drop.
Large-scale operation	Supports avoid dust formation.
	Oxides easily reduce to active metals.
	Feed pretreatment removes gaseous impurities (anticatalysts).
	Polymer or carbon deposits removed by regeneration in air.

Sabatier was awarded the Nobel prize for chemistry in 1912 and be presented his work in the book *Catalysis in Organic Chemistry*. This was translated into English by Professor E E Reid in 1922.[4]

3.1.2. The First Industrial Application of Nickel Catalysts

Sabatier did not extend his work to liquid phase hydrogenation, possibly because the condensation of liquid on the catalyst surface interfered with the reaction. Nevertheless, by 1902 the liquid phase hydrogenation of fatty oils had been introduced on an industrial scale.[5] Apart from nickel oxide the catalysts claimed in these patents included copper, platinum, and palladium, and were soon being supported on inert materials to increase activity.[6]

Although Sabatier and Senderens had hydrogenated oleic acid vapor to produce stearic acid, they did not extend this work themselves.[7] The appendix to Chapters 11 and 12 in their book describes early work to about 1916, by others who used nickel and palladium catalysts. They described the use of nickel supported on pumice, kieselguhr, asbestos, and wood charcoal.[8]

3.1.3. Ipatieff and High-Pressure Hydrogenation of Liquids

Ipatieff started investigating the hydrogenation of organic molecules at high pressure in Russia in about 1901. He knew about high-pressure equipment from his experience with explosives as a student in military school. After developing an interest in oxide catalysts, he began to work on liquid phase hydrogenation at pressures in the range 100–300 atm and temperatures above 250^0C. He used finely divided nickel and copper catalysts in a stirred reactor and followed the reactions by the change in hydrogen pressure.[9] He realized from his experiments with different grades of the finely divided catalysts that the surface area of a catalyst was important. The true significance of surface area was, of course, only realized later.

Ipatieff published a book in 1937 describing his work with catalysts.[10] He subsequently emigrated to the United States, worked with UOP, and continued to make tremendous contributions to the development of catalytic processes until the early 1940s. It was noted by Spitz in his very interesting book on the petrochemical industry[11] that Ipatieff frequently met and discussed catalysis with other catalyst researchers, including Sabatier, Caro, Willstatter, Bergius, Haber, and Nernst. These meetings obviously led to significant progress in catalyst and process development. Sabatier, for example, describes Ipatieff's use of nickel, iron, and copper catalysts in Chapter 12 of his own book.

3.1.4. Colloidal Platinum and Palladium Catalysts by Paal

While Sabatier and Ipatieff were experimenting with nickel and copper catalysts others were developing the use of finely divided platinum and palladium catalysts. The main objective was to achieve a range of more practicable hydrogenation procedures at low pressure in either the gas or the liquid phase.

Carl Ludwig Paal and others used colloidal platinum and palladium catalysts in liquid phase hydrogenation reactions at low temperatures.[12] Aromatic and unsaturated aliphatic compounds such as aldehydes or ketones were easily hydrogenated by either platinum or palladium. Aladir Skita collaborated with Paal in publishing a patent that described some of this work.[13] The colloidal metals could be stabilized by the use of albumen from egg whites, but they were not really practicable and it was difficult to separate them from products. An outline of an early colloidal catalyst preparation is given in Table 3.4.

3.1.5. Platinum and Palladium Black Catalysts by Willstatter

Fokin, who was also interested in precious metal catalysts, had earlier used platinum or palladium blacks to hydrogenate oleic acid at ambient temperature and much lower pressures than those used by Ipatieff.[14] Platinum or palladium black catalysts, which were finely divided metals containing some oxygen, were then

TABLE 3.4. Paal's Preparation of Colloidal Platinum Metals.

Step	Procedure
1	Method depended on use of suitable protective colloid such as protein, gum arabic, or starch to stabilize particles.
2	Paal prepared *protalbic acid* by dissolving egg albumin in sodium hydroxide solution and precipitation using either sulfuric or acetic acids. *Lysalbic acid* remained in solution and could be recovered either by evaporation, or by precipitation with ethanol.
3	Colloidal metal was obtained by adding sodium salts of the protalbic acid to a dilute solution of platinum chloride and reducing with hydrazine hydrate. The colloidal solution was dialyzed to remove electrolytes and then concentrated.

TABLE 3.5. Loew's preparation of platinum/palladium blacks

Step	Procedure
1	Dissolve platinic chloride in water and gradually add formaldehyde while cooling solution.
2	Slowly add caustic soda solution to neutralise the formic and hydrochloric acids.
3	Filter the finely divided platinum and wash until some colloidal particles pass the filter.
4	Age wet powder until particles are loose and porous and then wash again until filtrate is chloride free.

used for the hydrogenation of a whole range of organic materials by Willstatter, of BASF, who noted that some reactions were sensitive to the absence or presence of oxygen in the catalyst. Although Willstatter initially used an earlier catalyst preparation described by Oscar Loew,[15] he eventually evolved a better procedure himself.[16] Details of Loew's preparation are given in Table 3.5. Paal's method was improved on by Willstatter, who used caustic potash during the reduction step.

Continued developments in the use of precious metal catalysts for liquid phase reactions led to the introduction of more practical hydrogenation procedures using stirred or agitated reactors at lower pressures. Nevertheless, it was found that platinum black had a low and variable activity:

$$Pt\, Cl_4 + 2\, HCHO + 2\, H_2O \rightarrow Pt^0 + 2\, HCOOH + 4\, HCl \qquad (3.1)$$

$$2\, HCOOH + 2\, HCl + 4\, NaOH \rightarrow 2\, NaCl + 2\, HCOONa + 4\, H_2O \qquad (3.2)$$

Full details of the *methods* of Paal and Willstatter using precious metals and of Ipatieff using a range of base metal catalysts including nickel, iron, and copper are given in Chapters 11 and 12 of Sabatier's book. Many of the reactions concerned were eventually developed as industrial processes. Ipatieff investigated benzene hydrogenation with platinum black in 1912, though the current commercial process now uses nickel catalysts.[17]

Despite the increasing interest in precious metal catalysts, however, there were no significant industrial applications of high-pressure organic hydrogenation reactions until almost a generation after Sabatier and Ipatieff began their experiments. By that time several other important catalytic industrial processes based on high-pressure synthesis gas were being successfully introduced with a wide range of new catalysts. These included ammonia synthesis, coal hydrogenation, the Fischer–Tropsch reaction, and methanol synthesis.

TABLE 3.6. Preparation of Adams' Catalyst.

Step	Procedure
1	Fuse chloroplatinic acid with sodium nitrate to form brown platinum oxide.
2	Separate the oxide by washing with water, filtering, and drying. (Some residual sodiumcannot be washed from the catalyst.)
3	Reduce to Adams' platinum oxide by bubbling hydrogen through the reaction solution.

3.1.6. Adams' Platinum Oxide

The development of more active and reproducible precious metal catalysts continued in the 1920s when Roger Adams produced his platinum oxide catalyst by the method shown in Table 3.6.[18]

Since then Adams' platinum has been used in the pharmaceutical industry and in small-scale hydrogenation reactions. When industrial processes requiring precious metal catalysts were developed it was not economic to operate with high platinum concentrations. It was, therefore, necessary to reduce costs by supporting small amounts of the metal on a suitable diluent. Supports included alumina, asbestos, silica gel, and, most often, activated carbon. Products from these processes have included vitamins, cortisone, and dihydrostreptomycin.[19]

By 1930 Carleton Ellis, who worked on organic hydrogenation reactions and the chemistry of petroleum derivatives, was able to publish a book that included a literature survey of the recent developments in hydrogenation reactions and catalysts.[20]

3.1.7. Raney Nickel Catalysts

Progress in the use of precious metal catalysts for small-scale hydrogenation reactions, together with the increasing use of catalysts in new industrial processes, stimulated a much more practical interest in the development and commercial use of all types of catalysts. Improvements were based on using the most appropriate physical form of a catalyst for large-scale operation. Not surprisingly, it can be concluded that poor catalyst quality might explain why some of the early catalysts gave poor results. After the late 1920's, better quality control was introduced and a wide range of physical tests gradually became available.

Although most of the early experimental work on general catalysts was carried out by universities followed by industrial organizations, it was still possible for individuals to make significant contributions to process development. The invention of Raney nickel catalysts is a good example.[21]

Murray Raney was not a chemist, but he became interested in catalytic hydrogenation after he had designed a cottonseed oil hydrogenation unit for the Lookout Oil and Refining Co. For this process supported nickel catalysts, made

TABLE 3.7. Preparation of Raney Nickel.

Step	Procedure
1	Fuse equal parts of nickel and aluminum. Crush the alloy into suitable size granules or powder.
2	Extract aluminum by gradual addition of caustic soda solution. Cool mixture. If sodium aluminate hydrolyzes, dissolve alumina in more caustic soda.
3	Wash granules with water to remove all alkali and then with dry ethanol or inert solvent to remove residual water before storage under inert atmosphere.
4	Dry nonpyrophic Raney nickel containing 65% nickel has been used to hydrogenate benzene. It is briefly reduced in hydrogen before use.

on site, were used but they had irreproducible activities. This was probably due to the difficulties in synthesising a catalyst precursor that was consistently uniform in both chemical and physical properties, and then reducing this material to the metal in a consistent way so that the resulting active crystallites behaved similarly during catalysis. These problems led Raney to consider how he could make a better catalyst. He knew from his experience that when hydrogen was generated from a ferrosilicon alloy by treatment with caustic soda, sodium silicate and a fine iron oxide powder residue were formed. This suggested that if he made a nickel/silicon alloy and dissolved the silicon in caustic soda, he could make an active nickel oxide catalyst. A simple experiment using an alloy containing 50% nickel showed that nickel metal, not nickel oxide, was actually produced and that it was significantly more active than the supported nickel oxide catalyst that was being used to hydrogenate the cottonseed oil.[22] He delayed testing a nickel/aluminum alloy because aluminum was expensive, but did get around to patenting the procedure two years later.[23] The catalyst recipe is given in Table 3.7. Raney nickel is normally stored under a suitable liquid to prevent loss of activity.

Raney believed that his catalyst was active at low temperatures because it contained hydrogen. It is probable that the hydrogen evolved when the activated catalyst is heated arises from the reaction of residual water in the catalyst with aluminum.[24] The catalyst was soon being used to hydrogenate vegetable oil but at the time was not considered for other uses. Raney subsequently registered his name as a trademark for the catalyst and alloy powders.

In 1931 Homer Adkins came to the conclusion that Raney nickel was better than any other nickel catalyst then available for organic hydrogenation reactions, as well as being more convenient to use. He described the new catalyst in a 1932 paper,[25] and it was soon being widely used in other laboratories. Adkins was one of the first to study the catalyst extensively for a wider range of hydrogenation reactions.

Full details of the use of Raney nickel are given in Adkins' book and in the review by Lieber and Morritz.[26] It is interesting to recall that Adkins found the various Raney nickel catalysts described in the literature so different that he categorized them with a series of W numbers. The samples he prepared at different temperatures had variable aluminum and alumina content and variable stability when stored. An important feature was that the finished catalyst *contained* hydrogen and could be made by a consistent procedure to give the properties required by a particular operator. A *dry* form of Raney nickel is now used extensively to hydrogenate benzene by the industrial cyclohexane process licensed by IFP.[27]

Various Raney nickel, cobalt, and copper catalysts are still provided commercially by Grace Davison, who now owns the copyright, for use in both slurry reactors and fixed beds. These are often promoted with other metals, such as chromium and barium, and can be supplied in the form of powders, granules, or extrudates with a variety of pore sizes.[28]

Despite its usefulness in laboratory and small-scale hydrogenation procedures, Raney nickel was not immediately used for industrial hydrogenation processes. This was partly because of its relatively high manufacturing costs. In fact, from the late 1930s when reproducible nickel/silica catalysts became commercially available, it was no longer being used for vegetable oil hydrogenation. Since that time its use has been limited to the hydrogenation of ethylene oxide to glycol, dextrose to sorbitol and benzene to cyclohexane.

3.1.8. Nickel Oxide/Kieselguhr Catalysts

As the demand for organic chemicals began to exceed the supply from natural sources there was an increased industrial interest in the development of hydrogenation processes to saturate aromatic compounds and olefinic bonds.

Nickel oxide/kieselguhr catalysts had been used since just after Sabatier described his experiments. It has been noted that following his collaboration with Normann, Joseph Crosfield in the United Kingdom used supported nickel oxide industrially in 1910 to make soap. Supported nickel oxide catalysts were subsequently used in the United States as well.[29,56] Of course, these catalysts had to be reduced in hydrogen to form nickel metal before use. Supports such as kieselguhr or pumice were also known to improve catalyst stability and give longer life during use in large-scale processes such as natural oil hydrogenation. There were always problems in reducing the nickel oxide properly and most operators seem to have experienced difficulty in achieving reproducible results at a time when many small companies made their own catalysts in small quantities with little quality control.

Adkins investigated the preparation of nickel and copper hydrogenation catalysts in 1931 and attempted to optimize a nickel oxide/kieselguhr catalyst preparation. A typical method of production was to add sodium carbonate solution to a slurry of kieselguhr with a nickel salt solution and precipitate basic nickel carbonate. The mixed solid was then filtered, washed, dried, calcined at 400^0C, and pelleted with a lubricant such as graphite.

Reproducible catalyst performance could not be achieved unless the catalyst was carefully reduced at temperatures up to at least 400°C. It was finally concluded that the raw catalyst contained a complex mixture of nickel hydroxysilicates and kieselguhr.[30] With alkaline precipitation conditions it is possible to form a layer of *silica gel* on the kieselguhr surface that reacts with the nickel hydroxide/carbonate slurry as it is precipitated. This results in a layer of nickel antigorite ($Ni_3Si_2O_5(OH)_4$) on the support.[31] The same mineral has also been found following the reaction of nickel hydroxide with Pyrex glass under hydrothermal conditions.[32]

As the nickel antigorite forms, electron micrographs show that the kieselguhr/silica gel/nickel hydroxide structure changes. Silica plates develop as layers of hydroxyl groups from the nickel hydroxide brucite structure react with silica tetrahedra. Coenen suggests that the antigorite layer acts as a reactive surface that combines with a further layer of nickel hydroxide.[31]

The chemical combination of nickel oxide and support explains the need for high-temperature reduction of nickel oxide/kieselguhr catalysts before use and the high metal content required (see Table 3.10). A practical solution to the problem was to prereduce the catalyst at a high temperature and to stabilize the reduced nickel with air before use in industrial reactors. By the 1930s prereduction and stabilization were becoming standard procedures for catalysts used in fat-hardening and iso-octane operations.

The work by Coenen and others confirmed that the reduction of nickel hydrosilicates is inhibited by water in the lattice. This had led to the early problems in reducing the catalyst industrially because water forms continuously during reduction, not only as the nickel compounds are converted to metal but also as the remaining hydroxyl layers gradually decompose. Catalysts become active only after reduction at temperatures in the range 300–400°C, and even then a significant proportion of the nickel oxide is unreduced and the lattice still contains water.

Thermogravimetric analysis in air or inert atmospheres demonstrates that a typical catalyst gradually dries before the hydrates and residual carbonates decompose between 80–900°C. A typical catalyst reduces slowly between 240–500°C with a peak in water evolution at 350°C. However, small-scale water evolution is also noted between 200 and 250°C. The catalyst is only around 85% reduced at 500°C. These results are shown in Table 3.8.

Commercially, the prereduction of small batches of catalyst overcame the need for a high-temperature reduction in a reactor:

- Raw pellets were slowly heated to about 380°C for 72 h to decompose hydrates and residual carbonate from the structure.
- Calcined catalyst was cooled and then reduced in a stream of nitrogen containing 3–5% hydrogen as the bed temperature was slowly increased to 390°C. The temperature of hot-spots could be controlled by variations in the volume of hydrogen added.

TABLE 3.8. Reduction of Nickel Oxide/Kieselguhr Catalysts between 100 and 500°C.

Temperature (°C)	Wt % nickel oxide reduced		
	Typical catalyst	Prereduced	Plus 5% CuO
120		[a]	
160		12	[a]
180		15	10
200		20	20
250	5[a]	30	25
300	25	60	30
350	45		50
400	50–65		60–70
450	70–80		70–80
500	80–85		80–85
550	Still reducing		Still reducing

[a]Reduction begins

- Following reduction, the catalyst was again cooled in nitrogen and the reduced nickel was then slowly oxidized by adding oxygen to the circulating gas. The maximum catalyst temperature was kept below 250°C as the
  oxidation zone as monitored by the temperature profile moved through
  the catalyst bed. Reduced catalyst could also be stabilized by cooling in
  carbon dioxide to 25°C before adding air to the carbon dioxide until the
  bed temperature was stable.

Prereduced catalyst could be handled safely as it was being transferred to
the hydrogenation reactor and then easily reduced at temperatures in the range
180– 250°C before use. Table 3.8 compares reduction of a prereduced catalyst
and a typical catalyst.

Careful decomposition and reduction of nickel oxide/kieselguhr catalyst
produces small active nickel crystallites. Under normal conditions, when the
catalyst is not completely reduced, the nickel crystallites are supported by unreduced antigorite. It is important not to overreduce the remaining nickel compounds because high temperatures sinter the crystallites already formed. This is
illustrated in Table 3.9. Active metal surface area increases as the degree of reduction increases up to a temperature of about 400°C. Above this temperature,
metal surface area begins to fall as the increased degree of reduction is insufficient to balance the loss of metal area caused by sintering which leads to the
growth in the size of the metal crystallites.

Catalyst formulation was later changed by the addition of small amounts of
copper. This resulted in a product which could be reduced, at least partially,
below 200°C. The catalyst could therefore be reduced in the reactor, thus avoiding the cost of pre-reduction and the need for replacing equipment required for
pre-reduction.

TABLE 3.9. Nickel Metal Surface Area of Catalysts Reduced in Hydrogen.

| Temperature (^{0}C) | Metal area (m^2 g^{-1}) | | | |
| | NiO/kieselguhr catalyst | | NiO/CuO/kieselguhr catalyst | |
	3 h	5 h	3 h	5 h
150	—	—	2	5
200	—	—	8	10
250	5	8	20	26
300	25	38	50	60
350	67	73	75	82
375	76	80	80	82
400	78	78	78	79
450	~75	~75	~75	~75
500	~65	~65	~65	~65

Thermogravimetric analysis of the copper/nickel catalyst clearly indicates a significant difference from the original nickel catalyst:

- The copper/nickel catalyst contains more residual hydrate, hydroxyl and carbonate than the nickel catalyst.
- Undecomposed copper/nickel catalyst reduces very easily in hydrogen with two distinct reduction peaks. The first, between 180–280^{0}C, corresponds to the reduction of free nickel and copper oxides. The second, between 240–450^{0}C, corresponds to the single reduction peak of nickel catalyst.

The low-temperature peak explains the easy reduction of catalyst below 200^{0}C. Reduction details are shown in the last column of Table 3.8.

Thus, the replacement of some nickel with copper in the nickel/antigorite structure allows easier reduction of up to half of the nickel in the catalyst at a reasonably low temperature. The remaining nickel/antigorite still provides a support for the nickel and copper crystallites and has a structure similar to the original nickel catalyst. Table 3.9 shows that the nickel surface area, although higher in a copper/nickel catalyst up to about 300^{0}C, is the same for both catalysts at 400^{0}C. The addition of more than 5% copper can lead to rapid sintering during operation and thus, a shorter catalyst lifetime. The feedstock for most catalytic applications contains traces of sulfur compounds. This is absorbed by the catalyst up to levels of 16% and this also results in catalyst deactivation.

Table 3.10 gives the composition of typical nickel oxide/kieselguhr and copper-promoted nickel oxide/kieselguhr catalysts.

3.1.9. Nickel Oxide-Alumina Catalysts

The production of nickel oxide/kieselguhr catalysts illustrates that not only the composition but also the method of preparation of catalyst precursors determines

TABLE 3.10. Composition of Nickel Oxide Kieselguhr Catalysts.

Wt %	Typical catalyst	Catalyst plus 5% copper oxide
Nickel oxide	~68	~63
Copper oxide	—	5
Carbon dioxide	~ 2	~ 2
Hydrates	7–10	7–10
Silica	~20	~20
Impurities	From precipitation and kieselguhr	

industrial success. Most early nickel oxide/alumina catalysts were made by impregnation. This was not always satisfactory; however, particularly if operation was to be at a high temperature, since the nickel metal could react with the support.

Zelinsky, the Russian chemist, worked with nickel oxide/alumina catalysts for a number of hydrogenation reactions and was probably one of the first to describe co-precipitation of nickel oxide and alumina in 1924.[33] Since then many other nickel catalysts with alumina supports have been co-precipitated and used successfully in the production of synthesis gas, hydrogen, and town gas.

In the 1940's Feitknecht and others recognized that a particular form of blue-green basic nickel/aluminum carbonate could be prepared from mixed nickel and aluminum solutions under specific conditions.[34] The solid, as with the nickel oxide/ kieselguhr catalyst had the magnesia brucite structure, with part of the nickel layer replaced by aluminum and some of the hydroxyl groups replaced by carbonate.[35]

Altman showed that the formula of the nickel Feitknecht compound is $Ni_6Al_2(OH)_{16}(CO_3)4H_2O$. Nickel and aluminum can be replaced by certain other di-and trivalent metal ions. Since 1942, similar mixed metal precipitates were discovered during the development of copper oxide/zinc oxide/alumina catalysts and it is now recognised that they are similar in structure to natural *green rusts*.[36] It is important that during precipitation the nickel/aluminum atomic ratio in a Feitknecht compound be within the range 2:1–3:1, and carefully controlled conditions must be maintained in order to produce the most stable catalysts. For example, when alkali is added slowly to an acidic solution of nickel and aluminum nitrates to precipitate the basic carbonate, there is a gradual change in pH. The first solid to precipitate at low pH is alumina-rich, whereas the final precipitate, when the pH has increased, is nickel-rich. Precipitates with more uniform particle size and metal distribution form if the solutions are added as quickly as practicable with good mixing.[37] The precipitate should also be allowed to age at a reasonably high temperature for it to become more homogeneous.

After filtering and drying the Feitknecht compound is carefully decomposed at the lowest possible temperature to form a metastable mixture of nickel oxide and alumina ($Ni_6Al_2O_9$) that should not contain any free oxides. Thermogravimetric analysis shows that water of crystallization is lost between 150 and 210°C and that the hydroxyl and carbonate structure breaks down from 290 to 450°C.[38]

Subsequent reduction of the *mixed* oxide produces active nickel crystallites that have defects containing small particles of nickel aluminate. The nickel oxide content of the catalysts used for prereforming of synthesis gas or the catalytic rich-gas process (Chapter 9) will, therefore, be as high as 78–80%. Despite this, catalysts are very stable at temperatures up to 600°C and operate for long periods.

Other catalysts used for hydrocarbon steam reforming or methanation that have lower nickel content are prepared with the Feitknecht compound precipitated in the large pores of an inert support before it is decomposed.[39]

Catalysts prepared from Feitknecht compounds are analogous to the solid solutions of magnetite and alumina (Chapter 10), which, when reduced, give stable and active ammonia synthesis catalysts.

3.1.10. Copper Chromite Catalysts

The high pressure methanol process introduced by BASF in 1923 was based on the use of zinc/oxide/chromium oxide catalysts. The success of this process stimulated further work by others to make different catalysts for both methanol and higher alcohols. It was recognised quite early on that copper formulations were potentially very good catalysts, but that they were very prone to poisoning. It is interesting to note that most of the investigations were carried out by industrial organisations rather than universities, probably because of the need for high pressure technology that was not easily available to universities at the time. This led to most of the information being published, if disclosed at all, in patents rather than in scientific journals, with much of the early information being forgotten. The final BASF catalyst was a mixture of zinc oxide and chromic acid that was reduced before use.[40] Natta, working for Montecatini, produced a better precipitated zinc chromite catalyst with a relatively low chromium content.[41] While DuPont produced a precipitated zinc chromite catalyst containing a higher proportion of chromium, their patent (issued to Lazier)[42] described other chromites, including a copper chromite that was intended for use in higher alcohol production.[51] This early work led to the development of copper chromite catalyst.

When Adkins tried to modify the Lazier recipe and make a copper chromite hydrogenation catalyst, he found that an active black cupric oxide was produced instead of the red oxide claimed by Lazier.[43] Adkins and Folkers subsequently suggested modifications to the recipe, including the addition of barium, magnesium, or calcium oxides to stabilize the black oxide form, which was more active. A typical recipe and catalyst composition is shown in Table 3.11.

The investigations of Adkins and his colleagues confirmed that copper chromium catalysts were active for the hydrogenation of functional groups in

TABLE 3.11. Adkin's Copper Chromite Catalyst.

Preparation	Add ammonia to an orange solution of ammonium dichromate until the color changes to yellow (~pH 6.8).
	Mix with copper nitrate solution.
	Wash red-brown precipitate, dry and crush—$Cu(OH)(NH_4)_3CrO_4$.
	Calcine carefully until reaction stops and the color is black.
	Treat with dilute (10–15%) acetic acid to remove excess copper oxide (~10%).
	Filter, wash, dry, then crush powder and pellet.
Comment	Early catalysts contained about 1% MgO.
	Modern catalysts contain barium and/or manganese oxides. Manganese increases activity but reduces selectivity; barium increases selectivity and stabilizes catalyst by concentrating on the surface to prevent sintering and scavenge poisons.
	During calcination stages a proportion of the black trivalent chromium is oxidized to hexavalent chromium. This reaches a maximum at about 250–300^{0}C. At the final calcination temperature of up to 450^{0}C most of the undesirable hexavalent chromium has been rereduced to trivalent form. Only about 2–3 wt% remains. Care should be taken when reducing catalyst with high hexavalent chromium content because of the exothermic heat release.
	The catalyst is cheap and easy to produce,[43] and resists typical poisons more easily than nickel catalysts.
Typical catalyst composition	Copper oxide 35–37 wt%
	Chromium oxide 31–33 wt%
	Barium oxide 1–3 wt%
	Manganese oxide 2–3 wt%

esters, amides, aldehydes, and ketones under moderate conditions. It was shown that nickel catalysts were still preferred for the hydrogenation of olefinic bonds, aromatic rings, furane, pyridine, oximes, and cyanides, and nitro-compounds.

Adkins' copper chromite catalyst is still widely used today. Nickel chromite catalysts made according to a similar recipe were also used until recently as methanation catalysts. It is probable that as a result of environmental restrictions, future use of catalysts containing chromium will be limited.

3.1.11. Copper Oxide/Zinc Oxide Catalysts

Following the introduction of a copper chromite catalyst based on the DuPont recipe for zinc chromite, a further copper catalyst was developed from experimental work related to the high-pressure methanol synthesis process.[44]

Precipitated copper oxide/zinc oxide catalysts were more active for a range of reactions than zinc chromite but lost activity as the copper was poisoned by gaseous impurities in the synthesis gas. The two oxides were found to be *mutually promoting* in methanol synthesis because the mixture of very small crystallites was more active than the individual oxides.

TABLE 3.12. Copper Oxide/Zinc Oxide Catalyst.

Composition	Wt %
Copper oxide	32–33
Zinc oxide	63–64
Ignition loss at $900^{0}C$	< 3
Metal oxide impurity (Na_2O, Fe_2O_3, MgO, Al_2O_3)	< 0.15

Several catalyst formulations were originally described and confirmed high activity with an optimum copper oxide content in the range 30–40%. A composition corresponding to the formula $CuO \cdot 2ZnO$ (Table 3.12) was selected for industrial use in dehydrogenation, hydrogenation, and gas purification applications. Operating stability depended on careful washing of the precipitate, but by efficient control of production conditions, it was possible to obtain catalysts containing less than 1000 ppm of metal oxide impurities.

Catalyst activity depends on the surface area of metallic copper and the catalyst must be carefully reduced at a gas inlet temperature in the range 180–$200^{0}C$. The maximum temperature should be less than $230^{0}C$ to maximize the copper surface area. It was originally suggested that zinc oxide slowly reduced to form a solid solution of zinc and copper often described as α-brass. Some early experimental results are shown in Table 3.13.

Brass formation is not a practical problem because typical operating temperatures are less than the suggested onset of zinc oxide reduction. Differential thermogravimetric analysis simply shows that even with 100% hydrogen no zinc oxide reduction can be detected at temperatures up to $600^{0}C$. Copper oxide reduces sharply at approximately 200–$250^{0}C$. The deactivation attributed to zinc oxide reduction may have been due to poor temperature control and hot spots in the catalyst bed.

Because of the facile low-temperature reduction/reoxidation cycle, copper oxide/zinc oxide catalyst has often been used in gas purification as well as hydrogenation–dehydrogenation reactions (Table 3.14).

TABLE 3.13. Reduction of the Copper Oxide/Zinc Oxide Catalyst.

Thermogravimetric analysis	Direct reduction

1. 30–$100^{0}C$: loss of adsorbed water.
2. 200–$250^{0}C$: water evolution, 9 wt%.
3. 250–$650^{0}C$: no further reaction; no weight loss.
 	
Weight loss during reduction corresponds to reduction of copper oxide to copper and decomposition of hydrates (1.7 wt%).
	Extent of zinc oxide reduction estimated as water loss following copper oxide reduction.
	% Zinc oxide reduced:
	  1. 30–$150^{0}C$: nil
	  2. 150–$350^{0}C$: < 4 wt%
	  3. 350–$450^{0}C$: < 4 wt%
	  4. 450–$600^{0}C$: < 10 wt%
	No effort made to determine presence of -brass by X-ray diffraction. Results not confirmed by thermogravimetric analysis.

TABLE 3.14. Applications of the Copper Oxide/Zinc Oxide Catalyst.

Process	Isopropanol dehydrogenation[a]	OXO aldehyde hydrogenation	OXO aldehyde hydrofining
Capacity (tes year^{-1})	50,000	—	—
Reactor	Tubular	Adiabatic bed	Adiabatic bed
Catalyst volume (m^3)	3	—	—
Space velocity (h^{-1})	2000–2500 (GHSV)	1 (LHSV)	1–2 (LHSV)
Life	6 months	4–6 months	2 years
Temperature (^{0}C)	370–410	225–250	125–140
Pressure (atm)	1	250	50
Conversion (%)	90–92	99–100[b]	100
Yield (%)	95	100	—
Hydrogen (% theory)		130	130
Gas Purification			

			Concentration	
Impurity	Temperature (^{0}C)	Space velocity (h^{-1})	Inlet	Outlet
Oxygen	200	2000	1 vol %	< 10 ppm
Hydrogen	200	2000	1 vol %	< 10 ppm

Note: CuO/ZnO will also absorb hydrogen sulfide from gases and has been used to protect nickel catalysts from sulfur poisoning. It will absorb 10–12 wt% sulfur.
[a] Isopropanol azeotrope used as feed.
[b] Followed by hydrofining reactor using nickel catalyst to remove ~0.2% aldehyde.

Oxygen or hydrogen can be removed from inert gas streams by the addition of stoichiometric volumes of hydrogen or oxygen, respectively. Not surprisingly, the catalyst can also be used to remove traces of sulfur from gas streams. More than 10 wt% of sulfur can be absorbed by the catalyst at about 300^{0}C.

The main use of copper oxide/zinc oxide catalysts has been in dehydrogenation and hydrogenation reactions. These include the dehydrogenation of isopropyl alcohol to acetone as well as the hydrogenation of oxo-alcohols and fatty acid methyl esters. Although in many processes copper chromite catalysts are preferred to copper oxide/zinc oxide, the environmental problems involved in disposing of chromium wastes may reverse the situation.

Copper oxide/zinc oxide was the first catalyst to be tested in the low-pressure methanol synthesis process.[45] The relatively large copper and zinc oxide particles, the poor metal distribution, and the absence of a structural stabilizer led to rapid deactivation by poisons and thermal sintering. The problem was solved thanks to two significant changes. Firstly, improved versions of ternary catalysts based on copper/ zinc/alumina originally tested in the 1920s were developed.[46] These were followed by even better catalysts made by new precipitation techniques that produced Feitknechttype intermediates (see Chapter 10). Secondly, the purity of the synthesis gas increased dramatically, thanks to a change of feedstock from coal to naphtha followed later by natural gas, and the key problem of sulfur poisoning was largely solved.

3.2. HYDROGENATION OF FATS AND OILS

3.2.1. Process Development

The hardening of fats and oils became the first large-scale application of industrial hydrogenation catalysts during the period 1903–1908. This was only a few years after Sabatier began his work on hydrogenation in 1897. Patents were granted to Leprince and Siveke[47] in Germany and to Normann[48] in England for liquid phase hydrogenation processes using nickel catalysts. Normann, who worked at the Leprince and Siveke oil mills, developed the catalytic fat hydrogenation process in 1901 (Figure 3.1). No one in Germany was interested in the process, so in 1905 Normann was offered a job in England by Joseph Crosfield, who also purchased the rights for Normann's British patent and began to make soap using the process in about 1905. Because of problems in process development, Crosfield eventually sold the patent rights to Jurgens, a forerunner of Unilever, which also made edible fats and subsequently employed Normann. The Normann patent was then ruled invalid because he had not disclosed full details of the catalyst operation. Following legal action between Lever Bros. and Crosfield, who were still making soap with the process to meet an increasing demand, a further patent covering the use of a nickel oxide catalyst supported on kieselguhr was issued to Crosfield in 1910.[49]

Figure 3.1. Wilhelm Normann, Aufnahme, 1938.

These problems typify the difficulties in developing the early catalytic processes. Nevertheless, full-scale production was eventually very successful in many countries and led to the widespread use of edible butter substitutes as well as soap and candles. Hydrogenation removed the unpleasant smell of fats and allowed the use of fish and whale oils, which, until then, had only been useful in supplying glycerine. Crosfield was producing 100–150 tonnes of margarine a week from whale oil in1908 and 1000 tonnes a week by 1918 in plants in Bromborough and Port Selby. By the time Sabatier had published his book in 1922,[4] some 16 plants were operating in the United States making 92 brands of shortening! The first products to be made were lard substitutes, but soon vegetable shortenings became available, and products such as Crisco, Selex, and Fairco, melting at 33–37^{0}C became household names. It was not necessary to hydrogenate the oil completely and by mixing fully hydrogenated oil with untreated oil, the necessary consistency could be obtained using less hydrogen.

Experience led to the introduction of catalysts based on nickel nitrate and oxalate, followed by lactate or formate[50] as well as the original carbonates, all supported on *infusorial* earth, pumice, or even charcoal to increase activity. Reduction procedures were found to be important in obtaining the highest catalyst activity. Reoxidation of nickel before use had to be avoided. Mixed nickel oxide and copper oxide reduced more easily than nickel oxide alone.[51]

As Raney found in the 1920s, catalyst reproducibility was a real problem during a period when just about every small operator made his own catalyst. The ready availibility of Raney nickel supplied in an easily activated form and then the more reliable and active prereduced nickel catalysts provided by Harshaw[52] were a relief for producers and led to further developments in the process.

During the early period of development operating conditions evolved for the treatment of different fats and oils depending on the extent and type of unsaturation. More practical ways of mixing the oil and hydrogen were introduced and selective hydrogenation became more important.

3.2.2. Oil Hydrogenation

Glycerides are extensively used as butter and lard substitutes in foods, but they must be modified by hydrogenation before being used. This allows control of the melting point and removes unpleasant odors. Table 3.15 lists several important unsaturated fatty acids and the corresponding saturated derivatives.

Table 3.16 shows the most commercially useful vegetable oils with an indication of unsaturated or saturated acid content. Natural oils, or glycerides, contain a mixture of long-chain fatty acids randomly esterified with glycerol. All natural fatty acids have an even number of carbon atoms, usually C_{14}–C_{20}, but predominantly C_{16}–C_{18}. As many as three double bonds are present in some common fatty acids, all in the *cis* form and never conjugated.

TABLE 3.15. Saturated and Unsaturated Fatty Acids.

Acid	Carbon atoms	Melting point (^{0}C)
Saturated:		
capric	10	32
lauric	12	44
myristic	14	58
palmitic	16	63
stearic	18	70
arachidic	20	75
Monounsaturated:		
palmitoleic	16	
oleic	18	13–16
eichosenoic	20	
Diunsaturated:		
linoleic	18	5
eichosadioenic	20	
Triunsaturated:		
linolenic	18	−11

3.2.3. Fat Hardening Catalysts

Nickel catalysts are almost always used to hydrogenate natural oils. While palladium is also active and selective, it has usually proved to be too expensive. Copper catalysts are not active enough and, apart from being difficult to filter from the product, lead to toxicity problems. Low activity and quality control difficulties mean that copper cannot compete with nickel.

TABLE 3.16. Commercially Useful Vegetable Oils (Triglycerides).

Oil	Fatty acid content			Iodine value
	% Saturated	% Monounsaturated	% Di-unsaturated	
Soya bean	15 ($C_{16}C_{18}C_{20}$)	24 ($C_{18}C_{20}$)	61 (C_{18})	133
Rape seed	6 ($C_{16}C_{18}$)	58 (C_{18})	36 (C_{18})	118
Sunflower	11 ($C_{16}C_{18}$)	20 (C_{18})	69 (C_{18})	132
Palm kernel	51 ($C_{16}C_{18}$)	39 (C_{18})	10 (C_{18})	51–58
Coconut	92 ($C_{16}C_{18}$)	6 (C_{18})	2 (C_{18})	7–11
Maize (corn)	13 ($C_{16}C_{18}$)	25 (C_{18})	62 (C_{18})	125
Cotton seed	27 ($C_{16}C_{18}$)	19 (C_{18})	54 (C_{18})	108
Olive	16 ($C_{16}C_{18}$)	72 ($C_{16}C_{18}$)	12 (C_{18})	75–92
Compared with:				
Lard	41 ($C_{14}C_{16}C_{18}$)	47 (C_{18})	12 (C_{18})	62
Body fat	32 ($C_{16}C_{18}$)	47 (C_{18})	11 (C_{18})	

For many years after the fat hydrogenation process was introduced, manufacturers made their own catalysts when they were needed. Then, gradually, the nickel salt producers began to make the catalysts for operators. This improved quality and ensured more efficient operation. By 1928 the best catalyst supports were found to be kieselguhr or fuller's earth,[53] charcoal,[54] and complex silicates such as permutite.[55] Many other practical ideas were introduced such as:

- Protecting the prereduced catalyst with hardened fat before use.
- Adding more catalyst to combat the effect of recognized poisons.
- Regenerating oxidized catalyst by a second reduction.

It was concluded that the active catalyst was a nickel suboxide,[56] probably because of the difficulty in reducing nickel hydrosilicates in the catalyst.

The catalyst business gradually progressed until the 1930s during which time Harshaw, in the United States, became one of the principal suppliers. More reliable active catalysts supported on kieselguhr or silica alumina became available, backed by quality control and technical service.[52] Prereduced catalysts protected from reoxidation with solid fats were used almost exclusively. Catalysts were supported on a suitable inert material with large pores to provide a large accessible surface area for the reaction. The most important support material became kieselguhr, although some alumina and proprietary supports were also used. A typical catalyst composition is shown in Table 3.17.

Several different production methods are now standard:

- *Dry reduction*: Basic nickel carbonate is precipitated by adding sodium carbonate to a mixture of a nickel salt and a support at about 100^0C. During the precipitation nickel silicates are also believed to form, as well as basic carbonates, and thus makes it difficult to reduce all of the nickel to the metal but does provide a good support. The product is filtered, washed, and dried and then carefully reduced with a hydrogen–nitrogen mixture in a rotary calciner at 290–450^0C. The pyrophoric catalyst can then be mixed with a hardened oil that, when solidified, will prevent re-oxidation before use.

TABLE 3.17. Fat-Hardening Catalyst.

Composition	Property
Nickel	20–25 wt%
Kieselguhr	12–15 wt%
Hardened oil	Balance (mp 60^0C)
Bulk density	0.8 kg liter^{-1}

Notes: For *trans*-promoting hydrogenation reactions a specially sulfur-poisoned catalyst can be provided to achieve the maximum content of *trans*-isomers. Alumina and silica/alumina supports are also available.

- *Wet reduction*: Insoluble nickel formate is precipitated by adding sodium formate to a strong solution of a nickel salt. Alternatively, formic acid can be added to precipitated nickel hydroxide or carbonate. The precipitate is filtered and washed with minimum water to remove impurities and dried. Catalyst is suspended in dry saturated oil and slowly heated first to about 200^0C and then to about 250^0C. The hydrate first decomposes at up to 180^0C and finally the formate itself decomposes to produce finely divided nickel at about 200^0C. Nickel can be filtered from the mixture and suspended in fresh oil. The suspension forms flakes as it solidifies and the catalyst is ready for use.

- *Electrolytic precipitation*: Nickel hydroxide may also be precipitated onto a support from nickel anodes suspended in a stirred bath of 1% sodium chloride at pH 9–9.5. The catalyst is filtered, washed, dried, ground to the correct size, and dry-reduced before the addition of a hardened oil to protect it from oxidation. Before dispatch powdered catalyst is formed into flakes or shapes that can be easily added to the hydrogenator.

- *Raney nickel*: For some years after it was first introduced, Raney nickel was successfully used as a fat-hardening catalyst and provided a reproducible catalyst at a time when nickel catalyst production was unreliable.

3.2.4. Catalyst Selectivity

The melting point and the resistance of natural oils to oxidation depend on the unsaturation of the fatty acid component. For example, unstable linolenic acid, with three double bonds, must be selectively hydrogenated to linoleic or oleic acid before the oil is stable enough to be used domestically. On the other hand, the stearic acid content of a natural oil should not be increased unless a high-melting, hard product is required. Melting-point control is the most important factor in producing a selective catalyst.[57]

The hydrogenation process has, therefore, become popularly known as fat hardening. It converts oils to solids, with convenient softening points, that resist oxidation and contain *polyunsaturated* linoleic esters that are felt to be nutritionally useful. Most fats can be synthesized in the body, except for those containing linoleic and linolenic acids, so these are the essential fatty acids that must be provided with food.

As well as controlling the final product composition by selective, stepwise hydrogenation of the double bonds, it is important to control isomerization during the process. Double bonds in natural oils are always in the cis-isomer form, which leads to a higher melting point than in trans-isomers. Isomerization from cis- to trans-isomers is therefore generally undesirable. Double bonds in unsaturated fatty acids, which are always unconjugated, are separated by an active methylene group and, if possible, should not be isomerized to give a conjugated arrangement.

The ideal reaction would be the adsorption of the linolenic chain on the catalyst surface and hydrogenation of one double bond before desorption of the triglyceride molecule. When all of the linolenic acid in the triglyceride has been converted to linoleic acid, any further hydrogenation of linoleic to oleic acid would begin. The desired extent of hydrogenation depends on the melting properties required. The product should be solid at typical ambient temperatures yet melt in the mouth. This obviously varies in different climates.

It is not just the degree of hydrogenation that affects melting point, but also the nature of the isomers in the product. Unfortunately, the catalysts active for hydrogenation, also have activity for isomerization:

- The adsorbed double bond is rearranged rather than hydrogenated.
- The double bond migrates in either direction to form one of four possible positional cis- or trans-isomers.
- The isomers formed may also be hydrogenated to reach the thermodynamic equilibrium content of about 66% trans-isomers. Trans-isomerization can be suppressed or maximized to some extent by selecting appropriate operating conditions or using a sulfided catalyst.

Fortunately, polyunsaturated oils are preferentially adsorbed by the catalyst, compared with monounsaturated oils, and are therefore hydrogenated first with a selective catalyst. Moreover, when a conjugated diene does form, it is more reactive and is quickly hydrogenated.

A steady supply of hydrogen to the catalyst surface promotes hydrogenation rather than isomerization. Thus, when hydrogen is readily available, polyunsaturated oils are hydrogenated faster than conjugated chains can desorb. However, too much hydrogen on the catalyst surface is undesirable as it can lead to over-hydrogenation and lower selectivity. Selectivity is, therefore, controlled by a careful balance of operating pressure, stirring, hydrogen transfer, operating temperature, and the catalyst loading. Transisomerization increases as the monounsaturated content of the oil increases at high operating temperature.

3.2.5. Feed Pretreatment

The crude vegetable oils must be carefully purified before they are used. Free fatty acids are neutralized with alkali, while pigments and poisons, such as alkyl soaps, phosphatides, thioglucosides, and amino acids are *bleached* with fuller's earth. Oils are carefully filtered and dried to remove water, which can produce fatty acids by hydrolysis during hydrogenation and thus damage the catalyst.

3.2.6. Catalyst Operation

The catalyst is provided as solid flakes or droplets that contain prereduced nickel coated with a layer of solid fat that melts in the hot oil before reaction. The use of shapes retards reoxidation of the nickel and any problems associated with dust. Suppliers recommend the quantity of catalyst and the operating conditions required to provide the target melting point and iodine value for products. Most catalysts can be used with different vegetable oils to allow rapid switching from one product to another.

High selectivity towards linoleic acid is required for the production of edible oils or fats. This means that only one double bond in the triunsaturated linolenic glyceride is hydrogenated to give linoleic glyceride. Complete hydrogenation to the saturated stearate glyceride results in a product with a *fatty* taste. Furthermore, good selectivity to the linoleic glyceride also controls the texture of the fat produced, giving a uniform composition with sharper melting characteristics for products ranging from ice cream and salad dressings to soft margarines.

Trans-isomer selectivity is important in producing fats to replace cocoa butter in chocolate. A high *trans*-isomer content allows the chocolate to melt in the mouth. A high proportion of *trans*-isomers is obtained when sulfided nickel catalysts are used.

On the industrial scale, triglycerides are hydrogenated in large reactors and the reaction is often diffusion limited. Small catalyst particles are used to alleviate this limitation. However, the catalyst should still be easily removed from the product by a simple filtering procedure.

The pressure of hydrogen should be sufficient to enable hydrogenation of linolenate to linoleate and sequentially oleate to where necessary, with minimum isomerization of the remaining double bonds. The pressure of hydrogen should never be sufficiently high so that further hydrogenation to stearate occurs. Stirring within the reactor usually provides adequate mixing. If necessary, selectivity can usually be improved by using a different catalyst.[57] This may be either a

TABLE 3.18. Operating Conditions for Fat-Hardening Processes.

Process	Catalyst conditions
Low temperature	To remove triunsaturated acids and improve stability. 0.1–0.15% fresh nickel/oil. 110–120°C. 3–5 atm.
Iso or *trans* suppressing	For melting point 30–40°C. 0.05–0.15% fresh catalyst (selective). 160°C maximum. Up to 5 atm. Minimum *trans*-isomers
Normal	For a melting point higher than 40°C. 0.1–0.15% fresh nickel/oil (or more older catalyst). 140° rising to 180–200°C. Up to 3 atm.

Note: Hydrogenation proceeds to the melting point or iodine values required. Full recommendations are always available from the catalyst supplier.

less active older catalyst or a smaller dose of a more active catalyst with appropriate changes in the operating conditions. Typical sets of operating conditions are shown in Table 3.18.

3.2.7. Catalyst Poisons

Natural oils contain traces of sulfur, phosphorus, nitrogen, and oxygen compounds. These are usually removed to a reasonable level by suitable cleaning processes before the oils are hydrogenated. Where removing them completely is too expensive compared with the price of the catalyst, some poisoning has often been accepted by producers.

Oils can be pretreated with an old catalyst, no longer active enough for fat hardening, in a guard reactor to protect new catalysts. Where it is more convenient to use additional new catalyst to absorb the poisons, it has been estimated that the following proportions of the catalyst are equivalent to 1 ppm of poisons:

- Sulfur in thioglucosides 0.004%
- Phosphorus in lecithins 0.0008%
- Nitrogen in amino acids 0.0016%

In some cases rather more extra catalyst must be added, depending on how any poison is adsorbed. Poisons that are adsorbed on the catalyst surface rather than inside the pores make the catalyst less selective. If poisons are evenly distributed on the surface and in the pores then the catalyst selectivity can be restored at the expense of producing more isomers.

It is common for catalysts to be deliberately poisoned by sulfur to increase trans-isomerization and to provide products with sharper melting points. Up to 2–3% sulfur relative to the nickel content is added to trans-selective catalysts. However, the quantity of catalyst used must be increased because sulfided catalysts are less active.

3.3. FATTY ACID HYDROGENATION

Unsaturated fatty acids produced by the hydrolysis of triglycerides can be saturated by hydrogenation in batch or continuous processes. The catalysts already described for triglyceride hydrogenation can be used. Operation is at 160–180^{0}C and 25 atm pressure for vegetable fatty acids and up to 190–200^{0}C for tallow and fish oil fatty acids. The use of higher hydrogen pressures not only increases the rate of reaction and limits the attack of the acid on the catalyst but provides better reaction conditions. The catalyst *dose* is about 0.2% nickel catalyst to oil treated.

A continuous fixed bed process has been developed that uses a precious metal catalyst to hydrogenate vegetable oils and animal fats, as well as fatty acids.[58] The regenerable, easily recovered catalyst resists the effects of acid and avoids contamination of the products with nickel or copper soaps. Two beds are used. In the first, most of the feed is hydrogenated and in the second, hydrogenation to the required product specification is completed. Disadvantages of conventional nickel catalyst seem to have been overcome.

3.4. THE PRODUCTION OF FATTY ALCOHOLS

Long-chain alcohols are widely used as plasticizers or in the production of detergents. They are available from a variety of synthetic routes or the direct hydrogenation of natural fatty acids:

- Hydroformylation of olefins giving mixtures of normal and isoaldehydes that can be hydrogenated to alcohols.
- The partial oxidation of C_{12}–C_{14} normal paraffins to give secondary alcohols.
- The oligomerization of ethylene using aluminum alkyls, followed by oxidation and hydrolysis of the aluminum trialkoxides.
- Hydrogenation of the methyl or fatty alcohol esters of fatty acids obtained by the hydrolysis of natural oils or fats.

3.4.1. Natural Fatty Alcohols

Natural oils are hydrolyzed and the fatty acids separated by distillation. The acids are then hydrogenated to alcohols either as the methyl ester or as an ester with another fatty alcohol.

Catalysts used for the hydrogenation step are usually copper chromite formulations, although copper oxide/zinc oxide catalysts have also been used. The process accounts for about half of the copper chromite catalysts used commercially. Both acid group and double bonds in the long carbon chain are hydrogenated during the reaction, which produces a saturated alcohol. When an unsaturated fatty alcohol is required, a more selective zinc chromite catalyst may be used.

In commercial processes the copper chromite catalyst must be carefully promoted for use with different oils. Catalysts must resist the action of the acids being treated because colored metal soaps contaminate the products. To avoid dust formation the catalyst should also be strong enough to resist disintegration in the liquid reactants. Typical catalysts used are shown in Table 3.19.

TABLE 3.19. Catalysts for Fatty Alcohol Production.

	Slurry process	Fixed bed process	
Type	Copper chromite (wt%)	Copper chromite (wt%)	Copper oxide/zinc oxide (wt%)
Copper	36	33	CuO 33
Chromium	33	30	ZnO 67
Barium	0–2	8	—
Manganese	2–3	—	
Density	1.0–1.5 kg liter^{-1}	1.65 kg liter^{-1}	1.8 kg liter^{-1}
Surface area	35 m^2 g^{-1}	80 m^2 g^{-1}	45 m^2 g^{-1}

3.4.2. Catalyst Operation

Fatty alcohols are produced in either slurry or fixed bed processes.

- In slurry processes, the fatty acid is mixed in batches with a proportion of the fatty alcohol product to form fatty acid esters. The ester is then circulated through the reactor mixed with copper chromite catalyst powder and hydrogen. Fatty alcohol is removed from the system in a centrifuge that separates the catalyst. More than half of the catalyst can normally be reused, depending on the poisons present in the acid, and about 3–4 kg of catalyst are required per tonne of alcohol produced. With proper control of the acid concentration, the formation of hydrocarbon by-products can be minimized.

- Fixed bed processes can be used for the hydrogenation of fatty acid methyl esters. The methyl esters can be prepared directly from the fatty acid or by *trans*-esterification of the triglyceride with methanol. The hydrogenation is carried out in a bed of solid copper chromite catalyst, which usually loses activity after operating for 3–6 months. Copper oxide/zinc oxide catalysts have also been used.

Operating conditions for the two processes are shown in Table 3.20. A process for the direct hydrogenation of fats and oils to fatty alcohols and propanediol was developed by Henkel using a specially supported copper/chromium catalyst at 200°C and 250 atm pressure.

3.4.3. Reaction of Fatty Alcohols

- Fatty aldehydes are formed selectively by dehydrogenation of the corresponding fatty alcohol using copper chromite catalysts in slurries or fixed beds. Operation is at 250–350°C and a pressure of 1 atm or less to give an equilibrium conversion to about 30% aldehyde.

- α-Olefins can be formed by the dehydration of fatty alcohols with an acid catalyst at 300–350°C.

TABLE 3.20. Fatty Alcohol Production Processes.

Temperature (°C)	Slurry process 280–300	Mixed bed process 200–240
Pressure (atm)	300	60–250
Conversion (%)	~100	> 80
Catalyst use	2–5 kg catalyst.te alcohol^{-1}	0.25–0.75 h^{-1}
Hydrogen	Excess	20–100 × theory
Comment	Direct hydrogenation of the fatty acid with recycled fatty alcohol to produce ester as the first stage of reaction.	Hydrogenation of the methyl ester of the fatty acid. Ester produced directly from the fatty acid or *trans-esterification* of the triglyceride.

- Fatty amines are formed by the dehydration of the fatty acid ammonium salts to give nitriles, which are then hydrogenated to amines. Amines are also formed by ammination of fatty alcohols.

3.5. SOME INDUSTRIAL HYDROGENATION PROCESSES

3.5.1. Nitrobenzene Reduction

Nitrobenzene hydrogenation is the principal process for aniline production. Only relatively small quantities of aniline are used as such, the main demand being in the production of isocyanates required for polyurethane synthesis. Hydrogenation reactions can be carried out in the gas phase using tubular reactors and catalyst pellets. Reactions containing both liquid and gas phases can also be used, using catalyst powders. The catalysts that have been used include copper chromites, copper oxide or nickel oxide supported on kieselguhr, Raney copper, and nickel sulfide supported by alumina. All catalysts give good conversion with a high selectivity to aniline. During operation the conversion slowly declines and the catalyst must be regenerated after a few months. Deactivation is usually the result of carbon deposition from the thermal cracking of aniline.

Operation is carried out at 270–290°C and 1–5 atm of hydrogen with a hydrogen/nitrobenzene ratio of about 1:9. The copper chromite and nickel oxide/kieselguhr catalysts are made by the standard methods. Nickel sulfide catalyst is prepared by the method described by Allied Chemical and Dye Corp. A nickel oxide/alumina catalyst is prepared either by impregnating activated alumina with nickel nitrate followed by decomposition at 500°C or by co-precipitation of the mixed oxides. It is then sulfided by treatment with hydrogen sulfide at 450°C and the NiS reduced to Ni_2S_3 with hydrogen at 250°C. Oxygen-free gas should be used during the sulfiding and reduction steps. The product contains both Ni_2S_3 and NiS.

Deactivated catalyst is carefully regenerated in a mixture of air and steam at 300–400^{0}C. Hydrocarbons are purged from the system and the catalyst rereduced with hydrogen in steam before it is reused.

3.5.2. Benzene Hydrogenation

Large quantities of benzene are required throughout the world for a wide range of applications. A high proportion is hydrogenated to provide cyclohexane, an intermediate in the production of nylon fibers and resins.

The reaction involved is very simple and has been well known since Sabatier and Senderens reported on their experiments in 1901. They passed hydrogen saturated with benzene vapor at ambient temperature over a nickel catalyst at 180–200^{0}C. At this temperature an almost complete conversion of benzene to cyclohexane was achieved.[59] They made two important observations:

- Partially hydrogenated benzene derivatives were never found—only cyclohexane.
- Cyclohexane was dehydrogenated above 200^{0}C to give the reverse reaction. At higher temperatures benzene cracked to form methane and carbon.

Although Sabatier and Senderens claimed that other metals did not hydrogenate benzene, later work by Zelinsky showed that benzene was easily hydrogenated by platinum metals.

During the late 1800s it was realized that cyclohexane was identical with Caucasian petroleum. Subsequently, natural gas liquids became an important source of up to 85% pure cyclohexane. Even as late as 1968 some 20% of the cyclohexane used in the United States was obtained in this way, although the cyclohexane content was increased to about 98% by the isomerization of methylcyclopentane during fractional distillation.

Elsewhere benzene hydrogenation was increasingly used to provide 99.9% pure cyclohexane. Liquid phase hydrogenation at 40 atm pressure and temperatures in the range 170–230^{0}C is typical using supported nickel catalysts.[60] These conditions avoid the isomerization of cyclohexane to methycyclopentane. Benzene must be free from sulfur to avoid poisoning the catalyst, although, originally, short and uneconomic catalyst lives were common. Reaction temperature and exotherm can be controlled by evaporation of the product and dilution of the benzene feed with recycled cyclohexane. A process with two hydrogenation steps is currently favoured. Liquid phase reaction gives 95% benzene conversion and is followed by adiabatic vapor phase reaction to produce cyclohexane containing less than 100 ppm of benzene and methylcyclopentane.

In an alternative vapor phase process, a platinum catalyst is used in a tubular reactor at 30 atm and about 400^{0}C to give almost 100% selectivity. The catalyst used in this process is substantially more expensive than nickel oxide.

3.5.2.1. *Removal of Aromatics*

Environmental limits on the aromatic content of gasoline and diesel fuel have led to a further application of supported nickel hydrogenation catalysts. Benzene can be completely removed from light C_6 reformate or other similar streams by liquid phase hydrogenation, before blending into the refinery gasoline pool.

The catalysts contain more than 50% nickel oxide, supported on kieselguhr with some added alumina, and are prereduced and stabilized. This allows for rapid reduction in existing reactors. The process operates at the relatively low temperature of 80^0C with hydrogen pressures in the range 20–40 atm. A liquid space velocity of about 2.5 h^{-1} is required and hydrogen addition depends on the aromatics content of the feed being treated.

The aromatic-free product can be recycled to control temperature rise in the catalyst bed. Sulfur impurity in the feed gradually poisons the catalyst so that the inlet temperature must be gradually increased. Catalyst lives exceeding two years have been achieved.[61] The same catalyst can be used to *dearomatize* diesel fuel or white oils but is then operated at up to 200^0C and 125 atm hydrogen pressure with lower space velocity.

3.5.3. Hydrogenation of Phenol

Although most of the cyclohexanone used to produce adipic acid and ε-caprolactam has been made from cyclohexane it is also possible for phenol to be used. The original I. G. Farben process using phenol operated in two stages:

- Phenol was converted to cyclohexanol by hydrogenation in either the gas or liquid phase using a supported nickel oxide catalyst. Typical operating conditions were $140–160^0C$ and 15 atm with higher than 95% selectivity.
- The cyclohexanol was then dehydrogenated at $400–450^0C$ and atmospheric pressure using a copper oxide/zinc oxide catalyst. More than 95% selectivity at about 90% conversion was obtained.

A single-stage liquid phase process was subsequently developed by Allied Chemical[62] and Vickers Zimmer[63] using a selective palladium catalyst. More recently a single-stage gas phase process was introduced that uses a selective catalyst containing about 1% palladium supported on a calcium oxide/alumina mixture.[64] Almost complete conversion and greater than 95% selectivity is achieved at $140–170^0C$ and 1–2 atm. A relatively high calcium content (possibly in the form of calcium aluminate) is used to neutralise any acidic form of alumina, which would otherwise lead to catalyst deactivation via coke formation. Regeneration still remains a possibility, should the catalyst become deactivated.

TABLE 3.21. Early Sources of Ethylene.

Volume (%)	H_2	C_2H_2	C_2H_4	CO	Balance
Coke oven gas	52	0.15	2.4	8	Nitrogen, etc.
Acetylene plant off-gas	48	0.1	8.5	29	Methane, etc.
Refinery gas	10–20	0.1	5–15	0.1–1	Low-molecular-weight hydrocarbons
Ethane cracking	27	0.4	33	1	Ethane, etc.
Propane cracking	11	0.8	36	1	Propane, etc.

3.6. SELECTIVE HYDROGENATION OF ACETYLENES AND DIENES

The removal of acetylenes and dienes from steam-cracked olefins is a critical step in purification. Selective hydrogenation processes and catalysts have become more important as worldwide olefin production has increased in 1999 to more than 90 million tonnes of ethylene and almost 50 million tonnes of propylene. Demand for better catalysts with improved selectivity and longer operating cycles has grown as larger plants are built. Tighter product specifications have also been imposed now that more of the olefins produced are being converted to polyolefins.

Before the 1950s commercial ethylene was recovered from various off-gases, ethane or propane cracking and ethanol dehydration, as shown in Table 3.21. Various purification catalysts were used before ethylene production expanded, and as the feed gases usually contained sulfur the catalysts were often metal sulfides. In fact, sulfiding was actually necessary to improve selectivity and operating stability. Several early catalyst types are described in Table 3.22. Most of the Group-VIII metals are active and quite selective, but catalysts suffered from the need for frequent regeneration to remove polymers deposited during operation.

By the 1950s, as demand for ethylene increased, the existing catalysts were operating in newly designed small-capacity steam crackers. In the United States, where crackers were generally based on low-molecular-weight feeds, this meant that acetylene was hydrogenated in cracked gas containing hydrogen, often before sulfur was removed and the gas was dried. It was inevitable that polymers formed during operation and they became well known as *green oil*. A successful catalyst was thereafter judged not only on acetylene conversion but also on the ability to avoid green oil production and the operating *life* between regeneration.

The urgent demand for better catalysts intensified in the late 1950s as polyethylene production was developed and new plants in Europe began to use naphtha feeds. Larger single-stream ethylene plants needed better reliability and selectivity from more active catalysts. In the short term, better acetylene conver-

TABLE 3.22. Early Acetylene Hydrogenation Catalysts.

Year	Catalyst	Use
1931	Molybdenum disulfide supported on alumina.[65]	Acetylene hydrogenation in coke oven gas containing sulfur.
1931	Nickel oxide/chromium oxide supported on alumina.[67]	Selective hydrogenation in ethylene and hydrogen mixtures.
1940s	Palladium supported on silica gel.[66]	Acetylene plant off-gas—lifetime eight months.
1950+	Cobalt molybdate on alumina.[68]	Cracked gas streams containing hydrogen.
1950+	Nickel oxide supported on alumina and magnesia.[69]	Cracked gas streams containing hydrogen.
1955+	Nickel oxide/cobalt oxide/chromia on silica alumina.[70]	Cracked gas streams containing hydrogen.
1955+	Fused iron oxide with silica, magnesia, potash promoters.[71]	Acetylene hydrogenation in depropanizer overhead streams.
1955+	Palladium supported on γ-alumina.[72]	Used in tail-end guard beds following front-end nickel catalysts.

sion was achieved in existing plants by installing a guard reactor to remove traces of acetylene from the separated ethylene-ethane (C_2) stream. The new guard catalysts were prepared by supporting palladium on γ-alumina and a stoichiometric volume of hydrogen was added to react with the acetylene.[72] Of course, this did not solve the problem of polymer forming in either of the catalyst beds although the guard bed catalyst could operate for relatively long periods. Brief properties of some early catalysts are given in Table 3.23.

Since 1960 all new catalysts used for selective acetylene hydrogenation contain palladium supported on different forms of alumina and remove acetylene almost completely to achieve the much stricter specification demanded. The higher activity of palladium catalysts meant that smaller volumes of catalyst could be used at temperatures as low as 50–60°C.

TABLE 3.23. Operation of Some Early Acetylene Hydrogenation Catalysts.

	Cobalt molybdenum	Nickel cobalt chromium	Palladium alumina guard catalyst
Space velocity (h^{-1})	500–1000	1000–3000	1000–3000
Temperature inlet (°C)	175–315	120–200	60–120
Operating pressure (atm)	5–16	5–16	Plant design
Hydrogen concentration (%)	10–20	10–20	2–3 mol per mole acetylene
Acetylene inlet (%)	0.4–2.0	0.4–2.0	20–100 ppm
Outlet (ppm)	10–20	10–100	< 10
Ethylene loss (%)	1–3	1–3	Limited by hydrogen
Cycle time (months)	0.5–1	3–6	6–12
Life (years)	1–2	5	5–10

3.6.1. Acetylene Hydrogenation Process Design

Two different process configurations are now used to remove acetylene from ethylene. The choice of these depends on how the cracked hydrocarbon gases are separated following desulfurization and drying:

- If the demethanizer, which removes methane and hydrogen from the gases, is the first stage of gas separation, the acetylene removal reactor is placed before the ethylene-ethane (C_2 stream) splitter. A methyacetylene/propadiene removal reactor is also needed before the propylene-propane (C_3 stream) splitter. Sufficient hydrogen must be added to the C_2 and C_3 streams before they enter the catalyst beds. Spare reactors must be available to allow for regular regeneration of the on-line catalyst because green oil polymers form during the hydrogenation reactions. *Tail-end* hydrogenation was developed from the guard beds using a palladium/γ-alumina catalyst added to the early ethylene plants.[73]
- When either a depropanizer or de-ethanizer is the first stage of gas separation the acetylene can be hydrogenated in the mixed overhead streams, which contain up to almost 30% hydrogen. An advantage of this procedure, which uses several beds of a more selective palladium/α-alumina catalyst, has been that no spare reactor is required because green oils are not usually formed. Apart from the different catalyst used, *front-end* hydrogenation is based on the original acetylene removal designs.[74]

Both procedures work well. The choice between them is determined by the process supplied by the contracting company. However, the two types of palladium/alumina catalyst used are very different and are not interchangeable.

Operating problems with palladium catalysts have been associated with increasingly high volumes of acetylene in the process gas, which is a result of increased steam cracking severity to improve ethylene yields. Both front-end and tail-end reactors now include several adiabatic beds, with interbed cooling, to control reaction and remove the excessive heat evolved as acetylene is hydrogenated.

Significant hydrogenation of ethylene can occur if the gas is not efficiently cooled or the catalyst is not very selective. This is referred to as ethylene loss. Catalyst selectivity is also important to minimize the formation of green oil polymers, which wastes ethylene and causes operating problems. Some process designs have included tube-cooled adiabatic catalyst reactors to cope with high acetylene concentrations, but they have not been very popular.

3.6.2. Early Acetylene Hydrogenation Catalysts

3.6.2.1. *Sulfided Cobalt Molybdate*

A supported cobalt/molybdate catalyst, probably based on the ones developed in the 1930s, was one of the first types to be used in modern ethylene plants.[68] The front-end reactor was located in the compressor train after heavy hydrocarbons were removed but before sulfur removal or gas drying. The catalyst was, therefore, partly sulfided. Careful temperature control was required to limit ethylene loss. About 10% steam was added to cracked gas, which limited the temperature rise and improved selectivity. An unusual feature of operation was that a significant proportion of the acetylene was removed as a polymer. This decreased the potential temperature rise but meant that catalyst regeneration and subsequent reactivation was a routine procedure at intervals of 2–4 weeks and that a spare reactor was needed. To compensate for loss of activity the gas temperature was continuously increased throughout the operating cycle. Acetylene levels were reduced to about 10–20 ppm with 1–3% ethylene loss. Up to 50% of any butadiene present in the gas was also hydrogenated. The catalyst was replaced after 1–2 years.

The catalyst composition was 13.5 parts $Co(NO_3)_2 6H_2O$ and 10.5 parts MoO_3 (i.e., $CoO:MoO_3 = 1.0:1.6$) with 54.5 parts $Al_2O_3.H_2O$; 24 parts Portland cement and 16 parts Kentucky clay.

3.6.2.2. *Sulfided Nickel Oxide*

Following from early experience with a DuPont nickel oxide/silica alumina catalyst containing magnesia, and which was reduced and sulfided before use,[69] other nickel catalysts were later developed. Catalysts and Chemicals Inc. introduced a nickel cobalt/chromium catalyst supported on silica/alumina, which was used for several years in early ethylene plants.[70] It operated as a single bed, generally with a spare reactor, to remove acetylene from wet cracked gases containing sulfur compounds. Operating conditions depended on gas composition.

The addition of steam and, occasionally, sulfur compounds sometimes improved selectivity. Less polymer was generally formed than with cobalt/molybdate catalysts but regeneration at 375–425^{0}C was still essential at regular intervals of up to 3 months. Following regeneration the catalyst had to be re-reduced at up to 375–425^{0}C for 6–12 h. The sulfur content of the cracked gas treated could be as high as 25–50 grains per 100 standard ft^3 (~1000 ppm), although the operating temperature had then to be increased to compensate for the decreased activity of sulfided catalyst.

Acetylene content of product ethylene was claimed to be less than 10 ppm with only 1% ethylene loss. At this conversion all butadiene in the gas was also hydrogenated. At lower butadiene conversion the acetylene content in ethylene would rise to about 100 ppm. Many plants had problems in maintaining a low ethylene loss and found that the catalyst needed very frequent regeneration.

3.6.2.3. *Fused Iron Oxide*

ICI and several other operators used a fused magnetite catalyst promoted with magnesia, silica, and potash in modern naphtha steam crackers designed by Kellogg in the late 1950s. Up to 3000 ppm of acetylene could be reduced to less than 50 ppm in the sulfur-free depropanizer overheads containing 12% hydrogen. Catalyst activity declined after about 6 months as a result of polymer deposition. This was a low cost catalyst and was not regenerated but replaced as necessary.[71]

3.6.2.4. *Palladium Catalyst Guard Beds*

The nickel and iron hydrogenation catalysts were not able to meet the more stringent ethylene specifications required by the new polyethylene processes. Existing steam crackers therefore began to back up the front-end reactors, which produced ethylene containing 20–50 ppm acetylene, with a tail-end reactor.[72] The *guard* bed, located in the C_2 stream, contained a catalyst with less than 350 ppm of palladium on a suitable γ-alumina support. Up to 2–3 mol of hydrogen per mole of acetylene was added, and at 60–120^{0}C the outlet acetylene was reduced to less than 10 ppm. Any excess hydrogen was removed by increasing the operating temperature. The catalyst still needed regular regeneration to remove polymers and restore activity so that a spare reactor had to be available. The catalyst was often supplied ready for use in a small preloaded reactor.

3.6.3. Modern Acetylene Hydrogenation Catalysts

Until 1958 no ethylene plant had used a tail-end palladium catalyst to hydrogenate all of the acetylene formed in the steam cracker. This was an attractive possibility, however, and many of the large new US plants built in the 1960s were designed in this way. The less efficient front-end nickel and iron catalysts were soon obsolete. Several significant changes followed the use of tail-end catalysts:

- Two parallel reactors, one operating and the second regenerating, were placed before the C_2 splitter to remove acetylene. No front-end catalysts were used in these plants.

- Where necessary, particularly in naphtha steam crackers, the same type of tail-end system was used to hydrogenate methyl acetylene and propadiene in the feed to the C3 splitter.

In Europe four or five of the more modern 1950s steam crackers, based on naphtha feed, replaced the fused iron or nickel front-end catalysts with a new palladium catalyst using an α-alumina support.[72] Success in meeting the strict new acetylene specifications, while hydrogenating 95% of the methyl acetylene and forming no green oil, led to the use of this catalyst in many new ethylene plant designs.

3.6.4. Acetylene Hydrogenation Catalyst Preparation

Front-end and tail-end catalysts are both produced by relatively simple procedures in which palladium is impregnated onto the outside surface of an alumina support in a thin layer. Theoretically, in order to achieve the required selectivity, the support should be inert and take no part in the hydrogenation process.

Tail-end catalysts are usually made with suitable γ-alumina particles with a relatively small surface area,[72] with selectivity being controlled by the volume of hydrogen added to the olefin stream being treated. More selective catalysts have been developed to avoid excessive formation of polymers and the use of hydrogen ratios greater than two. The addition of a suitable Group-IB metal inhibited the oligomerization reactions that led to green oil formation. Selectivity was also improved by the addition of carbon monoxide to the hydrogen stream that adsorbed on the catalyst surface.

Front-end catalysts were produced from a suitable α-alumina support with carefully controlled surface area and pore volume.[75] The support could influence the hydrogenation reaction to give good selectivity with almost no polymer formation. It was found that the adsorption of carbon monoxide onto the catalyst surface inhibited ethylene hydrogenation in the presence of a few ppm of acetylene. Carbon monoxide was always present in the process gas.

Catalyst selectivity in front-end catalysts can also be controlled by the addition of a Group-IB metal when acetylene levels are high or carbon monoxide content is low.

3.6.5. Acetylene Hydrogenation Catalyst Operation

3.6.5.1. *Tail-End Acetylene Hydrogenation*

Acetylene is hydrogenated in the separated C_2 stream, which means that sufficient hydrogen must be added to the gas before the reactor. The theoretical amount for complete removal is 1-mol volume of hydrogen per mole volume of acetylene, giving 100% conversion to ethylene, but this has always been impossible to achieve. There is usually an ethylene loss associated with complete acetylene removal and in the 1960s at least three volumes of hydrogen were often required. Product ethylene contains less than 1 ppm acetylene.

In some of the early large ethylene plants, high ethylene losses led to the formation of several tonnes of green oil per day. This resulted in the need for extremely frequent catalyst regeneration. Good control of the catalyst temperature and the volume of added hydrogen was essential to give optimum and stable operation.

Much better catalysts now provide improved operation.[73] Hydrogenation can be controlled by adding traces of carbon monoxide to the hydrogen. Adsorbed carbon monoxide modifies the relative adsorption of acetylene and ethylene on the palladium and minimizes ethylene loss. The catalyst itself can also be made more selective by alloying the palladium with a further metal such as copper or silver.[76] This also affects palladium dispersion and the relative adsorption of acetylene and ethylene on the catalyst surface to improve selectivity. To minimize temperature rise catalyst suppliers recommend that one or more catalyst beds with intercoolers be used in each reactor, depending on the acetylene content of the C_2 stream:[73]

- One bed of catalyst is used when the acetylene content of the C_2 stream is less than 0.8%.
- Two beds of catalyst with intercooling are needed in each reactor if acetylene content is between 0.8 and 1.7%.
- Three beds of catalyst with intercooling are used in each reactor for acetylene content up to 2.5%.

The hydrogen ratio recommended is usually 1.5–2.0 in single or final beds, where the acetylene content is lowest and maximum conversion is needed, but only 0.8–1.3 in the first or second beds.[73] Temperature rise can be roughly calculated as 65^0C or 35^0C for the conversion of 1% acetylene to ethane or ethylene, respectively.

Reaction is controlled by increasing or decreasing the bed inlet temperature and changing the hydrogen ratio. Alternatively, if the reaction is not sufficiently selective, carbon monoxide can be added to the hydrogen. However, gas inlet temperature may then need to be increased if the preferential adsorption of carbon monoxide affects catalyst activity. The objective is to have about half of the total temperature rise in the top third of each catalyst bed. The bed inlet temperature should be increased as required to maintain conversion. Catalyst should be regenerated when the inlet temperature reaches 150^0C to avoid overhydrogenation. Bed intercoolers or cold recycle gas are used to control catalyst temperature.

With very high acetylene levels in the C_2 stream, the use of tubular reactors can provide good temperature control. An identical spare reactor should be available to allow continuous plant operation when the catalyst has to be regene-

TABLE 3.24. Tail-End Catalyst Operation.[a]

	Tail-end acetylene	Tail-end MAPD[b]
Space velocity (h[-1])	2000–8000	2000–4000
Inlet gas temperature (^{0}C)	20–150	50–150
Pressure (atm)	Depends on plant design	
C_2H_2 or C_3H_4 inlet (%)	0.3–2.8	Up to 6
Outlet (ppm)	< 1	< 5
Hydrogen ratio (molar)	0.9–2.2	1.1–1.8
Operating cycle (months)	3–18	4–24
Catalyst life	5–10	> 5
Feed gas composition	Mixed C_2	Mixed C_3

[a] Catalyst volume ~15 m^3 per 100,000 tonnes ethylene produced per year.
[b] Methyl acetylene and propadiene.

rated. Catalyst life is usually about 10 years. Operating conditions are summarized in Table 3.24.

3.6.5.2. *Tail-End Methyl Acetylene/Propadiene Hydrogenation*

Hydrogenation of methylacetylene and propadiene, its isomer (MAPD) in the separated C_3 stream is very similar to tail-end acetylene removal. The possible reactions are:

$$HC{\equiv}C{\cdot}CH_3 + H_2 \rightarrow CH_3{\cdot}CH{=}CH_2 \qquad (3.3)$$

$$HC{\equiv}C{\cdot}CH_3 + 2H_2 \rightarrow C_3H_8 \qquad (3.4)$$

$$H_2C{=}C{=}CH_2 + H_2 \rightarrow CH_3{\cdot}CH{=}CH_2 \qquad (3.5)$$

$$H_2C{=}C{=}CH_2 + 2H_2 \rightarrow C_3H_8 \qquad (3.6)$$

Two tail-end reactors are installed, with one operating until regeneration is necessary and the second ready for use. The maximum volume of hydrogen added should not usually exceed 2 mol per mole of MAPD. Even less may be used with a very selective catalyst, and there is often a significant propylene gain.

Green oil is not formed in large volumes, probably because C_3 molecules do not readily adsorb on the sites that allow oligomerization of ethylene in C_2 hydrogenation. However, occasional regeneration is required to clean the catalyst when the inlet temperature has been increased to 150^{0}C.

Each of the two reactors may have one or more catalyst beds, depending on the total MAPD in the feed gas. With regulated hydrogen addition, the temperature rise in each bed can be limited to less than about 40^{0}C. Temperature is controlled either by intercoolers between the beds or the addition of a cold recycle. The temperature rise for 1% MAPD converted to propylene is about 25^0C:

- One bed of catalyst is used when the MAPD content of the C_3 stream is less than 1.5%.
- Two beds of catalyst with intercooling are needed in each reactor if the MAPD is in the range 1.5–3.0%.
- With higher MAPD content, up to the maximum of about 6–7%, three beds are used with a combination of intercooling and cold gas recycle to control the temperature rise.

If the reaction is not selective and the required product specification is not achieved, performance can be improved by the addition of a few ppm of carbon monoxide to the hydrogen in a similar manner to acetylene hydrogenation. Operating conditions are summarized in Table 3.24.

3.6.5.3. Front-End Acetylene Hydrogenation

Acetylene can also be hydrogenated (Figure 3.2) in depropanizer or de-ethanizer overhead streams containing an excess of hydrogen in the range 10–30%. The palladium catalysts used are very selective and can provide a significant conversion of acetylene to ethylene rather than ethane. Green oil is not usually formed during operation, and the ethylene product contains less than 1 ppm of acetylene.[74]

Figure 3.2. Reactor in front-end acetylene hydrogenation unit. Reprinted with permission from Linde AG.

The catalyst is normally divided into several beds, with interbed cooling, to limit the temperature rise in each bed to less than 15–25^{0}C and to control the selectivity of the reaction. Temperature increases by 40^{0}C and 75^{0}C for every 1.0% of acetylene converted to ethylene or ethane, respectively, and the increase for converting 1.0% MAPD to propylene is 40^{0}C. The number of catalyst beds needed in a reactor can easily be estimated. Reaction is readily controlled by variation of the bed inlet temperature.

The first charges of front-end palladium catalyst were used in 1960 in three small ethylene plants with a combined capacity of 150,000 tonnes year^{-1}. These plants operated for six years with no regeneration, until they were eventually replaced by a larger, modern plant. The volume of palladium catalyst used in a single reactor with two beds was less than 15% of the iron catalyst it replaced, which had also needed a spare reactor. In other ethylene plants two reactors using a nickel catalyst were replaced by a single two-bed reactor with only 20% of the original volume using a palladium catalyst.[75]

For economic reasons the palladium catalyst has occasionally been used to hydrogenate all of the acetylene in a single bed. Performance was satisfactory although, with large-diameter reactors and shallow catalyst beds, there was occasionally a small ethylene loss. Table 3.25 gives an example of front-end catalyst operation.

Development of the front-end catalyst revealed the most important factor in achieving selectivity: carbon monoxide is adsorbed by the catalyst surface so that acetylene, methyl acetylene, and, less strongly, propadiene are adsorbed in preference to ethylene. Consequently, only small amounts of ethylene are hydrogenated at the bottom of the final bed when most of the acetylenes have been removed.

The need for carbon monoxide inhibition can be seen as the catalyst is commissioned. A hot spot may develop when cold gas enters the reactor and the overall temperature is much lower than normal operating levels. However, the catalyst surface in the vicinity of the hotspot is rapidly *deactivated* with carbon

TABLE 3.25. Front-End Catalyst Operation.

Conditions	Inlet bed 1	Exit bed 2
Overall space velocity (h^{-1})	5000	
Bed inlet temperature (^{0}C)	80–85	80–85
Temperature rise (^{0}C)	10–15	5
Acetylene (%)	0.25–0.3	< 1 ppm
Methyl acetylene (%)	0.18–0.22	< 100–200 ppm
Propadiene (%)	0.2–0.3	20% conversion
Carbon monoxide (%)	0.1–0.2	0.1–0.2
Catalyst life		> 7 years

Note: With increased acetylene content the number of beds increases. If the carbon monoxide content changes the inlet temperature must be adjusted. Catalyst volume required about 6 m^3 for 100,000 tonnes of ethylene produced per year.

monoxide. Fortunately carbon monoxide is always present in cracked gas, being formed on a reverse water gas shift reaction between carbon dioxide and steam on the nickel alloy tubes. This has led to the success of front-end palladium catalysts. Concentrations of carbon monoxide are usually in the range 500–5000 ppm, depending on the hydrocarbon feed and cracking conditions. Some operators have even used the catalyst with more than 1% carbon monoxide with no effect on acetylene removal efficiency. It is only necessary to increase the gas inlet temperature at higher carbon monoxide levels.

Although no liquid polymers are formed on front-end catalysts, some acetylene is dimerized to produce 200–400 ppm of butadiene that is not hydrogenated. Also, some organic material, usually forms on the catalyst surface and up to 1 wt% can be extracted from used catalyst samples. The only evidence of this during operation is the need to increase the operating temperature by a few degrees and the deposit most probably enhances the catalyst selectivity.

A further refinement in front-end catalyst formulation has been the addition of a second metal oxide.[76] The newest catalysts often contain less palladium than the original types, which achieved selectivity from the use of a special low-surface-area alumina. The new catalysts may possibly require regeneration after about two years.

3.6.6. Selective Hydrogenation of Pyrolysis Gasoline

There are other products, apart from ethylene, that form during the steam cracking of heavy feeds such as naphtha. About 20–30% of the naphtha cracked is recovered as a high-octane pyrolysis gasoline that contains more than 50% of mixed aromatics. This is significant because the extracted aromatics now provide more benzene than is available from naphtha platforming units. In modern large steam crackers it can also be economic to recover the styrene that is present. A typical pyrolysis gasoline composition is shown in Table 3.26. At

TABLE 3.26. Pyrolysis Gasoline Composition.

Analysis	Composition
Benzene, toluene, xylenes (vol%)	50
Ethyl benzene (vol%)	2
Styrene (vol%)	2
C_9 aromatics, alkylstyrenes, indenes (vol%)	15
C_5–C_7 paraffins, olefins, dienes (vol%)	25
Dicyclopentadiene (vol%)	6
Total sulfur (ppm)	300
Thiophene (ppm)	10
Gum (mg m^{-3})	< 100
Boiling range (^{0}C)	45–200
RON	100

least 50% of the worldwide ethylene steam crackers use more than 3 million tonnes per year of naphtha, so that up to 1 million tonnes per year of pyrolysis gasoline can be produced. Recovery as an aromatic gasoline fraction or an aromatic stream rich in benzene provides a significant contribution to steam cracking operating economics.

Steam cracking, of course, is designed to produce unsaturated hydrocarbons so pyrolysis gasoline contains dienes which must be removed by selective hydrogenation.

There are two ways in which raw pyrolysis gasoline can be treated to provide a useful product:[77]

- Low-temperature selective hydrogenation converts dienes to olefins, and alkyl benzenes such as styrene to ethyl benzene, in the liquid phase. Other unstable hydrocarbons, such as dicyclopentadiene, that form gums are also hydrogenated. At the same time α-olefins isomerize to higher-octane internal olefins.
- When aromatics are to be extracted a second, high-temperature hydrogenation step is needed. This operates in the vapor phase and converts most of the olefins to paraffins and removes the sulfur compounds. Aromatics remain unchanged.

3.6.6.1. *Catalyst Types*

A range of catalysts can be used for the selective hydrogenation step depending on the composition of the cracked gasoline:

- Palladium supported on γ-alumina and promoted with chromium oxide is active and selective provided that methyl mercaptan, disulfides, and thiophene do not exceed typical levels. Carbon monoxide in the hydrogen gas used may make the catalyst more selective.[73] The catalyst gradually deactivates owing to gum deposition and the effect of sulfur compounds but can be regenerated by burning the deposits in a stream of air at moderate temperatures. Several regenerations are possible before the catalyst must be replaced, usually after more than four years of use. Palladium is generally used when the pyrolysis gasoline contains low levels of gum.
- Nickel oxide supported on γ-alumina is also active and selective with typical pyrolysis gasolines.[77] It can be reactivated very easily at intervals of two to nine months by heating in hydrogen or regenerated by conventional burning of deposits in air when necessary. Replacement is required after about two years.
- Nickel oxide/tungsten oxides supported on alumina are more active as sulfides and can be used if pyrolysis gasoline contains relatively high levels of reactive sulfur compounds.[77]

3.6.6.2. *Catalyst Operation*

The catalyst is loaded into fixed beds and is reduced in hydrogen at 400^0C before use. Good mixing of the liquid pyrolysis gasoline with gas as it flows through the catalyst is essential and temperature is controlled by a cool recycle of treated gasoline. The flow rate is adjusted to obtain the necessary conversion at minimum operating temperature. The catalyst gradually becomes deactivated due to the deposition of gums, but conversion is maintained by increasing the operating temperature. Nickel catalysts are reactivated every few months by treatment with hydrogen alone at 400^0C. It is usual to regenerate the catalyst by burning off gum in air after two to three reactivations. Palladium catalysts must always be regenerated in air at 400^0C.

It is interesting to note that the nickel catalysts in particular are not rapidly deactivated by the small amounts of the mercaptans and disulfides that remain in pyrolysis gasoline. In fact thiophenes have the beneficial effect of retarding the hydrogenation of olefins while the dienes are selectively hydrogenated. Aromatics are not hydrogenated during the process, while paraffinic α-olefins are isomerized. Providing that the hydrogen does not contain more than about 2000 ppm of carbon monoxide, catalyst selectivity may also be enhanced. A range of operating conditions is shown in Table 3.27.

TABLE 3.27. Hydrogenation Conditions and Catalysts Used.

Variable	Value
Temperature	$40–200^0C$
Pressure	20–70 atm
Liquid space velocity:	
Fresh feed	1–8 h^{-1}
Feed plus recycle	Up to 30 h^{-1}
Hydrogen addition	80–250 m^3 H$_2$/m^3 fresh feed
Poisons:	
Mercaptans	< 120 ppm
Hydrogen sulfide	< 100 ppm in H2
Carbon monoxide	< 2000 ppm
Cycle time	2–18 months
Catalysts used:	0.5% Pd/0.5% Cr/alumina
	10% NiO/alumina
	5% NiO/20% WO$_3$/Al$_2$O$_3$

REFERENCES

1. P. Sabatier and J-B. Senderens, *Comptes Rendus* **128** (1899) 1173.
2. L. Mond, C. Langer, and F. Quirke, *JCS* **54** (1889) 296.
3. P. Sabatier, *Comptes Rendus* **134** (1902) 514, 639.
4. P. Sabatier, Nobel Laureate Address, *Revue Scientifique* **1** (1913) 289; *Catalysis in Organic Chemistry*, Van Nostrand Co, N York, 1922.
5. Leprince and Siveke, German Patent 141029 (1902); W. Normann, British Patent 1515 (1903).
6. J. Crosfield, British Patent 30282 (1910).
7. P. Sabatier and J-B. Senderens, French Patent 394957 (1907).
8. W. Normann, British Patent 1515 (1903); F. Bedford and C. E. Williams, British Patent 9142 (1908); E. Erdman, German Patent 211669 (1908)-Pumice; J. Crosfield, British Patent 30282 (1910)-Kieselguhr; K. Kaiser, US Patent 1004034 (1911); Kieselguhr; Schwoerer, German Patent, 199099 (1906)-Asbestos; S. B Ellis, US Patent 1060673 (1912)-Charcoal.
9. V. N. Ipatieff, *J. Russian Chem. Soc.* **36** (1904) 7861; *Berichte* **37** (1904) 2961.
10. V. N. Ipatieff, *Catalytic Reactions at High Temperature and Pressure*, Macmillan, New York, 1937.
11. P. H. Spitz, *Petrochemicals*, John Wiley, New York, 1988.
12. C. L. Paal, *Berichte* **40** (1907) 2201; **41** (1908) 805, 2273; German Patent 298193 (1918); *Chem. Centralbe* **2** (1917) 145.
13. A. Skita, German Patent 230724 (1904).
14. S. Fokin, *J. Russ. Phys. Chem. Soc.* **39** (1907) 607.
15. O. Loew, *Berichte* **23** (1890) 289.
16. R. Willstatter, *Berichte* **54** (1921) 121.
17. V. N. Ipatieff, *Berichte* **45** (1912) 3218.
18. R. Adams and L. B. Hunt, *Platinum Metals Review* **6** (1962) 150; R. Adams, V. Voorhees and R. L. Shriner, *Organic Syntheses*, Vol. 1, John Wiley, New York, 1932, p. 452.
19. W. H. Jones, *Platinum Metals Review* **2** (1958) 86.
20. C. Ellis, *Treatise on Hydrogenation of Organic Substances*, 3rd Edition, D. Van Nostrand, New York, 1930.
21. M. Raney, *Heterogeneous Catalysis, Selected Case Histories, ACS Symposium Series 222* (1983).
22. M. Raney, US Patent 1563587 (1925).
23. M. Raney, US Patent 1628190 (1927).
24. C. N. Satterfield, *Heterogeneous Catalysis in Industrial Practice*, 2nd Edition, Krieger Publishing Co., 1996.
25. Covert and H. Adkins, *JACS* **54** (1932) 4116.
26. H. Adkins, *Reactions of Hydrogen*, University of Wisconsin Press, 1932; E. Lieber and F. L. Moritz, *Advances in Catalysis*, Vol. 5, Academic Press, New York, 1953, p. 417.
27. J –F. Le Page et al., *Applied Heterogeneous Catalysis*, Editions Technip, Paris, 1987, p. 300.
28. L. D. Schmidt, *Catalysis of Organic Reactions*, Marcel Dekker, New York, 1995, p. 45.
29. J. Crosfield, British Patent 30282 (1910); US Patent 1004034 (1911).
30. J. J. De Lange and G. H. Visser, *Ingenieur* **58** (1964) 24; G. C. A. Schuit and L. L. Van Rijer, *Advances in Catalysis*, **10** (1958) p. 242.
31. J. W. E. Coenen, in *Preparation of Catalysts*, Vol. 2, Elsevier Scientific Publishing, Amsterdam, 1978, p. 89.
32. J. Longuet, *Comptes Rendus* **225** (1947) 869.
33. N. D. Zelinsky and W. Kommarewsky, *Berichte* **57** (1924) 667.
34. W. Feitknecht and M. Gerber, *Helv Chim Acta* **25** (1942) 131.
35. R. Altmann, *Chimia*, 99 (1990).
36. M. S. Spencer, *Top Catal.* **8** (1999) 259; A. M. Pollard, M. S. Spencer, R. G. Thomas, P. A. Williams, J. Holt, and J. R. Jennings, *Appl. Catal. (A): General* **85** (1992) 1.

37.   G. W. Bridger, *The Manufacture of High Activity Catalysts*, Council of Engineering Institutions Mac-Robert Award Lecture, November 25, 1975.
38.   D. C. Parkyns, I. J. Kitchener, C. Komodromos and N. D. Parkyns, in *Preparation of Catalysts III*, Ed. by G. Poncelet, P. Grange and P. A. Jacobs. Elsevier Science Publishers, Amsterdam, 1983, p. 237.
39.   K. B. Mok, J. R. H. Ross and R. M. Sambrook, in *Preparation of Catalysts III*, Ed. by G. Poncelet, P. Grange and P. A. Jacobs, Elsevier Science Publishers, Amsterdam, 1983, p. 291.
40.   I. G. Farbenindustrie, Oppau, Germany, *Fiat Report No 888*, August 1946, p. 7.
41.   G. Natta in *Catalysis*, Vol 3, Ed. by P H Emmett, Reinhold, New York, 1955, p. 349; G. Natta, *Chime Ind.* **35** (1953) 705.
42.   W. A. Lazier, US Patent 1746783 (1930); British Patent 301806 (1926).
43.   H. Adkins and H. Connor, *JACS* **53** (1931) 1092.
44.   P. K. Frolich, M. R. Fenske, L. R. Perry and N. J. Hurd, *JACS* **51** (1929) 187; *Ind. Eng. Chem.* **20** (1928) 698; *Ind. Eng. Chem.* **21** (1929) 109.
45.   W. Kotowski, *Chem. Tech.* **15** (1958) 976.
46.   D. H. Bolton, *Chem. Eng. Tech.* **41** (1969) 129.
47.   Le Prince and Siveke, German Patent 141029 (1902).
48.   W. Normann, British Patent 1515 (1903); German Patent 139457-later cancelled (1902).
49.   J. Crosfield, British Patent 30282 (1910).
50.   K. H. Wimmer and E. B .Higgins, British Patent 18282 (1912); French Patent 441097 (1913); F. Bedford and E. Erdmann, US Patent 1200696 (1916); H. Schonfeld, *Z. Angew. Chem.* **27**(2), 1 (1914).
51.   J. Dewar and A. Liebman, British Patents 12981, 12982 (1913); C. Ellis, US Patent 1156068 (1915).
52.   Engelhard Industries, *Performance Standards in Catalysts and Sorbents for Fats and Oils*, 1991.
53.   J. Crosfield, British Patent 30282 (1910); K. Kaiser, US Patent 1004034 (1911); G. H. Morey and Craine, US Patent 1232830 (1917).
54.   C. Ellis, US Patent 1060673 (1913); US Patent 1156674 (1915); M. H. Ittner, US Patent 1238774 (1917).
55.   Permutite Supports, British Patents 1358, 8452 (1915); US Patent 1256032 (1918).
56.   S. J. Green, *Industrial Catalysis*, Ernest Benn Ltd, London, 1928, p. 307.
57.   M. J. Breen and C. A. L Sicat, *Catalysis of Organic Compounds*, Dekker, New York, 1995, p. 37.
58.   Davy Process Technology, *Oils, Fats and Fatty Acid Hardening*, 1996.
59.   P. Sabatier and J-B. Senderens, *Berichte* **45** (1912) 3312; *Comptes Rendus* **132** (1901) 210.
60.   J-F. Le Page, *Applied Heterogeneous Catalysis*, Editions Technip, Paris, 1987, p. 300.
61.   *Oil Gas Journal*, 17 March (1997) 70.
62.   M. Taverna, *Hydrocarbon Processing* **49** (1970) 137.
63.   K. Kahr, *Ullman's Encyklopaedie de Technischen Chemie*, 1975, p. 96.
64.   Palladium Hydrogenation Catalysts, British Patents 1257609 (1971); 1332211 (1973); US Patents 4092360 (1978), 4203923 (1980).
65.   British Patent 359422 (1931).
66.   BASF, Germany, *Fiat Report No. 1107*, April 22 (1947).
67.   Howlett, Bowman and Wood, *J. Soc. Chem. Ind.*, March (1950) 69.
68.   Girdler Catalysts, *G54 Brochure*; H. W. Fleming, W. M. Keely and W. R. Gutmann, *Petroleum Refiner* **32** (1953) 138; Rediay, US Patent 2735897 (1956); Belgian Patent 567675.
69.   Barry/DuPont, US Patemt 2511453 (1950).
70.   R. E. Reitmeier and H. W. Fleming, *Chem. Eng. Prog.* **54** (1958) 48; Catalysts and Chemicals Inc, *C36 Brochure* (1958).
71.   *ICI Catalyst 36–1 Literature* (1957).
72.   H. C. Anderson, A. H. Haley and W. Egbert, *Ind. Eng. Chem.* **52** (1960) 901; Engelhard, US Patent 2927141(1960); Dow, US Patent 2802889 (1957).
73.   Sud-Chemie, *Selective Hydrogenation Catalysts in Steam Cracker Units* (1997).

74. ICI, *Selective Hydrogenation Catalyst Brochure* (1996).
75. W. Lam and L. Lloyd, *Oil Gas Journal*, March (1972) 27; US Patent 3116342 (1975).
76. Kataleuna, *Oil Gas J. Special*, Sept 27 (1999) 56.
77. J-F. Le Page, *Applied Heterogeneous Catalysis*, Editions Technip, 1987, p. 329.

4

OXIDATION CATALYSTS

Oxidation catalysts were among the first to be described and then developed industrially. Because of the energy evolved, oxidation processes were originally known as catalytically induced combustion. Some of the earliest catalytic oxidation reactions used commercially are shown in Table 4.1. This list could also include the Deacon and the Claus processes, which were described in Chapter 2. Subsequently, nitric acid and formaldehyde were produced on a large scale by catalytic oxidation processes. In most early processes, once a reasonable catalyst had been developed, production was limited only by demand and the availability of efficient equipment.

Complete combustion of organic materials to form carbon dioxide was, of course, well known! By 1920 the *partial combustion* of organic chemicals was being investigated and selective catalysts were gradually developed to control the reactions taking place. Two important processes to produce maleic anhydride and phthalic anhydride from the benzene and naphthalene in coal tar were among the first to be developed commercially.

Phthalic anhydride, produced in Germany as early as 1916 and in other parts of the world in the 1920s, was used initially in the synthesis of indigo dyes. At first naphthalene was oxidized by chromic acid or oleum but, by a convenient accident, it was found that mercury catalyzed the oxidation reaction. Later work by BASF in Germany and H. D. Gibbs and C. Condover in the United States developed catalysts for the vapor phase oxidation reaction.

Oxidation of benzene to provide maleic anhydride was developed at the same time, but because maleic anhydride was available as a by-product from phthalic anhydride production, the reaction did not become important comer-

L. Lloyd, *Handbook of Industrial Catalysts*, Fundamental and Applied Catalysis,
DOI 10.1007/978-0-387-49962-8_4, © Springer Science+Business Media, LLC 2011

TABLE 4.1. Introduction of Catalytic Oxidation Processes.

Date	Process
1901	Ammonia oxidation to nitric acid.
1905–1910	Methanol oxidation to formaldehyde (copper or silver catalysts).
1914–1923	Sulfur dioxide oxidation to sulfuric acid (contact process).
1916–1924	Naphthalene oxidation to phthalic anhydride.
1931	Ethylene oxidation to ethylene oxide (Union Carbide plant 1937).
1933	Benzene oxidation to maleic anhydride.
	Methanol oxidation to formaldehyde (iron molybdate catalyst).
1946–1950	Orthoxylene oxidation to phthalic anhydride.
1962–1974	Butene/butane oxidation to maleic anhydride.
1957–1959	Ethylene oxidation to acetaldehyde.
1950s	Propylene oxidation to acrolein/acrylic acid.
1960s	Propylene ammoxidation to acrylonitrile.

cially until larger quantities of maleic anhydride were required for the production of unsaturated polyester resins.

4.1. NITRIC ACID

In 1839 Kuhlmann described ammonia oxidation to produce nitrogen oxides for nitric acid production using a platinum sponge catalyst at 300^0C.[1] At the same time he was also granted a patent for the oxidation of sulfur dioxide and used the process in his factory at Loos.[2] He was apparently unaware of the Phillips patent granted in the United Kingdom, but he attempted to make sulfuric acid with a platinum catalyst.

A second ammonia oxidation patent was granted to T. J. Smith, for T. du Motay, who heated ammonia and air with, for example, manganates, permanganates, dichromates, and plumbates in a closed vessel at 300–500^0C.[3] Low yields of nitrogen oxides and nitrates were recovered. This patent was not very significant but it did lead to later attempts covering the use of nonplatinic catalysts, including bismuth oxide/copper oxide by Bayer[4] and iron oxide/bismuth oxide, or a rare earth, by BASF.[5] BASF emphasized the usefulness of bismuth oxide as a promoter.[6] Later, in the 1920s and the 1950s, further attempts were made to commercialize the use of cobalt oxide.[7] Although the catalysts were often very active compared with platinum, they deactivated quickly and large volumes were needed for industrial applications. Apart from a very early plant in Leverkusen, in Germany, operated by Friedrich Bayer between 1914 and 1918, none has been used on a large scale.[8] The catalyst Bayer used was not known precisely but consisted mainly of iron oxides, with promoters such as chromium, manganese, and bismuth. A 12-ft diameter bed of catalyst (5–10 mm granules), 6-in deep, was used and gave an overall efficiency of 80–85% at 700–850°C. Iron oxide/bismuth oxide/manganese dioxide mixtures were studied as ammonia oxidation catalysts in the 1940s and were found to produce up to 80% yields of nitrous oxide between 300–400°C, but nitric oxide was formed at temperatures above 400°C.[9]

The most significant early developments came when Professor Ostwald in Leipzig began his experiments on ammonia oxidation and published his results in 1902.[10] The application for a German patent was disallowed because of Kuhlmann's earlier patent. Ostwald had developed his interest following the encouragement of Professor Pfeffer at Bonn in response to Sir William Crookes' address to the British Association in 1898. In 1909 Ostwald was awarded the Nobel Prize for Chemistry for his process, which was of vital importance in the production of fertilizer. He used platinum catalysts and obtained the best results with a coil of platinum foil as the catalyst at a very high linear gas velocity and removing the products from the tube as quickly as possible. He went on to define appropriate operating conditions in a pilot unit that gave 85% conversion and started a production unit at Bockum in May 1906 that produced 300 kg day^{-1}. The life of 50 g of catalyst was 4–6 weeks.

As a result of this success a second plant was built by 1908 giving a 53% yield and producing 3 tonnes day^{-1}. Although the Ostwald process had the disadvantage of using a large amount of platinum and had poor temperature control,[11] more or less the same conditions were used for about 30 years, but with many improvements in the plant design and the form of the catalyst.

In 1911 Karl Kaiser introduced preheating of the air up to 300–400°C before passing it through four platinum gauzes in a square reactor. The gauze was a 1050 mesh of 0.06-mm wire that was alloyed with traces of palladium or iridium and produced 1.5 tonnes of ammonia per square foot per day with a life of three months. This corresponded to 90–92% conversion in plants operating in Kharkoff, Russia, and in England.[12]

Caro and Frank, who had been granted several patents by 1914,[13] developed the first process to be used on a large scale. The first plant was built at the Bayerische Stickstoffwerke and was later engineered by BAMAG (Berlin Anhaltische Maschinenbau AG), who built 30 plants. The nitrogen oxides produced were initially used in the sulfuric acid lead chamber process but were eventually used to make all the nitric acid needed by Germany in the later stages of World War I. Following developments by BAMAG, three autothermal platinum gauzes of 80 mesh/in with 0.0026-in wire were used at 650–750°C with 10% ammonia in air and gave a 6-month lifetime. About 0.75 tonnes day^{-1} of ammonia per square foot of gauze was produced at 92% conversion in a plant operated at Hoechst by Meister, Lucius, and Bruning.[14]

American Cyanamid operated the first US plant in 1916 and the Air Nitrates Corporation was formed in 1917 to build 700 nitric acid units and produce

Figure 4.1. A typical modern plant for the manufacture of nitric acid. Reprinted with permission from Uhde GmbH.

110,000 tonnes of ammonium nitrate per year.[15] Experimental units using the Caro and Frank process had been used in the United States since 1916 and the converter design, using a cylindrical gauze that had been improved by Parsons and Jones was one of the best, being cheap to construct and operate.[16] Conversion was 94% with 10–11% ammonia in air and the efficiency was greater than that of the Hoechst plant.

By the time the German and American plants were operating on a large scale, the layout of ammonia oxidation units was fairly well established (Figure 4.1):

- Platinum was used by all of the major producers in the form of a gauze. The mesh dimensions and wire used between the days when Ostwald and Kaiser first used a gauze and the present are summarized in Table 4.2. The small amounts of rhodium and iridium used in Germany were probably added to increase the strength of the fine platinum wires during both the drawing process and operation. The first patent for platinum/rhodium alloys as used today was granted to DuPont in 1928.[17]

- Fresh gauze was not very active and it took a little time for the surface to activate and the plant to achieve full output. The weight of platinum used per tonne of acid produced was less than in the original plants.

TABLE 4.2. Ammonia Oxidation Catalysts.

Process	Catalyst	Form of catalyst	Where used
Ostwald	Platinum	Foil: 2-cm strip (50 g) rolled up and used in nickel tube.	Gerthe, Westphalia (1909) Vilvoorde, Belgium Angouleme, France Landis, *Chem Met Eng* **20**, (1919) 471. *Iron and Coal Trades Rev.* (23 May 1913)
BASF	Iron with 3–4% bismuth (could include rare earth)	Mixed oxide layers 4 in on perforated plates.	British Patent 138488 (1914)
Kaiser	Platinum	Gauze: 4 layers (first use of gauze catalyst). Maybe containing small traces of Pd or Ir.	Spandau, Berlin (1912) *Chem. Zeit.*, (1916) 14
Frank and Caro	Platinum	Gauze: early testing with 80 mesh/in. using 0.0026-in diameter wire. Later use of 80 mesh/in using 0.06-in diameter wire. Three layers used (330 g).	Experimental units in US. Fairlee, *Chem. Met. Eng.* **20** (1920) 6. Plant at Hoechst, Germany Partington, *J. Soc. Chem. Ind.* **46** (1921) 185R.
US plants	Platinum	Gauze: 4 layers. Flat sheets 13 in wide by 113.5 in, 80 mesh/in rolled into 9-in diameter tube (16.5 oz).	American Cyanamid, 22,500 tons (100% acid) per year by 1919. Parsons, *J. Ind. Eng. Chem.* **11** (1919) 541. Taylor, *J. Ind. Eng. Chem.* **11** (1919) 1121.
Modern plants	Platinum/rhodium	Gauze: 3–5 layers wire 0.075 mm.	

- Optimum operating conditions at atmospheric pressure were established as greater than 800 $\overset{\circ}{}$C, with 12% ammonia in air, and conversion in well-designed plants exceeded 90%.

The chemistry of the first stage in the overall process, reaction (4.1), is the exothermic oxidation of ammonia to nitric oxide and water. The many reactions involved in the overall process to nitric acid may be simplified into three equations: the burning of ammonia to nitric oxide, reaction (4.2); the oxidation of nitric oxide, reaction (4.3); and the reaction of dinitrogen tetroxide to give nitric acid, reaction (4.4):

$$NH_3 + 2 O_2 \rightarrow HNO_3 + H_2O \qquad (4.1)$$

$$4 NH_3 + 5 O_2 \rightarrow 4 NO + 6 H_2O \qquad (4.2)$$

$$2 NO + O_2 \rightarrow N_2O_4 \qquad (4.3)$$

$$2 N_2O_4 + 2 H_2O + O_2 \rightarrow 4 HNO_3 \qquad (4.4)$$

As demand for nitric acid increased, efforts were made to improve the process by operation at higher pressures, which would allow the use of smaller equipment with lower capital costs. High pressure could also provide higher acid concentrations with more efficient absorption and an increased rate of reaction in converting nitrogen oxides to nitric acid. High-pressure operation was made possible when chromium and chromium/nickel alloy steels replaced ceramic materials.

The oxidation of ammonia, however, was less economic at high pressure than at atmospheric pressure because the burner temperatures had to be increased in order to achieve the same selectivity. This led to shorter catalyst life as the metal gauzes deteriorated more rapidly and the loss of platinum became uneconomic. The 90% platinum/10% rhodium alloy introduced by DuPont solved this problem by reducing platinum loss by 50% and also improving selectivity.[17] DuPont also found that the loss of platinum was proportional to the oxygen content of the gas mixture. It was realized that the surface of the catalyst etched as it was activated and the wires were covered by tinsel.

Bimetallic gauzes not only improved the physical performance of the catalyst to give a longer life but also increased the selectivity to more than 94%. There has been little change in operation since the 1930s, except that larger plants have been built. This has required better plant design and improved temperature-resistant materials to support the larger-diameter gauzes used.

4.1.1. The Ammonia Oxidation Process

Three types of ammonia oxidation processes are now used with absorption at atmospheric (AOP), intermediate (IOP), and high (POP) pressures. The intermediate-and high-pressure plant burners can also be operated at low pressure by incorporating gas compression before the absorber. Simplified flow sheets for typical ammonia oxidation plants are shown in Figures 4.2, 4.3 and 4.4. Operating conditions for the different designs are given in Table 4.3. In general the burner efficiency is in the range 95–98% and the process chosen depends on the economic requirements of individual operators. Inlet gas to the burner is preheated up to about 300°C with an ammonia concentration of less than 12% to avoid explosive mixtures and to ensure efficient operation at high conversion. The gauze temperature, which is usually in the range 850–950°C, depends on the ammonia content, the preheating temperature, and the gas rate. At higher temperature ammonia can be fully oxidized to nitrogen and operating efficiency is

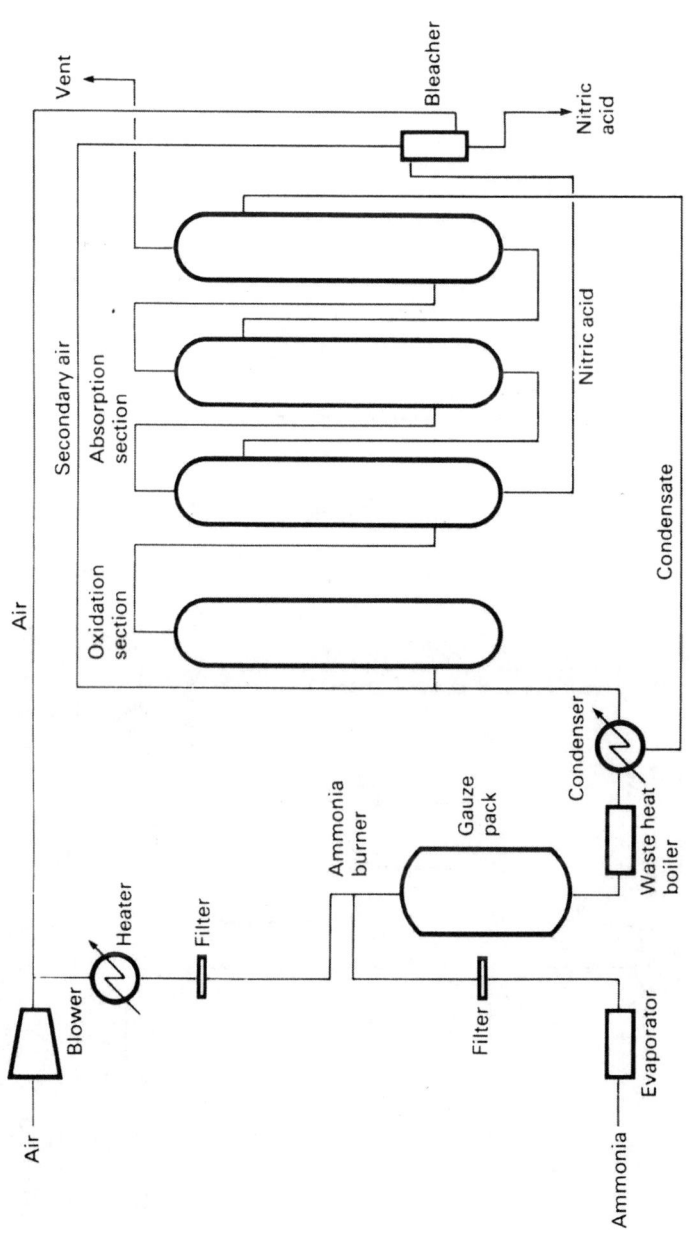

Figure 4.2. Simplified flow sheet for typical atmospheric-pressure nitric acid plant. Reprinted from *Catalyst Handbook*, 2nd ed., by kind permission of M. Twigg.

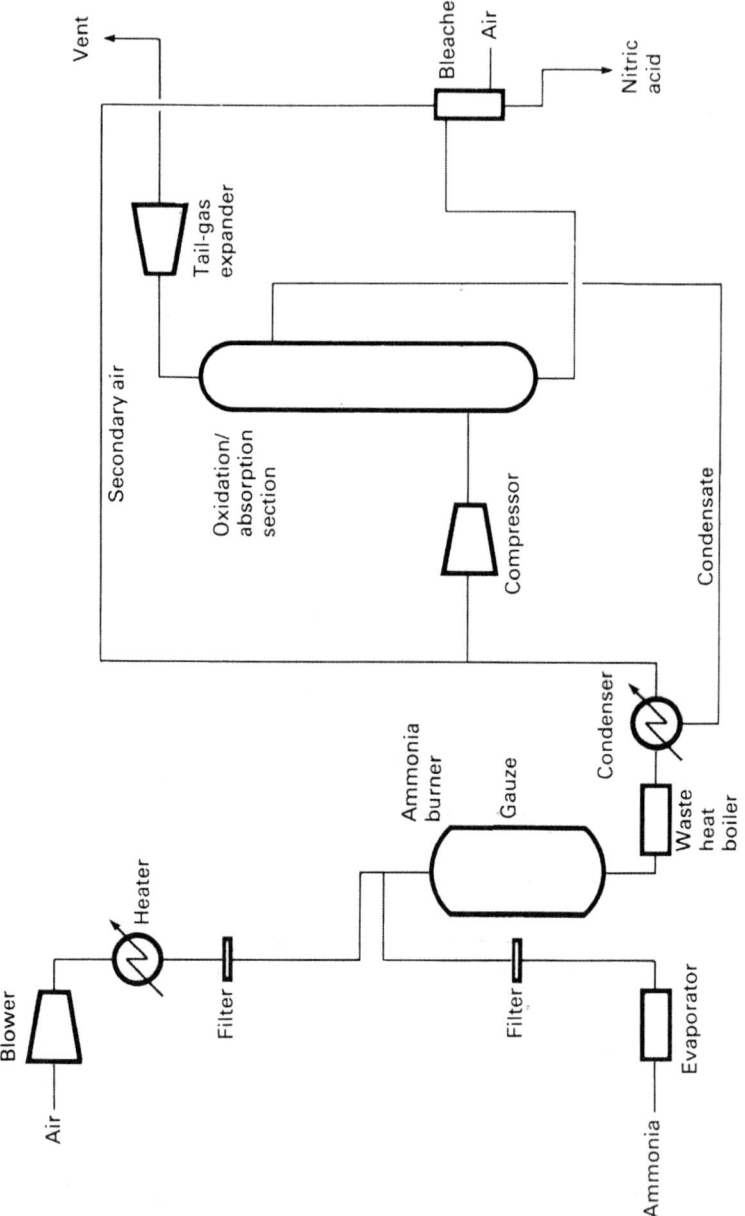

Figure 4.3. Simplified flow sheet for typical dual-pressure, medium-pressure nitric acid plant. Reprinted from *Catalyst Handbook*, 2nd ed., by kind permission of M. Twigg.

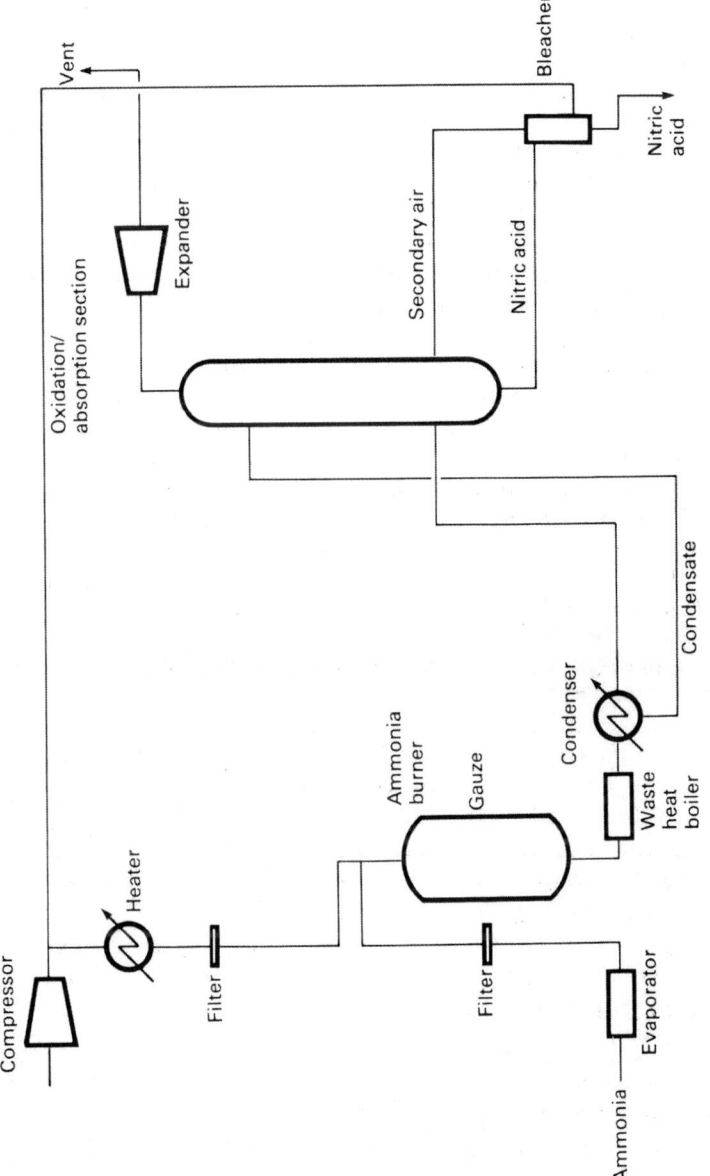

Figure 4.4. Simplified flow sheet for typical single pressure, high-pressure nitric acid plant. Reprinted from *Catalyst Handbook*, 2nd ed., by kind permission of M. Twigg.

TABLE 4.3. Modern Nitric Acid Plant Design.

	Atmospheric pressure (AOP)	Intermediate pressure (IOP)	High pressure (POP)
NH_3 concentration (vol%)	12	12	11
Gauze temperature (^{0}C)	810–850	810–850	870–890
Burner pressure (atm)	1	1	4
Absorber pressure (atm)	1	3	6
Number of gauzes	3–6	10–20	35–45
Gauze diameter (m)	3–5	3–5	3–5
Life of gauze (months)	8–12	8–12	4–6
Pt loss (g/tonne HNO_3)	0.05	0.05	0.1
Acid strength (%)	49–52	55–69	60–62
Conversion efficiency (%)	97–98	97–98	96–96.5

Note: Preheat temperature up to 300^{0}C. Ammonia concentration < 12% to avoid explosive limits. Catalyst gauze supported on high-chrome steel mesh. Burners can be operated at absorber pressure in IOP/POP plants.

therefore reduced. Nitric oxide is thermodynamically unstable, and can revert to oxygen and nitrogen at elevated temperatures. Following the burner (the oxidation reactor), the nitric oxide produced is cooled rapidly to a temperature below 130°C so that much of the water formed in the elevated pressure process condenses, and the nitric oxide is not decomposed. Torn or poorly packed gauzes in the pad allow ammonia to pass through the pad unchanged and allow subsequent reaction with nitric oxide to produce nitrogen.

4.1.2. Catalyst Operation

The optimum pad thickness increases with operating pressure and 3–6, 10–20, and 35–45 gauzes are used in atmospheric, intermediate, and high pressure plants. Pads must be carefully fitted to avoid tears or creases and laid flat on a support of high-chrome steel mesh (Figure 4.5). New gauzes are free of lubricants and iron contamination, but old gauzes can also be reused provided that they contain no holes and have been carefully cleaned and reactivated. Dust and other contamination is carefully removed and the gauze *pickled* in hydrochloric acid to remove iron oxide scale.

Low-activity new gauze is always packed below used gauze because it does not reach full activity for several hours. The smooth surface of the wire becomes activated by the development of crystallites, which increases the surface area of platinum and exposes the more active crystallographic planes. This does, however, weaken the wires as platinum migrates and vaporizes to condense on lower layers of gauze or even be lost altogether from the pad. Platinum loss increases during high-pressure operation. The reasons for the deactivation of platinum/rhodium gauzes are not properly understood although the practical consequences are all too clear (Figure 4.6).

Figure 4.5. Installation of platinum gauze pad in a nitric acid plant–checking for wrinkles. Reprinted from *Catalyst Handbook*, 2nd ed., by kind permission of M. Twigg.

During atmospheric pressure operation, once the platinum has become active by this crystallization process the gauze gradually deactivates as platinum is lost as a volatile platinum oxide. Loss per tonne of nitric acid produced is proportional to the gauze temperature and increases from 50–100 mg of platinum, at atmospheric pressure and 800°C, to about 400 mg of platinum, at 8-atm pressure and 900°C.[18] Gauzes are usually changed when about 5% of the metal has been lost and the rhodium content has increased to more than 12%.

Increased rhodium content, particularly at the surface of the wires, results in deactivation of the platinum surface as rhodium oxide accumulates.[19] The oxide is insoluble in acid and catalyst activity can only be restored by decomposing the oxide in air or nitrogen at temperatures exceeding $1000^{0}C$, or by reduction in hydrogen. Rhodium can diffuse back into the metal during a long, high-temperature annealing procedure or operation at a lower pressure.

4.1.3. Platinum Recovery

Platinum loss has been reduced by 25–50% when base metal gauzes have been incorporated into the pad as they help to disperse the heat of reaction and reduce local hot spots.[20] A further improvement in process economics was the recovery

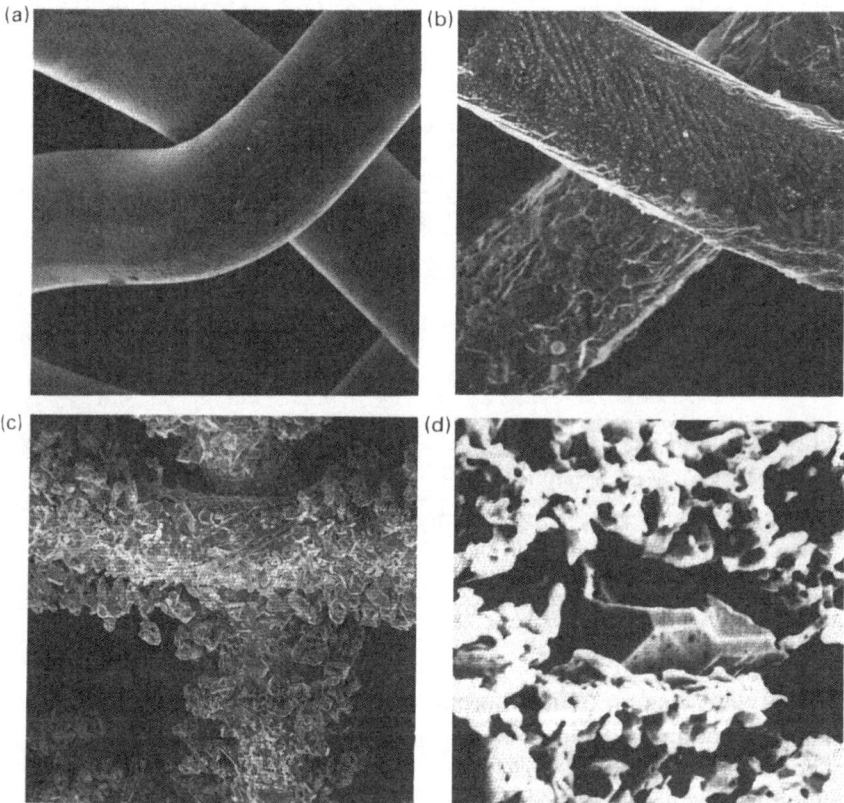

Figure 4.6. Scanning electron microscope (SEM) images of gauzes. (a) New unactivated gauze, showing draw marks (X200); (b) partially activated gauze, showing etching (X200); (c) well-activated gauze, showing excrescences (X20); (d) POP gauze with Rh_2O_3 crystals covering active alloy surface (X2500). Reprinted from *Catalyst Handbook*, 2nd ed., by kind permission of M. Twigg.

of some of the platinum lost from the catalyst during operation. Platinum sludge formed in the absorber was relatively easy to separate. About 60% of the platinum could also be recovered from the burner gases with a suitable filter, although this introduced a pressure drop in the system.

Since about 1970 a new recovery system using a palladium *getter* gauze has been developed.[21] This absorbs up to 80% of its weight of platinum as a platinum/palladium alloy but does lose about 0.33 g of palladium for every gram of platinum collected. Up to 20% gold must be added to the palladium to provide physical strength. Several getter gauzes can be used to increase platinum recovery to about 70%. A disadvantage of the procedure is the expense of recovering platinum and the other metals from spent getter gauzes.

Since 1920 the platinum required to produce a given amout of nitric acid has fallen to less than 30% of the earlier levels a result of using catalysts with higher activity, greater selectivity, and longer operating lives. It is unlikely that better catalysts will be developed that can replace platinum/rhodium gauze.

4.2. FORMALDEHYDE

Formaldehyde has been well known as a disinfectant and preservative from early times and was originally obtained in low yields from special lamps by burning *wood alcohol*. As further practical applications were introduced, larger quantities were supplied from the partial combustion or selective oxidation of methanol.

Hoffmann developed flameless combustion of methanol in 1867. He used a platinum coil as a catalyst that glowed red hot as the methanol dehydrogenated and produced some formaldehyde.[22] A commercial plant was designed by Trillat in 1889 to convert a methanol/air mixture into formaldehyde using a platinized asbestos catalyst.[23] Trillat subsequently showed that other catalysts could also be used, such as oxidized copper at 330^0C, although platinum at 200^0C was most effective. Yields of about 50% formaldehyde were produced and he claimed that the addition of 20% steam to the gases improved performance.

During an investigation into the dehydrogenation and dehydration of alcohols, Sabatier and Maille dehydrogenated methanol over a number of metal oxides. In general the methanol was *reformed* to give carbon oxides and hydrogen, although some metals including copper did produce formaldehyde. They concluded that the Trillat process proceeded with a methanol dehydrogenation step followed by the reaction of the hydrogen formed with the excess oxygen present:

$$2 \, CH_3OH \rightarrow 2 \, HCHO + 2 \, H_2O \tag{4.5}$$

$$2 \, H_2 + O_2 \rightarrow 2 \, H_2O \tag{4.6}$$

The overall reaction is exothermic, and is a good early example of oxidative dehydrogenation. In reality, this is an example of selective oxidation:

$$2\ CH_3OH + O_2 \rightarrow 2\ HCHO + 2\ H_2O \tag{4.7}$$

Orlov studied the production of formaldehyde using a wide range of catalysts and published a summary of his conclusions.[24] He began by extending Sabatier's conclusions and recommended the use of a copper catalyst. A plug of platinised asbestos was incorporated into the process gas stream before the copper gauze; this acted as an *ignition pellet*, raising the temperature of the process gas before it came into contact with the main catalyst bed. Le Blanc and Plaschke showed that a silver catalyst was preferable to copper and worked out what they considered to be the best operating conditions.[25] Bouliard[26] and Le Blanc[27] both recommended the use of silver supported on asbestos.

Several commercial formaldehyde plants were operating after 1900 using both copper and silver catalysts. These included the Formal plant, Cote d'Or, France, which used long copper tubes,[28] and theFHMeyer plant, Hanover-Heinholz, based on Orlov's work, with a copper reactor and copper or silver gauzes operating at 450–500°C.

It has been suggested that the practical similarities between methanol and ammonia oxidation led to the more rapid development of platinum gauzes for nitric acid production during the 1914–1918 war.[29] Large-scale production of formaldehyde did not become important until the demand for phenol-formaldehyde plastics developed in the 1920s. With so little information available on the production of formaldehyde, it is more likely that the experience gained from ammonia oxidation in the wartime plants was then applied to the manufacture of formaldehyde.

By the 1930s, Adkins, working with the Bakelite Corporation, introduced a mixed oxide catalyst for the direct oxidation of methanol.[30] During the development he found that pure molybdenum oxide gave about 60% conversion to formaldehyde at 400°C, although activity fell after 12–24 h to 30% conversion. Pure iron oxide, on the other hand, was not selective and produced only carbon dioxide. However, a mixed iron/molybdate catalyst converted more than 90% of the methanol to formaldehyde. Operation was relatively stable and by 1952 DuPont had built a plant using iron/molybdate in a process similar to that described in Adkins' patent.[31]

Several reviews of the commercial formaldehyde processes then available had been published by 1953 and gave summaries of the operating conditions used.[31–33] Simplified flow sheets for both processes are shown in Figures 4.7 and 4.8, and a photograph of a metal-oxide catalyzed plant is shown in Figure 4.9. Typical catalyst properties are given in Table 4.4.

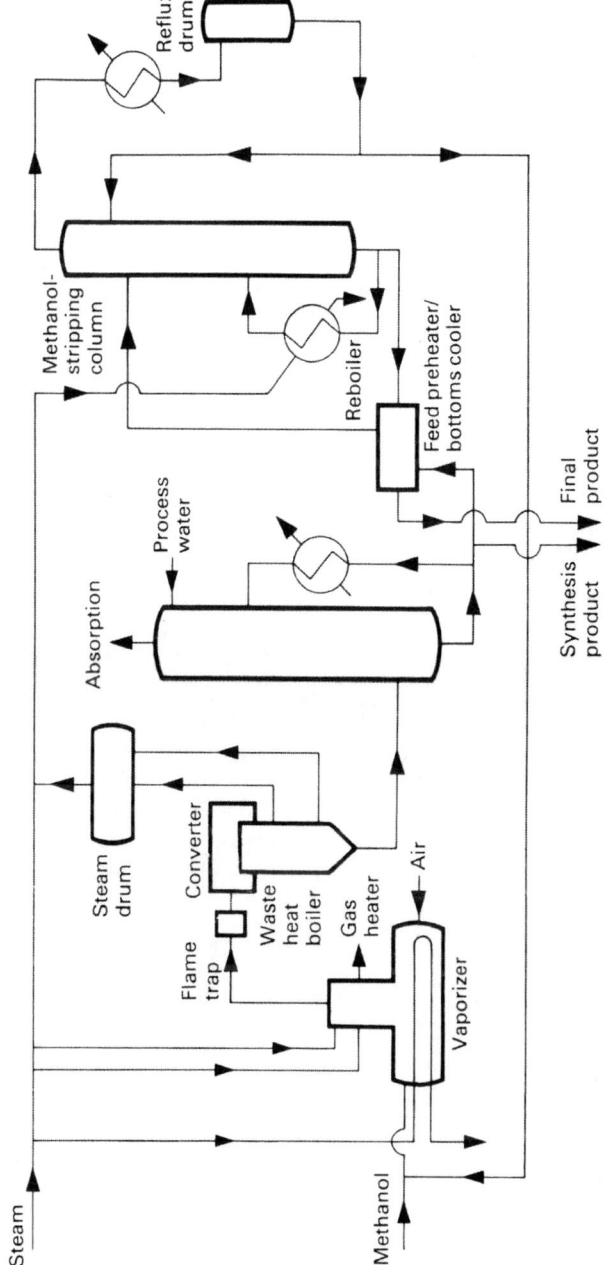

Figure 4.7. Simplified flow sheet of a typical silver-catalysed formaldehyde process. Reprinted from *Catalyst Handbook*, 2nd ed., by kind permission of M. Twigg.

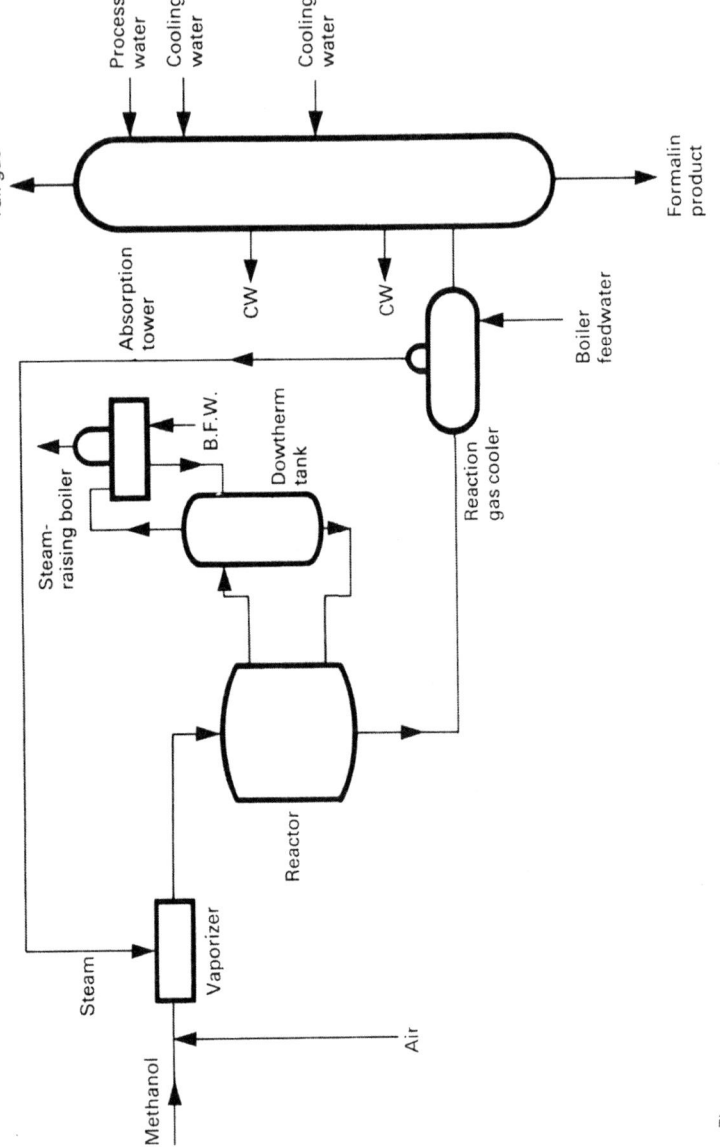

Figure 4.8. Simplified flow sheet of a typical metal-oxide-catalysed formaldehyde process. Reprinted from *Catalyst Handbook*, 2nd ed., by kind permission of M. Twigg.

Figure 4.9. Plant for the manufacture of formaldehyde by metal-oxide-catalysed process. Photograph by Stan Erismann, Perstop Formex.

TABLE 4.4. Silver Granule and Iron Molybdate Catalysts.

	Iron molybdate	Silver granules
Catalyst size	Rings 4.5 x 4.5 mm x 2 mm hole	Gauze or 0.5–5 mm granules
Composition	$Fe_2(MoO_4)_3$ No free Fe_2O_3 Slight excess MoO_3	Pure silver
Bulk density	0.7–0.9 kg liter^{-1}	—
Surface area	11 $m^2 g^{-1}$	Geometric
Pore volume	0.3 ml g^{-1}	Small

4.2.1. Silver Catalyst Operation

As demand for formaldehyde increased during the 1930s, processes using silver gauze or a layer of small silver granules as a catalyst were used almost exclusively. The oxidative dehydrogenative process operates with a methanol-rich air mixture, and the methanol concentration must be greater than the upper flammability limit. The feed gas must therefore contain more than 37% methanol and the oxygen content is inevitably less than the stoichiometric amount required for 100% conversion of the methanol. Unreacted methanol is therefore recovered and recycled.

Reactors consist of a 4-m wide catalyst bed, filled with a 3-cm layer of 0.5- to 5-mm silver granules or a pack of silver gauzes. Space velocity through the bed is as high as 300,000 h^{-1} to minimize formaldehyde loss. Process gas at 560–680^{0}C must be rapidly cooled in a waste heat boiler immediately it leaves the reaction zone to avoid thermal decomposition of the product.

Before use, the catalyst is activated in situ by the chemisorption of oxygen onto the silver surface. Oxygen, in the atomic state, then reacts with methanol to produce surface methoxy and hydroxyl species. Further methanol can react with these surface hydroxyl species to give more methoxy groups and free water. It has been suggested[34] that the surface methoxy groups decompose to formaldehyde and hydrogen, suggesting the presence of equal amounts of formaldehyde and hydrogen in the process gas stream as is passes through the reactor. An alternative, or even simultaneous possibility is for surface methoxy groups to react with adjacent chemisorbed oxygen atoms giving formaldehyde and surface hydroxyl species. Low oxygen coverage is necessary to avoid further oxidation to formate ions and, ultimately, carbon dioxide. Silver is a more selective catalyst than copper because it adsorbs less oxygen. A small volume of steam added to the feed can repress carbon formation, but its concentration is limited to maintain the final product formaldehyde concentration.

The product contains up to 50% formaldehyde after distillation. By recirculating the vent gases, the operating temperature can be reduced and the methanol conversion increased. This results in an increase in the concentration of formaldehyde in the product to 55%. Methanol conversion is, however, limited to 70% in a single reactor. If a second reactor and additional air is added to give a higher methanol conversion, distillation can be avoided, although yields are reduced.

4.2.2. Mixed Oxide Catalyst Operation

The oxidative dehydrogenation process is highly exothermic and the reactor temperature must be controlled to maintain selectivity. The reactor consists of many tubes cooled by the circulation of a heat transfer fluid such as Dowtherm. The catalyst tubes have a very small diameter to improve heat transfer and the catalyst is usually supplied as rings to minimize pressure drop. Despite these precautions there is a higher temperature zone within the reactor that moves gradually to the bottom of the tube as the catalyst ages. The process operates just above atmospheric pressure and the required space velocity is in the range 8000–10,000 h^{-1}. The methanol content is always below its lower flammability limit in air about 7%, and the methanol is completely converted. One of the main problems with this design is that there can be up to 40,000 catalyst tubes in a plant producing 35,000 tonnes year^{-1} of pure formaldehyde.[35] This makes loading and discharging catalyst a long job!

Although the stoichiometric ratio of molybdenum to iron in ferric molybdate is 1.5, the maximum activity is obtained at an atomic ratio of 1.7. However, the presence of free ferric oxide in the catalyst is known to reduce considerably the selectivity of the catalyst to the formation of formaldehyde. For this reason, excess molybdenum is usually added to the catalyst formulation to maximize the yield of the product. The optimum ratio is about 2.0.[35]

Some catalyst producers have made catalysts with a higher molybdenum-iron ratio between 2.5–4.5, and the very high molybdenum content does affect the surface area and physical strength of the catalyst. Molybdenum oxide is quite volatile at high temperatures and the catalyst activity gradually decreases so that the high temperature zone, the position in the bed of maximum conversion, gradually moves down the tube. The excess molybdenum is lost first, so that there is no immediate loss in activity. Catalyst dust, however, deposits on active catalyst at the bottom of the tubes, resulting in an increase in pressure drop.

The catalyst is prepared by precipitation from solutions of ferric chloride and ammonium molybdate. The precipitate may not be homogeneous, with significant variations within a single batch. Hydrothermal aging of the precipitate may be necessary to provide a more uniform composition. Precipitation of the catalyst as a gel provides a more uniform ferric molybdate composition.

Additives such as chromium or cobalt oxides can stabilize the catalyst. In Table 4.5, catalyst compositions and operating conditions in modern formaldehyde processes are shown.

4.3. ANDRUSSOV SYNTHESIS OF HYDROGEN CYANIDE

Hydrogen cyanide is used in a number of industrial processes, including the manufacture of methylmethacrylate (MMA). The elegant synthesis of MMA by John Crawford of ICI in 1932 was made possible by the Andrussov process, which was introduced in the early 1930s.[36] The polymerization of MMA produces the plastic, Perspex, the trademark registered by ICI in November 1934. This acrylic was used in the manufacture of the lightweight canopies required for the Spitfire fighter plane, which first flew in 1936 and which was widely used during

TABLE 4.5. Formaldehyde Production with Silver and Iron Molybdate Catalysts.

	Silver	Iron molybdate
Operating temperature (^{0}C)	560–680	230–280 (hot zone 330–380)
Space velocity (h^{-1})	300,000	8000–10,000
Operating pressure (atm)	1.2–1.5	1.2–1.5
Catalyst bed	3-mm deep; 4-m diameter	Up to 40,000 cooled tubes
Operating cycle (years)	1–1.5	Up to 1
Feed methanol in air (vol%)	37	7
Conversion (%)	70	90–95
Selectivity (%)	98	96–97
Poisons	Iron carbonyl, chloride, sulfur, sodium, heavy metals.	

the war. Perspex is known as Lucite in the United States and as Plexiglas in Germany.

The Andrussov Process for the manufacture of hydrogen cyanide by ammoxidation of methane is now widely used:

$$CH_4 + NH_3 + 1.5 \, O_2 \rightarrow HCN + 3 \, H_2O \qquad (4.8)$$

Air is the usual oxidant and while natural gas is the most common source of hydrocarbons, by-product gases from other units such as an ethylene plant can also be used.

The original catalyst, still widely used, is essentially the same as that developed for ammonia oxidation. Leonid Andrussov had previously worked out the reaction mechanism for ammonia oxidation with Bodenstein in 1926 and investigated the process for many years.[37]

Andrussov's catalyst was a platinum/rhodium gauze that contained up to 10% rhodium, although he did originally think that 2–3% iridium was better. A typical reaction mixture contained 11.2% ammonia, 11.7% methane, 15.6% oxygen, and 61.2% nitrogen, with traces of ethane.[38] Reaction was adiabatic with a hot spot in the range 900–1100^{0}C, at a linear velocity of 2–4 ft s^{-1}. This gave an ammonia conversion in the range 60–65%. The actual temperature and conversion depended on the composition of feed gas.

Nowadays, while reaction conditions are more or less the same, seven to ten layers of gauze are used to ensure that the gauze is uniformly heated, and to avoid carbon deposition. The process operates at about 70% conversion and requires rapid cooling. As in ammonia oxidation, the surface of the platinum alloy is etched as it is activated.[39] Metal foil catalysts have often been formed from expanded sheets

Catalysts prepared by supporting platinum/rhodium on an inert support have also been used in shallow beds.[40] These catalysts are stronger at high temperature than gauzes. Supports include zirconia, beryl, and silicon carbide. Variations in the quality of solid supports can, however, lead to changes in performance. Supported catalysts also need pretreatment to roughen the surface. Platinum/rhodium supported on beryl, when used with a platinum/rhodium gauze, has been claimed to have a longer life than either catalyst used alone, giving yields of HCN in the range 68–70% for at least 60 days.[41]

Catalyst activity falls by 4–5% during the operating life as a result of losing active metal, tearing of the gauzes, poisoning, poor gas distribution, or thermal degradation of the support. Performance can be improved by filtering the feed gas and removing any organometallic compounds that may be present in the feed. Heavy metals, arsenic, and phosphorus are common permanent poisons, but poisoning due to high levels of sulfur is reversible.

4.4. HOPCALITE CATALYSTS FOR CARBON MONOXIDE OXIDATION

Some early catalysts are no longer used in the application for which they were developed. They do, however, often have an important bearing on subsequent catalyst and process development. One important example is the use of Hopcalite catalysts during the 1914–1918 war. Carbon monoxide, highly toxic and not easily detectable, was generated inside tanks, and during the firing of naval cannons and machine guns. There was, therefore, an urgent demand for efficient catalysts that could oxidize carbon monoxide at ambient temperature for use in gas masks.

The pre-1817 experiments of Davy and Erman,[42] followed by those of Fletcher,[43] had shown that coal gas could ignite over platinum or hot iron wire at low temperatures giving *flameless combustion*. In 1902 the work was continued by Bone,[44] and during the 1914–1918 war several oxides were identified that promoted combustion at temperatures below 20^0C. These included copper oxide, manganese dioxide, silver oxide, and cobalt oxide, as well as palladium metal. Mixed oxides were even more active, particularly if a promoter such as ceria was included.

The composition of the catalysts used in respirators by British and American soldiers was as follows:

- British: Mixed $CuO/MnO2$ plus 1–5% cerium oxide.[45]
- US Chemical Warfare Service: CuO 30%, MnO_2 50%, Co_2O_3 15%, AgO 5%.[46]

The US catalyst was known as Hopcalite 1 and, later, the more active Hopcalite 2 containing 60% MnO2, 40% CuO, was introduced.[47]

The oxidation of carbon monoxide is strongly exothermic, and even at carbon monoxide concentrations as low as 2%, the heat produced made the purified gas too hot to breathe. A *cooler* containing low-melting sodium thiosulphate was, therefore, incorporated into the gas mask, the latent heat of fusion of sodium thiosulphate being sufficient to cool the air to an acceptable temperature. Catalysts were sensitive to moisture so a replaceable desiccant was also needed to dry the air entering the respirator.

As with later oxidation catalysts, Hopcalites were prepared by mixing freshly precipitated oxides. Silver was added by impregnating the oxides with a silver salt and precipitating the oxide with alkali.

A full review of carbon monoxide oxidation has been given by Katz,[48] who noted that the development of Hopcalites had produced very active multicomponent catalysts. The silver and manganese oxides used were often nonstoichiometric and able to lose or gain oxygen, depending on the temperature and pressure. He also suggested a probable reaction mechanism involving the oxidation of adsorbed carbon monoxide by lattice oxygen and the subsequent adsorption and activation of molecular oxygen by the catalyst to regenerate the lattice. It follow that surface defects and weaker oxygen bonds were important for the exchange of surface ions and electrons. Huttig had already reported at a Faraday Society meeting that oxygen ions on the surface of Cr_2O_3 and other oxides are mobile. At the same meeting Taylor, Rideal, and Garner discussed the relationship between chemisorption and heterogeneous catalysis.[49]

The academic interest in mixed oxides and catalysts between 1930 and 1950 clearly promoted the understanding and development of oxidation catalysts. Many improved formulations had, of course, been gradually developed by largely empirical methods for organic oxidation processes from the 1940s, and these led to the redox reaction mechanism proposed by Mars and van Krevelen in 1954.[50]

These developments would undoubtedly have taken place in the fullness of time without the need for wartime gas masks, but this is just another good example of how general theories and improved industrial catalysts develop from early, often random, experiments.

4.5. PHTHALIC ANHYDRIDE

The oxidation of organic compounds to useful products was not reported extensively until the 1920s. Before then, any hot combustible material mixed with air passing over a catalyst produced mainly oxides of carbon.

A number of low-activity catalysts that could control oxidation to some extent were eventually identified. These were metal oxides from Groups V and VI, such as vanadium, molybdenum, and tungsten, particularly when mixed with phosphoric, arsenic, or boric acids. Oxidation was controlled by choosing the appropriate temperature and contact time. At first it was difficult to achieve reasonable selectivity when dealing with exothermic reactions.

A review of the work up to 1920 was produced by Weiss and Downs,[51] of the Barrett Company, who were among the first to investigate the catalytic oxidation of naphthalene. Weiss remained an active consultant until at least 1946.

4.5.1. Naphthalene Oxidation

Although BASF are said to have produced phthalic anhydride by oxidizing naphthalene as early as 1916,[52] the first serious investigations were described by H. D. Gibbs and his associates at the US Bureau of Chemistry. Gibbs and Condover were granted a large number of patents from 1917 for the production of phthalic anhydride from naphthalene.[53]

Further patents were also granted to the Seldon Company,[54] Wohl,[55] Weiss and Downs[56] and Craver (Barrett Co).[57] It is interesting to note that much of the published information first became available in the form of patents. Gibbs and Condover did, however, summarize their work in a number of papers and specified the use of both vanadium and molybdenum oxide catalysts.[58] Tests showed that vanadium pentoxide could produce yields of up to 85%, whereas molybdenum trioxide required higher temperatures and, only gave yields up to 50–60%. Tungstic oxide was not very active. It was found that fused vanadium pentoxide gave the best results when it was supported on a range of low-surface-area materials such as kieselguhr, pumice, asbestos, or even metallic aluminum.[59] Craver recommended a mixture of 65% vanadium pentoxide and 35% molybdenum trioxide with traces of manganese dioxide or copper oxide.[57]

Operating temperatures were in the range 400–450°C with contact times less than 0.5 s, although reaction did begin in the temperature range 270–280°C. By 1928 *large* quantities of phthalic anhydride were being made commercially in the United States, Germany, and the United Kingdom.[60] It was recognized that tubular reactors or a number of shallow adiabatic beds should be used to control the hot spot that developed in the catalyst.[61] Temperature was controlled to maintain selectivity by cooling adiabatic beds with a cold air quench or in tubular reactors by heat exchange with a suitable liquid. Surprisingly, adiabatic catalyst beds were preferred until the 1940s, although Downs did investigate the use of square *tubes* in a 3-ft diameter vessel.[62] The catalyst was cooled by liquid mercury surrounding the tubes, the mercury boiling point being controlled by changes in the pressure of the bath. Tubular catalytic reactors cooled by eutectic salt mixtures were also developed, but generally the use of adiabatic catalyst beds continued.

Up to about 1945 the typical naphthalene oxidation catalyst was fused vanadium pentoxide, sometimes combined with molybdenum trioxide, on an inert support. At that time US production of phthalic anhydride was probably less than 60,000 tonnes year^{-1} and catalyst quality was not very important.

Over a period of time, the introduction of alkali sulfates to moderate the reaction led to some improvements in selectivity.[63] The developments may have been derived from studies on the vanadium catalysts used commercially in the oxidation of sulphur dioxide. It was discovered that the alkali metal sulphates and pyrosulphates formed complexes with vanadium pentoxide. In 1944 the success of the Badger/Sherwin Williams fluid bed process using naphthalene as feedstock, confirmed the activity and stability of these catalysts. Improved operation with a richer feed gas and better temperature control gave 90% selectivity at almost 100% conversion. Typical operating conditions at the time are shown in Table 4.6.

4.5.2. Orthoxylene Oxidation

As the petrochemical industry developed orthoxylene became widely available as a feedstock for phthalic anhydride. In the 1940s, Chevron became the first company to manufacture phthalic anhydride commercially by the oxidation of *o*-xylene, obtained as a by-product from a hydroforming plant. As *o*-xylene became available from platformers and demand for phthalic anhydride increased, it became the major feedstock. About 90% of the phthalic anhydride used in 1990 was produced from *o*-xylene.

TABLE 4.6. Phthalic Anhydride Production.

Conditions	Feed/process variation		
	A: Processes using naphthalene feed		
Feed quality	Pure	Impure	Pure
Temperature (^{0}C)	350–400	400–550	350–380
Air/feed	20/1	20/1	8.5/1
Contact time (s)	< 0.5	< 0.5	3–20
Selectivity (%)	80–85	60–70	90
		(plus ~10% maleic anhydride)	
Reactor	Tubular	Tubular	Fluid bed
Catalyst	10% V_2O_5/1% K_2SO_4 on silica or alumina	10% V_2O_5/1% K_2SO_4 on silica or alumina	9% V_2O_5 15% K_2O 23% SO_3 53% SiO_2
Surface area (m^2g^{-1})	~1	~1	30–35
	B: Processes using *o*-xylene feed.		
Temperature (^{0}C)	375–410		
Air/feed	18/1		
Contact time (s)	< 0.5		
Selectivity (%)	75–80		
Reactor	Tubular		
Catalyst	0.4% V_2O_5/9.6% TiO_2 (promoters Al, Zr, phosphates) supported on cordierite		

BASF and then von Heyden introduced new processes using tubular reactors with either naphthalene or o-xylene as feed. This design proved to be extremely successful in Europe during the 1950s and is still widely used. Scientific Design also introduced a similar, successful tubular reactor design in North America. The Badger/Sherwin Williams benzene catalyst was not selective when used with o-xylene feed and the fluid bed process soon became obsolete. A new catalyst was soon being used for o-xylene oxidation. About 10 wt% of mixed vanadium pentoxide and titania (anatase) was supported on a rugged cordierite support. Alumina and phosphates were also included as promoters. The form and quantity of vanadium used was claimed to control the catalyst activity, and by using more than one catalyst the hot-spot temperature could be minimized. Catalyst shape evolved from granules to spheres and the preferred shape is now small rings. Operation at 375–400^{0}C gave a selectivity of about 75–80%. Precise temperature control allowed use of the maximum o-xylene/air ratio.

The inclusion of titania was believed to inhibit the desorption of the many intermediates involved in the overall reaction[65] and is now a key component of the preferred o-xylene oxidation catalyst.

Titanium dioxide catalysts were first described in the 1940s and 1950s, when mixed oxide catalysts were being investigated and used in a number of oxidation reactions. Mixtures of vanadium pentoxide with titanium dioxide gave better operation and longer life as phthalic anhydride demand increased. An early catalyst that did not sinter and clearly increased the stability of vanadium pentoxide was described in a patent as $TiO(VO_3)_2$.[66] At about the same time vanadium pentoxide/phosphorous pentoxide *mixtures* were also being developed for use in maleic anhydride processes.

A common feature of the new vanadium catalysts, including those used in sulfuric acid production, was the need to reduce pentavalent vanadium by reaction with hydrochloric or oxalic acid solutions before the active compounds that improved catalyst performance were formed. It was well known, especially from sulfuric acid catalysts, that tetravalent vanadium also formed during operation.

Titanium dioxide in the form of anatase provides a suitable surface on which the vanadyl ions can react with hydroxyl groups. This forms pseudotetrahedral groups giving a *theoretical monolayer*. The layer is a two-dimensional sheet corresponding to the formula $VO_{2.5}$, which despite the stoichiometry, is believed to contain both strongly bound tetravalent and pentavalent vanadium. Up to about 10% vanadium pentoxide can combine with titanium dioxide, depending on its surface area, and any vanadium pentoxide that is not part of the monolayer forms crystals on the catalyst during the final stages of catalyst preparation. Both of the valence states take part in the oxidation of o-xylene by a typical redox mechanism.[67]

4.6. MALEIC ANHYDRIDE

4.6.1. Benzene Feedstock

Benzene oxidation using a vanadium pentoxide/pumice catalyst was first studied at the time that the phthalic anhydride process was being developed. Weiss and Downs discovered that maleic anhydride was formed in significant amounts.[51] They concluded that the maleic anhydride was produced via benzoquinone as the intermediate. The yields of maleic anhydride were not high with the unselective vanadium or molybdenum oxide catalysts being tested at that time.

The first commercial production unit was built by the National Aniline and Chemical Company, part of the Barrett Company, in 1933. However, most of the maleic anhydride used at that time was supplied from phthalic anhydride plants, which produced about 5–10% as a by-product. Demand for maleic anhydride was still low during the 1940s. At that time two typical catalysts were available. One contained 12% vanadium pentoxide/4% molybdenum trioxide supported on α-alumina, while the other contained 10% vanadium pentoxide moderated with less than 1% of lithium sulfate/sodium sulfate, also supported on α-alumina. The alkali sulfate moderated catalyst was, however, sensitive to sulfur poisons in the benzene feed.

A mixed oxide catalyst containing 13% titanium dioxide plus molybdenum trioxide and tungstic oxide supported on low-surface area α-alumina was developed by DuPont for butene-2 oxidation[68] in 1952.

In general, catalysts were prepared by dissolving and reducing the oxides in concentrated hydrochloric acid, adding the corundum granules and evaporating the liquid before calcining to decompose the chlorides. Oxalic acid was often included to act as glue, enabling the catalyst to become more firmly fixed to the support.

By 1955, Montedison had published sales literature to describe their MAT 5 catalyst which was still based on supported vanadium pentoxide/molybdenum trioxide.[69] The composition of MAT 5 is shown in Table 4.7. A life of between two and three years was claimed, depending on the poisons present in the benzene used, before the pass yield fell from 72% to about 65%. Decreased selectivity may also have depended on the amount of molybdenum lost at the higher operating temperature required to maintain conversion.

4.6.2. *n*-Butene Feedstock

Prior to 1960 benzene was the only feed used to produce maleic anhydride. From 1962, however, the Petrotex Chemical Corporation began to use *n*-butene feed in its plant in Houston, Texas.[70]

TABLE 4.7. Benzene Oxidation Catalyst for Maleic Anhydride Production. (MAT-5 catalyst produced by Montecatini.)

Vanadium pentoxide (wt%)	8
Molybdenum oxide (wt%)	4
Alumina (wt%)	88
Sodium oxide	0.12 atoms Na per atom of Mo
Shape (mm)	5 × 5 pellets
Porosity (ml g^{-1})	0.2
Conversion (%)	95–100
Operation	Tubular reactor
Temperature (^{0}C)	350–400
Pressure (atm)	2.5
Contact time (s)	0.5–1.0
Molar pass yield (%)	New 72
	1.5 year 70
	2.5 year 65
Catalyst life	~ 3 years
Productivity	Up to 1900 kg MA per kg catalyst

Note: Early naphthalene and benzene oxidation catalysts often contained alkali metal promoters. Commercial benzene oxidation catalysts were shown to contain a β-bronze phase (Na$_2$O·V$_2$O$_4$·5V$_2$O$_5$ or Na$_2$O·MoO$_3$·5V$_2$O$_5$) with other mixed oxide compounds such as V$_9$Mo$_6$O$_{40}$. The β-bronze could possibly stabilize the other active compounds and limit loss of molybdena during operation. M. Najbar, *Preparation of Catalysts IV*, , Elsevier, Amsterdam, 1987, p. 217.

The *n*-butene feed was supplied by the dehydrogenation of *n*-butane and the plant began the trend to develop oxidation processes using aliphatic petrochemical hydrocarbons. The main incentive, of course, was to use surplus C$_4$ hydrocarbon from steam cracking. This not only used a cheap by-product gas, but the reaction was less complicated because the straight-chain C$_4$ molecule contained fewer carbon atoms than aromatic benzene.

The mixed vanadium pentoxide/phosphorous pentoxide (with niobium, copper, lithium promoters) catalyst used by Petrotex was, at the time, a further step change in the catalyst types used for hydrocarbon oxidation.[71] It also eventually contributed to a better understanding of the catalyst structures used in oxidation reactions. The catalyst must have evolved from the accumulated experience obtained with a variety of mixed oxide catalysts and had a composition similar to that shown in Table 4.8. Distillers patented a molybdenum trioxide/phosphorous pentoxide catalyst,[72] and the Atlantic Refining Company took out a patent for a vanadium pentoxide/phosphorous pentoxide catalyst specifically for butene-2 oxidation.[73] The vanadium pentoxide catalyst gave higher yields.

TABLE 4.8. Operating Conditions for *n*-Butene/*n*-Butane Oxidation.

Catalyst	10–20% active phase (40% V_2O_5 : 60% P_2O_5)	
Support	α-alumina or fused silica	
Typical operating conditions	*n*-Butene	*n*-Butane
Temperature (°C)	390–400	390–400
Feed concentration (vol%)	~2	~3
Conversion (%)	~100	~80
Selectivity (%)	~50	~50
Reactor	Tubular	Tubular
Contact time (s)	0.5	0.5
Hot zone (°C)	450–500	~450
Catalyst life (years)	4	4

Note: Dissolve ammonium metavanadate in phosphoric acid. Heat to reduce vanadium and precipitate catalyst. Atlantic Refining Co., US Patent 2773838 (1957).

As new mixed oxide catalysts were introduced for industrial oxidation processes in the period from 1950 to 1965 it was difficult to recognize a pattern of performance. The relationship between the original catalysts used to produce phthalic and maleic anhydrides and the new efficient catalysts being developed for acrolein and acrylonitrile processes is easier to see now substantial research, both industrial and academic, has led to a better understanding of the reactions. The vital piece in the catalyst jigsaw puzzle came when Mars and van Krevelen recognized that mixed oxide catalysis generally proceeded by a redox mechanism. Hydrocarbons reacted with lattice oxygen from one oxide, which was subsequently reoxidized by oxygen supplied by the second oxide.[74]

An interesting practical feature of all mixed oxide catalysts is the very simple preparation from the appropriate ingredients. Maleic anhydride catalyst is prepared as follows:

- Dissolve vanadium pentoxide in concentrated hydrochloric acid and reflux to reduce V^{5+} to V^{4+} forming a blue-green solution.
- Add phosphoric acid with an organic solvent such as isobutanol to precipitate the catalyst.
- Add support and evaporate to dryness.
- Calcine the catalyst to remove hydrochloric acid and chloride.
- Promoters such as lithium, zinc, or molybdenum may also be added as appropriate to the original vanadium solution.

$(VO)_2P_2O_7$

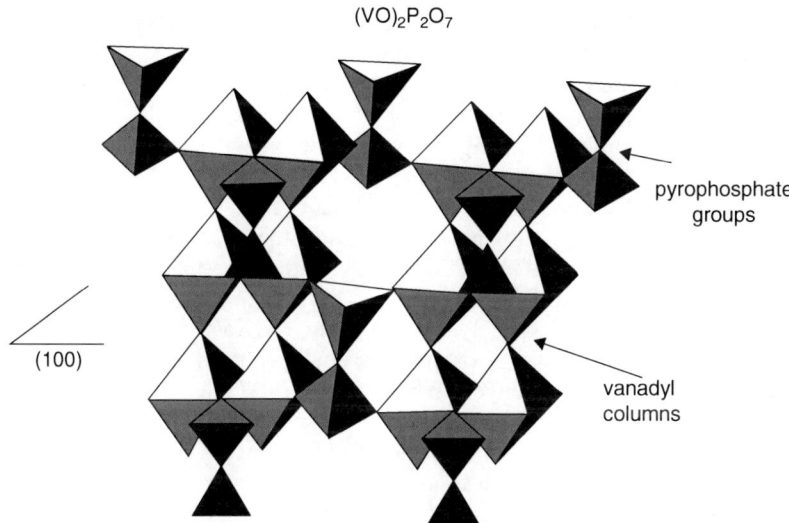

Figure 4.10. Bulk structure of $(VO)_2P_2O_7$.

The active catalyst precursor has the composition $VOHPO_4\cdot0.5H_2O$. During the initial stages of reaction the precursor decomposes in two steps of dehydration, at 375^0C and 480^0C, to give the catalyst $(VO)_2P_2O_7$. The bulk and surface structures of $(VO)_2P_2O_7$ are shown in Figures 4.10 and 4.11.

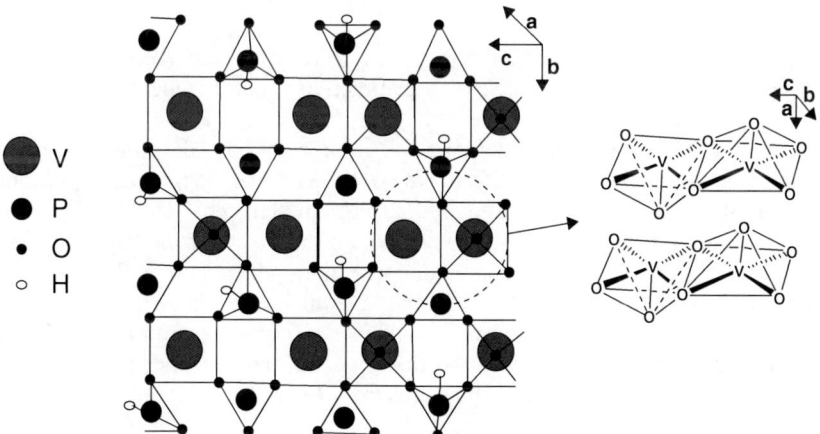

Figure 4.11. Idealized surface structure of $(VO)_2P_2O_7$.

The catalyst precursor decomposes at a lower temperature under reaction conditions, and the catalyst is more active when normal or isobutanol is used as the organic solvent. Tests under reaction conditions show that maleic anhydride begins to form at 388^0C, when the precursor decomposes. A final step of crystallization takes place as maximum conversion is reached at $390–400^0C$. The active catalyst has the same particle morphology as the precursor when examined by scanning electron microscopy.[75]

The original Petrotex plant was reconverted back to a benzene feed in 1967 suggesting that they may not have had access to the latest catalysts. Early patents claimed a pass yield of about 50% but operational results seem not to have been published.[76] Other plants, operated by BASF[77] and Bayer,[78] also used *n*-butene feed and proprietary catalysts from about 1969 and functioned until the 1980s. The Bayer plant used a range of gradually improved catalysts and increased the yield of maleic anhydride from 59 lb in 1975 to 70 lb in 1981, based on 100 lb of feed containing 75% butenes.[79] Catalysts in these plants were also thought to be based on vanadium pentoxide/phosphorous pentoxide compositions with some titania and a steatite support.[80] There was a move to use fluid bed reactors in 1970 by Mitsubishi Chemical Industries, Ltd. A mixed C_4 stream was used as a feed and only the butenes were converted to maleic anhydride. No catalyst details were published, although a vanadium pentoxide/phosphorous pentoxide formulation must have been used.[77,81]

4.6.3. *n*-Butane Feedstock

Monsanto and Alusuisse both introduced processes using *n*-butane as feed in 1974. The overall reaction is:

$$C_4H_{10} + 7/2\,O_2 \rightarrow C_4H_2O_3 + 4\,H_2O \tag{4.12}$$

Despite the need to operate at a higher temperature with a relatively low conversion and lower yield, *n*-butane was more attractive than *n*-butene because it was cheaper. It is now being used more widely.

Since 1980 several plant operators and contractors have developed fluidized bed reactors to oxidize *n*-butane feed. One of the first was Badger, using a Mobil catalyst, at the Denka plant (originally Petrotex) in Houston.[82] Badger estimated that the fluid bed butane process would require only 64% of the capital investment needed for a conventional plant. Since then several other fluidized bed processes have been introduced, including a Sohio/UCB plant near Cleveland, Ohio[83] and a BP/Mitsui Toatsu plant in Japan. The Alusuisse/Lummus Crest process began pilot plant operation in Italy in 1983 and has since been licensed to several companies throughout the world from 1987.[84]

By injecting butane and air separately, fluidized beds allow the use of an inlet concentration of butane that is higher than the flammable limit. The same catalysts are used as in *n*-butene oxidation. Operating conditions for the oxidation of *n*-butane in the conventional Alusuisse fluid bed and the DuPont circulating fluid bed process described in the next Section are given in Table 4.9.

4.6.4. *n*-Butane Oxidation in a Circulating Fluidized Bed

A significant technological advance was made by DuPont in 1986 when maleic anhydride was produced from *n*-butane in a *circulating fluidized bed*.[85] In this process, *active* catalyst reacts with butene to give maleic anhydride, itself becoming reduced in the process. The overall reaction is:

$$C_4H_8 + 3\,O_2 \rightarrow C_4H_2O_3 + 3\,H_2O \qquad (4.13)$$

Reduced catalyst is then reoxidized with air, in a separate regeneration reactor, to regenerate the active form. This innovation followed the successful introduction of conventional fluidized bed operation by Alusuisse and other companies in 1983. Physical circulation of a fluidized bed of catalyst particles, or microspheres, is an unusual technology and has been developed commercially only for the fluid catalytic cracking of heavy gas oils and the SASOL version of the Fischer–Tropsch Synthol process. Success depends not only on an active and selective catalyst but also on the resistance of the catalyst to attrition during the transfer from the reactor to the regenerator and back again.

The DuPont process is based on a practical application of the Mars and van Krevelen redox mechanism in a reducing fluidized bed reactor followed by an oxidizing regenerator.[86] During reaction, lattice oxygen from the fluidized catalyst selectively converts an *n*-butane feed to maleic acid with almost no carbon dioxide formation. Then, when the fluidized bed of the reduced catalyst passes through the regenerator, it is reoxidized with air before re-entering the reactor.

TABLE 4.9. Oxidation of butane to maleic anhydride in fluid beds

	Alusuisse	DuPont
Design	Conventional fluid bed with separate air and feed injection	Circulating fluid bed reaction with no oxygen in reactor and separate regenerator with air addition
Operating temperature (^{0}C)	~ 350	~ 350
Butane concentration in reactor (vol%)	4	100
Operation	Feed recovered	Feed recycled
Conversion (%)	~ 60	#~i75
Catalyst life (years)	> 1	> 1
Product separation	Organic solvent	Water

By separating the catalyst reduction and oxidation stages of the redox process, and avoiding the possibility of explosion, there are no practical restrictions on the concentration of butane during reaction or to the concentration of air during the reoxidation of oxygen-depleted surface sites. During operation $(VO)_2P_2O_7$ catalyst is, therefore, always in the optimum form, with the average valency of the vanadium component slightly greater than four. Operating temperature is similar to that of other conventional processes, although the pressure is somewhat higher as a result of having to circulate a fluid bed. Lower water production in the reaction zone makes product recovery easier at 30^0C. Throughput can be increased by 10% with 20% higher yield.

The catalyst microspheres are approximately 40–150 m in diameter. Adequate strength for use in the circulating fluid bed is obtained by spray drying a slurry of small, 0.5–2 m, $(VO)_2P_2O_7$ particles with 5% of polysilicic acid at pH 3. During this process silica forms a hard porous shell over the surface of the catalyst particles.

4.7. ETHYLENE OXIDE

Ethylene oxide was first produced by BASF and Union Carbide in 1916 and 1923, respectively, via the chlorhydrin route:

$$CH_2=CH_2 + HOCl \rightarrow HOCH_2CH_2Cl \qquad (4.14)$$

$$HOCH_2CH_2Cl \rightarrow \underset{O}{CH_2\!-\!CH_2} + HCl \qquad (4.15)$$

This process used chlorine that could not easily be recovered, so attempts were made to develop a catalyst for direct oxidation The reaction is

$$C_2H_4 + \tfrac{1}{2}O_2 \rightarrow (CH_2)_2O \qquad (4.16)$$

Lefort[87] was successful in 1930 and he subsequently assigned his patents to the Union Carbide corporation, which commissioned the first large-scale direct oxidation plant at South Charleston, West Virginia, during 1938–1939. The success of this plant led Scientific Design to develop an air oxidation process in 1953,[88] followed by Shell, using oxygen, in 1955.[89] Operating conditions for both processes are given in Table 4.10.

Modern processes operate reactors with the catalyst packed into 1- to 2-in tubes that are cooled by suitable heat transfer liquids:

- The air oxidation process uses a reaction mixture containing 4% ethylene and 6% oxygen, with the balance being nitrogen. This is passed through

TABLE 4.10. Ethylene Oxide Production by Air and Oxygen Processes.[a]

	Air	Oxygen
Ethylene concentration (vol%)	20	15–20
Oxygen concentration (vol%)	6	6
Carbon dioxide (vol%)	5	30
Balance	Nitrogen	Methane
Operating temperature (^{0}C)	260–280	230
Pressure (atm)	20	10–20
Conversion (%)	10–15	10–15
Selectivity (%)	70+	80+
Contact time (s)	1	1
Inhibitor	$C_2H_4Cl_2$	$C_2H_4Cl_2$
Reactor	Tubular	Tubular
	Several thousand tubes	

[a]The oxygen process is now the most widely used.

the catalyst tubes and then the product absorber. Ethylene and air are added to maintain the reactor inlet composition so a purge is necessary to remove inerts. The purge gas passes through a second reactor to recover the residual ethylene, which can be as much as 20% of the initial feed. The second purge reactor may take purge gas from several main reactors.

• When using pure oxygen the reaction mixture can contain about 15–20% ethylene, 6% oxygen, and 15–30% methane, with the balance being an inert gas such as carbon dioxide. After the product has been absorbed the carbon dioxide/methane mixture is recirculated, with appropriate makeup of ethylene and oxygen. No purge is necessary. To maintain the correct feed gas composition, carbon dioxide is simply stripped from the circulating gas and vented. There is no need for a second reactor to recover ethylene. A very brief review of early industrial patents was given by Hucknall.[90]

• Since the Hucknall review, which dates back to 1974, there have been many substantial developments in the ethylene oxide process arising from separate contributions from companies such as Union Carbide, Shell and ICI. These improvements include the addition of ppm quantities of chloride to the process gas stream, the addition of alkali metal with potassium and to a lesser extent, caesium being preferred, and advances in the preparation of the support involving the addition of hydrofluoric acid and acetic acid prior to calcinations to control the morphology of the α-alumina. This sequence in process developments resulted in an improvement in selectivity from around 65% to about 83%, thus reducing significantly the amount of carbon dioxide formed, particularly in fresh catalysts. As the catalyst aged and activity fell, the operating temperature had to be raised slightly to maintain conversion, and this resulted in a gradual and ongoing fall in selectivity. When the selectivity eventually fell to about 78% after a period of two to three years, it became more economic to replace the catalyst.

- The addition of NO_x to the process gas was found to increase selectivity even further, and with careful selection of the methods of preparation of the support, the addition of the silver and potassium components and the catalyst pre-conditioning procedures, it was possible to achieve selectivities in the range 90–93% It is interesting to observe that in the earlier process, the combustion of ethylene to carbon dioxide generated so much heat that the overall process, including product recovery and purification, was a nett exporter of steam, that is, by-product energy. In contrast, as only a small amount of ethylene is wasted at the high selectivities achieved in the NO_X promoted process, the process is a net importer of energy. However, combustion of ethylene to fuel a boiler is hardly the most economic way to provide energy!!

4.7.1. Catalyst

The same catalyst can be used in both air and oxygen processes. In the early catalysts about 10–15% silver was deposited on a low-surface-area support such as α-alumina or silicon carbide. The nature of the support is critical. Low surface area α-alumina is normally used. Any residual acidity in the support promotes the isomerisation of ethylene oxide to acetaldehyde, a key intermediate in the formation of carbon dioxide, and hence low selectivity. All γ-alumina must be converted to α-alumina during calcination. Silver lactate solution plus an alkaline earth lactate could be used to coat the preformed support.[91] Metallic silver formed as the catalyst was dried and then calcined. Suitable supports include small spheres or rings with a pore volume about 0.5 ml g^{-1}. Up to 2% of an alkaline earth promoter such as barium may have been added to the catalyst. Further details are shown in Table 4.11.

TABLE 4.11. Early Ethylene Oxide Catalyst Composition.

Composition	Silver:	10–15%
	Support:	α-Al$_2$O$_3$, surface area < 1m^2 g^{-1}
	Promoters (based on silver):	Alkaline earth, e.g., BaO, 1–2%
		Alkali salt, e.g., Cs$_2$O, 0.005–0.05%
Shape	Spheres:	0.125 in diameter
Porosity		20–24%
Life	Up to 6 years	

The high cost of ethylene is such that even quite small improvements in catalyst selectivity offer substantial savings to the ethylene oxide producer. High selectivity is, therefore, commercially very valuable to the catalyst producer. The patent situation regarding ethylene oxide catalysts is so congested and complex that it is extremely difficult for the companies to guarantee adequate protection for their work, or even to recognize and prove that infringement of their rights has taken place. Consequently, specific details of the preparation of current commercial catalysts are therefore almost never available in the open literature.

4.7.2. Operation and Reaction Mechanism

During operation of early catalysts it was usual to limit the conversion of ethylene both to achieve maximum selectivity and to control the evolution of heat. A short contact time was required and conversion was limited to about 8–10%. The ethylene content of the feed gas was close to the lower flammability limit.

The reaction mechanism was suggested by Sachtler et al. to proceed by adsorption of molecular oxygen directly onto the silver surface, where it reacts with ethylene:[92]

$$[Ag] + O_2 \rightarrow [Ag]O_2 \tag{4.17}$$

$$[Ag]O_2 + C_2H_4 \rightarrow C_2H_4O + [Ag]O \tag{4.18}$$

The adsorbed atomic oxygen that remains on the silver surface does not react with ethylene to produce more ethylene oxide but instead forms oxides of carbon:

$$6 [Ag]O + C_2H_4 \rightarrow 6 [Ag] + 2 H_2O + 2 CO_2 \tag{4.19}$$

The catalyst surface is, therefore, continuously oxidized by oxygen and then reduced by ethylene during operation. Only six out of every seven ethylene molecules form ethylene oxide, so that the maximum selectivity allowed by the Sachtler mechanism is only 85.7%.

The Sachtler mechanism did not take into account the critical role of chloride, and with high selectivities now being regularly and reliably reported, it clearly no longer satisfied the latest information, particularly when NO_x is used as a moderator in the feed gas, and selectivities in excess of 90% are achieved. Further contributions to this debate were clearly required by the industry and academy alike.

Van Santen[93] proposed an alternative mechanism in which molecular oxygen on the silver surface first dissociates to give atomic oxygen species. He suggested that only weakly bound oxygen atoms interact with ethylene to give the ethylene oxide precursor. This occurred preferentially when both oxygen and chloride atoms competed for the same silver site. The role of the alkali metal, potassium, was believed to stabilise an oxychloride ionic species.

Silver is in the form of spherical particles that can change size during operation. Particles smaller than 0.1 μ sinter and larger particles seem to break up as the size stabilizes. The equilibrium silver particle size is reached faster by the addition of the alkali metal promoter to the catalyst and is less than 1 m, the typical diameter[94] of the pores with the alumina.

The reaction of ethylene with adsorbed oxygen atoms to form carbon dioxide is inhibited by adding a few parts per million of ethylene dichloride to the process gas. The chlorine atoms cover a significant proportion of the silver surface, and the rate of oxygen adsorption is reduced. This can deactivate the catalyst. Moreover, at high levels of chloride in the process gas, the catalyst eventually becomes *chlorided* and thus deactivated. The alkaline promoter in the catalyst can *store chlorine* and help control deactivation, but it can also lead to a small decrease in selectivity. The effect of chlorine on silver can be reversed by the addition of methane or ethane to the feed.[95]

The beneficial effect of ethylene dichloride was discovered accidentally at Union Carbide in the 1940s when it was a trace impurity in the air at Charleston, West Virginia. Careful work determined that occasional bursts of ethylene dichloride, present with more than 200 other impurities, improved operation, and it has been added as a promoter in controlled quantities ever since.[96]

Typical operating selectivity varies in the ranges 65–75% and 70–83% for the air and oxygen processes, respectively.

4.7.3. Applications of Ethylene Oxide

Demand for ethylene oxide has increased rapidly in the second half of the 20th century as ethylene has been produced by steam cracking and more importantly new applications for ethylene glycol have been introduced. About 60% of the total ethylene oxide produced is converted to ethylene glycol by hydrolysis in an excess of water at temperatures exceeding 140^0C and about 20–30 atm pressure. About 1% of sulfuric acid can be added as a catalyst. Selectivity is only about 90% as diethylene glycol and small amounts of glycol ethers also form during the reaction. Increasing amounts of polyvinyl alcohol are formed at low water concentrations.

The earliest important use of ethylene glycol was as antifreeze in automobile engines, but it had also been used earlier, converted to the dinitrate, as a component in low-freezing dynamite. Nowadays it is used mainly in the production of polyesters.

Ethylene oxide is used in a variety of other applications such as nonionic surfactants, ethanolamines, and glycol ethers.

TABLE 4.12. Mixed Oxide Catalysts.

Catalyst/Process	Composition
Bismuth phosphomolybdate	Phases present: α-$Bi_2O_3 \cdot 3MoO_3$ β-$Bi_2O_3 \cdot 2MoO_3$ (most active) γ-$Bi_2O_3 \cdot MoO_3$
Uranyl antimonate	Phase 1 (active) USb_3O_{10} Phase 2 (inactive) $USbO_5$
Acrylonitrile	Bi, Co, Ni, Fe, Mo oxides. Bi_2O_3 (10%), CoO (5%), NiO (2.5%), Fe_2O_3 (10%), MoO_3 (30%), SiO_2 (40%)
Isobutene to methacrolein	Bi, Mg, Co, $Ni_{5.5}$,$Ti_{0.4}$, Mn, $P_{0.1}$, Mo_{12} oxides [US Patent 3928462 (1975)]
Propylene to acrolein	$Bi_{0.1-4}$,$Co_{0.5-7}$, Ni_{10-15},$Fe_{0.5-7}$, La_{0-4}, $(K,Rb,Cs)_{0.01-5}$, Mo_{12}, O_{35-85} [US Patent 3959384 (1976)]
Ammonia oxidation to nitrous oxide	Iron oxide/bismuth oxide [Von Nagel, *Zeit Electrochem* **3** (1980) 754] Iron oxide/bismuth oxide/molybdenum oxide [Zawadski, *Trans Faraday Soc Disc*, No. 8 (1950) 140]

Note: Mixed oxide catalysts are now known to be extremely complex and the large number of different structures discovered can be very active catalysts.[107] Thomas and Thomas suggest that bismuth oxide can be a solvent for many other metal oxides. Others have tried to establish that a layer structure may form with the most active component in the surface of particles.

4.8. A REDOX OXIDATION MECHANISM: MARS AND VAN KREVELEN

The introduction of new vanadium pentoxide/phosphorous pentoxide and vanadium pentoxide/titania catalysts to meet the increasing demand for maleic anhydride and phthalic anhydride was a huge step forward in catalyst development. It is hard to tell in retrospect whether the discoveries were made by trial and error or as part of a focused research effort. Even the catalyst suppliers do not admit to or remember a sudden breakthrough. Nevertheless the understanding of catalyst design and improved physical testing coincided with the escalating demand for petrochemicals to produce better catalysts. During this period a number of apparently random experimental results that had been accumulating since the 1920s gradually began to make sense (see Table 4.12).

From experience in providing oxidation catalyst samples similar to those produced by Petrotex and von Heyden it seems almost certain that tentative formulations were developed before a theory evolved. There was certainly very little prior disclosure of catalysts and only a few patents before the 1960s, although active catalysts could be produced very easily. As the multitude of later patents for acrolein and acrylonitrile catalysts showed, *mixed* oxides was an appropriate description.

From their experiments on the oxidation of naphthalene and presumably other published information, Mars and van Krevelen realized that the reaction could be explained by a *redox* mechanism.[74] They suggested that two stages were involved:

- Initially naphthalene vapor reacted with the oxide catalyst to provide products while the catalyst was reduced.
- Reduced catalyst was then reoxidized with molecular oxygen before the reaction proceeded further.

The catalyst used in their experiments to determine the reaction kinetics was reported to be similar to the Davison fluid bed phthalic anhydride catalyst, and results confirmed that the redox reaction was independent of the hydrocarbon pressure.

Mars and van Krevelen did not describe the form of oxygen taking part in the redox procedure. It has since been shown by isotopic labelling experiments that the hydrocarbon is oxidized with oxygen from the lattice of the actual catalyst. During the oxidation of propylene to acrolein with oxygen-18 over a typical bismuth molybdate catalyst, the acrolein and carbon dioxide initially formed contain only oxygen-16. It appears that the bismuth molybdate catalyst can apparently use most of its lattice oxygen before it is completely inactive. Naturally, in commercial oxidation, the oxygen vacancies in the lattice are quickly replaced by dissociated molecular oxygen. Catalysts usually achieve maximum activity when the surface is partly reduced.

It is clear from subsequent developments since Mars and van Krevelen made their initial suggestion, that sites of higher activity are formed by *cation pairs* relative to single cations as in, for example, partially reduced vanadium pentoxide, or cupric oxide. At least two oxides are required although neither may be active or selective enough alone. One part within the *cation pair* forms the active intermediate that is then oxidized with oxygen ions provided by the other part. Mixed oxides can exist as different crystalline phases and it is important to select the particular structure with the highest catalyst activity for the required reaction.

4.9. ACROLEIN AND ACRYLONITRILE

Acrylonitrile has been an important organic intermediate since the 1930s, when it was used in a copolymer for synthetic butadiene rubbers in Germany. While Buna-N was not as successful as the similar Buna-S, made with styrene, acrylonitrile is now widely used in the production of fibers, and ABS resins. Early production routes were based on the reaction of ethylene oxide or acetylene with

TABLE 4.13. Early Propylene Oxidation Catalysts.

Company	Composition	Reference
Shell	Cu_2O/Al_2O_3	US Patent 2451485 (1948)
		British Patent 778125 (1948)
Sohio	BiPMo/SiO$_2$	British Patent 821999 (1959)
Knapsack	BiFePMo/SiO$_2$	British Patent 908655 (1963)
Distillers	SbSn/SiO$_2$	British Patent 906328 (1963)
	CoTe^a Mo/SiO$_2$	British Patent 878803 (1960)
Shell	NiTeMo/SiO$_2$	French Patent 1335423 (1964)
	TeMo/SiO$_2$	Frrench Patent 1342963 (1964)
	TePMo/SiO$_2$	French Patent 1342962 (1964)
Socony	Tellurium included	US Patent 2669586 (1953)
	Tellurium included	US Patent 2653138 (1953)
	BiFeNiCoPKMo/SiO$_2$	Japan Patent 6906246 (1969);
		6906245 (1969)
	TeWVAsSn or Sb/SiO$_2$	Japan Patent 6908990 (1969)
	TeWZnPCd or g/SiO$_2$	Japan Patent 6808806 (1968)

^{a}Toxic tellurium oxide is volatile in operation and is not now widely used commercially.

hydrogen cyanide,[97] rather than by heterogeneous catalytic. Later, DuPont produced acrylonitrile on a small scale by oxidizing propylene with nitric oxide using a silver oxide/silica catalyst.

The first use of oxide oxidation catalysts for the production of acrolein from propylene with a cuprous oxide/silica formulation was described by Shell in 1948. This followed an Allied Chemical Company patent describing the potential production of acrylonitrile from propylene. As demand for these products increased during the 1950s, other, more efficient, catalysts based on mixed oxides were developed. The best early catalysts are listed in Table 4.13.

It was soon realized that commercial units would be based on the use of fluidized beds, which were then being introduced in refineries to produce gasoline. The most successful fluid bed process was introduced by Idol of Sohio in 1960 and used a bismuth phosphomolybdate catalyst supported on silica.[98] Knapsack described a bismuth phosphomolybdate catalyst containing iron for acrylonitrile production in 1962.[99] This rather more complex mixed oxide formulation, $Fe_7Bi_2Mo_{12}O_{52}$, also supported on silica, foreshadowed improvements in the 1970s and the introduction of multicomponent catalysts. Since then several *generations* of improved catalysts have been introduced, e.g., by Sohio, as the reaction mechanism has been better understood.

4.9.1. Manufacture of Mixed Oxide Catalysts for Acrolein and Acrylonitrile

The original bismuth molybdate catalyst described in the early patents was supported on silica and prepared by a relatively simple procedure:[100]

- Aqueous solutions of molybdic acid and bismuth nitrate were added, successively, to an aqueous silica sol and the mixture acidified with nitric acid.
- The solution was evaporated to dryness and heated to about 540^0C for 16 h. The solid was ground to a fine powder for use as catalyst in a fluid bed reactor.

A more active bismuth phosphomolybdate (Table 4.13) was prepared simply by adding an appropriate volume of phosphoric acid to the initial solution. A typical catalyst composition was claimed to be $Bi_9PMo_{12}O_{52-55} \cdot 2SiO_2$. The same catalysts could be used to produce both acrolein and acrylonitrile.

During full-scale operation it was found that *whiskers* of molybdenum oxide could form on the catalyst surface in the presence of steam. This led to caking of the particles as well as loss of molybdenum. The problem with catalyst caking is that the fluid bed ceases to operate correctly and temperature control is severely impaired. This results in significant loss in selectivity. By ensuring that the ratio of bismuth to molybdenum was greater than 2:3 the loss could be controlled. An upper bismuth-tomolybdenum ratio of about 3:4 was fixed[101] because bismuth was expensive.

The *second generation* Sohio catalyst, introduced during the mid-1960s, was an antimony oxide/uranium oxide mixture ($UO_3 \cdot 2Sb_2O_3$).[102] A successful iron oxide/antimony oxide catalyst containing some tellurium oxide was subsequently developed by Nitto, a Sohio licensee, in Japan.[103]

In 1972, a *third generation* Sohio catalyst with a number of promoters was introduced to limit by-product acetonitrile formation. It had a general formula $(Ni,Co)_8(Fe^{3+})_3BiMo_{12}O_{60}$, was also supported on silica, and was made by the original recipe. Other promoters were later included in smaller proportions. Many other mixed oxide catalysts for the oxidation of propylene to acrolein had been investigated since the late 1940s. Apart from the bismuth molybdates, other formulations were based on cuprous oxide, antimony and tin oxides, tellurium and tin oxides, and other combinations. The methods of preparation ranged from the ball milling of oxides and the silica support to the mixing of metal salts and support in an acid solution followed by evaporation to dryness and grinding to the required particle size. In most cases conversion increased as the proportion of support increased to about 70–80%, after which the product yield declined. Tellurium is no longer used as a catalyst component because the oxide is toxic and disposal of deactivated catalysts is no longer allowed.

4.9.2. The Acrylonitrile Process

The Sohio fluidized bed acrylonitrile process has proved to be extremely successful and provides most of the worldwide acrylonitrile demand. During operation the catalyst temperature is controlled by circulating water through vertical tubes within the catalyst bed.

The catalyst is very hard and almost no attrition losses are experienced. Any dust forming is filtered from the quench water. As the inventory declines, makeup catalyst can be added or, if activity falls, an appropriate volume can be withdrawn and replaced. Molybdenum losses could be replaced during operation or the catalyst could be discharged and milled to remove lumps. *Whiskers* of molybdenum oxide grew on the catalyst particles causing either the loss of molybdenum or caking during the earlier days of operation of the process. This problem was solved by optimisation of the bismuth/molybdenum ratio. There do not seem to be any serious operating problems and catalyst can be used almost continuously for up to 10 years. By-products, such as acetonitrile, and particularly hydrogen cyanide are recovered and used commercially. Typical process conditions are shown in Table 4.14.

4.9.3. Reaction Mechanism

Catalysts used for the oxidation or ammoxidation of propylene must have an appropriate structure to give selective operation. The *first generation* Sohio catalysts had active sites that consisted of adjacent bismuth and molybdenum atoms in the mixed oxide lattice.

Similar mechanisms were suggested for both reactions in which the initial hydrogen abstraction from a propylene molecule adsorbed on a molybdenum atom was the rate-determining step.[104] Simplified descriptions are as shown next.[105]

TABLE 4.14. Acrylonitrile Process Conditions.

Plant capacity	50,000–150,000 tonnes.year^{-1}		
Catalyst volume	1–2 tonnes per 1000 tonnes ACN		
Contact time (s)	5–8		
Yield (%)	Bi/P/Mo 65+	U/Sb 70+	Multicomponent 75+
Fluid bed reactor:			
Conversion (%)	> 98		
Selectivity (%)	> 80		
Temperature (^{0}C)	450		
Pressure (atm)	1–2		
Feed composition (vol%):			
Propylene	8		
Air	80		
Ammonia	10		
Water	2		
By-products (wt%) based on acrylonitrile:			
Acetonitrile	2–4%		
Hydrogen cyanide	14–18%		

Acrolein formation:

- α-Hydrogen abstraction from propylene by the bismuth atom forms a surface hydroxyl group and an adsorbed allyl intermediate.
- Reaction of the allyl intermediate with a surface oxygen atom and a second hydrogen abstraction forms acrolein and leaves a lattice oxygen vacancy.
- Acrolein desorbs.
- The oxygen vacancies are filled by the dissociation of adsorbed molecular oxygen at the molybdenum atom, which provides mobile oxygen ions to restore the active bismuth site.
- Adjacent hydroxyl groups dehydrate, forming water and once again, restoring the active site.

Acrylonitrile formation:

- The molybdenum atom is activated by reaction with ammonia, which produces a molybdenum di-imido species before propylene is adsorbed.
- α-Hydrogen abstraction from propylene by the bismuth atom then forms the allyl intermediate.
- The allyl intermediate is converted to an organic nitrogen intermediate by reaction with the di-imido species and a second hydrogen abstraction to leave a lattice oxygen vacancy.
- Acrylonitrile is produced by a third hydrogen abstraction and desorbs.
- The oxygen vacancies are filled by molecular oxygen adsorbed and dissociated at the molybdenum atom to produce mobile oxygen ions.
- The molybdenum atom is again activated by ammonia.

The ammoxidation process operates at a higher temperature than that required for the production of acrolein. This is probably a consequence of the different reactivity of ammonia towards the catalyst, and the difference in reactivities between oxide and imido groups. A similar range of operating temperatures was used during early experiments on the catalytic oxidation of ammonia to produce nitric oxide for nitric acid manufacture. In this work, however, nitrous oxide was the initial product.[106]

It is important to have a high concentration of bismuth atoms at the catalyst surface, to generate the optimum structure, for forming active sites, $-Bi-Mo-Bi-Mo-$. The optimum bismuth/molybdenum ratio for practical operation is in the range 2:3 to 3:4, and this is a composition close to $\alpha-Bi_2O_3 \cdot 3MoO_3$, $[Bi_2(MoO_4)_3]$. This has a type of defect-Scheelite structure, $Bi_{2/3}[]1/3MoO_4$[107] and is probably more selective than either $\beta-Bi_2O_3 \cdot 2MoO_3$ or $\gamma-Bi_2O_3 \cdot MoO_3$, which may result as molybdenum oxide is lost from the catalyst structure during operation. All catalysts active for acrylonitrile production are mixtures of oxides that when used alone are either inactive or unselective for the ammoxidation reaction.

The *second generation* Sohio catalyst was a uranium antimonate (USb_3O_{10}).[108] This was more active and selective than the earlier bismuth phosphomolybdate and has been described as *Phase 1*. Active sites in the layer structure were also defect-Scheelite structures containing uranium-antimony *cation pairs*. Catalysts containing $USbO_5$, or *Phase 2*, were less selective.

Third generation Sohio catalysts were also based on bismuth molybdates and contained nickel, cobalt, iron, and minor amounts of other promoters. It has been suggested that the Fe^{2+}/Fe^{3+} redox couple facilitates the adsorption and activation of molecular oxygen at the catalyst surface.

4.9.4. Partial Oxidation of Propane

The selective oxidation of propane is more difficult than that of propylene due to its low reactivity and there has been little success in developing a viable process.

Moro-Oka and his colleagues tested bismuth vanadophosphates doped with a number of promoters such as silver.[109] A catalyst with the composition $Bi_{0.85}V_{0.54}Ag_{0.01}Mo_{0.45}O_4$ gave a conversion of only 13%, but 67% selectivity to acrylonitrile at 500^0C and 3000 h^{-1} space velocity using oxygen. While yields were reasonable, the conversion was not high enough for a commercial process. Although the concentration of propane in feed gas was considerably higher than the propylene in the conventional acrylonitrile process, the recycle costs were still too high. A propane ammoxidation process patent was granted to Harris,[110] of Davy Process Technology, in 1973 for the use of a vanadium antimonate catalyst. BP Amoco have tested the process on a large scale,[113] but far detailed information on operation has not yet been published. This process was developed jointly in conjunction with ICI. Excellent selectivities were achieved, but in a similar situation to the Moro-Oka work, the process suffered from low conversions.

Since then Centi has also examined catalysts based on vanadium antimonates.[111] Conversion as high as 60–80%, but only 35–40% yields, were obtained with a VSb_5WO_x catalyst supported on alumina. Bowker confirmed Centi's conclusion that the reaction proceeds in two steps.[112] Propane is first dehydrogenated to propylene, which is then ammoxidized to acrylonitrile by the well-established reaction mechanism.

4.9.5. Acrylic Acid

Acrylic acid was originally produced by reacting ethylene oxide with hydrogen cyanide and later by the carbonylation of acetylene, using a nickel tetracarbonyl catalyst promoted with copper halide. It is now generally produced by the oxidation of propylene using mixed oxide catalysts in processes that also provide acrolein as a useful co-product.

It is common practice to use two reactors. The first reactor contains a conventional bismuth phosphomolybdate catalyst and is used to convert propylene to acrolein. The second reactor contains a selective vanadium molybdate catalyst promoted with tungsten, nickel, manganese or copper,[114] to convert acrolein to acrylic acid. Fixed bed tubular reactors are used in both stages. Typical operating conditions are shown in Table 4.15.

4.9.6. Oxidation of Isobutene

Isobutene can be oxidized to methacrolein or methacrylonitrile using the same catalysts as for propylene oxidation. The reactions are:

$$C_4H_8 + O_2 \rightarrow CH_2{:}CH(CH_3)CHO + \tfrac{1}{2} H_2O \qquad (4.20)$$

$$C_4H_8 + 5/4\, O_2 \rightarrow CH_2{:}CH(CH_3)CN + 5/2\, H_2O \qquad (4.21)$$

This process avoids the by-product ammonium sulfate formed from the acetone cyanohydrin route.

4.10. OXIDATIVE DEHYDROGENATION OF n-BUTENES TO BUTADIENE

The oxidative dehydrogenation of normal butenes in the presence of steam can produce a higher conversion to butadiene than typical dehydrogenation processes.[115] Added oxygen not only converts the hydrogen being formed to water but also prevents carbon formation, so that catalyst regeneration is not required. Petrotex introduced the Oxo-D process in 1965. It operates at a temperature in the range 575–600^{0}C with a steam ratio of up to 12 to control selectivity.

TABLE 4.15. Acrylic Acid Production from Propylene in Two Steps.

Process	Two tubular reactors in series	
	First reactor	Second reactor
Temperature (°C)	330–370	260–300
Pressure (atm)	1–2	1–2
Conversion (%)	~95	~95
Selectivity (%)	#	85+
Catalyst	$(Bi_2O_3)_x$ $(Fe_2O_3)_y$ $(MoO_3)_z$ silica support	$(V_2O_5)_x$ (W,Fe,Ni,Mn,Cu) $(MoO_3)_y$ silica support

Note: Steam added to propylene/air mixture with interbed cooling.

About 65% butene conversion is achieved at more than 95% selectivity.[116] Phillips also developed the O-X-D process, which operated under similar conditions. Both processes use fixed beds of catalyst. Oxidative dehydrogenation with metal oxide catalysts involves a redox mechanism and the catalyst used should not be reduced irreversibly during the cycle.

In early patents, it was claimed that various phosphate catalysts, including calcium/nickel phosphate and bismuth/molybdate, were active. A variety of other catalysts was also introduced, including the zinc and magnesium ferrites developed by Petrotex.[117] The best ferrite catalyst included chromium to prevent excessive catalyst reduction.[118] Phillips described tin oxide catalysts that were promoted with 4% phosphate and 1% lithium as well as the bismuth/phosphate catalysts that were promoted with boron or lithium compounds.

Several units using these processes were built during the late 1960s but closed down before 1980 when butadiene became available as a byproduct from the steam cracking of hydrocarbons.

REFERENCES

1. F. Kuhlmann, *Annalen.* **29** (1839) 281; French Patent 11331 (1839).
2. F. Kuhlmann, French Patent 11332 (1839).
3. T. J. Smith and T. du Motay, British Patent 491 (1871).
4. Beyer & Company, British Patent 18594 (1903).
5. BASF British Patent 13848 (1914).
6. BASF Norwegian Patent 26691 (1916); US Patent 1211394 (1916).
7. Scott, *Ind. Eng. Chem.* **16** (1924) 74; Scott and Leech **19** (1927) 170; ICI, *Chem. Eng. News.* **48**, May 25 (1970); C. and I. Girdler, *Chem. Eng.* **77**, No. 14 (1970) 24.
8. F. Beyer, *Chem. Met. Eng.* **24** (1921) 305, 347.
9. J. Zawadski, *Disc. Faraday Soc.* **8** (1950)140.
10. W. Ostwald, British Patents 698 (1902), 8300 (1902), 7909 (1908); French Patent 317544 (1902); US Patent 858904 (1902).
11. W. S. Landis, *Chem. Met. Eng.* **20** (1919) 471.
12. K. Kaiser, *Chem. Zeit*, 14 (1916); German Patent 271517 (1910); British Patent 20325 (1910); US Patent 987375 (1911); Partington, *J. Soc. Chem. Ind.* **37** (1918) 337R.
13. N. Caro and A. Frank; German Patents 286991, 303822, 304269 (1914); Schuphans *Chem. Met. Eng.* **14** (1916) 425.
14. J. R. Partington, *J. Soc. Chem. Ind.* **40** (1921) 185R.
15. E. J. Pranke, *Chem. Met. Eng.* **19** (1918) 396; Fairlee, *Chem. Met. Eng.* **20** (1918) 6.
16. Parsons and Jones, US Patent 1321376; Parsons, *J. Ind. Eng. Chem.* **11** (1919) 541.

17. S. L. Handforth and J. N. Tilley, British Patent 306382 (1928); *Ind. Eng. Chem.* **26** (1934) 1287.
18. A. E. Heywood, *Plat. Met. Rev.* **17** (1973) 118.
19. N. H. Harboard, *Plat. Met. Rev.* **18** (1974) 97.
20. G. R. Gillespie and R. E. Kenson, *Chem. Tech.* (1971) 627; British Patents 1347491, 1471327.
21. A. E. Heywood, *Plat. Met. Rev.* **26** (1982) 28; Connor, *Plat. Met. Rev.* **11** (1967) 60; Holzmann, *Plat. Met. Rev.* **13** (1969) 2.
22. Hofmann, *Berichte* **11** (1878) 1685.
23. A. Trillat, *Oxydation des Alcohols*, (1901); *Bull. Soc. Chem.* **27** (3) (1902) 797; **29** (1903) 35; French Patent 199919 (1889); German Patent 55176 (1889).
24. E. I. Orlov, *Formaldehyde*, Barth, Leipzig (1909).
25. Le Blanc and Plaschke *Zeit Electrochem.* **17** (1911) 45.
26. Bouliard, French Patent 415501 (1910).
27. Le Blanc German Patent 228697 (1910); French Patent 418349 (1910).
28. Morel, *J. Pharm. Chem.* **21** (1905) 177.
29. S. J. Green, *Industrial Catalysis*, Benn, London, 1928, p. 385.
30. V. E. Meharg and H Adkins, US Patent 1913405 (1923); H. Adkins and W. R. Peterson, *J. Am. Chem. Soc.* **53** (1931) 1512.
31. R. N. Hader, R. D. Wallace, and R. W. McKinney (du Pont), *Ind. Eng. Chem.* **44** (1952) 1508.
32. L. F. Marec and D. A. Hahn, *Catalytic Oxidation of Organic Compounds in the Vapor Phase*, Chem. Catalogue Co, New York, 1932.
33. J. F. Walker, *Formaldehyde*, Reinhold, New York (1st Ed.), 1953; (3rd Edition) 1964.
34. I. E. Wachs and R. J. Madix, *J. Catal.* **53** (1978) 208.
35. J. F. Le Page, *Applied Heterogeneous Catalysis*, Editions Technip, Paris, 1987, p. 311.
36. L. Andrussov, German Patent 549055 (1932); US Patent 1934838 (1933); *Angew. Chem.* **48** (1935) 593.
37. L. Andrussov, *Angew. Chem.* **39** (1926) 321; **40** (1927) 166; **41** (1928) 205, 262; **48** (1935) 593; *Berichte* **59** (1926) 458; **60** (1927) 536, 2005; **71** (1938) 76.
38. L. Andrussov, *Chem. Ing-Tech.* **27** (1955) 469; *Berichte* **60** (1927) 2005; *Bull. Soc. Chim.*, (1951) 45–50.
39. L. Andrussov, *Chem. Ing-Tech.* **25** (1953) 697; *Plat. Me.t Rev.* **22** (1978) 131.
40. A. B. Stiles, US Patent 2726931 (1955); D. R. Merrill and W. A. Perry, US Patent 2478875 (1948).
41. US Patent 2831752 (1958); British Patent 785657 (1957).
42. H. Davy and Erdman, *Phil. Trans.* **107** (1817) 77.
43. Fletcher, *J. Gas Light* **1** (1887) 168.

44. W. A. Bone, *J. Roy. Soc. Arts* **62** (1914) 787, 801, 818.
45. S. J. Green, *Industrial Catalysts*, p. 113, Benn, London, 1928.
46. A. B. Lamb, W. C. Bray and J. C. W. Frazer, *Ind. Eng. Chem.* **12** (1920) 213.
47. W. A. Whitesell and J. C. W. Frazer, *J. Am. Chem. Soc.* **45** (1923) 2841.
48. M. Katz, *Advances in Catalysis*, Vol. 5, Academic, New York. 1953, p. 177.
49. *Disc. Faraday Soc.* **8** (1950) 215.
50. J. Mars and D. W. van Krevelan, *Chem. Eng. Sci.* **3** (1954) 41.
51. H. G. Weiss and C. R. Downs, *Ind. Eng. Chem.* **12** (1920) 228; **15** (1923) 965.
52. K. Weissermel and H. J. Arpe, *Industrial Organic Chemistry,* (2nd Edition) VCH, Weinheim, 1993.
53. Gibbs and Condover; US Patent 1284887-8; 1285117; 1288431; 1303168 (1918/19); British Patents 119517-8; 14150-1 (1917).
54. Seldom Company, British Patent 170022 (1920).
55. A. Wohl, British Patent 145071 (1920).
56. H. G. Weiss and C. R. Downs, US Patents 1374965; 11374720-2; 1377534 (1921).
57. A. E. Craver, US Patent 1489741 (1924).
58. Gibbs and Condover, *Ind. Eng. Chem.* **11** (1919) 1031; **14** (1922) 120.
59. Gibbs, US Patent 1458478 (1923).
60. S. J. Green, *Industrial Catalysts*, Benn, London, 1928, p. 122.
61. Kusama, *J. Chem. Soc. Jap.,* **44** (1923) 605.
62. C. R. Downs, US Patents 1589632; 1604739 (1926).
63. *BIOS* Report 936, p 6.
64. W. O. Fugate, US Patent 2698330 (1954); British Patent 702616 (1954).
65. G. C. Bond, *J. Catal.* **116** (1989) 531.
66. P. Schoen and N. V. Zoon, Dutch Patent 64720 (1949).
67. G. Cavani, F. Centi, Parrinello, and F. Trefiro, *Preparation of Catalysts IV*, Ed. by B. Delmon, P. Grange, P. A. Jacobs and G. Poncelet, Elsevier, Amsterdam, 1987, p. 227; *Chem. and Eng. News*, April 10 (1995) 37.
68. DuPont, US Patent 2605238 (1952).
69. R. J. Sampson in *Catalysis, Science and Technology*, Vol. 8, Ed. by Anderson and Boudart, Springer-Verlag, Berlin, 1987, p. 49.
70. *Hydrocarbon Processing*, Nov. (1980) 149.
71. Petrotex US Patents 3255211-3; 3288721 (1966); British Patent 1095223 (1967).
72. Distillers, US Patent 2649477.
73. Atlantic Refining Company, US Patent 2773838.
74. D. W. van Krevelan, thesis, Delft University, Excelsior, The Hague (1958).
75. G. J. Hutchings, A. Desmartin-Chomel, R. Oliev and J-C. Volta, *Nature,* **368,** March 3 (1999) 41.

76. R. J. Sampson in *Catalysis, Science and Technology*, Vol. 8, Ed. by Anderson and Boudart, Springer-Verlag, Berlin, 1987, p. 54.
77. Trevida and Culbertson, *Maleic Anhydride*, Plenum, New York, 1982.
78. H. Heller, G. Lenz and R. Thiel, *Inst. Chem. Eng. Symp. Ser.*, No. 50 (1977) 121.
79. *Hydrocarbon Processing*, Nov. (1977) 180; Nov. (1981) 180.
80. BASF, German Patent 1443452 (1970); German Offen 2030201 (1971); British Patent 1154148 (1969).
81. Mitsubishi Chemical Industries Ltd., *Chem. Eng. Econ. Reporter*, Oct. (1982) 25.
82. Badger, *Chem. Market Reporter*, Dec. 8 (1980) 3; Schaffel. et al., *Erdol. Kohle* **36** (1987) 85.
83. Sohio/UCB *ECN*, Sept. 20 (1982) 25; *Chem. Week,* Oct. 6 (1982) 31.
84. Alusuisse/Lummus Crest *ECN*, May 30 (1983) 22.
85. DuPont *Chem. Eng. News*, March (1989) 35.
86. DuPont *Chem. Eng. News*, April (1995) 20.
87. T. E. Lefort, French Patent 729952 (1931); US Patent 1998878 (1935).
88. Scientific Design, British Patents 711601; 721412.
89. Shell, British Patents 754493; 638319; van Oosten, *J. Inst. Petrol.,* **46** (1960) 347.
90. D. J. Hucknall, *Selective Oxidation of Hydrocarbons*, Academic, London, 1974, p. 10.
91. US Patent 2477435 (1949); British Patent 2043481 (1980).
92. P. A. Kilty and W. M. H. Sachtler, *Catal. Rev.—Sci. Eng.* **10** (1974) 1; S. Carra and P. Forzatti, *Catal. Rev.—Sci. Eng.* **15** (1977) 1.
93. R. A. van Santen and H. P. C. E. Kuipers, *Adv. Catal.* **35** (1987) 265; R. A. van Santen, *Proc. 9th Int. Conf. Catalysts*, Chemical Institute, Canada, 1988, p. 1152.
94. C. N. Satterfield, *Heterogeneous Catalysis in Industrial Practice* (2nd Edition), Krieger Malabar, 1996, p. 282.
95. McKim and Cambron, *Can. J. Res.* **B27** (11) (1949) 813.
96. G. H. Law and H. C. Chitwood, US Patent 2194602 (1940); US Patent 2279469 (1942).
97. M. A. Dalin, I. K. Kolchin and B. R.Serebryakov, *Acrylonitrile*, Technomic, Westport, Conn., 1971.
98. J. D. Idol US Patent 2904480 (1959); J. L. Callahan, J. J. Szabo and B. Gertuser, US Patent 3186955 (1966); British Patent 821999 (1958).
99. British Patent 908655 (1962).
100. J. D. Idol US Patent 2904580 (1959).
101. J. L. Callahan, R. W. Foreman, and F. Veatch, US Patent 3044966 (1962).
102. Sohio, US Patents 3198750 (1965); 3308151 (1969).
103. Nitto Chemical Industries, Co. Ltd., Japanese Patent 7103438 (1971); German Offen 1811063 (1969).

104. R. K. Grasselli and Burrington, *Adv. Catal.* **30** (1981) 133; R. A. Sneider and Hill, *Catal. Rev.—Sci. Eng.* **31** (1989) 43; R. K. Grasselli, Burrington, and Lartisak, *J. Catal.* **63** (1980) 239.

105. R. K. Grasselli, *Heterogeneous Catalysis: Selected American Histories*, Ed. by Davis and Hettinger, ACS Symposium Series, No. 222,1983, p. 317; *Appl. Catal. A: General* **136** (1996) 205.

106. P. H. Emmett, *Catalysis*, Vol 7, Reinhold, New York, 1960, p. 294.

107. J. M. Thomas and W. J. Thomas, *Principles and Practice of Heterogeneous Catalysis*, VCH, Weinheim, 1997, p. 345.

108. R. K. Grasselli and J. L. Callahan, *J. Catal.* **14** (1969) 93.

109. Y. Kim, W. Ueda, and Y. Moro-Oka, *New Developments in Selective Oxidation*, Elsevier, Amsterdam, 1990, p. 491; *Appl. Catal.* **70** (1991) 175, 189.

110. N. Harris, British Patent 1336136 (1973).

111. G. Centi, F. Trifiro, R. K. Grasselli, and E. Patane, *New Developments in Selective Oxidation*, Elsevier, Amsterdam, 1990, p.515; *Ind. Eng. Chem. Res.* **31** (1992) 107; *Catal. Today* **13** (1992) 661.

112. M. Bowker, C. R. Bicknell, and P. Kirwin, *Appl. Catal. A: General* **136**, (1996) 205.

113. B. P. Amoco, *Chem. Eng. News*, Sept. 23 (1996) 18.

114. German Offen 2056614 (1972).

115. K. W. Furman and G. W. Hearne, US Patent 2991320 (1961).

116. *Hydrocarbon Processing*, Nov. (1978) 131; R. J. Rennard and W. L. Kehl, *J. Catal.* **21** (1971) 282; Massoth and Scarpiello, *J. Catal.* **21** (1971) 294.

117. Petrotex, US Patents 3607966 (1971); 3666687 (1972).

118. Phillips Petroleum, US Patents 3580969 (1971); 3686346 (1972); 3501547 (1970).

5

CATALYTIC CRACKING
CATALYSTS

5.1. INTRODUCTION

Catalytic cracking is one of the most important processes in a modern refinery. It is the most economic way to convert low-value crude oil fractions into more valuable products and it has been described not only as the heart of the refinery but also as the garbage can![1] Although the process was originally developed as a gasoline producer it also supplies large volumes of gaseous hydrocarbons that are used in alkylation plants and as petrochemical feedstock. Furthermore, domestic fuel oil is an important by-product.

The complexity of catalytic cracking units and the number of catalysts used has significantly increased in the 60 years since the first practical process was introduced by Eugene Houdry in 1936. Not only did the original *fixed* catalyst beds give way to *moving* beds but, more significantly, the development of *fluidized* beds and active zeolite catalysts led to greatly improved process designs with higher production capacity.

The first feed to be used in catalytic cracking units was virgin gas oil. However, from the 1970s on, cheaper residual fractions are also used as the cost of crude oil increased. Demand for higher octane ratings, particularly as lead-free gasoline was introduced, led to improvements in catalyst formulation. Later, when residual fractions were added to catalytic cracker feeds, more active catalyst matrices were needed together with additives to absorb poisons and control sulfur emissions.

L. Lloyd, *Handbook of Industrial Catalysts*, Fundamental and Applied Catalysis,
DOI 10.1007/978-0-387-49962-8_5, © Springer Science+Business Media, LLC 2011

5.2. PROCESS DEVELOPMENT

The Houdry catalytic cracking process began semicommercial operation at the Sun Oil Company, Marcus Hook, Pennsylvania, refinery in 1936 with a capacity of 2000 barrels per day (bpd). The first large-scale unit soon followed at the same location in 1937 with a capacity of 15,000 bpd of heavy gas oil.

5.2.1. Fixed Beds

The main problem with this process is that the catalyst rapidly becomes deactivated due to the deposition of coke, and therefore it needs to be regenerated in a separate stage. To achieve continuous operation, three fixed-bed, tube-cooled reactors each containing a clay catalyst were used. One reactor was operated for up to 10 min until the catalyst was deactivated by coke. At the same time the deactivated catalyst in the second reactor was regenerated with air. The catalyst in the third reactor had already been regenerated and so it was ready for further use. The cycle time was typically 15 min, depending primarily on the rate of carbon deposition. The cracking reaction is endothermic and requires an input of heat. The catalyst bed was heated to about 450^0C by passing a molten eutectic mixture of sodium nitrite and potassium nitrate through the cooling tubes. The oxidative regenerative procedure is highly exothermic and the catalyst bed needs to be cooled to about 500^0C before use. This was achieved by passing the same, cooled eutectic mixture through the cooling tubes of the catalyst bed. Thus, the heat generated during the regeneration procedure was used to supply the heat required for the cracking reaction. The original clay catalysts had a life of up to 18 months before they were permanently deactivated and replaced.

Houdry catalytic cracking units (Figure 5.1) produced better-quality gasoline of higher octane rating, with fewer unsaturated compounds, than thermal cracking units. They were an immediate success and made an important contribution to the high-octane aviation gasoline requirements during World War II. By 1944 US capacity had grown to 24 units processing 330,000 bpd of feed. A synthetic silica/alumina catalyst was developed by Houdry and first used in July 1940. The new catalyst was superior to clays because it had a uniform chemical composition and formed less coke. Although marginally less gasoline was produced at a given conversion, it was of better quality with a higher octane rating.

The fixed bed design was soon replaced by more convenient processes with continuous circulation of the catalyst from the reactor to the regenerator and then back to the reactor. The new crackers had the advantage of using smaller vessels with less heat loss. They were also more flexible to operate because the catalyst itself acted as the heat transfer medium.

Figure 5.1. Early Houdry catalytic cracking unit at Marcus Hook, Pennsylvania.

5.2.2. Moving and Fluidized Beds

Fluid catalytic cracking (FCC), introduced in May 1942 by the Standard Oil (NJ) Company, used small particles of silica/alumina catalyst to convert vaporized gas oil feed into lighter products. The process design required fluidized catalyst to flow from the reactor to a regenerator and back again in a continuous stream. This was possible because the catalyst in both the reactor and the regenerator formed well-mixed fluidized phases, like a frothing liquid. This *fluidisation* was achieved by passing a stream of vapour, hydrocarbon during reaction and air during regeneration, upwards through the fine catalyst particles so that the particles were suspended on a cushion of vapour.

Thermofor catalytic cracking (TCC) introduced by Mobil in 1943, fluid catalytic cracking (FCC) introduced by Exxon, and several other similar processes used moving or fluidized beds of strong catalyst particles. Catalyst was withdrawn continuously from the bottom of the reactor and lifted in buckets or by an air stream to the top of a regenerator, or kiln, after the residual hydrocarbons had been stripped out with steam. Catalyst was then returned to the reactor after regeneration. There was a limit to the capacity of moving bed processes

Figure 5.2. FCC units 1, 2, and 3 of Standard Oil at Baton Rouge, LA.

which depended on the rate at which the catalyst and the feed could be transferred. Circulation rates were increased to a certain extent by using the air lifts rather than buckets to move the catalyst but, eventually, as larger-capacity cracking units were required, the process became obsolete. Existing units, however, were still operating even into the 1990s. Both processes were being widely used by the end of World War II with a total capacity of more than 500,000 bpd of feed.

Production was significantly increased and only minor changes required in the processing equipment when the more active zeolite catalysts were introduced between 1962 and 1964. Subsequently, modified catalysts with various additives were developed. These were required to accommodate a wider range of feeds, which included some residual fractions with high molecular weight hydrocarbons and higher levels of impurities, and to meet stricter environmental controls.

Although FCC unit designs were extensively modified following the introduction of zeolite catalysts, the basic flow sheet remained the same. In fact most of the existing units were simply revamped to use the new catalyst and to increase production. The second and third FCC units (Figure 5.2) which started at Baton Rouge in June 1942 with a capacity of 17,000 bpd, were still in operation more than 50 years later with a combined up-rated capacity of 188,000 bpd.[2]

During production the vaporized, high-boiling feed is cracked in the fluidized bed of catalyst to produce a mixture of lighter hydrocarbons. The catalyst is quickly deactivated by the deposition of coke. The catalyst is then separated from products in a stripping section and transferred to a regenerator, where coke is burned in a stream of air. The regenerated catalyst then leaves the regenerator, is mixed with fresh feed and recirculated. Modern plants operate with *riser-only-cracking* and reaction takes place in the original transfer line between the regenerator and the old reactor. The original reactor serves only as a *disengager* to separate the catalyst from the products. This allows more efficient operation with the active zeolite catalyst and minimizes the catalyst volume, or inventory, needed in modern FCC units.

Any catalyst dust formed by attrition and lost in the cyclones has to be replaced at regular intervals. It is also necessary to replace a small proportion of the circulating *equilibrium* catalyst to compensate for gradual permanent deactivation and maintain conversion. About 3% of fresh catalyst is added to a unit on a daily basis to maintain the necessary catalyst inventory. The whole operation is continuous and a unit may be operated for several years without shut down. An important feature of the process is that heat transferred from the regenerator to the reactor by the hot catalyst as a heat transfer agent is an integral part of the energy balance.

The clay-based catalysts used in the early cracking units were of low activity and of poor thermal and structural stability. High recycle rates of uncracked feed and the severe coke deposition at low-space-velocity operation limited output. Regeneration temperatures were limited to below $600^{0}C$ not only because of metallurgical restrictions but also to avoid catalyst deactivation. This meant that the volume of regeneration air was restricted and regenerated catalyst still contained 0.6% coke. The flue gas was a mixture of carbon dioxide with some carbon monoxide because of incomplete carbon combustion. When catalyst fluidization in the regenerator was not uniform, after-burn was a regular problem as carbon monoxide reacted with oxygen, causing possible ignition and leading to excessive temperatures. To minimize catalyst damage during these temperature runaways, coolers and water sprays needed to be installed. Catalyst coolers were used to control the regenerator temperature, and this allowed additional air to be passed into the regenerator which, in turn, resulted in lower levels of residual coke in the regenerated catalyst. Lower levels of carbon in the catalyst led to higher conversions in the reactor, the use of increased feed rates and hence greater production capacity. Figure 5.3 shows an Orthoflow Resid FCC converter.

When higher-activity catalysts consisting of zeolites incorporated in a silica-alumina matrix were introduced, first to TCC units in 1962 and then to FCC

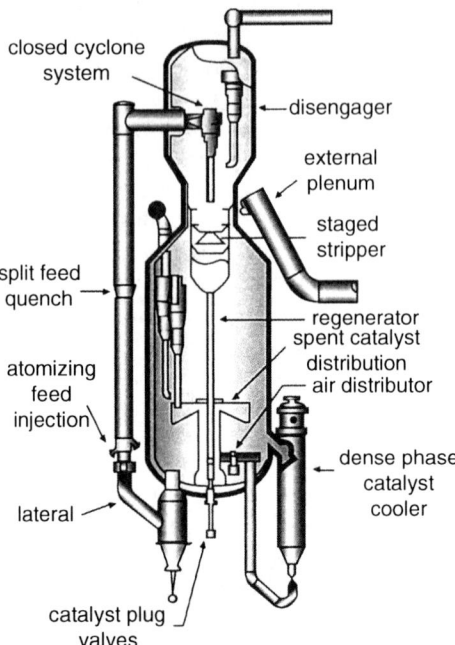

Figure 5.3. Outline flow sheet of FCC plant.

units in 1964, it became possible to make a number of further process improvements:

- Increased catalyst activity and higher conversion meant that the volume of recycled feed was significantly decreased.
- This led to increased production by using more fresh feed to maintain the same reactor throughput.
- At the same time more cracking reactions took place in the riser, so that eventually the reactor was redundant.
- Unit revamps to increase throughput were possible.
- Air flow to the regenerator was decreased because less coke formed on the catalyst and regeneration temperature could be increased to 700–750^{0}C, because of higher thermal stability of the catalyst. This gave more complete combustion to carbon dioxide.

Some of the more important developments in catalytic cracking catalysts and the additives used are shown in Table 5.1.

TABLE 5.1. Developments in Catalytic Cracking Catalysts.

Year	Catalyst development
1936	First use of FCC clay catalysts in Houdry plants.
1940	Synthetic silica/alumina powder catalyst.
1942	Fluid bed catalyst in Standard Oil PCLA No 1.
1943	Moving bed bead catalysts used in Mobil Thermofor process.
1948	Spray-dried microspheroidal catalyst introduced by Davison.
1955	High-alumina silica/alumina microspheroidal catalyst.
1959	Semisynthetic silica/alumina plus kaolin.
1962	X-and Y-zeolites with matrix beads introduced by Mobil. Soon followed by spray-dried microspheres in 1964.
1964	First development of ultrastable Y-zeolite.
1964	Rare earth exchanged zeolite catalysts.
1973	Silica sol binders used in high-zeolite catalysts by Davison.
1975	Octane catalysts developed by Davison.
1981	Alumina sol binders.
1986	Residue catalysts with high-cracking activity matrix.
1990+	Nickel and vanadium passivation by catalyst matrix.
	Additives used in FCC Units:
1976	Platinum catalyst for carbon monoxide combustion in regenerator.
1976	Nickel passivation with antimony compound.
1982	Sulfur oxide transfer spinel additive.
1982	Vanadium traps introduced.
1983	Nickel passivation with bismuth compound.
1983	Shape-selective cracking ZSM-5 *octane* additive.
1980s	Coke-selective *deep bottoms* cracking additive.

5.2.3. Catalyst Regeneration and Carbon Monoxide Combustion

5.2.3.1. *Catalyst Regeneration*

Catalyst regeneration is an important part of the FCC process. It removes coke from deactivated catalyst and provides heat to maintain operation. As catalyst circulates through the regenerator, coke burns in air and the catalyst is regenerated for further use. Hot regenerated catalyst carries the heat around the unit to vaporize feed and maintain temperature in the riser during the endothermic cracking reaction. Heat from the regenerator is also used in other ways. Steam is generated to strip hydrocarbons from catalyst returning to the regenerator and heat recovered from flue gas is used to preheat combustion air.

About 70% of the combustion heat is absorbed by the catalyst during regeneration, with the remainder leaving the regenerator in flue gas or as heat loss. The endothermic cracking reaction absorbs 10–25% of the heat circulated by the catalyst while a further 70–80% is needed to heat the feed to the reaction temperature. About 5% of the heat is lost or recovered for other purposes. The heat balance in the FCC unit is critical to the economy of its operation.

Typical coke is a mixture of high molecular weight, hydrogen-deficient hydrocarbons, containing about 6–7% hydrogen. Any variation in the hydrogen content affects the heat of combustion and the heat balance. The ratio of "regeneration air" to the amount of hydrocarbon/coke on the spent catalyst controls the combustion reaction and this influences both the final temperature and the ratio of carbon dioxide to carbon monoxide in the exhaust gas. When the stripping stage is inefficient and more hydrocarbon passes with the catalyst into the regenerator bed, the temperature rises and the flue gas composition changes.

Early FCC units were made with carbon steel internals and unstable catalysts, which restricted the regenerator temperature to less than 600°C and left a substantial volume of carbon monoxide in flue gas. This was not only toxic but led to operating difficulties. After-burning occurred when air did not mix properly with the catalyst, leading to hot spots in the regenerator. It was then necessary to install steam injectors to control bed temperature and "boilers" in the flue gas line to burn residual carbon monoxide. Two-stage regenerators were also designed to avoid catalyst damage. More stable catalysts allowed an increase of regeneration temperature to 650°C. Finally, when stainless steel internals were introduced, regenerator temperature were increased to about 750°C. This allowed the addition of more air which resulted in almost complete combustion.

5.2.3.2. Carbon Monoxide Combustion Promoter

Early attempts by Mobil to minimize after-burning in TCC units led to the addition of chromium oxide to their Durabead catalyst to oxidize carbon monoxide, but this unfortunately also decreased cracking selectivity. Mobil then introduced a platinum/alumina additive in 1976 to control carbon monoxide combustion in the regenerator.[3] Platinum was added either as a component of the cracking catalyst or in separate particles. Complete combustion of carbon monoxide was achieved by adding the equivalent of 0.5 ppm of platinum to the catalyst inventory.

The use of platinum additives provides efficient heat transfer in the dense phase and controls temperature runaways in the dilute phase, even when excess oxygen is used. Either complete or partial carbon monoxide combustion is now possible, depending on the unit requirements, simply by controlling the air rate to the regenerator and using the additive. Improved regenerator operation at higher temperatures gives less residual coke on regenerated catalyst, which improves activity in the riser. Improved heat transfer in the regenerator with fewer hot spots lessens hydrothermal catalyst deactivation.

5.2.4. Equilibrium Catalyst

During operation of an FCC unit several factors influence catalyst performance. Catalyst activity declines rapidly. This is mainly the temporary effect of coke formation and this activity loss can be restored by regeneration. Permanent deactivation also takes place as thermal and hydrothermal sintering of the zeolite leads to dealumination of the crystal structure. Metal impurities in the feed also affect performance and deactivate the zeolite. Although the zeolite deactivates quickly the catalyst matrix generally retains its activity for longer periods. The physical movement of fluidized catalyst between the reactor and the regenerator causes loss of the catalyst by attrition to dust, which leaves the system through a series of cyclones.

To compensate for deactivation and poisoning the low-activity catalyst is regularly withdrawn from the circulating inventory, which is replaced with fresh catalyst. Replacement is usually about 1–3% of the total inventory every day. This is very substantial. A replacement rate of 3% a day, an FCC unit with an inventory of 200 tons of catalyst, more than 2000 tons of catalyst are replaced every year. The replacement rate is based on the need to maintain a constant operating activity. The catalyst inventory has a significant age distribution. Generally, about half of the total catalyst is less than 3 weeks old and accounts for more than 70% of the total activity.

FCC units operate continuously and are hardly ever closed down to change the catalyst inventory completely because of the lost production this would involve. When it is necessary to use a different catalyst type, it would take a long time to remove the old catalyst at the small makeup rate needed to maintain activity and to replace physical losses. At 3% replacement per day it takes eight weeks to change 80% of the inventory. However, because the freshest catalyst contributes most to conversion and overall yield, almost the full effects of a catalyst change are noticeable after 50–60% replacement. Catalyst producers provide information that enables operators to estimate the time taken for changing catalysts to take effect.[4]

It is usual to check the properties of equilibrium catalyst (E Cat) to maintain the replacement rate at the optimum level and to review any potential process problems.[5] The following tests are typical:

- Micro-activity tests (MAT) are used to measure activity at a constant level and determine the appropriate catalyst makeup rate.
- Catalyst surface area measurements of the zeolite and matrix components help to analyze deactivation mechanisms and provide a rapid assessment of activity.
- Chemical analyses of known catalyst poisons, such as vanadium, sodium, and nickel, also allow control of catalyst makeup rates to maintain activity.

- Equilibrium catalyst attrition index and average particle size distribution (APS) indicate changes in the rate of catalyst attrition. Further analysis of APS for any catalyst that is carried forward into the fractionator, present in the slurry, or which leaves the unit via the regenerator stack can identify problems associated with catalyst quality or cyclone operation. Problems include operation at greater than design feed, catalyst rates or cyclone maloperation. APS is also important in predicting the fluidization properties of the catalyst inventory.

As well as the need for routine analyses of equilibrium catalyst, regular checks on all batches of fresh catalyst are carried out to check the consistency of particle size, the attrition index, and the activity of the catalyst added to a unit. Excess steam deactivates the catalyst and causes abnormal attrition. High air velocity or maldistribution of air in the regenerator increases catalyst attrition and leads to variations in the carbon content of the equilibrium catalyst.

Troubleshooting is extremely important in maintaining optimum operation of FCC units. The correlation of trends in equilibrium catalyst properties with operating data can quickly identify potential problems. These can often be confirmed by a number of useful nondestructive tests. For example, radioisotope tracer experiments can measure vapor velocities and catalyst flow patterns throughout the unit.[6] It is possible to show catalyst distribution within the riser, the regenerator, and the cyclones as well as the stripper. Cracks in internal cyclones or blockages in the stripper can been quickly identified. Not surprisingly, maldistribution of catalyst and vapor has often been confirmed in risers and regenerators.

5.2.5. Reaction Mechanism of Catalytic Cracking Reactions

Catalytic cracking proceeds via a carbenium ion mechanism that provides a higher yield of more useful products than thermal cracking reactions. These products include more hydrocarbons in the gasoline boiling range, with high proportions of branched paraffins, olefins and aromatics to increase octane numbers. Carbenium ions can form in a number of ways, but the main initiation processes are acid catalyzed by both Brønsted and Lewis acid sites. This involves either the removal of a hydride ion from a saturated hydrocarbon (Lewis site) or the addition of a proton to an olefin or aromatic nucleus (Brønsted site). It has also been suggested that at high temperatures the addition of a proton to a paraffin can form a pentacoordinated carboniun ion, which undergoes β-scission or loses hydrogen to form a carbenium ion. Carbenium ions are extremely reactive and have short lifetimes but take part in all of the catalytic cracking reactions.

The typical feeds to catalytic cracking units shown in Table 5.2 are mixtures of paraffins, naphthenes, alkyl chain substituted aromatics, and more complicated molecules. These undergo the series of complex cracking reactions, includ-

TABLE 5.2. Typical FCC Unit Feed and Operating Conditions.

Feedstock	Operation	
	Vacuum gas oil	Atmospheric residue
API gravity	25.5	22.4
Sulfur (wt%)	0.7	0.8
Nickel (ppm)	0.4	3
Vanadium (ppm)	0.6	3.5
Conradson carbon (wt%)	0.2	4.0
Conversion (vol%)	86	76
Fuel gas (wt%)	4.4	4.6
Total C_3 (vol%)	13.9	10.4
Total C_4 (vol%)	18.6	13.2
C_5+ gasoline (vol%)	62.6	57.4
LCO (vol%)	6.8	11.6
Slurry (vol%)	6.9	11.9
Coke (wt%)	5.6	7.5
Reactor temperature (^{0}C)	535	535
Regenerator temperature (^{0}C)	720	720

ding isomerization, carbon–carbon bond β-scission and hydrogen transfer, which are summarized in Table 5.3. Coke is formed from those hydrocarbons which do not readily crack and in modern units, coke contains about 6–7% hydrogen. It can also contain significant amounts of sulfur.

Table 5.3 Catalytic Cracking Reactions.

Hydrocarbon	Initial products	Further products
Paraffins	Branched paraffins and olefins mainly in C_3–C_{10} range.	Olefins crack and isomerize and are also saturated by hydrogen transfer to give paraffins. Olefins also cyclize to naphthenes.
Naphthenes	Crack to olefins. Dehydrogenate to cyclic olefins. Isomerize to smaller rings.	Further dehydrogenation to aromatics, by hydrogen transfer.
Aromatics	Alkyl groups crack at ring to form olefins. Dehydrogenation and condensation to polyaromatics.	Further dehydrogenation and condensation forms coke.
Typical products (approximate)	Light gas (3%) LPG (17%) Naphtha (52%) LCO (16%) HCO (5%) Coke (5%)	H_2, CH_4, C_2H_6, C_2H_4 C_3H_6, C_3H_8, C_4H_8, C_4H_{10} Light 40^0–110^0C/Heavy 110^0–220^0C Jet fuel 220^0–340^0C, Kerosene, diesel, heating oil Recycle. Higher than 340^0C.

Note: During typical operation to produce gasoline from gas oils the hydrocarbons that can enter the zeolite structure crack into smaller molecules ranging from C_5 up to a boiling point of about 110^0C. *Dry gas*, i.e., C_2 and lower, forms from thermal cracking and secondary reactions. Light-cycle oil results from matrix cracking. Heavy-cycle oil is recycled.

5.3. CATALYST DEVELOPMENT

As soon as the automobile industry became established, it was recognized that straight-run gasoline available from refineries could not satisfy potential demand. Statistics shows that by the 1920's, the demand from an increasing number of automobiles could only be met by the use of thermal cracking processes.[7] Attempts were soon started to develop more efficient and economic catalytic cracking processes. Sabatier had tested various metal oxides as catalysts to crack petroleum fractions; subsequently several patents were issued in Germany for processes based on clay catalysts, which were not successful. The McAfee process, which was developed by Gulf and operated from about 1915, used an aluminum chloride cracking catalyst. Despite petroleum yields of 35–48%, the process was uneconomic by 1929 and not widely used, mainly because the catalyst could not be recycled and more efficient thermal cracking processes had been developed.[8]

The major problem in developing a reliable catalytic process for cracking gas oils was the rapid deactivation of catalysts by coke deposition. It was not until Eugene Houdry began his work around 1927 that there was any significant progress. Houdry showed that cracked gasoline was better than thermal gasoline, and he was able to remove the carbonaceous residues from his catalyst by regeneration in a stream of air. More significantly, however, he demonstrated that certain clays were both active and economic catalysts because they retained activity during regeneration.

After large-scale pilot plant testing, in cooperation with the Vacuum Oil Company from 1931 to 1933 and Sun Oil from 1933 to 1937, the first full-scale catalytic cracking unit began operation in 1937.

The first clays selected for testing by Houdry were the acid treated materials originally used as adsorbents to purify lubricating oils. The most active clay was supplied by the Pechelbron Oil Refining Company of San Diego. Bentonite clays were being tested by 1933 and the Filtrol Company supplied catalyst pellets for the first large-scale unit in 1937. It is significant that synthetic silica/alumina compositions, similar to the natural products but of different chemical structures, have continued to be the most successful catalysts, although the chemical and physical properties have been considerably developed. Clay catalysts gave variable performance, which was improved by the use of synthetic silica/alumina powders in Houdry plants from about 1940. These catalysts contained no metal impurities and produced better-quality gasoline with increased octane numbers, and more light gas and less coke were produced. While the fluid bed catalyst used by Standard Oil in its development work was based on acid-treated clays, a more suitable silica/alumina catalyst was developed by Davison for the full-scale plant.

Further improvements continued and spray-dried *microspheroidal* particles of silica/alumina were introduced in 1948. These gave better activity and selectivity, together with more stable performance. Also, the more regular shape improved fluidization properties and decreased attrition losses. A strong, active high-alumina content silica/alumina catalyst was introduced in 1955. This was more active and stable than earlier catalysts with the same composition and provided more resistance to the poisoning effects of metal impurities present in feed.

5.3.1. Natural Clay Catalysts

The cracking catalysts used in the Houdry fixed bed process were based on the commercially available clays used in pilot plant tests. Acid-treated bentonites were found to have an acceptable activity and could be easily regenerated. Bentonite clays are formed from volcanic ash and contain up to 90% of the mineral montmorillonite. Montmorillonite has a three-sheet lattice structure consisting of a central layer of alumina octahedra sandwiched between two layers of silica tetrahedra. About one in six aluminum atoms is substituted isomorphously with a magnesium atom. All metal atoms are linked, within and between the layers by oxygen atoms. Some oxygen atoms in each of the alumina octahedra also form hydroxyl groups.

This structure has a negative lattice charge for every magnesium atom that has replaced an aluminum atom, and the mineral has base exchange properties. Iron also replaces some of the aluminum atoms in the lattice.

When the bentonite clays are treated with acids, up to 80% of the aluminum can be extracted from the montmorillonite lattice together with most of the magnesium and iron. No silica is dissolved during the extraction process, but it is probable that some may be peptized to form an active amorphous phase with alumina. This increases the surface area and pore volume of the catalyst. Typical analyses of commercial catalysts shown in Table 5.4 indicate that sulfuric acid was a common activating agent.[9]

Impurities affect the performance of cracking catalysts. For example, when iron atoms are sulfided they are displaced from the montmorillonite lattice. Iron sulfide is oxidized during regeneration and subsequently catalyses the dehydrogenation of hydrocarbons in the reactor to form gas and coke. Bentonite from some particular locations could not be used as catalysts because of high iron content.

Activated kaolinite and halloysite/endellite clays were also used as cracking catalysts. The double-layer lattice of kaolinite consists of alternating tetrahedral silica and octahedral alumina layers and the halloysite/endellite structures have interlamellar water layers. There is very little cation exchange with natural clays of this type but after heating at 600^0C to dehydrate and destroy the lattice, alumina can be extracted with acid to give catalysts comparable to activated montmorillonite. Kaolin-based cracking catalysts were therefore used to replace montmorillonite types, mainly because they did not suffer from the deactivating

TABLE 5.4. Composition and Properties of Some FCC Catalysts, 1947–1967.

	Super Filtrol[a]	Low alumina	High alumina	Semisynthetic
Composition (wt%):				
SiO_2	66.6	Balance	Balance	Prepared as blend of
Al_2O_3	15.4	>12.5[c]	>24.5[c]	silica/alumina base
MgO	4.3	—	—	with activated kaolin
Fe_2O_3	2.3	0.05[c]	0.05[c]	$[Al_2Si_2O_5(OH)_4]$. Low-
CaO	2.2	—	—	er activity but better
TiO_2	0.4	—	—	resistance to metal poi-
SO_3	3.0[b]	0.5	0.5	soning.
Ignition loss	3.8	(~12)	(~12)	
Average particle size (μm)		55–65	55–65	55–65
Bulk density (kg liter^{-1})		0.43–0.47	0.45–0.50	0.48–0.52
Surface area (m^2 g^{-1})		480–550[d]	400–460	250–300
Pore volume (ml g^{-1})		0.6–0.95	0.6–0.95	—
Pore diameter (nm)		7[d]	8	9.5

[a]Produced from acid extracted natural bentonite.
[b]From sulfuric acid extraction—present as anhydrite ($CaSO_4$).
[c]On loss free basis.
[d]Other catalysts up to 650 m^2 g^{-1} with 4-nm pore diameter.

effects of iron. Kaolin is still widely used as part of a catalyst matrix and, after further calcination, is even more important because it can be converted to Y-zeolite for in situ catalysts.[10]

5.3.2. Synthetic Silica Alumina Catalysts

The composition of acid-treated natural clay catalysts was more or less duplicated in the synthetic catalyst formulations. The main advantage of synthetic catalysts was a reproducible composition with few impurities known to cause deactivation. Both silica/alumina and silica/magnesia formulations were used in early tests but, despite good activity, silica/magnesia catalysts were unstable and difficult to regenerate.

The first fluid bed unit used a newly developed synthetic silica/alumina catalyst supplied by Davison (Figure 5.4) and performed more reliably than clay catalysts under the demanding operating conditions.[11] Although synthetic catalysts gave only a marginal improvement in conversion and selectivity, they were more stable and resistant to attrition than clays in fluid bed operation. Properties of silica/alumina catalysts are shown in Table 5.4.

5.3.3. Preparation of Synthetic Catalysts

Silica/alumina catalysts were carefully prepared by precipitating alumina from an aluminum salt at pH 7 onto a freshly prepared silica hydrogel. The precipitate

Figure 5.4. Grace FCC Catalyst : (a) zeolite synthesis; (b) spray dryer; (c) calciner; (d) finished catalyst; and (e) the end user, the FCCU. Photographs reprinted with permission from Grace Davidson Refining Technologies.

had to be carefully washed before being spray dried. It has been reported that about 15,000 gal (US) of water were required to produce 1 ton of finished catalyst.

Early silica/alumina catalysts contained 10–13% Al_2O_3, which, at the time, appeared to be optimum. Increased alumina content had little effect on activity and produced gasoline with a lower octane number, while forming more coke. Powdered silica/alumina catalysts were used until spray-dried microspheroidal particles were introduced in 1948.

Eventually, by improving the dispersion of alumina within the silica, increased activity was achieved with 25% alumina in the finished catalyst.[12] This formulation, with a larger pore volume, gave the same product distribution and coke yield but was stronger and more stable. High-alumina catalysts also provided better resistance to metals poisoning. It is interesting that a *semisynthetic* spray-dried silica/alumina catalyst containing a kaolin clay was also produced. Semisynthetics, which were stronger and more poison resistant than silica/aluminas, were the forerunners of the matrix used in zeolite catalysts.

5.4. ZEOLITE CATALYSTS

The new TCC catalysts containing zeolite introduced by Mobil in 1962 resulted in an immediate increase in the gasoline yield. Compared with amorphous silica/alumina catalysts, zeolites were much more active and formed less coke. The new high-activity catalysts were used in FCC units during 1964. Improved performance and high activity indicated significant potential for improvements in plant design and efficiency. The activity of pure zeolite was so high that only about 10% could be incorporated with a matrix in the new catalyst for use in existing units.[13]

Despite the dilution, high-activity zeolite catalysts achieved almost 100% *riser* cracking compared with only 15–20% with silica/alumina catalysts. This made it possible to redesign FCC units with full *riser* cracking and so avoid the usual overcracking in the *dense* phase of the reactor. Increased conversion and higher throughput led to a reduction in the volume of heavy cycle oil produced and recycle rates could be decreased. Thus, production capacity was further increased with little capital expenditure.

The driving force behind the development of zeolite catalysts by Mobil was said to be the additional potential profit of $1 million a year from the production of 1% extra gasoline. In fact, by 1968, it was estimated that refineries in the United States were saving $1 million a day through the use of zeolites and that the total capital savings, by expanding production from older plants and delaying capital expenditure, was $300 million![14] These remarkable statistics showed not only how important zeolite catalysts would be to the refining industry, but also emphasized the large scale of gasoline production by the FCC process.

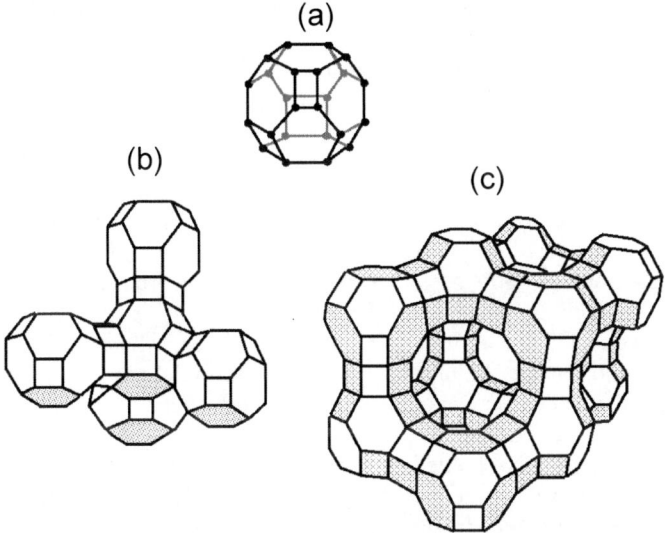

Figure 5.5. Zeolite Y structures showing silicon-silicon and silicon-aluminium interconnections only : (a) single sodalite cage ; (b) tetrahedral interconnection of sodalite cages ; and (c) extended structure showing the supercage.

5.4.1. Commercial Zeolites

Zeolites are crystalline silica/aluminas with a high surface area resulting from a regular pore structure of cavities connected by channels. Synthetic X-and Y-zeolites, which were first developed by Union Carbide and resemble naturally occurring faujasite, both have the same well-defined three-dimensional system of cavities and channels, despite having a different silicon/aluminum ratio.[15] The channels are large enough to allow most molecules in typical vacuum gas oils, such as isoparaffins, naphthenes, and aromatics, easy access. Both Y-zeolite and X-zeolite, which is cheaper to produce, were used in early cracking catalysts. However, only Y-zeolite is used in modern catalysts because it is more stable under reaction conditions.

The zeolite framework consists of SiO_4 and AlO_4 tetrahedra. These are joined, through bridging oxygen atoms, to give hollow, truncated octahedra known as *sodalite cages*, with alternating six-and four-sided faces formed from rings of tetrahedra. Sodalite cages are connected at four of the eight six-sided faces to form smaller hexagonal prisms. Within the resulting three-dimensional framework of tetrahedra, channels connect large cavities, which are known as *supercages*. Each supercage is surrounded by ten sodalite cages connected by hexagonal prisms (Figure 5.5). The dimensions and electronic properties of zeolite structures account for their use as cracking catalysts, and hydrocarbon conversion takes place within the supercages.

Sodium Y-zeolite forms cubic crystals, with each unit cell containing eight sodalite cages. The unit cell contains 192 silicon and aluminum atoms. The minimum silica/alumina ratio for fresh Y-zeolite is at least 3 (equivalent to 48 aluminum atoms in the unit cell). However, it can be much higher due to the loss of aluminum atoms or dealumination during preparation or regeneration. The bond length of the Al-O bond is 0.171 nm while that of the Si-O bond is only 0.164 nm. Thus, any increase in the silica/alumina ratio results in a shrinkage of the unit cell, as shown in Table 5.5. This can be calculated from X-ray diffraction measurements. The unit cell size (UCS) for sodium Y-zeolite with a silica/alumina ratio of 5 (32 aluminum atoms) is 2.46 nm, whereas if it could be completely dealuminated the unit cell would shrink to 2.42 nm. Exchange with rare earth ions, which will be explained later, has the effect of stretching the unit cell size at any silica/alumina ratio.

Each supercage in Y-zeolite has four circular openings, or ports, each 0.74 nm wide, which allow suitable size molecules, such as those shown in Table 5.6, to enter the supercage, which is 1.3 nm in diameter. The sodalite cages are too small to take part in the cracking reactions, but during exchange reactions they can accept rare earth ions, which thus affect the zeolite properties. Acid site density and the total acidity of the zeolite, which can be equivalent to that of a strong acid, is proportional to the number of negatively charged aluminum atoms in the framework. The strength of individual sites does, however, decrease with increasing site density. The silica/alumina ratio and the distribution of aluminum atoms in the sodalite cages can, therefore, affect catalyst performance. Lowenstein's rules, which are summarized in Table 5.7, states that no two adjacent tetrahedral sites can be occupied by aluminum atoms but otherwise distribution is random.

Each zeolite type has a typical silica/alumina ratio related to the crystal structure. It is possible, to increase the silica/alumina ratio however, by removing aluminum atoms from the Y-zeolite framework, with no effect on crystallinity. This, of course, modifies catalyst performance by changing the nature and

TABLE 5.5. Shrinkage of Zeolite Unit Cell as Silica/Alumina Ratio Increases.

SiO_2/Al_2O_3	Si/Al	UCS (nm)[a]
5	2–5	2.46
10 (≡HY)	5	2.445
20 (≡USY)[b]	10	2.435
40	20	2.425
Calculated for crystalline SiO_2		2.42

[a]Measured by X-ray diffraction: Si–O bond length 0.164 nm; Al–O bond length 0.171 nm.
[b]USY-zeolite contains mesopores (3–6 nm) in the zeolite formed during dealumination.

TABLE 5.6 Hydrocarbon Molecular Diameter.

Hydrocarbon	Size (nm)
n-Paraffins	0.45
Methyl paraffins	0.57
Dimethyl paraffins	0.63
Benzene	0.63
Toluene	0.63
Cyclohexane	0.65
1,2,4-trimethyl benzene	0.69
1,2,4,5-tetramethyl benzene	0.69
1,3,5-trimethyl benzene	0.78
Pentamethyl benzene	0.78

number of acid sites. Y-zeolites with silica/alumina ratios in the range 5–80 are available commercially.

Pentasil zeolites, such as ZSM-5, are often used with octane catalysts. They are different from Y-zeolite because the SiO_4 and AlO_4 tetrahedra form five-membered rings. The rings are linked in the framework to form two types of channels. Straight, elliptical channels, 0.51×0.58 nm, run parallel to the axis of the unit cell, and sinusoidal channels, 0.54×0.56 nm, connect at right angles to the straight channels. Strong acid sites are located where channels intersect and the diameter of the cavity formed is 0.9 nm. The silica/alumina ratio of ZSM-5 is variable in the range 10–1000, depending on the synthesis and subsequent treatment. ZSM-5 is commercially available with silica/alumina ratios in the range 30–300.

TABLE 5.7 Lowenstein's Rules and Y-Zeolite Acidity.

Crystalline Y-zeolite is formed from framework silicon and aluminum atoms in tetrahedral coordination with oxygen atoms and linked to other tetrahedra. Certain rules apply:

- Each aluminum atom has four silicon atoms as *nearest neighbors* (NN).
- None of the aluminum atoms is linked directly.
- The four silicon *nearest neighbors* are connected to a total of nine other silicon or aluminum *next nearest neighbors* (NNN), which are limited in number by the way sodalite and hexagonal cages are linked around the supercage in the unit cell.
- There are 192 silicon or aluminum atoms in each unit cell. For a typical silicon-aluminum ratio of 5 there are about 55 aluminum exchange sites.
- After sodium exchange (~70% Na^+ in sodalite and ~30% Na^+ in hexagonal cages) with protons, which form hydroxyl Brønsted acid sites, the aluminum atoms become acid sites.

Aluminum atom sites with no next nearest neighbors have less electronic interaction with other sites and more acidity—maximum acidity of individual sites is reached with about 9–12 aluminum atoms in the unit cell.

TABLE 5.8 Composition and Properties of Commercial Zeolites.

Zeolite	Composition (empirical)	Port size (nm)
X-zeolite	$Na_2O.Al_2O_3.2\text{-}3SiO_2.8H_2O$	0.74
Y-zeolite	$Na_2O.Al_2O_3.3\text{-}6SiO_2.9H_2O$	0.74
ZSM-5	$Na_2O.Al_2O_3.5\text{-}100SiO_2.7H_2O$	0.54

Typical properties and the composition of zeolites used in FCC catalysts are listed in Table 5.8, with the molecular diameters of relevant hydrocarbons given in Table 5.6.

5.4.2. Production of Zeolites

Most zeolites are synthesized from proportional mixtures of sodium aluminate and sodium silicate and, in some cases, colloidal silica. The procedure is different to that for the preparation of silica alumina catalysts because zeolites form under alkaline conditions as the silica/alumina co-gel crystallizes in the presence of hydroxyl ions. The zeolite type formed depends on the proportion of silicate and aluminate in the solution and the reaction temperature and pressure. The time taken for zeolite crystals to form can range from a few hours to several days. Seeds or templates are often added to induce formation of the appropriate crystalline product. *Knowhow* is very important, and precise details for a specific preparation are not always published.

Y-zeolite is relatively easy to produce requiring only the addition of finely divided silica to seed the crystallization of the saturated solution. The method originally used by Filtrol, the company that supplied the first catalysts to Houdry, was to produce pure Y-zeolite by the conversion of kaolin clay. The first step was to form metakaolin by dehydrating the kaolin above 600^0C. The dehydrated kaolin then contained activated alumina that could be extracted with acid to give the appropriate silica/alumina ratio. Following extraction the clay was aged in caustic soda solution until pure Y-zeolite crystals formed.[16] The success of the procedure depended on the use of pure kaolin and careful temperature control.

An improved synthesis of Y zeolite within a matrix of kaolin was introduced by Engelhard. Kaolin was calcined to about 1000^0C to produce a silica/alumina spinel, but without forming mullite. The spinel contained less active alumina that meta-kaolin. The spinel mixture was then slurried with kaolin and *seed* particles before being spray-dried to form micro-spheres. These are then initially aged for some time in caustic solution and then heated to 80^0C during which time the Y-zeolite crystallizes. The supernatant liquid from the crystallization stage still contains sodium silicate, and the strength of the kaolin matrix can be increased if the wet micro-spheres are flash dried. The matrix is thermally stable, has large pores and is strong enough to be used without binders.

ZSM-5 is produced by reacting a solution/suspension of silica in N-tetrapropylammonium hydroxide, with aqueous sodium aluminate solution to form a gel. The gel is then heated in an autoclave at 150°C for 5 to 8 days during which time, the zeolite crystallizes. After being filtered, washed, and dried, the crystals are calcined to decompose the organic cation. The zeolite can be ion exchanged to give H-ZSM5.[17] The silica/alumina ratio varies depending on the template used and the reaction conditions selected.

5.4.3. Formation of Active Sites by Ion Exchange

Freshly synthesized X-and Y-zeolites contain up to 13% sodium oxide, with about 60–70% in supercages, with the balance in the sodalite cages and hexagonal prisms. As prepared, both zeolites have only a limited catalytic activity and produce gasoline typical of those produced by thermal cracking. Acid sites can, however, be easily generated by ion exchange. Bronsted acids destroy the zeolite framework and it is necessary to exchange sodium with ammonium ions. The ammonium zeolite can then be converted to the hydrogen zeolite by a thermal treatment. Normally with two ammonium exchanges the sodium content of fresh zeolite can be reduced to less than 2% sodium oxide. Both HX and HY zeolites provide FCC catalysts with higher cracking activity than silica/alumina, but are unstable and rapidly lose activity as the residual sodium *poison* migrates from the framework to the supercages.

Removal of sodium from the early Y-zeolites remained an urgent priority for catalyst manufacturers who were attempting to increase gasoline yield and obtain catalysts with higher thermal stability. Experimental work with Y-zeolite had confirmed the benefits of a catalyst with low residual sodium content, but repeated exchange with ammonium salts was too expensive to be used commercially. Consequently the Y-zeolite produced commercially was not sufficiently stable.

Early development work by Mobil had shown that the exchange of sodium with higher-valency ions, particularly those derived from rare earth metals, increased stability. Rare earth–exchanged NaY-zeolite (REY-zeolite) was not as easily dealuminated by steam and high temperatures in the regenerator as HY-zeolite. Consequently, catalysts manufactured from REY-zeolite were soon being used to maximize gasoline production in most FCC units.

Initial exchange with rare earth and ammonium chloride removes sodium ions from the supercages. On calcinations, the rare earth oxides and hydroxides decompose and migrate into the sodalite cages, where they can exchange for more sodium ions. During calcination some of the rare earth ions are converted to cationic polynuclear hydroxy complexes that provide additional acid sites.[18] Residual sodium ions in the supercages can then be exchanged with ammonium chloride and calcined to decompose the ammonium ions. The quantity of rare earth that can be introduced depends on the extent to which the Y-zeolite has been dealuminated and thus the remaining number of aluminum atoms remaining in the unit cell. A fully exchanged REY-zeolite contains up to about 18% of rare earth oxides.

REY-zeolite catalysts produce higher yields of gasoline than the early silica/ alumina catalysts although the octane level is lower. This is due to increased hydrogen transfer between naphthenes and olefins which produces a mixture of aromatics and paraffins. The low-octane paraffins are not compensated by the high-octane aromatics and the motor octane number (MON) is quite low while the research octane number (RON) is not greatly affected. This was not significant until lead-free gasoline was introduced, and up until then REY-zeolite catalysts were a great success.

5.4.4. Use of Zeolites in Catalytic Cracking

The use of cracking catalysts containing zeolite increased rapidly following their introduction in 1962, and by 1972 they were being used in at least 90% of the FCC units in the United States.[19] Gasoline production was increased, with a lower role of recycle of the residual heavy oil. Gas oil feed rate could, therefore, be increased. Zeolites were produced with more consistent quality and were relatively more stable, producing less coke during operation than synthetic silica/alumina catalysts.

Some early zeolite catalysts contained X-zeolite, probably because it was cheaper to manufacture, but experience soon showed that Y-zeolite was more

TABLE 5.9. FCC Catalyst Composition, 1965–2000.

	1965–1970	1975–1990	1980–2000
Catalyst	Gasoline	Gasoline, octane	Octane, residue, reformulated gasoline.
Zeolite	X/Y-zeolite, REY-zeolite	REY-zeolite, REUSY-zeolite	REUSY-zeolite
Content	5–10%	25–30%	30–40%
Matrix[a]	Kaolin	Kaolin	Alumina (major), silica/alumina, cerium-pillared clays, acid treated meta kaolin.
Binders	Silica/alumina, peptized alumina	Silica sol	Aluminum chlorhydrol, peptized alumina.
Properties[b]	Active/unselective, strong.	Inactive/selective, strong.	Large pore size, active, strong.

[a]Plank and Rossini introduced Matrix USP 3271418 (1966).
[b]Components have small size and are milled to ~ 2 μm APS.

stable under typical operating conditions. The most stable and successful catalysts in producing high yields of gasoline were made from REY-zeolite and these became the standard in refineries.

Most existing units were unable to make full advantage of the higher activity of pure zeolite catalysts, and the initial catalysts consisted of 5–10% zeolite, supported within a matrix. This led to the use of many different formulations so that individual operators could make the best use of available equipment. From the 1970s, when legislation required that lead additives were phased out of gasoline and residual fractions were cracked together with gas oil, an even wider range of catalysts became available. Typical examples are described in Table 5.9.

5.4.5. The Catalyst Matrix

The matrix has a most important role in FCC catalysts and must be carefully formulated depending on the feed being treated. Early catalysts needed a matrix because undiluted zeolite was too active for use in existing units and was immediately deactivated as coke deposited on the surface. Zeolites would never have been used without a matrix. In 1964 the matrix was simply a diluent and binder to form porous particles strong enough to resist attrition while circulating continuously between the reactor and the regenerator. The hot matrix also acted as a heat sink that both circulated heat and effected vaporization of the feed to sustain the endothermic reaction.[20] These requirements could be provided by kaolin, often used with amorphous silica/alumina, and a suitable binder. Alumina sols made by dissolving pseudoboehmite in a monobasic acid, such as formic acid, had previously been used to bind silica/alumina catalysts. Other binders now include silica sols and aluminum chlorhydroxide:[21]

- The first matrix materials, based on the early catalysts, were really nonselective catalysts that promoted the cracking of heavy feed molecules and formed some coke. They have now been further developed.
- More selective catalysts were needed to produce gasoline of a higher octane rating, and a new more active matrix, tailored to improve octane rating, was required. During the period while these new catalysts were introduced, older production units were also being updated to allow more reaction to take place in the *riser* section of the plant. Increased activity was achieved by including a higher proportion of zeolite in the catalyst. This caused the catalyst particles to be weaker, and stronger binders were required in the formulation. Silica sol became the most favored binder for high-zeolite/kaolin catalysts. These new catalysts were less prone to coke deposition so the process could be operated under more severe conditions. This led to an increase in both the rate of production and the octane number of the resulting gasoline.

- The nature of the matrix continued to be changed to cope with ever increasing amounts of cracker residues added to the gas oil feedstock. Better porosity was needed to allow larger residue molecules access to the active matrix pores, where they could crack into smaller fragments that, in turn, could enter the small zeolite pores. Alumina, and kaolin, fired at high temperature and then acid extracted were successful components of an active matrix. The alumina in the matrix could also absorb poisons from the residues, such as the metals nickel and vanadium and some sodium compounds.

More effective binders are now used to give sufficient strength to the new high-zeolite catalysts required for the processing of feedstocks containing high levels of recycled heavy ends.

5.5. OCTANE CATALYSTS (CATALYSTS TO INCREASE OCTANE RATING)

Despite the better conversion and increased yield of gasoline with REY-zeolite catalysts it was found that octane levels were lower because fewer olefins were produced. This was not really a problem for a number of years because gasoline quality could be improved by either increasing cracking severity or using anti-knock compounds such as tetraethyl lead (TEL) and tetramethyl lead (TML). About 3-ml TEL per US gallon could increase the octane number of catalytic cracked gasoline by five to six points.

The 1970 Clean Air Act, introduced by the US Environmental Protection Agency, included regulations requiring the control of automobile emissions. Lead compounds used to increase the octane rating of gasoline were to be phased out from 1973. All new cars, starting with 1975 models, would be fitted with catalytic converters and use unleaded gasoline.

These new constraints on gasoline formulation focused attention on the decreased octane levels of gasoline produced with REY-zeolite cracking catalysts and the *octane dip*, or low octane number, of C6–C10 paraffins. To compensate, the aromatic content of the gasoline pool was increased from about 20% in 1973 to almost 40%, but the need for new *octane* catalysts was soon an important objective for refiners and catalyst producers. Octane catalysts require a suitable zeolite that can limit the hydrogen transfer reactions that convert olefins to paraffins.

Despite the ever increasing demand of gasoline in the 1970s, the *work horse* REY-zeolite was still widely used until octane number became a serious problem. As late as 1972, out-of-date catalysts were still used in low-yield plants by operators not wishing to pay more than the minimum price for catalysts. About 90% of cracking units were using zeolites, but around 20% was the relatively unstable, yet cheaper, X-zeolite![19] The remaining 10% of the catalysts used were improved versions of silica/alumina catalysts dating from the 1940s. The new environmental requirements changed this situation and by 1979 X-zeolite was no longer being produced for cracking catalysts.

5.5.1. Hydrothermal Dealumination of Y-Zeolites

Improvements in plant design and better operation had allowed the zeolite content of cracking catalysts to be increased from about 10% before 1972 to 15–20% by 1975. The higher zeolite content resulted in improved conversion and gasoline yield. Matrix composition had also assumed a greater importance, not only to increase the strength of catalyst particles but also to reduce the overall cost by using cheap, yet stronger, fillers.

The need for more thermal and hydrothermal stability of the zeolite became progressively greater as they were operated under increasingly severe conditions. It was already known that higher stability could be achieved by increasing the silica/alumina ratio and by lowering the sodium content. Nevertheless, it was very difficult to produce directly Y-zeolite with a silica/alumina ratio greater that six, and efforts to increase this ratio had led to the first de-aluminated catalyst. This was first marketed as an ultrastable Y (USY)-zeolite catalyst in 1964, but was not readily accepted because of its lower activity compared with REY-zeolites. It did, however, produce gasoline with more olefins and a higher octane number. USY-zeolites also contained less sodium than Y-zeolites. The manner in which crystal form and sodium content of Y-zeolite changes with rare earth exchange and calcination is shown in Table 5.10.

TABLE 5.10. Dealuminated Y-Zeolite Properties.

Zeolite type	UCS (nm)	SiO_2/Al_2O_3	Na_2O (wt%)	Surface area ($m^2\,g^{-1}$)
NaY	2.468	5	13	900
First exchange:				
NH$_4$Y	2.473	5	2.5–2.8	925
REY	2.468		2.5	
CREY	2.465		0.3	
RENH$_4$Y	2.470		2.5	
CREY	2.454		0.3	
HY	2.450	5	2.5	730
Second exchange:				
NH$_4$Y	2.453	5	0.2–0.3	750
USY	2.435	15	0–0.3	
USY(1)	2.428	30	< 0.05	780
USY(2)	~ 2.425	60	< 0.05	780
USY(3)	~ 2.423	80	< 0.05	780
RENH$_4$Y	2.450		< 0.3	
REUSY	2.430		< 0.1	

Note: Crosfield catalyst zeolite Y (October 1989). Zeolyst form ZC (September 8, 1998).

Equilibrium representations of USY and REY zeolites

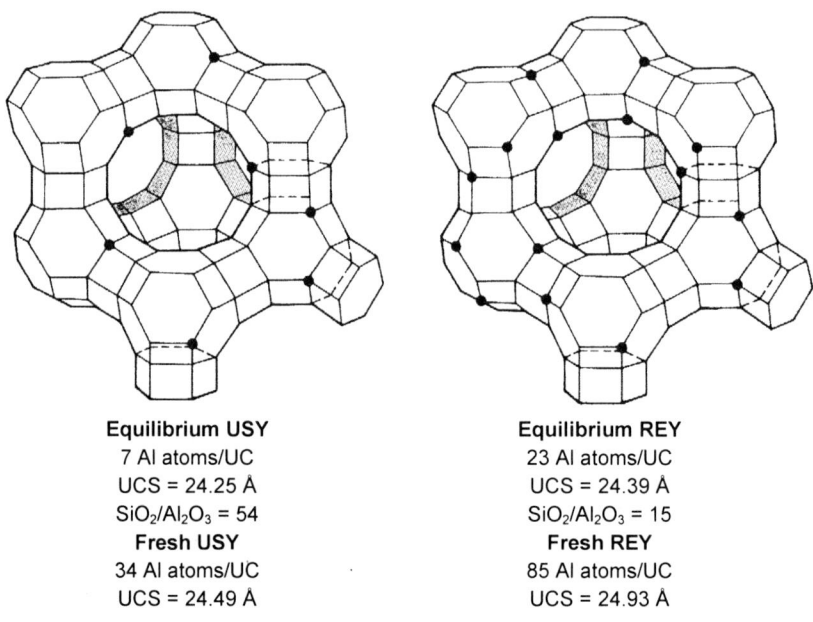

Equilibrium USY	Equilibrium REY
7 Al atoms/UC	23 Al atoms/UC
UCS = 24.25 Å	UCS = 24.39 Å
$SiO_2/Al_2O_3 = 54$	$SiO_2/Al_2O_3 = 15$
Fresh USY	**Fresh REY**
34 Al atoms/UC	85 Al atoms/UC
UCS = 24.49 Å	UCS = 24.93 Å

Figure 5.6. Dealumination of Y zeolite. Reproduced with permission from Grace Davidson Refining Technologies.

The framework was dealuminated (Figure 5.6) to give a higher silica/alumina ratio and sodium was simultaneously removed from the sodalite cages by high-temperature treatment of wet NaY-zeolite filter cake, following ammonium exchange to produce HY-zeolite.[22] The first USY-zeolites were more stable but less active than Y-zeolites because they continued fewer, through stronger, acid sites (Figure 5.7). Although the unit cell size decreased slightly as a result of dealumination there was no loss of crystallinity and typical feed hydrocarbons were still converted within the supercages. Developments continued and a number of other high-silica Y-zeolites were eventually introduced, in particular, by calcining ammonium-exchanged Y-zeolite in a flowing steam atmosphere at 750°C.[23] Aluminium atoms are progressively removed from the zeolite framework by increasing the temperature and time during which the HY-zeolite is steamed. The unit cell size gradually decreases and eventually the zeolite framework collapses.

High-silica Y-zeolites, referred to as either USY-or HSY-zeolites, have considerably improved thermal/hydrothermal stability. In addition to the original

Figure 5.7. Scanning electron micrograph (SEM) of dealuminationed Y zeolite (20 000 x, 130% relative crystallinity). Reproduced with permission from David Rawlance/Crosfield Limited.

pores and channels, USY-zeolites also have large pores in the size range 25–60 Å, which are formed during dealumination. Some of the defect sites formed by hydrolysis of the silicon–aluminum bonds are filled by the migration and insertion of noncrystalline silica to give a more stable silica framework. The displaced alumina, however, remains trapped within the zeolite pore structure and is known as nonframework alumina (NFA). Although defect-free Y-zeolite can be made with a silica/alumina ratio up to 12 by using tetraalkylamine hydroxide templates at high temperature and pressure, the procedure is expensive and not widely used.[24]

5.5.2. Chemical Dealumination of Y-Zeolites

High-silica Y-zeolites can also be prepared by chemical extraction of alumina despite the practical problems involved. De-aluminated Y-zeolite may be stabilised by replacement of aluminium atoms in the zeolite structure by some silicon atoms from the solvent, or, depending upon the preparative procedure, the zeolite may simply be left in a hydrogen-deleted form.

In practice the only chemically produced HSY-zeolite to be used on a large scale was made by reaction of NH4Y-zeolite with ammonium fluorosilicate

(AFS) under controlled, low-pH, conditions.[25] Aluminum in the zeolite frame-work is removed as ammonium fluoroaluminate and replaced by silicon:

$$NH_4Y\text{-zeolite} + (NH_4)_2SiF_6 \rightarrow AFSY\text{-zeolite} + (NH_4)_3AlF_6 \qquad (5.1)$$

High-silica AFSY-zeolites have a silica/alumina ratio of about 12 and good hydrothermal stability. The low active site density, however, gives a low intrin-sic activity requiring a higher proportion of the zeolite in a catalyst. AFSY-zeolite is expensive to use because of high production costs and the need to dis-pose of toxic effluent. The main advantage in chemical dealumination is that the formation of non-framework alumina is avoided. However, AFSY-zeolite is rarely used, because of the other difficulties involved.

Nonframework alumina can also be removed from USY-zeolite catalysts by solution in ammonium nitrate, at pH 2.5, and washing out the aluminum nitrate formed. This does lead to further dealumination and, although defects remain, the zeolite is still very crystalline.[26] The NFA-free catalysts have increased sta-bility, produce higher-octane gasoline, have better coke selectivity, and are widely used. Most catalyst suppliers can provide dealuminated USY-zeolite with reduced nonframework alumina.

Cracking catalysts containing more than 25% of USY-zeolite to compensate for lower activity became available in 1976, although they were not widely used until the addition of lead to gasoline was almost completely abandoned. The lead phase-out coincided with the introduction of steaming processes for making more stable USY-zeolite.

5.5.3. Increasing Octane Number

The development of ultrastable Y-zeolite catalyst led to the production of gaso-line with higher olefin content and increased octane number. However, the need for improved catalysts continued because the zeolite was not sufficiently stable and the motor octane number did not rise as much as the research octane num-ber. As discovered previously with other zeolites in the 1960's, partial exchange of USY-zeolite with rare earth (REUSY) gave better stability as well as activity and provided more branched hydrocarbons and aromatics. Both motor octane number and gasoline production could thus be increased.

Fully exchanged REUSY-zeolite, containing about 7% rare earth oxides, produced more gasoline but unfortunately also promoted more hydrogen transfer from naphthenes to olefins, which decreased the octane number. By selecting an appropriate level of rare earth exchange a compromise between octane number, gasoline yield, and catalyst stability could be achieved. *Octane-barrel* catalysts, therefore, maximized the production of gasoline consistent with an acceptable octane level.

5.5.4. Shape Selective Cracking

The use of a small pore ZSM-5 co-catalyst with REUSY catalysts increases the octane number of gasoline by a process known as shape-selective cracking.[27] Straight-chain C_6–C_{10} olefins produced by normal cracking reactions and which are the precursors of low-octane paraffins are selectively cracked by ZSM-5 catalysts.

So-called *centre cracking* produces a C_4-C_5 olefin fraction which rapidly isomerises to isobutene and isoamylene. These products were converted to methyl tertiary butyl ether (MTBE) and tertiary amyl methyl ether (TAME) by reaction with methanol to produce octane-enhancing additives for use in reformulated gasoline. Propane and n-butane are also produced. Fresh ZSM-5 also cracks C_7^+ paraffins until the acid site density decreases. Eventually, olefin cracking activity declines but isomerization activity is retained. Regular addition of fresh ZSM-5 is therefore required to maintain the shape-selective activity.

The introduction of shape-selective ZSM-5 additives in 1975 resulted in an increase in the gasoline octane number by two to four units within a few days.[28] This was weeks faster than changing to a more selective octane catalyst because of the long time internal needed to change significantly the composition of the catalyst inventing at normal replacement rates, see Table 5.11. ZSM-5 additives were also used with cheaper REY-zeolite catalysts to increase octane levels with little change in operating conditions.

Although ZSM-5 additives have the disadvantage of decreasing gasoline yield by up to 2%, there is no change in dry gas, heavy oil, or coke production. The gasoline loss was more than compensated for by using the by-product propylene and *n*-butylene in an alkylation unit. By 1993, ZSM-5 additives were being used in 20% of the world's FCC capacity, although very little commercial

TABLE 5.11. Time Taken to Change Catalyst Inventory.[4,5]

Days after addition begins	% *New* catalyst at 2% addition
14	20
17	30
22	40
30	50
44	60
60	70
105	90

Note: Catalyst producers provide information on the time taken to change an existing catalyst. With a relatively low daily replacement rate of 2% it can take 20 days to change 40% of the old catalyst inventory. Because the zeolite in an FCC catalyst is rapidly deactivated, more than 50% of the cracking activity is supplied by catalyst less than 20 days old. Older catalyst still has some activity in the matrix, which converts heavier fractions.

operating data had been made available. ZSM-5 is usually added to an FCC unit on a daily basis with makeup catalyst. It is supplied in the form of particles containing up to 15% ZSM-5 on an inert matrix. As little as 3% of the additive may be needed in the catalyst inventory to achieve the required improvement in gasoline octane. M-5 additives do not form coke and are not poisoned by basic nitrogen because of the lower acid site density. Similarly the ZSM-5 framework is extremely stable and so there is less deactivation than with typical Y-zeolites.

5.6. RESIDUE CRACKING CATALYSTS

Some heavy residual fractions from petroleum refineries have been added to FCC unit feeds since the late 1970s, when crude oil prices rose sharply. The processing of these more intractable fractions has partly compensated for the higher oil price without the need to process more crude. Although some new units have been designed to be able to cope with 100% residue, it is more usual to add about 30% of residue to gas oil feeds in conventional units (see Table 5.2).

The economic benefit of using cheaper residues was offset to a certain extent, by the need to use more expensive catalysts and increased catalyst makeup rates. The proportion of residue which can safely be added to the feed is usually limited by the impurity levels in the residue and the need to maintain a satisfactory conversion, since use of residues generally leads to a reduction in conversion. These problems have been partly overcome by the gradual introduction of better catalysts and the use of metal passivation additives. More effective additives continue to be developed.

5.6.1. Residual Feeds

Residues include coker gas oil, atmospheric and vacuum residues, lube extracts and deasphalted oils. They differ in chemical composition from vacuum gas oil. Typically residues contain a wide range of hydrocarbon types including polynuclear naphthenes, high-molecular-weight aromatics, and asphaltenes. They also contain significant amounts of metal, sulfur, and nitrogen compounds. Some of the higher-boiling residual components are not completely vaporized during the cracking process and nonvolatile Conradson carbon coke levels are increased. Operation seems to be reasonably acceptable despite the potential problems!

The larger molecules found in residues have diameters in the size range 2.5–10.5 nm, with molecular weights ranging from 1000 to 100,000. This compares with typical gas oil feeds, boiling in the range of about 150^0–600^0C, with molecules in the size range 1.0–2.5 nm and average molecular weights of less than 400.

Consequently, there are significant differences in FCC unit operation when residue is added to normal feed. Conversion falls and less gasoline is produced, as shown in Table 5.2, and the catalyst-to-oil ratio must rise as coke yields increase. The coke also has a different composition relative to that produced from normal feed not only because of the higher Conradson carbon levels and high-boiling compounds, which are absorbed by the catalyst particles, but also from the dehydrogenation activity of the metal impurities, which leads to polymerization reactions and *contaminant coke* formation.

5.6.2. Residue Catalyst Formulation

Catalysts used to crack gas oil to which residue has been added are generally quite similar to octane catalysts. They incorporate high-silica USY-zeolites, exchanged with rare earths and supported on a high-activity matrix. To compensate for the more demanding reaction and regeneration conditions, the manufacturing processes for USY-zeolite have been optimized to give a more crystalline and thermally stable structure which produces less coke. It is necessary, however, as shown in Table 5.9, to compensate partly for rapid zeolite deactivation by including up to 40% REUSY-zeolite in the catalyst formulation.

The matrix used contains a higher proportion of active alumina or silica/alumina together with natural or activated clay filler. It is important for the matrix to have carefully controlled large pores because of the more important role it plays in cracking the larger molecules present in residual feeds. The composition of the catalyst is set by the need to achieve a balance between zeolite and matrix activities. This balance depends on the operating conditions of the process and the nature and quantity of residue to be mixed with the gas oil. The higher sulfur content of residual fractions, which increases the sulfur oxide emission in regenerator flue gas, often requires the use of further additives to meet statutory limits.

5.6.3. Coke Formation

The heat balance in an FCC unit is complex and depends on the combustion of coke in the regenerator. Coke formation on the catalyst must be carefully controlled when the feeds contains residue. Impurities such as organic nickel, vanadium compounds, and Conradson carbon lead to increased coke deposition and this affects the rest of the unit. It is necessary to passivate the metals with additives and dilute or hydrotreat the residue.

The cracking reaction is endothermic and so requires an input of heat. This heat is provided by combustion of residual coke on the catalyst in the regenerator. In a heat-balanced unit, the level of coke deposition is controlled so that the

TABLE 5.12. Distribution of Delta Coke with Gas Oil and Residue Feeds.

	Gas oil (Con-carbon 0.3 wt%) Wt% coke	Gas oil/residue (Con-carbon 5 wt%) Wt% coke
Catalytic	0.52	0.40
Contaminant	0.12	0.25
Feed	0.04	0.45
Catalyst to oil	0.12	0.12
Delta coke (total)	0.8	1.22
Wt% feed	5.6a	6.7

a Heat balanced cat/oil = 7

amount of coke deposition is in balance with the amount of heat needed to sustain reaction. A coke-selective catalyst makes more gasoline for a given coke content in the regenerator. The quantity of coke deposited on the catalyst varies with each catalyst type.

Only part of the coke on spent catalyst is burnt in the regenerator. The difference between the amount of coke on spent and regenerated catalyst is referred to as delta coke (Δ coke) and is usually expressed as a percentage. At steady state, this equates to the overall amount of coke formed per pass. Coke yield is the percentage of feed that is converted to coke. A useful measure of Δ coke is the coke yield divided by the corresponding catalyst to oil ratio.

When cracking residue with coke selective REUSY-zeolite catalysts, the low Δ coke allows more flexible operation by increasing conversion and gasoline selectivity at a constant coke yield. Several different types of Δ coke are deposited during the reaction:

- Catalytic coke forms on acid sites by cracking or polymerization of feed and depends on the catalyst type.
- Catalyst/oil coke consists of residual very high molecular weight hydrocarbons that are not stripped from catalyst before regeneration.
- Contaminant coke forms as a result of the dehydrogenation and polymerization reactions catalyzed by metal impurities in the feed.
- Feed coke results from Conradson carbon, asphaltenes, or other high molecular weight compounds that do not crack in the riser.

About 1.2–1.4% Δcoke forms on *residue* catalysts compared with about 0.7– 0.8% on *gasoline* catalysts. The distribution of Δcoke for both types of feed is shown in Table 5.12. Most of the increase is associated with contaminant and feed coke.

5.7. RESIDUE CATALYST ADDITIVES

The addition of residual fractions to gas oil feed results in an increase in the impurity content of the equilibrium FCC catalyst and causes a decrease in activity. Metal impurities exist as porphyrin complexes which crack and deposit metal residues on the catalyst surface, causing catalyst deactivation. The most serious effects on catalyst performance result from nickel and vanadium compounds. Sodium can also deactivate acid sites on the catalyst, but the effect is generally reduced by desalting crude oils and by absorption of small amounts of sodium on the matrix. Sulfur compounds in the feed contaminate products and regenerator flue gas.

Efforts have been made to develop additives that limit the effects of impurities in the feed. However, a practical way to counter the effect of metal impurities has been to increase the withdrawal and replacement of equilibrium catalyst with fresh catalyst, despite the increase in cost. Some of the ways in which metals and other impurities can be managed in modern FCC units are as follows:

- When possible use a feed with low metal content. This can mean hydrotreating the feed, which is expensive and does not remove all the metal impurities. It may also be possible to use a more metal-tolerant catalyst.
- Add more fresh makeup catalyst to maintain activity. This is generally used for high-vanadium content feed but is expensive.
- Use an uncontaminated equilibrium catalyst (E cat) as replacement to maintain activity. It is difficult to obtain supplies of E cat, which is usually of variable quality and may only have a short lifetime.
- Catalyst containing high metal impurity levels may be flushed from the inventory using cheaper low-activity catalyst. This does, however, dilute activity.
- Use vanadium traps or nickel additives. These are not yet widely used or very efficient, but integral traps for FCC catalysts are being developed. These will dilute an active catalyst but are beneficial, particularly when the feed contains a high level of vanadium.
- Magnetic separation and catalyst demetallization procedures to segregate or revive contaminated catalyst have good potential but more development is needed.
- The use of carbon monoxide combustion additives and sulfur transfer additives help to reduce coke formation and sulfur emissions.

5.7.1. Nickel Additives

Nickel compounds deposited on the catalyst surface are oxidized to nickel oxide in the regenerator. Despite the potentially high nickel oxide concentration in equilibrium catalyst this does not affect zeolite activity. Nickel oxide does, however, reduce to nickel in the reaction zone and catalyses the dehydrogenation of hydrocarbons to produce hydrogen and coke. Increased volumes of hydrogen in dry gas lead to compressor limitations. Additional coke deposited on the catalyst increases the regenerator load and affects the unit heat balance.

Nickel can be absorbed as an aluminate in a large pore matrix. It is also possible to use additives that limit dehydrogenation by forming stable nickel compounds. One successful additive is an antimony trisdipropyl dithiophosphate solution.[29] Optimum nickel/antimony ratios are claimed to be in the range 1.5–5, depending on operating conditions and the quantity of nickel deposited. About 96–98% of the antimony remains on the catalyst or leaves the reactor with catalyst fines. Environmental metal limits for catalyst fines in the United States and Europe are currently 4000 ppm nickel and 1500 ppm antimony (with 7000 ppm vanadium), which may indicate a typical nickel/antimony ratio of about four. The additive can decrease hydrogen formation by 40–60% with a corresponding reduction in contaminant coke. Nickel passivators are generally needed if hydrogen production exceeds 60–75 scf/barrel of feed and it has been estimated that about 50% of US FCC plants with nickel in the feeds use this procedure.

Because antimony is on the EPA list of hazardous chemicals it is sometimes replaced by a similar bismuth additive. Solutions of this additive contain 28 wt% bismuth and act in the same way.[30]

5.7.2. Vanadium Additives

Vanadium has a more severe effect on FCC catalyst than nickel. Its most serious effect is to cause irreversible deactivation of the zeolite. It also has significant activity for dehydrogenation (25–30% that of nickel) and this contributes to both coke deposition and hydrogen formation.

When vanadium (IV) porphyrin complexes deposit on the catalyst matrix they are cracked and form vanadium pentoxide in the regenerator. Vanadium pentoxide melts at 680^0C, and forms a liquid phase that diffuses through the catalyst. Experiments have shown that by heating mixtures of vanadium pentoxide and FCC catalysts at 700^0C, vanadium infiltrates the catalyst particles within about 15 min. It has been shown by differential thermal analysis, that the zeolite structure is destroyed when the vanadium pentoxide melts in the temperature range 630^0–660^0C. The effect of vanadium is to shrink the zeolite unit cell by a dealumination mechanism, and this appears to accelerate the typical aging of zeolite catalysts. Vanadium also reacts with rare earths in the sodalite cages forming vanadates that destroy the rare earth stabilizing bridges.

Sodium compounds accelerate vanadium deactivation of FCC catalysts. Under typical operating conditions, sodium hydorxide and vanadium pentoxide form mixed oxide phases that melt at temperatures as low as 525^0C. The mixed salts can dissolve alumina from the zeolite and matrix during regeneration, but both alumina and vanadate redeposit as the catalyst temperature falls in the reaction zone.

Early attempts to control vanadium deactivation involved the addition of amorphous alumina to the cracking catalyst matrix. This was not particularly suitable because alumina increased coke formation and led to wider dispersion of nickel impurities. Hydrodesulfurization of FCC feeds is useful; it not only removes sulfur, but the desulfurisation catalyst also adsorbs a significant proportion of the metal porphyrins.

The easiest procedures to compensate for vanadium poisoning are to increase the zeolite content of the catalyst or to replace a larger proportion of the catalyst inventory every day. Those options are expensive and when equilibrium catalyst contains more than about 5000 ppm of vanadium it is more cost effective to use a *vanadium trap*. The role of the trap is to stop the migration of vanadium and therefore to prevent deactivation of the zeolite. The trap must not interfere with the cracking reaction and must maintain the vanadium at a lower oxidation state with a high melting point.

Early traps contained butyl tin compounds, and reduced the deactivation effect of vanadium by 30% and the dehydrogenation activity by 50%[31] as soon as it entered the reactor.[28] Tin, however, poisoned FCC catalysts if used in large quantities and had no effect on the vanadium already on a catalyst. More recently, traps derived from basic alkaline earth oxides, such as strontium titanate, were claimed to reduce zeolite deactivation by 90%.[32] Unfortunately, in full-scale operation, these traps were poisoned by sulfur oxides forming very stable and intractable sulfates, and are not often used.

The most effective traps currently available are based on supported rare earth oxides, such as those of cerium, plus promoters and absorb significantly more vanadium than the catalyst. The trap may be added to the matrix or used as separate particles if the quantity required would excessively dilute the zeolite content or affect the strength of the FCC catalyst. Intercat V-trap additive has been shown to absorb 17 times more vanadium than a FCC catalyst.[33] Many suppliers now provide catalysts with integral metal traps. For example, Engelhard Ultrium catalyst traps vanadium on a magnesium-based component and the activity of the nickel-based contaminant is reduced by agglomeration onto the surface of the catalyst particle. *Millenium* catalyst absorbs porphyrin molecules onto a surface alumina compound, where they are immobilized.[34] Demetallizing processes, which remove nickel and vanadium from spent FCC catalysts so that they can be recycled, have been developed but are not yet widely accepted.[35]

5.7.3. Sulfur Oxides Transfer Additives

Sulfur emissions from FCC units cause to atmospheric pollution problems and refineries have to control the sulfur oxide (SOX) content of regenerator flue gas to comply with local or national restrictions.

TABLE 5.13. Distribution of Feed Sulfur in Products.

Wt% sulfur	Feed	H₂S	Gasoline	Light-cycle oil	Heavy-cycle oil	Coke
Gas oil	0.7	0.29	0.05	0.20	0.14	0.03
Gas oil + 10% vac bottoms	1.0	0.38	0.06	0.22	0.21	0.14
Gas oil + 20% vac bottoms	1.3	0.51	0.07	0.25	0.24	0.25

In 1978 the South Coast Air Quality Management District (SCAQMD) of California announced that the limit on FCC flue gas SOX emissions would be 130 kg SOX per 1000 barrels of feed by 1981. This was then reduced to 60 kg SOX per 1000 barrels of feed by 1987. A further proposal, in 1990, suggested that emissions should eventually be lowered to 6-kg SOX per 1000 barrels of feed. The Federal Environmental Protection Agency, while establishing no limits, has specified the options available to reduce SOX emissions. These are the use of flue gas scrubbing, feed hydrodesulfurization, SOX additives, or low sulfur feeds. In Europe SOX emissions were included in an overall refinery sulfur emission limit and not treated separately. The sulfur content of reformulated gasoline has also been restricted by the EPA.

Most FCC unit feeds contain sulfur compounds, which become distributed among the reaction products as shown in Table 5.13. When residues are added to the feed, the coke contains a significantly increased amount of sulfur, which is oxidized in the regenerator, and becomes the SOX component of the flue gas. Although most sulfur compounds in the feed can be removed by hydrotreating, the more refractory sulfur compounds that deposit in the coke are less easily hydrogenated. Residual sulfur oxides as well as particulates can be removed from the flue gas by scrubbing. However, both options are expensive to install and operate. For this reason, it is usually more economic to use a sulfur transfer additive. Additives absorb sulfur oxides in the regenerator, forming reactive sulfates that are reduced to hydrogen sulfide on returning to the reactor. Hydrogen sulfide in light gas is then removed in an existing downstream sulfur recovery unit.

Additives must rapidly absorb sulfur dioxide and trioxide as it is produced during catalyst regeneration. The resulting sulfate is then reduced in the reactor and stripper to regenerate the additive and continue the cycle. The reactions taking place are shown schematically in Table 5.14.

It was observed during the 1970s that high-alumina catalysts partly fulfilled these requirements. However, sulfur oxides in the flue gas were only decreased by about 20% because aluminum sulfate decomposes at a relatively low temperature of 580°C. Better absorbents were then investigated. Cerium oxide supported on alumina improved the absorption of sulfur trioxide, but performance

TABLE 5.14. Reactions Taking Place During Sulfur Transfer.

Location	Reaction
Regenerator	$2\ SO_2 + O_2 \xrightarrow{\text{CO combustion additive}} SO_3$
	$MO(SOX\ additive) + SO_3 \xrightarrow{700\text{-}760^0C} MSO_4$
Reactor	$MSO_4 + 4\ H_2 \xrightarrow{500\text{-}550^0C} MO + H_2S + 3\ H_2O$
	or $MSO_4 + 4\ H_2 \xrightarrow{500\text{-}550^0C} MS + 4\ H_2O$
Steam stripper	$MS + H_2O \xrightarrow{\text{reactor outlet}} MO + H_2S$

was still affected by partial decomposition of the sulfate as in the high-temperature regenerators. Better results still were obtained with cerium oxide supported on magnesia, which absorbed sulfur trioxide to form magnesium sulfate. This was stable in the regenerator but did not reduce completely to hydrogen sulfide in the reactor.

A solid solution of about 10% cerium oxide in the lattice of magnesium aluminate spinel proved to be a more effective sulfur transfer agent. It is important, however, that the spinel does not contain any free magnesium oxide because this will be converted to magnesium sulfate during service. Magnesium sulfate is not reduced to hydrogen sulfide at temperatures below 730^0C, which is somewhat greater than the temperature of the reactor. The addition of a chromium component gives an increase in the conversion of sulfur dioxide to sulfur trioxide in the regenerator.

Magnesia/alumina spinels are not prepared by precipitation for this application because it is difficult to wash the gelatinous precipitate free of cations, and any residual sodium causes the deactivation of zeolites. Suitable additives have to be prepared either by mixing extremely small particles (2–5 μm) of magnesia and alumina in water or mixing magnesium acetate with alumina sol and spray drying the mixtures at 750^0C. This produces the desired magnesia-rich spinel with no free magnesium oxide. The spinel can be impregnated with cerium and chromium nitrate solutions before final calcination.

Commercial trials with the additive in full-scale FCC units have seen a reduction in SOX emissions by 50–80%. Since the first commercial trial of SOX transfer additives was reported in 1984 about 40% of US refineries have tested additives to eliminate SOX from effluent gas. Several different additives are commercially available but despite the relatively good results, not many refineries use them on a regular basis unless there is an economic or political reason to do so.

Sulfur oxide transfer additives work more effectively if a combustion promoter such as platinum is used to oxidize sulfur dioxide to sulfur trioxide more efficiently. More additive is required when a unit is operating under less oxidizing conditions and the coke is only partially converted to carbon dioxide.

Metals do not poison SOX additives and reduction of the spinel surface area does not seem to affect sulfate formation. Silica does, however, deposit on the surface and decreases sulfur pickup. Normal life of a SOX additive is about 28 days.

5.7.4. Bottoms Cracking Additive

When residual fractions boiling above 450⁰C are cracked on a high-activity matrix, the matrix is quickly deactivated by the high levels of metal impurity and Conradson carbon. A bottom-cracking additive can be added to reduce coke deposition when using a high proportion of residue in the feed. If necessary, a vanadium metal trap can also be included.

Use of between 5–10% of a cracking additive can reduce the amount of heavy-cycle oil by converting it to more valuable products and minimizing thermal cracking. The additive does not contain any zeolite and is basically a stable, high-activity 30% silica/50% alumina material with a suitable binder and additives such as rare earth oxides.

5.8. REFORMULATED GASOLINE

The FCC process has been upgraded continuously since it was introduced in 1942. Two of the original three units built by Standard Oil of New Jersey at Baton Rouge, Louisiana, in 1942–1943 were still operating well after 1992.[2] Combined production of PCLA 2 and 3 had been increased from 34,000 bpd of gas oil in 1943 to 188,000 bpd in 1992. PCLA 1 was dismantled in 1963 after expansion of the original 15,000 bpd capacity to 41,000 bpd. Many similar successful improvements have been made at other refineries.[36] Table 5.15 shows how the developments in the FCC process and catalysts made this possible. It was a dynamic and flexible process. Further demands were made when reformulated gasoline was introduced, first in the United States and then in other parts of the world.

The US Clean Air Act (CAA) of 1970 required that automobile exhaust emissions be regulated to meet new environmental standards. From 1975 all new models were to be fitted with catalytic combustion converters to reduce levels of carbon monoxide and unburnt hydrocarbons in the exhaust. This led to the phase-out of lead additives in gasoline between 1973 and 1996. To compensate for the loss of octane rating of the gasoline more reformate and alkylate needed to be added. Octane catalysts were also developed for FCC units and the aromat-

TABLE 5.15. Gradual Revamping of FCC Units to Improve Performance.

Operating variable	1945–1955	1965–1975	1985–1995
Increased feed	100%	200%	300%
Catalyst type	Silica alumina	Early zeolites/RE exchange	REUSY-zeolites/ active matrix
Oil conversion (wt%)	55–60	67	76
Gasoline produced (vol%)	40	50	56
Reactor type	Catalyst bed	Riser	Riser
Regeneration (^{0}C)	565	600	700
	—Use of better steels and combustion additives→		
Catalyst losses (tonne day^{-1})	2.5	< 2	< 0.5
	—Improvement in cyclones and stronger catalysts→		
Regenerated catalyst (wt% coke)	0.5	< 0.05	< 0.05
	—Increasing regenerator temperature→		
Carbon monoxide in flue gas	7%	< 0.1%	< 50 ppm
	—Better combustion→		
Sulfur emission (ppm)	500–1500	300–900	300–600

ic content of gasoline was doubled from 20% to about 40% by using more reformate.

Amendments to the Act in 1989 resulted in changes in the actual composition of gasoline and became part of the CAA in 1990. The purpose of the changes was to lower the toxic and ozone-forming volatile emissions from gasoline evaporation during the summer months. In phase 1 of the summer gasoline program, Reid vapor pressure (RVP) was set to a maximum level according to locality. This meant using less butane, adding even more alkylate and including an oxygen-containing compound such as methyl tertiary butyl ether (MTBE) to premium gasoline. The use of *oxygenates* to promote cleaner burning, to limit carbon monoxide formation and to increase octane levels.

Phase 2 of the amendments was instituted in 1992 and later became part of the CAA. This demanded even more stringent conditions on exhaust emissions, together with a further reduction of RVP. From November 1992 the oxygenated gasoline program required that up to 2.7 wt% oxygen be added to gasoline during the winter months in areas not meeting the carbon monoxide reduction standard. From January 1995, the reformulated gasoline program demanded further reductions in the ozone pollution levels.

The 1990–1997 Federal Phase 1 *simple model* for reformulated gasoline was:

- RVP 7.1 psi (south); 8.0 psi (north)
- Benzene 0.95 vol%
- Oxygen 2.1 wt%
- Olefins 8.5 vol% (less than 1990 base line)
- Sulfur 130 ppm (less than 1990 base line)

The complex model from January 1998 was the same, apart from changes in the baseline gasoline properties.

Significant changes to the unit operation of FCC units have been made so that the gasoline pool complies with these new regulations. Three modifications to the catalysts were required to achieve those new standards:

- Isobutylene and isoamylene production was increased by using up to 10% ZSM-5 additive to provide feed for MTBE and TAME production.
- The volumes of isobutane, propylene, and butylenes was also increased to provide feed for alkylation units.
- USY-zeolite with a very low level of rare earth exchange and an active matrix was used to increase gasoline olefins and to decrease aromatic formation by limiting hydrogen transfer.

These changes decrease gasoline yield but give better coke selectivity. The rare earth content of the zeolite and the proportion of ZSM-5 added above the 10% level are usually optimized as required to meet a given product range.[37] This is necessary because the usual operating limit is the wet gas compressor capacity available to separate LPG. In order to comply with the sulfur regulation it may also be useful to hydrotreat the feed to a FCC unit. This has the additional benefit of reducing the nickel, vanadium, and Conradson carbon contents.

When using typical high rare-earth zeolite catalysts, a lower metal content in the feed significantly increases conversion, giving higher yields of CPG and fewer olefins. However, the gasoline produced has a low octane rating. Conversely, when using low rare-earth catalyst with a ZSM-5 additive, conversion can be higher with the production of more isobutylene and propylene which can be used to increase alkylation and MTBE capacity.

There is even more potential to improve gasoline yield as well as LPG olefins such as isobutene by using a catalyst that incorporates the properties of a ZSM-5 additive. This type of catalyst may also provide better conversion of light-cycle oil and residues because it has larger pores than ZSM-5.[38]

When the gas oil feed contains significant residue, it is necessary to control carefully the level of metal impurity on the equilibrium catalyst. The various metal passivation technologies introduced and the other additives used can affect the unit operation and this must also be taken into account.

REFERENCES

1. Grace Davidson, *Guide to Fluid Catalytic Cracking*, Part 1, 1993, p. 46.
2. A. D. Reichle, Fluid Catalytic Cracking Hits 50 Year Mark on the Run, *Oil Gas J. Special*, May (1992) 18, 47.
3. L. Rheaume, R. E. Ritter, J. J. Blazec and Montgomery, *Oil Gas J.* **74**(20), (1976) 103; **74** (21) (1976) 66; US Patent 4064639 (1977); A. B. Schwartz US Patents 4072600, 4088568, 4093535, 4107032 (1978).
4. R. G. McClung, Calculation of Catalyst Inventory Changes, *Engelhard Catal. Rep.*, EC6758 (1995).
5. A. F. Sweezey, Troubleshooting FCCU Operation Using E Cat Trends, *Engelhard Catal. Report*, EC6911 (1996); Grace Davison, *Guide to Fluid Catalyst Cracking*, Part 2, Ch. 8, 1996.
6. Tracerco, *Radioisotope Studies on FCC Units*, July, 1992.
7. *American Petroleum Institute*, McGraw-Hill, Inc., New York, 1925.
8. A M McAffee, *Ind. Eng. Chem.* **7** (1915) 737; C. Thompson, *Since Spindletop: Gulf's First Half Century*, 1951, p. 45.
9. L. B. Ryland, M. W. Tamele, and J. M. Wilson, *Catalysis*, Ed. by Emmett, Vol. 7, Reinhold, New York, 1960, pp. 1–92.
10. W. L. Hayden and F. J. D. Zierzanowski, US Patents 3402996 (1968), 3624718, 3657154 (1972).
11. Grace Davison, *Guide to Fluid Catalytic Cracking* Pt. 1, (1993), p. 11; *Chem. Eng.* **58**(11) (1951) 218.
12. B. H. Loper, *Oil Gas J.* **53**(51) (1955) 115.
13. C. J. Plank and E. J. Rosinski, US Patents 3140249, 3140251, 3140253 (1964), 3271418 (1966).
14. Heterogeneous Catalysis, Selected Case Histories, Ed. by B. H. Davis and W. P Hettinger, (C. Plank, *The Invention of Zeolite Catalytic Cracking Catalysts*, p 253), ACS Symposium Series 222 (1983).
15. R. M. Milton, US Patent 2882244 (1959).
16. P. A. Howell, US Patent 3390958 (1968).
17. R. J. Argauer and G. R. Landolt, US Patent 3702886 (1972).
18. P. K. Maher and C. V. McDaniel, US Patent 3402996 (1968).
19. D. P. Burke, *Chemical Week*, November 1, (1972) 24; March 28 (1979) 48.
20. J. D. Danforth, *Advances in Catalysis*, Vol. 9, Academic Press, New York, 1959, p. 558.
21. G. M. Woltermann, J. S. Magee, and S. D. Griffiths, Commercial preparation and characterization of FCC catalysts, in: *Fluid Catalytic Cracking: Science and Technology, Studies in Surface Science and Catalysts*, Ed. by J. S. Magee and M. M. Mitchell, Elsevier, 1993, p. 118.
22. P. K. Maher and C. V. McDaniel, US Patents 3293192 (1966); 3449070 (1969).

23. G. J. Kerr, J. N. Miale and R. S. Mikovsky, US Patent 3493519 (1970); P. E. Eberly and H. E. Robson, US Patent 3591488 (1971); J. Ward , *J. Cat.* **18** (1970) 348.

24. G. M. Woltermann, J. S. Magee, and S. D. Griffiths, Commercial preparation and characterization of FCC catalysts, in: *Fluid Catalytic Cracking: Science and Technology, Studies in Surface Science and Catalysts*, Ed. by J. S. Magee and M. M. Mitchell, Elsevier, 1993, p. 116.

25. D. W. Breck and G. W. Skeels, US Patent 4503023 (1985).

26. G. J. Kerr, J. N. Mialle, and R. S. Mikovsky, US Patent 3493519 (1969).

27. S. J. Miller, US Patents 4309276 (1982); 4416765 (1983).

28. C. D. Anderson, *Proceedings 9th Iberoamerican Symposium of Catalysts,* Lisbon, July, 1984.

29. R. W. Bohmer, D. L. Mckay and K. G. Knopp, *NPRA Annual Meeting*, March 1997, AM-97–29,

30. P. Ramamoorthy, A. R. English, J. V. Kennedy, L. W. Jossens and A. S. Krishna, *NPRA Annual Meeting*, March 1988, AM-88–50.

31. A. S. Krishna, *NPRA Annual Meeting*, March 1984, AM-84–51.

32. D. J. Rawlance, K. Gosling, L. H. Stahl and A. P. Chapplein *Preparation of Catalysts V* , C Poncelet, P A Jacobs, P Grange and B Delmon, editors, Elsevier Science Publishers B V Amsterdam (1991), p. 407.

33. Commercialization of the Chevron FCC Vanadium Trap, *Intercat Symposium*, Jakarta (1944); L. M. Magnabosco and K. J. Demmel, Intercat High Activity FCC SOX Reduction Agent, *International Symposium on Advances in FCC*, Division of Petroleum Chemistry, ACS (August 22–27 1993).

34. G. Woltermann, *Engelhard Catalyst Report*, EC-6946 (1996).

35. F. J. Elvin and S. K. Pavel, Demetallization Commercial Results, *NPRA Meeting*, March 17 (1991), AM-91-40; *Oil Gas J.*, July 19 (1991) 22.

36. Bienstock, Pramel, Ladwig, and Patel, *AIChE Spring Meeting*, March 28, (1993); R. E. Evans and G. P. Quinn, *Studies in Surface Science & Catalysts*, Vol. 76, Chapter 15, Elsevier, 1993; *Katalystics Symposia*: J. A. Celestinos, A. Melo Gonzales, and L. Lopez Oraquieta, *Pemex*, May 18 (1983) and W. Bannke, *Deutsche BP,* May 2 (1984).

37. T. E. Johnson and A. A. Avidan, *NPRA Annual Meeting*, March (1993); *C&EN,* September 2 (1993) 33.

38. *Oil Gas J.*, May 1, (1995) 98.

6

REFINERY CATALYSTS

6.1. THE DEVELOPMENT OF CATALYTIC REFINERY PROCESSES

The conversion of distillates into more useful products and the use of off-gases were among the major early developments in the refining industry. Hydrogenation of aromatics in kerosene with nickel catalysts was being tested around 1906–1910, as a method of improving burning qualities.[1] As gasoline for automobiles became the most important refinery product during the period from 1900 onward, the need for increased production and better quality dominated the industry. Fears of future shortages led to the realization that better use had to be made of the crude oil available. Development of thermal cracking processes increased refinery gasoline yields and it was found that unsaturated hydrocarbons in the gasoline improved engine performance.

As conventional spark ignition internal combustion engines developed and compression ratios were increased, problems with pre-ignition of the gasoline/air mixture caused *knocking*. The concept of octane rating evolved to compare different gasolines and to produce high octane antiknock additives. The octane number was determined by comparing gasoline with a mixture of isooctane and normal heptane. This emphasized the fact that branched-chain molecules were desirable gasoline components, whereas straight-chain paraffins all have a low octane number.

The *gasoline pool* in a refinery is now based on the most efficient available blends of high-octane components. The proportion of crude oil treated in catalytic processes and the average contribution of these processes to the US *gasoline pool* are given in Table 6.1. From the 1920s until the 1970s tetraethyl lead, the

L. Lloyd, *Handbook of Industrial Catalysts*, Fundamental and Applied Catalysis,
DOI 10.1007/978-0-387-49962-8_6, © Springer Science+Business Media, LLC 2011

most efficient antiknock additive discovered was added to improve the octane rating of gasoline.

Gasolines derived by cracking were always more valuable than those from straight-run naphthas, and Shell decided that heavy naphtha could probably be converted into a better-quality gasoline by cracking. The thermal reforming process was introduced by UOP in 1931.[2] While higher-octane gasoline was produced, as expected, large volumes of waste gases containing olefins were also formed and this reduced the yield. Up to this point no catalysts were used in gasoline production. A search then began for suitable ways to convert fuel value olefins to high-octane liquid hydrocarbons with boiling points in the desired gasoline range.

In 1932, UOP showed that C_3/C_4 olefins separated from the thermal cracker off-gas could be polymerized to dimers with a phosphoric acid catalyst. This gave *polymer gasoline* with an octane number in the range 83–87. Later, when the *Catpoly* process became more widely used, propylene and butylenes were separated to provide the optimum catalyst conditions for the dimerisation of each olefin separately. The C_4 dimer could be hydrogenated to provide the even more useful iso-octane for aviation gasoline. It was clear that low-molecular-weight olefins were to become important building blocks, and in 1935 Phillips introduced a cracking process to produce propylene and butylene from the corresponding paraffins.

TABLE 6.1. Proportion of Crude Oil Treated by Catalytic Processes and Contribution to the Gasoline Pool in 1996/1997.

Process	Percent crude treated		US gasoline pool (%)
	US	World	
Hydrotreating	54	35	—
Catalytic cracking	33	17	30–35
Catalytic Reforming	21	14	30–32
Hydrorefining	11	11	—
Hydrocracking	9	5	—
Alkylation	7	2	10–15
Isomerization	4	2	4–5
Polymerization	0.4	0.2	1
			(Naphtha 5)
			(Butanes 5)
			(MTBE 2–5)
Total crude (10^6 bpd)	16.54	81.55	
Or million tones year^{-1}	827	4078	
		includes US	
Refineries	154	756	

In the meantime Gulf and Shell had started to recycle off-gas from thermal reformers with the heavy naphtha feed to the process. The Gulf co-reforming and Shell polyforming processes provided up to 5% greater yields of higher-octane gasoline,[2] probably by an acid-catalyzed alkylation mechanism. Further development work by Shell, started in 1928, showed that the isobutylene in mixed C_4 hydrocarbons dissolved in cold 60% sulfuric acid. Then, after heating the acid, an oily layer containing about 65% of di-isobutylene separated. Iso-octane was produced by hydrogenation of the di-isobutylene and Shell delivered its first tank load of the product in 1935.[3] This *cold acid* process was almost immediately replaced by a *hot acid* process in which all butylenes were dissolved at elevated temperatures and polymerized. The yields could be doubled, but the octane level fell from 97/98 to 96/97.[4] Cold and hot acid polymerization presumably led to the alkylation process introduced by Shell and Anglo-Iranian Oil (now BP) in 1938.[5] High-octane gasoline was produced by an alkylation reaction between butenes and isobutane using cold concentrated sulfuric acid as the catalyst. Thus, from 1938 onwards, an alkylation process was available to produce isooctane directly, without the need for a hydrogenation step. A second financial benefit was the economic use of olefins and an increased yield from combining a paraffin with the olefin molecule. Alkylation capacity was rapidly installed despite the capital cost of the new plant, and the dimerisation of olefins became unattractive.

One early limitation in the use of alkylation processes was a shortage of isobutane in refineries. In some refineries, however, further supplies of isobutane could be extracted from natural gas. Where this was not possible, the butane in off-gas could be isomerized to isobutane using a new process which used an acid aluminum chloride catalyst. Shell, for example, operated an isomerization pilot plant at Pernis, in Holland, from 1939.[6]

A major disadvantage of the alkylation process was the accumulation of waste sulfuric acid which had no value unless it could be reused by chemical works close to the refineries. This led, in 1942, to the development of the Phillips process in which hydrofluoric acid replaced sulfuric acid, and could be recovered relatively easily by distillation.[7]

In this chapter, the use of catalysts in various major refinery processes is described. Catalytic cracking processes and catalysts are the subject of Chapter 5.

6.2. POLYMER GASOLINE

Low-molecular-weight polymers can be obtained by processing refinery off-gas containing mixed C_3/C_4 hydrocarbons. Various products are obtained, depending on the gas composition and operating conditions, but the process has generally been used to make polymer gasolines in the C_6–C_9 range. This is an acid-catalyzed reaction and phosphoric acid has been the preferred catalyst since the 1930's. The catalyst is prepared by extrusion of a paste made from phosphoric acid and kieselguhr. The concentration of the acid within the pore structure is critical to the performance of the catalyst. Some water is essential for the reaction; it is required to provide sufficient protons to initiate the reaction. If there is too much water, then the hydrogen ion concentration is too low to sustain the reaction at a satisfactory rate. Where there is too little water, the catalyst activity is too high, and this leads to the formation of polymers that block the pores of the catalyst bed.

When water is removed from a strong aqueous solution of phosphoric acid, (H_3PO_4), some dehydration of the acid takes place, and some pyrophosphoric acid is formed:

$$2 \; H_3PO_4 \rightarrow H_4PO_7 + H_2O \tag{6.1}$$

So-called *100% acid* is a mixture comprising 88% orthophosphoric acid, 6% pyrophosphporic acid and 6% water. The optimum catalyst for the alkylation reaction contains *102%* phosphoric acid and therefore contains an even higher concentration of pyrophosphoric acid.

There is a small loss of water during reaction, and this must be replaced continuously by addition of water to the feed gases to maintain catalyst performance. The specifications for early polymerization catalysts are given in Table 6.2. Total phosphoric acid (wt%) is the amount of acid that has combined with the support and that cannot be extracted with cold water. The specifications of two typical dehydrogenation catalysts are given in Table 6.3

The olefin polymerization reaction is exothermic and the reactor design depends on the olefin content of feed gas. For polygasoline production the olefin content is usually in the range 50–60% and a tube-cooled reactor is used. Temperature control is important because phosphate esters can form at temperatures below about 150^0C, while at temperatures above about 230^0C, thermal cracking or gum forming reactions are likely. In both cases the catalyst loses activity (Table 6.4).

TABLE 6.2. Early Polymerization and Dehydrogenation Catalysts.

Isobutylene polymerization (solid phosphoric acid) catalyst	
Total phosphoric acid (wt%)	65–75
Short acid combined with support (wt%)	15–25
Acid concentration (mol%)	102
Extrusion size (mm)	6
Bulk density (kg liter^{-1})	0.95–1.05

TABLE 6.3. Butane Dehydrogenation Catalyst.

	A	B
Chromic oxide (wt%)	12	45
Magnesium oxide (wt%)	2	—
Calcium oxide (wt%)	—	2
Potassium chromate (wt%)	—	1
Alumina	Balance	Balance
Pellet size (mm)	3	3
Bulk density (kg.liter$^{-1)}$	0.95–1.05	0.95–1.05

The extent of polymerization of the reactants is determined by the operating pressure. With reaction conditions of 70 atm pressure and a temperature of 160–200^{0}C, the product boils in the range 77–220^{0}C, whereas lower molecular weight product boiling in the range 36–230^{0}C is produced at 35 atm and 190–240^{0}C. The operating conditions reported in 1946 for a plant producing isobutylene and di-isobutylene are shown in Table 6.5. In many wartime plants di-isobutylene was hydrogenated to provide high-octane aviation fuel.[9]

Propylene can be polymerized to produce the tetramers and higher olefins, and these were used until the 1960s for detergent alkyl benzene production. Fixed bed reactors were used with lower olefin concentrations of 20–25% in the feed gas. During the period from 1939 to 1945 alkylated benzenes were used to increase the octane number of aviation fuel. In the United States, in 1942, Shell

TABLE 6.4. Di-Isobutylene and Polymer Gasoline Processes.

	Di-isobutylene	Polygasoline
Reactor design	2×2 m^3 tubular reactors in series	Adiabatic beds for < 30% C_3H_6 Tubular reactors for 50–60% olefins
Liquid space velocity (h^{-1})		0.8 (high conversion) 3.5 (low conversion)
Temperature (^{0}C)	140–170	160–230
Pressure (atm)		30 (low conversion) 70 (high conversion)
Feed composition (vol%)	— C_3H_8 (1) C_4H_8 (25) C_4H_{10} (74)	C_3H_6 (22) C_3H_8 (60) C_4H_8 (5) C_4H_{10} (13)
Conversion (%)	90	90

TABLE 6.5. Isobutylene Process Conditions.

	Isobutylene
Catalyst volume (m^3)	2 × 3.6
	Two reactors in parallel—one operating
Temperature (^{0}C)	Inlet 550–575
	Outlet 530–560
Pressure (atm)	6
Feed composition (vol%)	i-C$_4$H$_{10}$ (70)
	n-C$_4$H$_{10}$ (29)
	C$_3$H$_6$ (1)
Recycle (141 tonnes day^{-1})	C$_4$H$_{10}$ (97)
(vol%)	C$_4$H$_8$ (3)
Cycle time (h)	1–1.5

produced cumene by the alkylation of benzene with propylene, while ICI used the same process to alkylate benzene with butylenes in the United Kingdom. Both processes used the phosphoric acid/kieselguhr catalyst; details of the ICI operation are given in Table 6.6. Cumene later became important as the major source of phenol and acetone.

Polymerization processes are very exothermic and a high recycle of paraffins is needed to control the temperature rise. In benzene alkylation processes there should be a large recycle of benzene partly for the same reason, but also to suppress polyalkylate formation.

TABLE 6.6. Butylbenzene and Cumene Production.

Benzene alkylation product	Butylbenzene		Cumene	
Catalyst volume (m^3)	9.4		—	
Inlet temperature (^{0}C)	263		250	
Space velocity (h^{-1})	8.6 te C$_4$		—	
	18.6 te C$_6$H$_6$			
Inlet pressure (atm)	53		35	
Feed (wt%)	i-C$_4$H$_8$	17	C$_3$H$_6$	80
	n-C$_4$H$_8$	8	C$_3$H$_8$	20
	C$_4$H$_{10}$	75	C$_6$H$_6$	Excess
	C$_6$H$_6$	216		
	H$_2$O	~ 1%		~ 1%
Selectivity—aromatics (%)	92–3		91–92	
Production (tonnes.day-1)	4		500	

6.3. ALKYLATION

The purpose of alkylation is to combine propylene, butenes, and pentenes with isobutane, in the presence of an acid catalyst, to produce branched paraffins for blending in premium gasoline (Figs. 6.1 and 6.2). The first commercial process, using sulfuric acid, was introduced by Shell in 1938. The only other commercial process, which uses hydrofluoric acid (HF), was developed by Phillips Petroleum and introduced in 1942. The use of both processes has increased gradually since 1938 with the use of hydrogen fluoride gradually overtaking sulfuric acid, mainly because there are no spent acid disposal problems. The nature of this change is shown in Table 6.7. The introduction of reformulated gasoline in 1994 led to an increased level of interest in alkylates, because they have high octane ratings and are free from benzene and aromatics. However, sulfuric acid is once again finding favor as the preferred catalyst, because of the potential environmental and safety problems associated with the use of hydrogen fluoride.

Figure 6.1. Alkylation plant near Port-Jérôme, France. Reproduced with permission from ExxonMobil Research and Engineering Company.

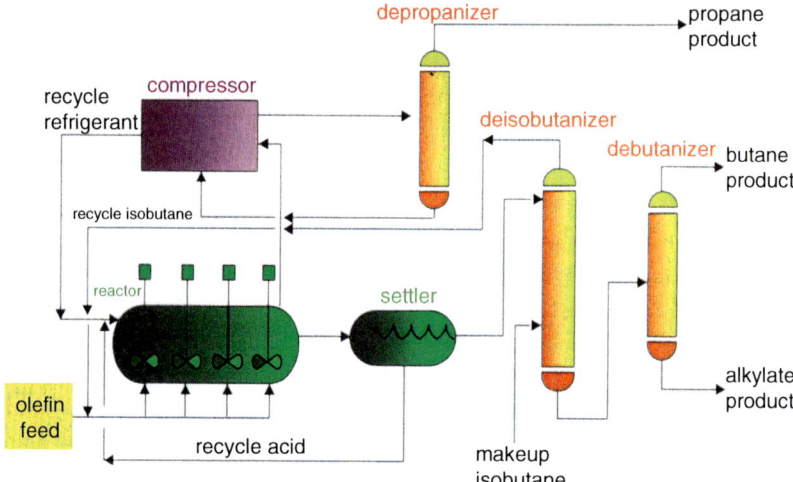

Figure 6.2. Outline flow sheet of alkylation plant. Reproduced with permission from ExxonMobil Research and Engineering Company.

TABLE 6.7. Importance of the H_2SO_4/HF Alkylation Processes in the United States.

Year	% H_2SO_4	% HF	Total production (million bpd)
1938	100	—	Small
1946	80	20	0.17
1957	75	25	0.27
1978	60	40	1.00
1986	50	50	1.05
1993	45	55	1.07
1998	50	50	1.10

TABLE 6.8. Sulfuric Acid and Hydrogen Fluoride Alkylation.

	H_2SO_4	HF
Temperature (°C)	5–10	21–38
Isobutane/olefin feed ratio (vol)	5–8	10–14
Olefin space velocity (vol olefin/vol acid)	0.2–0.3	
Contact time minutes	20–30	10–15
Acid strength (%)	90–92	85–95
Acid in emulsion (%)	50–60	25–60
Acid used (lb barrel^{-1})	13–30	0.1–0.3
Alkylate from C_4 feed (%):		
Trimethyl pentane		73
Dimethyl hexane		10
Heavy alkylate C_9^+		8
Light ends C_7^-		8

6.3.1. Liquid Acid Processes

All commercial plants operate with a high proportion of isobutane in the circulating gas and conversion depends on the residence time of the gas in the reactor. Typical feeds include olefins derived from catalytic cracking and, more recently, the refnates from MTBE or TAME units. Typical operating conditions are summarized in Table 6.8. Some of the acid catalyst must be removed on a continuous basis for recovery as it becomes diluted with inactive hydrocarbons. Sulfuric acid must be reused in other processes or recovered off-site for resale, a process which costs more than fresh acid. Hydrogen fluoride is usually recovered on site and reused.[10]

6.3.2. The Mechanism of Alkylation with an Acid Catalyst

The generally accepted mechanism for the alkylation reaction is as follows:

- Initiation. Butylene combines with a hydrogen ion to form a secondary butyl carbonium ion:

$$CH_3CH{=}CHCH_3 + H^+ \quad \rightarrow \quad CH_3CH_2\overset{+}{C}HCH_3 \qquad (6.2)$$

- Isomerism. The secondary butyl carbenium ion can then either isomerize to form a tertiary butyl carbonium ion:

$$CH_3CH_2\overset{+}{C}HCH_3 \quad \rightarrow \quad \begin{matrix} CH_3 \\ | \\ CH_3\overset{+}{C} \\ | \\ CH_3 \end{matrix} \qquad (6.3)$$

or react with an isobutane molecule to give a butane molecule and a tertiary butyl carbonium ion:

$$CH_3CH_2\overset{+}{C}HCH_3 + CH_3\overset{\overset{\displaystyle CH_3}{|}}{\underset{\underset{\displaystyle CH_3}{|}}{C}}H \rightarrow CH_3CH_2CH_2CH_3 + CH_3\overset{\overset{\displaystyle CH_3}{|}}{\underset{\underset{\displaystyle CH_3}{|}}{\overset{+}{C}}} \qquad (6.4)$$

- Addition. The tertiary butyl carbonium ion reacts with a butylene molecule to form branched C_8^+ trimethyl pentene carbonium ions:

$$CH_3\overset{\overset{\displaystyle CH_3}{|}}{\underset{\underset{\displaystyle CH_3}{|}}{\overset{+}{C}}} + CH_2{=}CH_2CH_2CH_3 \rightarrow CH_3\overset{\overset{\displaystyle CH_3}{|}}{\underset{\underset{\displaystyle CH_3}{|}}{C}}\,CH^+\,CHCH_2CH_3 \qquad (6.5)$$

- Propagation. The C^+ carbonium ion reacts with an isobutane molecule by hydrogen transfer to give the C_8 paraffin and to generate a tertiary C_4^+ carbonium ion that continues the sequence of alkylation reactions.

A number of side reactions can also take place such as:

- The decomposition of a C_8^+ carbonium ion to the C_8 olefin and a hydrogen ion. This is likely at either a low isobutane/olefin ratio or low acid strength.
- Reaction of a C_8^+ carbonium ion with another C_4 olefin molecule to form a C_{12}^+ carbonium ion. This can react with an isobutane molecule to give a C_{12} paraffin, but in the presence of excess olefin, further chain growth can take place.

6.3.3. Liquid Acid Operating Conditions

During operation hydrocarbons are emulsified in acid, ensuring that the olefin is not in excess, and reacts quickly. Operating pressure is adjusted depending on the boiling point of the hydrocarbon mixture and the need to dissolve the hydrocarbons. The space velocity of the mixed olefins and the catalyst is regulated, depending on the octane level of the product needed and operating temperature. Octane levels increase if the acid strength is low, which means that more spent acid has to be removed from the system. In sulfuric acid alkylation the optimum acid strength decreases with propylene, butylene, and pentylene feeds.

6.3.4. Processes Using Solid-State Acid Catalysts

Several solid acid catalysts were tested in the laboratory for the alkylation of ethylene with isobutane following the introduction of the sulfuric acid alkylation process.[11] These were mostly derived from aluminum chloride or boron trifluoride and were never used in the full-scale production of alkylates.

Operators have always hoped that a successful solid acid catalyst would be discovered because of the problems and hazards involved in handling large quantities of sulfuric and hydrofluoric acids. This situation reached a climax in 1990, when California refineries were faced with the possibility of having to close down units using HF catalysts. From about that time, potential solid-state acid catalysts were investigated by catalyst producers and several have been tested in pilot units.[12] In the meantime, although modifications were introduced to HF units and additives were used to reduce the volatility of the acid, the process has been less popular than the sulfuric acid process.[13] The short-term alternative of converting HF units to the use of sulfuric acid was much more expensive than the use of additives.

Suitable solid acids have been difficult to find. Those used for benzene alkylation are not sufficiently acidic for economic operation and, also, the olefin can polymerize on the catalyst surface. This blocks the active sites and prevents adequate hydrogen transfer. Frequent regeneration of the catalyst is not economic. The use of antimony pentafluoride slurried with the hydrocarbons is one of the options that has been investigated in recent years.[12]

The best solution appears to be the use of an almost insoluble liquid catalyst held within the pores of a suitable inert support. Supported liquid catalysts are well known and can be used with a continuous catalytic regeneration system similar to that developed for catalytic reforming processes. Haldor Topsøe has successfully tested trifluoromethane sulfonic acid in this way since 1993 with a variety of olefin feeds.[14] No formal regeneration was necessary apart from periodic removal of some catalyst for reimpregnation and the recovery of dissolved acid from the alkylate. Both catalyst and support are, therefore, recirculated. The small quantity of polymeric by-products formed (acid soluble oil) appears to be less than that formed in the sulfuric acid process, but slightly more than in the HF process.

6.4. HYDROTREATING

Sulfur impurities in products manufactured from crude oil products are undesirable because hydrogen sulfide, sulfur dioxide, etc., formed during product use. For many years it was possible to obtain acceptable quality gasoline and kerosene by selecting low-sulfur, or *sweet*, crude oils. *Sour* crude oils contain dissolved hydrogen sulfide, mercaptans, organic sulfides, thiophenes, and elemental sulfur in varying amounts. These could be *sweetened* by a number of chemical processes. High-sulfur crude oils were more difficult to desulfurize and the chemical and solvent extraction processes were combined with or replaced by relatively cheap and more efficient catalytic processes that could also remove gum-forming compounds from cracked gasolines.

For a short time from 1946 bauxite or fuller's earth was used without the addition of hydrogen.[15] It was found that sulfides and mercaptans reacted with impurities in the bauxite and that this together with some mild cracking of hydrocarbons produced sufficient hydrogen to hydrogenate thiophenes. It was soon realized that the catalytic desulfurization was actually a mild, selective hydrogenation process that did not saturate aromatics.[16] During the 1950s, cobalt/molybdate catalysts supported on bauxite or Fuller's earth were used. These were similar to the catalysts developed for coal hydrogenation and which were also used to desulfurize steam reforming feeds. The new catalysts were most effective when hydrogen was added to the feed. This also had the effect of reducing the deposition of carbon, and allowed for longer operating cycles before regeneration was necessary. More effective cobalt/molybdate catalysts were developed using γ-alumina as support. The activation step for the catalyst involved the formation of metal sulfides, and when the catalyst was pre-sulfided before use, it was found that light distillates, kerosene and even crude oils could be treated effectively with these catalysts.[18] Operating conditions depended on the boiling range of the fraction being treated. Catalyst temperature was usually limited to about 400^0C in order to avoid excessive carbon deposition while total pressure was increased from 300–500 psig for low-boiling distillates and up to 700–1000 psig for higher-boiling or cracked feeds. Liquid space velocity was usually up to 8 h^{-1}, with a hydrogenn/oil ratio of about 1000 scf of hydrogen per barrel of feed for low-sulfur distillates. Lower space velocities, in the range from 0.5–3 h^{-1}, with hydrogen/oil ratios up to 10,000 scf per barrel, needed to be used for higher-boiling residues. In the hydrotreating of heavy feeds, more carbon was deposited by thermal cracking than in the hydrotreating of lighter feeds. Catalyst regeneration was required after operation for less than 24h.

The use of hydrodesulfurization became more widespread as catalytic naphtha reforming processes were introduced. The operation of platinum catalysts needed an increasingly strict sulfur specification for the naphtha, and as a bonus, the cheap by-product hydrogen from the reforming process could be used to hydrotreat other refinery product streams.

By the early 1960s, hydrotreating capacity in the United States had increased to 2.5 million bpd, with catalyst lifetimes averaging about 5 years. The use of hydrotreating was extended to kerosene, gas oil, and vacuum gas oils as government regulations on sulfur emissions became more stringent and as better cobalt molybdate catalysts became available. By the late 1970s, when atmospheric and vacuum residues were also being desulfurized, US hydrotreating capacity had increased to about eight million bpd. By 1999 US capacity exceeded 10 million bpd, with more than 35 million bpd being treated worldwide.

6.4.1. What Is Hydrotreating?

The term hydrodesulfurization was used to describe processes that removed sulfur compounds from crude oil fractions by reaction with hydrogen. As the processes evolved to include nitrogen and oxygen removal, together with the hydrogenation of aromatics and olefins, the group of processes became known as *hydrotreating*. Hydrotreatment simply results in the conversion of organic sulfur, nitrogen, and oxygen compounds to hydrocarbons and hydrogen sulfide, ammonia, or water, respectively. At the same time olefins and aromatics may be converted to saturated hydrocarbons without any cracking of the hydrocarbons. When high-boiling crude oil fractions are hydrotreated under more severe conditions a proportion of the heavy molecules may crack as impurities are removed. For this reason it is now common to refer to processes that crack a relatively small proportion of the light feeds while sulfur or other impurities are removed as hydrorefining.

Deliberate reduction in the molecular weight of heavy feeds was introduced with the hydrocracking process (Section 6.5). This operates at high pressure, and heavy gas oils and residual fractions are converted to gasoline or other desirable products. Obviously there can be an overlap in defining processes where some precracking is an advantage to refineries. When an overall description is required, the term *hydroprocessing* is often used.

6.4.2. Hydrotreating Processes

In the early hydrotreating processes, sulfur compounds were removed from the light hydrocarbon fractions used in gasoline by hydrogenation over cobalt/molybdate catalysts to produce hydrogen sulfide and a saturated hydrocarbon. Around the same time, it was found that nickel/molybdate catalysts were more active for the hydrogenation of nitrogen compounds to ammonia and a hydrocarbon while also giving some saturation of olefins and aromatics.

In modern refineries both cobalt/molybdate and nickel/molybdate catalysts are now widely used in the purification of various crude oil fractions. These include:

- Straight-run naphthas, used as feedstock for catalytic reforming and steam reforming processes. They must contain less than 1 ppm of sulfur and nitrogen to avoid poisoning platinum or nickel catalysts.
- Cracked gasoline, to hydrogenate undesirable sulfur and nitrogen compounds as well as olefins.

- Middle distillates such as diesel fuel, kerosene, jet fuel, domestic heating oil, and other gas oils, to remove sulfur for environmental reasons. Hydrotreating is also used to increase the smoke point or cetane number by hydrogenating aromatic components.
- Vacuum gas oils, used as catalytic cracker or hydrocracker feeds, to remove sulfur, nitrogen, and metal impurities.
- Atmospheric and vacuum residues, to remove as much sulfur as possible to provide low-sulfur fuel oils. It is also used to hydrogenate asphaltenes and porphyrins to reduce both Conradson carbon and metal contents.

As the boiling point and the specific gravity of the fractions increase, more severe hydrotreating operating conditions are needed. A lower space velocity and more extensive hydrogen recycle are needed to limit deactivation of the catalyst by deposition of coke. The catalyst must be regenerated after shorter intervals and discarded more often than when using light fractions.

A very brief summary of typical operating conditions is shown in Table 6.9 although these can vary significantly for different crude oils and blends containing cracked fractions.

6.4.2.1. Catalyst Production and Operation

Hydrotreating catalysts are usually produced by impregnating preformed γ-alumina supports with aqueous solutions of ammonium molybdate. Particles are then dried and calcined. Molybdenum oxide forms a uniform surface layer on the alumina by reaction with the surface hydroxyl groups. It is important that the surface area of the alumina is consistent with the amount of molybdenum oxide required for the hydrotreating catalyst specification (typically 1 wt% $MoO_3 \equiv 12$ $m^2 g^{-1}$ area).[19]

Table 6.9. Typical Operating Conditions.

	Fraction			
	Naphtha	Gas oil	Vacuum gas oil	Atmospheric residue
Boiling Range (^{0}C)	66–200	240–380	350–560	560+
Liquid space velocity (h^{-1})	6–10	2–4	1–2	0.2–0.5
Hydrogen/oil ratio	60	240	350	7500
Temperature (^{0}C)	280–320	340–360	360–400	400–420
Pressure (atm)	10–25	20–40	30–60	80–100
Regeneration interval	1–3 years	1–2 years	1 year	0.5–1 year
Sulfur content feed (%)	0.05–0.15	1–2	2–3	4–5
Sulfur content product (%)	< 1 ppm	0.1–0.3	0.2–0.4	0.6–0.8
Nitrogen removal (%)	99	45	40–45	45

The molybdenum-treated alumina is then impregnated with either cobalt or nickel nitrate solutions and again dried and calcined. Several reactions can take place during the calcination step. At temperatures in the range 400–500°C, cobalt and nickel nitrates decompose to form oxides on the molybdena/alumina catalyst surface. The deposition of cobalt and nickel in contact with the molybdenum on the catalyst surface is important in forming active sites. As the temperatures are increased both cobalt and nickel oxides react to form aluminates close to the alumina surface. Cobalt aluminate ($CoAl_2O_4$) produces the characteristic blue color of cobalt/molybdate catalysts. The nickel and cobalt ions migrate into the alumina particles as the temperature is increased to 650–700°C, forming bulk aluminates. These are not active catalyst precursors and the temperature of calcinations must be carefully limited to avoid their formation.

A typical support can be prepared by mixing boehmite [AlO(OH)] and water to form a smooth paste, often adding a small volume of dilute nitric acid to peptize the alumina to give increased strength. The paste is extruded to form small-diameter, relatively long, spaghetti-like cylinders, rings, and other shapes, or the boehmite can be granulated to produce spheres. The supports are dried and then calcined at up to 600°C to produce γ-alumina. It is common practice to add up to 1% of silica, often as montmorillonite or bentonite, to stabilize the γ-alumina at high operating temperatures and to act as a lubricant during extrusion. Addition of phosphoric acid appears to produce phosphate groups on the alumina surface that inhibit coke formation as well as to improve molybdenum dispersion.[21] Finally, small amounts of potash can be added to reduce surface acidity and to control coke formation. Typical catalyst specifications are given in Table 6.10 and some catalyst are shown in Figure 6.3.

When using heavy feeds, particularly residues, the catalyst pores at the top of the bed soon become blocked with metals and high molecular-weight hydrocarbons present in the feed. This reduces catalyst activity and the catalyst must be regenerated. In an attempt to overcome the problem and to increase the cycle time, so-called *bimodal* catalysts were introduced containing both large and small pores. Small particles of cellulose, combustible polymers, or carbon were mixed with the boehmite and burned out during the final calcination step of the formation of the γ-alumina support. By using different techniques it was possible to generate pores shaped like bottles which could contain large molecules, and large pores which gave access to small subsidiary pores. These special supports extended the operating cycle before the catalyst had to be regenerated or discarded.

6.4.2.2. Catalyst Handling

The catalyst is supported by a layer of relatively large inert spheres, and must be packed carefully into the reactor. *Dense-loading* procedures to increase the packing density of the catalyst are now available that use a rotating disk at the

Figure 6.3. Typical hydrotreating catalysts. Reproduced with permission from Haldor Topsøe A/S.

end of a hose to distribute the catalyst. The packing density of spheres, extrudates, and trilobe shapes can be increased by 8%, 12%, and 20%, respectively. An advantage of the *dense-loading* technique is that the catalyst layer is always flat as loading proceeds. Catalyst particles, which are very long and thin, lie horizontally and this improves the distribution of liquid residue throughout the bed, resulting in better control of temperature distribution and conversion. When a catalyst is packed manually into a converter, the surface layer is rarely completely flat, but usually has rather a convex or concave orientation. Process liquids then run preferentially down some channels, resulting in poor catalyst and plane performance.[22] In the design for modern hydrotreating processes, up to six different catalyst types in a two-or three-bed reactor system may be used. The idea is to have large diameter particles at the top of the reactor to cope with solids and impurities in the feed. Different catalysts having different compositions, shapes and activities can also be used.

A layer of ceramic balls is placed on top of the catalyst bed to help improve distribution of the feed and to catch some of the solid particles introduced by the

Table 6.10. Typical Hydrotreating Catalyst Specification.

	Cobalt molybdate	Nickel molybdate
Composition		
Cobalt oxide %	4–5	—
Nickel oxide %	—	4–6
Molybdenum oxide %	12–20	15–20
Phosphorus %	Up to 2	Up to 2
Alumina	Balance	Balance
Impurities SO_4, Na_2O, Fe_2O_3	1%, 0.06%, 0.6%, respectively	1%, 0.06%, 0.6%, respectively
Properties		
Loss on ignition 500°C (wt%)	2–4	2–4
Bulk density (kg liter^{-1})	0.5–0.7	0.5–0.7
Surface area (m^2 g^{-1})	200–300	200–300
Pore volume (ml g^{-1})	0.5–0.7	0.5–0.7
Mean pore radius (nm)	4.5–5.0	4.5–5.0
Shapes	**Size (mm)**	
Extrudates	1.6, 1.3, 0.8 diameter × 3–5 long	
Rings	5 × 5, 3 × 3	
Trilobes	1.6, 1.3 × 5	
Quadrilobes	1.6, 1.3 × 5	
Used catalyst	Before regeneration	After regeneration
Sulfur (wt%)	Up to 10	< 0.5
Carbon (wt%)	8–12	0.1–0.3
Loss on ignition (wt%)	17–20	< 0.5

flow of the reactants. In older plants, wire mesh baskets were often placed on top of the catalyst bed and filled with pebbles. These were part of the vessel design to catch solids and to distribute liquid feed, and were known as *trash* baskets. Although still in use, better filtering of the feed and graded catalyst layers with the larger catalyst particles at the top of the beds have largely replaced the use of baskets.

6.4.2.3. *Activating the Catalyst*

Hydrotreating catalyst is activated by presulfiding with hydrogen sulfide. This is achieved by circulating sulfur compounds in a stream of hydrogen at temperatures too low to reduce the oxides to metals. Small molybdenum disulfide platelets (truncated hexagons) that incorporate cobalt or nickel ions at the plate edges are formed to produce the active CoMoS or NiMoS sites. The surface area of the alumina support is critical to achieve the appropriate distribution of molybdenum, and the optimum ratio of nickel or cobalt to molybdenum is approximately 1:4. In the case of cobalt molybdate, if the atomic ratio exceeds 30:1, cobalt sulfide (Co_9S_8) crystals form on the molybdenum disulfide in preference to the active mixed sulfide. The subsurface cobalt/aluminate does not sulfide. The same mechanism probably applies to nickel/molybate and the metal ratio should be controlled to avoid the formation of nickel sulfide (Ni_3S_2).

Catalyst can be presulfided in the reactor in a controlled manner using the following procedure:

- The reactor is freed from moisture by purging with dry air, before purging with nitrogen to remove oxygen. Hydrogen is then circulated at the operating pressure of the process and at 175^0C.
- Feed containing 1–2% sulfur is circulated and the temperature is increased to $200–250^0C$ to start the sulfiding process.
- Some of the sulfur in the feed is incorporated into the catalyst as metal sulfides. The level of sulfur in the feed leaving the catalyst bed reaches a constant level when this stage has come to quasi-equilibrium. The temperature is then increased to 300^0C and the process repeated, giving further sulfiding of the catalyst. Normal operation can be started after this stage.

Presulfiding normally takes 10–20 h to complete. Sulfur uptake depends on the metal oxide content of the catalyst used but is normally in the range 7–10%. Good gas distribution through the catalyst bed is very important to sulfide evenly all of the catalyst. *Pre-sulfurised* catalysts were introduced to avoid problems associated with poor gas distribution and temperature control. They also had the added bonus of a quicker and simpler start-up procedure. The early pre-sulfurised catalysts were prepared by treating green catalyst with elemental sulfur to fill the pores within the catalyst particle. The active sulfides were formed subsequently, when the catalyst was heated in a stream of hydrogen in the reactor. This procedure led to a significant temperature rise during start-up, which resulted in the vaporization and subsequent deposition of sulfur in downstream equipment.

Improved pre-sulfurised catalysts containing polysulfides or metal oxysulfides were later developed to overcome these problems.[24] Polysulfides still react rapidly with hydrogen during conversion of the cobalt, nickel, and molybdenum oxides to sulfides and produce a temperature rise in the catalyst bed. On the other hand, metal oxysulfides react slowly over a much wider temperature range and do not produce an exotherm.

Presulfurized catalysts can be commissioned satisfactorily with a flow of hydrogen in the temperature range $150–300^0C$. There is no significant temperature rise in the catalyst bed and water evolves gradually between $200–300^0C$.

6.4.2.4. *Catalyst Operation*

The early processes used to hydrotreat naphthas only required single beds, but modern plants treating heavier feeds are more complicated. Several catalyst beds with a variety of catalysts are usually required.[25] Interbed cooling with cold gas quench or liquid feed may also be necessary. Hydrotreating reactions are exothermic and cooling is essential to avoid catalyst damage.

Different catalysts are used in separate beds, because several different reactions are taking place at the same time. Top beds operate as guards that remove metals from vacuum gas oils and residues, as well as promoting sulfur and nitrogen removal reactions. Guard catalysts consist of large particles which have large pores to absorb metal porphyrins and Conradson carbon. The large particle size is needed to limit any increase in pressure drop caused by trapped impurities. There is usually an improvement in hydrotreating performance if the molybdenum and nickel content of these catalysts are high. Pressure drop is still the most significant operating problem.

A selection of the different sulfurous and nitrogenous impurities in a typical feed and the reaction sequences involved in their removal is shown in Table 6.11.

Coke is gradually deposited onto the catalyst surface during operation, even when using naphtha and middle distillates as feed, and this leads to a reduction in catalyst activity and an increase in pressure drop throughout the reactor. The plant may need to be shut down for catalyst regeneration to restore activity and to reduce pressure drop. Some liquid feeds do not vaporize completelyunder operating conditions and uneven liquid distribution through the beds also leads to poor performance.

6.4.2.5. *Catalyst Regeneration*

A typical procedure to regenerate a hydrotreating catalyst in the reactor is as follows:

- Isolate feed to the reactor and purge all volatile organic material from the catalyst with hydrogen.
- Purge hydrogen from the reactor with steam and increase temperature to 350^0C.
- Add air to the steam, keeping the maximum bed temperature to 400°C. There is a localized hot zone passing through the bed as the residual carbon is burnt to carbon dioxide, and as the hot spot passes through the bed, increase the temperature to 450^0C.
- Maintain temperature at 450^0C until no carbon dioxide is detected at the bed outlet and then purge residual oxygen from the reactor with steam.

TABLE 6.11. Hydrotreating Reactions.

Impurity	
	Hydrodesulfurization (HDS)
Mercaptans	$RSH + H_2 \rightarrow RH + H_2S$
Sulfides	$RSR + 2\,H_2 \rightarrow 2\,RH + H_2S$
Disulfides	$RSSH + 3\,H_2 \rightarrow 2\,RH + 2\,H_2S$
Thiophene	$+\,4\,H_2 \rightarrow C_4H_{10} + H_2S$
Benzothiophene	$+\,3\,H_2 \rightarrow$ $+ H_2S$
Dibenzothiophene	$+\,2\,H_2 \rightarrow$ $+ H_2S$
	Hydrodenitrogenation (HDN)
Amines	$RNH_2 + 2\,H_2 \rightarrow RH + NH_3$
Pyrrole	$+\,4\,H_2 \rightarrow C_4H_{10} + NH_3$
Pyridine	$+\,5\,H_2 \rightarrow C_5H_{12} + NH_3$
Indole	$+\,4\,H_2 \rightarrow$ $+ NH_3$
Quinoline	$+\,4\,H_2 \rightarrow$ $+ NH_3$
Carbazole	$+\,4\,H_2 \rightarrow$ $+ NH_3$

Note: Olefins are also saturated during its hydrotreating process.

The catalyst can be reused following regeneration provided that it has not been deactivated by metals such as vanadium, nickel, sodium from the feed or iron scale. Deactivated catalysts can contain between 10 and 30% of metal deposits, which may be recovered for economic or environmental reasons. Oxidation of cobalt or nickel sulfides to sulfate, during in situ regeneration, can gradually lead to deactivation of the catalyst.

With the large volumes of catalyst now used it has become usual to regenerate catalyst externally in special rotary kilns to save time and energy.[26] In situ regeneration can often take up to 20 days. External regeneration can be quicker and gives better temperature control and gas circulation, so that regenerated catalyst can be cleaner and almost as active as new catalyst. At the same time, discharged catalyst can be separated into cobalt or nickel types and different sized particles.

Apart from the possible need to hold a spare catalyst charge, external regeneration is more economic. However, as most refineries operate several hydrotreating units, it is usual to cycle regenerated catalysts from the less severe naphtha hydrotreating to middle distillate and finally to residue treatments.

External regeneration in the rotary kiln requires two stages:

- Sulfur is removed from the catalyst at 200–250°C by gentle oxidation of sulfides to sulfur dioxide in a flow of air. Temperature control is important to avoid complete oxidation of sulfide to sulfate at high temperatures.
- Carbon can then be oxidized in air at 400°C, taking care to limit the oxygen content of the hot gas and avoid temperature runaway, which would deactivate catalyst.

The operating cycle time before regeneration is needed varies considerably depending on the quality of the feed being used, as shown in Table 6.9. The type and size of catalyst is also important. For example, a naphtha hydrotreater can probably operate for at least two years before regeneration, whereas diesel oil and residue hydrotreaters only operate for a maximum of about 350 and 250 days, respectively.

6.5. HYDROCRACKING

Hydrocracking is a process for the conversion of heavy gas oils, including low-value fractions not suitable for catalytic cracking, into light gases, gasoline, jet fuel, and diesel oil. The process operates at a high hydrogen pressure (60–200 atm) and moderate temperatures in the range 350–450°C. Dual-function catalysts are used to combine the hydrogenation activity of hydrotreating catalysts with an acidic cracking support.

The process was introduced in modern refineries by the Union Oil Company (Unocal) in 1964,[27] but dates back to the 1920s when coal hydrogenation processes were developed in Germany and the United Kingdom. Gas oil was hydrocracked by Standard Oil at their Baton Rouge Refinery, during the 1930s using the German catalysts. The process produced aviation gasoline during World War II in both the United States and the United Kingdom.

The nickel catalysts supported on silica/alumina introduced by Chevron[28] in the low-temperature isocracking process in 1959 were used to upgrade distillates and to produce low-molecular-weight isoparaffins for high-octane gasoline. When the Union Oil Company used a more acidic zeolite support in 1964 the process became more versatile and a wider range of high-boiling-point feeds including middle distillates could be converted to high-quality products.[29] Organic nitrogen and sulfur compounds present in these feeds were also hydrogenated, forming ammonia and hydrogen sulfide, and removed from the products. Since the 1970s there has been a continuous improvement to the process and catalysts, as lower-quality feeds have been upgraded.

6.5.1. Hydrocracking Processes

Low-value distillates, including heavy-cycle oils from FCC units, thermal and coker gas oils, and other heavy-vacuum gas oils, are cracked to produce naphtha, jet fuel, and diesel oils. The reaction mechanism is the same as in catalytic cracking and some aromatic products are also hydrogenated.

There are two steps in the hydrocracking process. Following hydrotreatment to remove organic nitrogen and sulfur compounds, the feed is cracked. The main hydrocracking reactions in the second step which proceed via a carbenium ion mechanism are as follows:

- Dehydrogenation of long-chain paraffins; isomerization and cracking of the resulting olefins at the third carbon atom; hydrogenation of olefins to isoparaffins.
- Stepwise ring hydrogenation and hydrocracking of polycyclic aromatics.
- Dealkylation of alkyl aromatics and naphthenes.
- Saturation and cracking of aromatics.
- Hydrogenation of coke precursors formed during the hydrocracking reactions.

The proportion of different products formed depends on the catalysts used. Weak hydrogenation catalysts with a strongly acid support produce low-boiling gasoline with a high iso-to-normal paraffin ratio. Catalysts with high hydrogenation activity and low support acidity provide middle distillates. The hydrogen added for the hydrogenation reactions suppresses coke formation, which would quickly deactivate the catalyst. Catalysts operate for 2–6 years before being regenerated or replaced.

Feed and hydrogen are circulated through the catalyst at the conditions selected for the range of products required. Reaction temperatures in the range $250–500^0C$ and pressures in the range 60–200 atm have been used. Hydrotreating and hydrocracking reactions are exothermic and the catalyst is loaded into several beds so that temperature can be controlled by the addition of a cold hydrogen quench. Several different process designs have been used since 1959.

TABLE 6.12. Hydrocracking Processes.

Feed	Aromatic cycle oil, heavy gas oil, thermal/coker gas oil
Product	LPG, gasoline, middle distillates
Single-stage process/with recycle:	
One or two beds, relatively low sulfur and nitrogen in feed	
Feed temperature (^{0}C)	350–425
Pressure (atm)	60–200
Conversion (%)	40–70
Aromatics (vol%)	~15
Catalyst (sulfided)	Hydrotreating plus ammonia tolerant hydrocracking types.
Two-stage process:	
Two reactors, nitrogen and/or sulfur compounds can be removed between beds.	
First stage:	
Feed temperature (^{0}C)	350–425
Pressure (atm)	60–200
Conversion (%)	40–50
Catalyst (sulfided)	Early units: nickel molybdate/alumina; high nitrogen conversion; little cracking. Later units: nickel tungstate/alumina; High nitrogen conversion up to 40% cracking.
Second stage:	
Feed temperature (^{0}C)	~260–375
Pressure (atm)	60–200
Conversion (%)	40–50 (better aromatic conversion than single stage)
Aromatics (vol%)	~2
Catalyst	Early units: sulfided nickel molybdate/silica alumina catalysts. Later units: Catalysts with high zeolite content.

These can be divided into two different groups,[27] and details are given in Table 6.12.

6.5.1.1. *Single-Stage Processes*

In the early hydrocrackers, the feed was hydrotreated in a separate unit to remove organic nitrogen compounds, which are catalyst poisons. A nickel/tungsten/silica/ alumina hydrocracking catalyst was then used in a single vessel operating in the range 350–450^{0}C. Conversion was in the range 40–70%, and unconverted feed was recirculated after fractionation. These units operate with relatively low levels of impurity in the feed, but recycle gas contains hydrogen sulfide. However, early units still operating now use the new zeolite-based catalysts to treat higher-boiling feeds.

Design was improved by combining the two stages. Feed from the hydrotreating catalyst passed directly to the hydrocracking catalyst for conversion
and unconverted feed was recycled back to the hydrocracking section. Hydrotreating and hydrocracking catalysts were used so both feed and hydrogen
passed directly from one reactor to the other. This meant that after hydrotreatment, the hydrogen still contained ammonia and hydrogen sulfide when it entered the hydrocracker. This was usually acceptable because ammonia did not
poison the zeolite-based catalysts being used at the time.

6.5.1.2. *Two-Stage Processes*

In two-stage processes mainly hydrotreating and some cracking takes place in
the first reactor at $350-425^0C$. The cracked products are then fractionated before
the unconverted feed passes to the second reactor which contains hydrocracking
catalyst. Recycled gas is also scrubbed to remove ammonia and hydrogen sulfide
if palladium-based hydrocracking catalysts are used in the second reactor. Nickel molybdate or tungstate catalysts must be sulfided before use, so there is no
need to remove hydrogen sulfide from recycle hydrogen. Hydrocracking catalysts can operate with or without sulfur at $260-375^0C$. Feeds with high nitrogen
content can therefore be hydrocracked at a lower second-stage temperature and
higher space velocity than in single-stage processes.

By using a high-activity hydrotreating catalyst in the first reactor, with a zeolite catalyst in the second, overall conversion can be increased to 50–80%. Unconverted feed is still recycled, but a purge of about 5–10% may be needed to
remove accumulated polynuclear aromatics from the circulating feed. These can
cause deactivation of the catalyst, and foul the reactors. Typical compounds include coronene (seven fused rings) and ovalene (ten fused rings), both of which
are not very soluble. They are brightly colored and have been called the *red
death!*[30]

6.5.1.3. *Once-Through Process*

More recently a very flexible hydrocracking process has been operated, based on
the single-stage process.[27] Conversion of the feed can range between 40–90%,
but unconverted feed is not recycled. Feeds including heavy-vacuum gas oils,
heavy-cycle oil, coker gas oil, and deasphalted oil are cracked to produce high-
quality middle distillates and naphtha for catalytic reformers. Residue from the
fractionator is used as catalytic cracker and ethylene plant feed, lube oil stocks,
or high-quality fuel oil. Organic nitrogen is almost completely removed from the
product and sulfur content is very low. Moderately active zeolite catalysts are
used.

6.5.2. Hydrocracking Catalysts

It is important that hydrocracking catalysts have both high cracking and hydrogenation activities. Typical catalysts are generally made with an acid support containing either nickel/tungstate or palladium as the hydrogenation catalyst. Typical examples are shown in Table 6.13. All nickel/molybdate and nickel/tungstate catalysts can only work effectively in the form of sulfides in hydrotreating and hydrocracking reactions. For this reason the hydrogen sulfide content of recycle hydrogen must be controlled at an appropriate level. Ammonia is normally scrubbed from recycle gas.

6.5.2.1. *Acid Supports*

The first acid support to be used in hydrocracking catalysts was an amorphous silica/alumina, but following the successful application of zeolites in catalytic cracking processes in 1964, this has generally been replaced by Y-zeolite. The most important feature of the support, apart from the high acidity needed to crack high-molecular-weight hydrocarbons, is high thermal stability to withstand the process conditions. Pore size is not as critical when using heavy feed for hydrocracking as is the case for hydrotreating, provided that the surface area is sufficiently high. Many different support formulations have been patented, with many of them similar to FCC catalysts. All are produced as extrudates, held together with binders such as peptized alumina.

TABLE 6.13. Hydrocracking Catalysts.

Design	Composition
First stage/single stage	
Catalyst	5–10% NiO/10–20% MoO$_3$/1–2% P$_2$O$_5$
Support	Alumina containing varying amounts of silica
Porosity (cm^3 g^{-1})	0.4–0.6
Shape	Extrudates Rings
Second stage	
Catalyst	5–10%NiO/10–20% MoO$_3$ 5–10% NiO/10–20% WO$_3$ Palladium 0.5%
Support	Amorphous silica/alumina 0–80% Ultrastable Y-zeolite 20–80% Mixed zeolite/silica/alumina (zeolite 0–100%) Binders such as peptized alumina (0–20%)
Porosity (cm^3 g^{-1})	0.4–0.6

Zeolites, often combined with amorphous alumina or silica/alumina and a suitable binder, are now the most widely used acidic component to hydrocrack heavy feeds. Hydrogen-exchanged and dealuminated Y-zeolites are normally preferred. Zeolites are more active than amorphous materials because of their uniform and acidic structure. Less coke is formed during operation, resulting in less deactivation of the catalyst so that operational life can be longer than six years. With nitrogen-free feed the zeolite is active at lower operating temperatures and heavier feeds can be cracked.

Modification of the silica/alumina ratio of the Y-zeolite used in catalytic cracking catalysts can also improve its hydrocracking performance. It was found that dealuminated Y-zeolites give high diesel selectivity, whereas untreated Y-zeolite provides a hydrocracking catalyst with high selectivity for gasoline.

A hydrocracking catalyst support can therefore be tailor-made depending on how the catalyst is to be operated and the product type needed. To obtain maximum conversion to gasoline at the lowest operating temperature, up to 80% Y-zeolite and about 20% of a peptized alumina binder is used as support. High middle-distillate conversion is obtained at higher operating temperatures with a support containing about 10% dealuminated Y-zeolite, 70% alumina or silica/alumina in varying proportions, and 20% of the binder. The appropriate Y-zeolite and the proportion to be used in the catalyst can be selected for the products, as required.

6.5.2.2. Hydrogenation Catalysts

Sulfided nickel/molybdate catalysts supported on a suitable γ-alumina are used for first-stage hydrocracking reactions and they have a higher nickel molybdate content than typical hydrotreating catalysts. Sulfided nickel/tungstate or palladium, supported on zeolite or silica/alumina, is used in the second hydrocracking stage. Palladium catalysts are active if the feed contains residual sulfur compounds but are not active for the hydrogenation of benzene rings. Hydrogen sulfide need not, therefore, be removed from the recirculating hydrogen unless it is necessary to remove all aromatics.

6.5.2.3. Catalyst Preparation

Nickel/molybdate hydrocracking catalysts are made by impregnating preformed supports with solutions of nickel and molybdenum salts.[31,32] The addition of ammonia or phosphoric acid to the solutions before impregnation is claimed to simplify the procedure and to improve activity of the catalyst.

Palladium catalysts are made by ion exchanging the zeolite with a solution of palladium chloride containing ammonia followed by careful washing and drying.[31,32] The tetramine complex can be decomposed in air at temperatures above 800^0C to produce palladium metal in the form of finely divided crystallites. Direct reduction with hydrogen can lead to agglomeration of the palladium crystals, which are inactive. The zeolite is extruded with a binder and any other necessary component of the support.

6.5.2.4. Catalyst Activity

Catalysts are activated before use. Nickel catalysts are carefully presulfided either before or after loading to the reactor with the same organic sulfur compounds and procedures used for hydrotreating catalysts. Palladium catalysts are briefly activated in hydrogen at about 350^0C, taking care not to overheat the bed, so that the crystallites do not sinter.

Silica/alumina hydrocracking catalysts are severely poisoned by ammonia. The feed must therefore be hydrotreated and hydrogen recycle must be scrubbed to eliminate as much organic nitrogen and ammonia as possible. Zeolites are slightly less affected by typical ammonia levels and are widely used in the more efficient two-reactor single-stage process.[27] Catalysts should, however, contain more zeolite if the ammonia content is variable. The active metal content must also be increased to achieve the same conversion in the presence of ammonia. The cracking activity of the support and the hydrogenation activity of the metal component must therefore be carefully optimized to obtain maximum conversion with different operating conditions and feeds.

6.5.2.5. Catalyst Reactivation

A typical catalyst can operate for a number of years, and any slow loss of activity can be compensated for by a gradual increase in process operating temperature until the reactor temperature limit is reached. Regeneration is then possible, either in the reactor or externally, by burning accumulated coke in dilute air. This restores the activity of most nickel/molybdate or nickel/tungstate catalysts, which must then be resulfided before further use.

Deactivated palladium supported on zeolite can also be regenerated, and the catalyst must also be reactivated before being reused. This is because palladium in the zeolite framework sinters. By treating the catalyst with an excess of ammonium hydroxide solution, however, the agglomerated palladium dissolves to form a tetramine complex. Calcination in air then decomposes the complex and the original palladium distribution and activity are restored.[31]

The first Y-zeolite hydrocracking catalysts contained residual sodium ions in the sodalite cages that were mobile during operation and they entered the supercage. This led to a loss of cracking activity. Treatment of the zeolite with an ammonium salt solution removed the mobile sodium ions and restored acidity. The redistribution of palladium with ammonia solution could be combined with an exchange of sodium ions to rejuvenate the catalyst in one step.[32] This was done before reactivation by burning off the carbon deposits.

6.6. CATALYTIC REFORMING

Catalytic reforming is a key refinery process. It improves the octane rating of virgin naphthas and light distillates so they can be used in gasoline formulations. The process has also become an important source of aromatics for use in petrochemical production.[33]

Low-octane, sulfur-and nitrogen-free paraffins and naphthas in the boiling range 80–160^{0}C are converted to higher octane iso-paraffins and aromatic compounds by a sequence of dehydrogenation, dehydrocyclisation and isomerization reactions. Hydrocracking reactions are less important, but can also convert the higher-molecular-weight paraffins to more useful products, but with a reduction in the yield of C_5^+ products.

When gasoline is the main product required, it is usual to remove the C_5–C_6 straight-run naphtha cut before reforming the 100–160^{0}C fraction. This is because separate isomerization of the C_5–C_6 paraffins (Section 6.10) gives an overall improvement in octane number. If aromatics are required, a lower-boiling fraction, in the range 80–120^{0}C, is reformed. The heavy naphthas are not reformed because coke forms more readily and quickly leads to deactivation of the catalyst.

The first naphtha-reforming catalysts were used during World War II in a process known as hydroforming and were based on molybdenum or chromium oxides supported on alumina.[34] At the time, the new process provided high-octane aviation fuel and the toluene needed for the production of explosives. The decreased demand for these products after the war meant that the seven operating plants were idle by 1952. In any case the preferred catalyst (9% MoO_3/alumina) was neither active enough nor selective enough to give a really economic performance and needed frequent regeneration. Vladimir Haensel, who invented the platinum reforming catalyst,[35] recalled that during and after the wartime period, UOP was investigating a hydrocracking process that used desulfurized kerosene as feed. The work involved an analytical procedure for six-membered rings, based on the catalytic dehydrogenation of naphthenes with 3% platinum supported on carbon. Haensel tells how this led to the treatment of virgin naphthas containing sulfur with a similar platinum catalyst supported on silica. Operation at up to 450^{0}C and 500 psig, with 5 mol of hydrogen recycle per mole of feed, showed not only that the virgin straight-run naphthas were reformed to give a high-octane gasoline product but that sulfur poisoning was avoided. At the same time, however, the silica support and others such as silica/alumina promoted hydrocracking reactions. Further investigations produced an economic platinum/alumina catalyst with a platinum content in the range 0.3–0.7%.[36] The catalyst proved to be particularly effective when the alumina was prepared from aluminum chloride rather than the nitrate. This led to the treatment of supports with chloride or fluoride to improve acidity—a practice that has continued ever since. The new UOP process was announced at the Western

Figure 6.4. The first UOP catalytic reforming plant at the Old Dutch Refinery, Muskegon, Michigan, USA. Reproduced with permission from UOP, LLC.

Petroleum Refiners' Association meeting in San Antonio, Texas, in April 1949. The Old Dutch Refinery at Muskegon, Michigan, was immediately converted to catalytic reforming (Figure 6.4). A new era of refinery processing had begun.

Both the platinum component and the acidic alumina support play an important part in the *dual-function* catalysts, which, with the recirculation of hydrogen, operate for several years, with periodic regeneration to remove coke deposits. The first platinum catalysts have now been improved by the addition of a second metal component that enhances activity and increases the period before regeneration is needed. The first *bimetallic* catalyst, containing platinum and rhenium oxide, was introduced by Chevron in 1969 and considerably improved reformer operation.[37] Soon after this breakthrough was announced, other catalysts containing iridium[38] (1972, Exxon) and tin[39] (1973, UOP), were also introduced.[38] Alumina is still the most economic support but its acidity must always be controlled by the careful incorporation of chloride during production and further additions to the feed during operation.

The dehydrogenation of naphthenes to aromatics results in the production of large volumes of extremely valuable, co-product hydrogen. This has been vital for the development of refinery hydrodesulfurization.

6.6.1. Naphtha Reforming Reactions

The principal reforming reactions are shown in Table 6.14. The isomerization of *n*-paraffins proceeds via a dehydrogenation step on the platinum component of the catalyst to an *n*-olefin intermediate. The *n*-olefin then migrates to the acidic alumina site and isomerizes to the iso-olefin. Finally the iso-olefin is hydrogenated to the isoparaffin by the platinum. This mechanism requires good dispersion of platinum on the highly acidic, chloride-treated alumina support.

Similar mechanisms apply to the isomerization and dehydrocyclization of *n*-hexane to methyl cyclopentane and then benzene, as shown in Table 6.15.

6.6.1.1. *Reformer Operation*

The combined sequence reforming reaction is endothermic, and there is a significant temperature decrease across the catalyst bed. Heat has to be supplied continuously to maintain operation, so catalyst is loaded into a number of beds or reactors with interbed heating of the reaction mixture. The process is complicated because of the different reactions and temperature changes taking place in each bed. Variable heat balances were compensated in early plants by loading increasing volumes of catalyst in each bed from the top to the bottom of each reactor. An alternative arrangement was to load several reactors with equal volumes of catalyst while having a similar reactor in reserve to replace any operating reactor needing regeneration.

Three different reactor systems have evolved since the process was first introduced in 1952:

Table 6.14. Naphtha Reforming Reactions.

Reaction	Heat balance

Dehydrogenation:
    Naphthenes to aromatics
    (e.g., methylcyclohexane to toluene)
	Extremely endothermic

Dehydroisomerization: (e.g., methyl cyclopentane to benzene)	Endothermic
Isomerization: Normal paraffins to branched paraffins	Mildly exothermic
Dehydrocyclization: Normal paraffins to aromatics	Endothermic
Hydrocracking: Normal heptane to C_4/C_3 and C_6/C_1 paraffins	Exothermic

TABLE 6.15. Catalytic Rreforming of *n*-Hexane and iso-Hexanes.

| iso-hexane |
| ↑↓ |

iso-hexene

(iso-hexyl carbenium ion)

↓↑

| n-hexane |
| ↑↓ |

n-hexene

(n-hexene carbenium ion)

↕

methylcyclopentane

↑↓

methylcyclopentene

↑↓

methylcyclopentadiene

←Isomerization on acidic alumina→

cyclohexane

↑↓

cyclohexene

↑↓

cyclohexadiene

↑↓

benzene

← Hydrogenation and dehydrogenation on platinum catalyst →

Note: From Mills, Heinemann, Milliken and Oblad, Ind. and Eng. Chemistry, 1953.

- The semi-regenerative process operates continuously over long periods of up to 9–12 months. There are usually three reactors in series that contain about 15–20%, 30–35%, and 50–60% respectively of the catalyst's charge, depending on the feed used. Operation is in the range 10–25 atm for different feeds and reforming severity. When the catalyst activity has decreased, resulting in lower liquid yields and more hydrocracking activity, the catalyst is regenerated in situ, using air at about 8 atm. The catalyst can usually be regenerated up to ten times before the performance becomes uneconomic and it has to be replaced. Reaction conditions are summarized in Table 6.16.

- Cyclic processes generally operate with four reactors although as many as six are used if frequent regeneration is necessary. All reactors contain equal volumes of catalyst and operate at about 480^0C and 20 atm, with a temperature decrease of 25^0–30^0C across each bed. Low hydrogen/feed ratios can be used. One reactor is kept in reserve so that it can be switched to replace any on-line reactor requiring regeneration, so giving continuous operation. Cyclic processes operate at low pressure to give high yields of a high-octane product, but with relatively high coke laydown. The interval between switching reactors to regenerate the catalyst can be rather short, for example, after processing only about eight barrels of feed per pound of catalyst used, with catalyst regeneration either daily or perhaps after only a few days. This complicates the process engineering because alternating hydrogen and oxygen streams have to be provided. Catalyst life is also relatively short as a result of the frequent regenerationneeded and can be less than about 200 barrels of feed per pound of catalyst.

- Contin uous catalyst regeneration (CCR) processes were introduced in 1971 to avoid the shut-down intervals necessary in semi-regenerative reformers. The reformers have four beds, either stacked vertically or placed side by side horizontally, so that the catalyst can be moved continuously from one catalyst bed to another and ultimately, into the regenerator (Fig. 6.5). After regeneration, the catalyst is reduced before being returned to the first reactor to begin the sequence again. The average residence time of a catalyst in a CCR unit has decreased since the process was first introduced. The first design in 1971 recycled the catalyst charge about 12 times a year. By 1980 the recycle rate was 50 times a year, but as operating conditions have become even more severe, the rate can now be as high as 150 times a year.[40] Despite this, the catalysts used have lasted for more than seven years. Operating conditions are shown in Table 6.17.

TABLE 6.16. Semi-Regenerative Catalytic Reforming.

	Catalyst (vol %)	Temperature (^{0}C)		RON out	Carbon end of cycle (%)
		Inlet	outlet		
Reactions					
Reactor 1	15–20	500	430	65	3–10
Dehydrogenation					
Dehydroisomerization					
Reactor 2	30–35	500	470	80	5–15
Dehydrogenation					
Dehydroisomerization					
Some hydrocracking					
Dehydrocyclization					
Reactor 3	50	500	495	95–100	8–20
Hydrocracking					
Dehydrocyclization					
Conditions					
Feed (m^{3}h^{-1})				85–90	
Boiling range (^{0}C)				~100–160	
H$_2$/HC				4–5	
LHSV (h^{-1})				1–3	
Pressure (atm)				10–25	
Recycle gas (% H$_2$)				80	
Yield reformate (wt%)				70–80	
Hydrogen (wt%)				2–3	
Feed/products		Feed		Products	
Paraffins (%)		45–55		30–50	
Naphthenes (%)		30–40		5–10	
Aromatics (%)		5–10		45–60	

Platforming converts aliphatics into aromatics

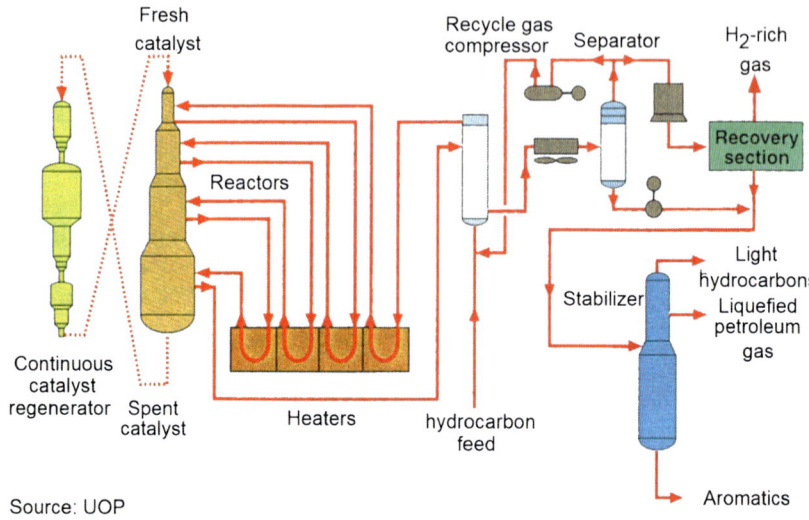

Source: UOP

Figure 6.5. Modern catalytic reforming plant with continuous catalyst regeneration (by permission of UOP).

TABLE 6.17. Continuous Catalytic Regeneration.

	1971	1990
Throughput (bbl/day)	12,000	25,000
Pressure (atm)	9–10	3–4
Hydrogen/hydrocarbon	2–3	2–3
Regeneration (lb h^{-1})	200	4500
Regenerations per year	12	150
Liquid space velocity (h^{-1})	1.6	1.6
Catalyst charge (tonnes)	60	110
RON	95	102
Temperature (^{0}C)	450	510
Yield H$_2$ (wt%)	3	4
C$_5^+$ (vol%)	78	82
Aromatics (vol%)	54	58

The CCR process has been very successful and is now used extensively throughout the world. The main advantage is that the coke content of catalyst before regeneration is as low as 5–6%, despite the severe operating conditions at hydrogen pressure as low as 3 atm. Yields of high-octane product and hydrogen were significantly increased. Old catalyst can also be withdrawn from the reactor and replaced with fresh catalyst when activity declines. High yields are favored by low pressure and low hydrogen circulation, but more coke is formed. Despite the higher capital cost of CCR reformers the higher yields and more flexible operation make the process very popular.

By the late 1990s more than 900 reformers were operating throughout the world.[41] The semi-regenerative process was being used in 60–70% of them, with continuous regeneration processes being used in 15–20%.[42] Cyclic processes were only used in about 10–15% of the reformers. Since 1990, however, almost all of the new reformers being built or designed use the continuous regeneration process. As larger-capacity reformers have been built, axial flow through the catalyst beds has been replaced by radial flow because this configuration allows for better gas distribution and reduced pressure drop.

Catalyst operation in the catalytic reforming process is influenced by feed composition, hydrogen ratio, and operating conditions. The benefits obtained by any changes in design always have to be balanced with certain disadvantages. For example, although straight-run naphthas boiling at up to about 200^0C can be used in catalytic reforming processes, the use of lower-boiling naphthas gives better results. More coke forms on the catalyst as high-molecular-weight hydrocarbons are reformed, and it is estimated that above 190^0C every 13^0C increase in the boiling range of the naphtha can reduce the period before regeneration by 35%.

Sulfur impurities in naphtha poison platinum hydrogenation activity and nitrogen impurities neutralize acid sites. Both poisons have to be carefully removed before the naphtha can be used. Despite this need to remove sulfur compounds from reformer feeds, it is also necessary to selectively poison the active platinum in new or regenerated catalysts by adding about 0.25% of sulfur to the catalyst in a sulfiding procedure using 100–200 ppm H_2S in the feed. This limits the extent to which the exothermic hydrogenolysis reactions that deposit coke and lead to sintered catalyst can occur. Without this sulfur conditioning, the yield of product can be reduced by 1% and the run length reduced by up to 15–20%. Unlike chloride, sulfur should not be added continuously during operation and the normal sulfur specification of naphtha must be applied.

Since the first catalytic reformers were used in the 1950s the hydrogen/ hydrocarbon mole ratio and the reformer operating pressure have both been gradually decreased. These developments resulted from improvements in the alumina support, the use of bimetallic catalysts, and finally the introduction of the low-pressure, continuous catalyst regeneration processes. The trends in operation are shown in Table 6.18.

TABLE 6.18. Trends in Reformer Severity.

Year	H$_2$/hydrocarbon	Space velocity (h^{-1})	Pressure (atm)
1950s	8–10	2–3	25–35
1960–1970	5–8	2–3	15–25
1970–1990	2–5	1–2	3–15

These changes have resulted in improved operation, not only by reducing gas compression costs, but also by increasing gasoline yields and hydrogen production from the dehydrogenation reactions, in combination with a lower degree of the hydrocracking reaction. Unfortunately, some hydrogen is lost because it also reacts (beneficially) with coke precursors, so removing them and decreasing the rate of coke deposition. The use of low hydrogen pressure slows down this reaction, and ultimately leads to more coke deposition and the need for more frequent regeneration. It is necessary, therefore, to balance the advantage of higher yields against increased catalyst deactivation in processes that allow continuous catalyst regeneration. Modern catalytic reformers operate at liquid hourly space velocity in the range 1.0–2.0 h^{-1}. The lower limit is a compromise between an increase in undesirable hydrocracking and a decrease in beneficial dehydrocyclization reactions. These conditions have no effect on either dehydrogenation or isomerization reactions, which still reach equilibrium. The inlet temperature to the catalyst bed must be increased as the catalyst is either deactivated or if a higher octane product is required. Operating temperatures also depends on the feed composition, hydrogen ratio, and the need to control the temperature decrease in all beds. Typical temperatures are shown in Tables 6.16 and 6.17.

An essential requirement of reformer operation is the addition of chlorine compounds to the feed to maintain the necessary chlorine content and hence the acidity of the alumina support. This requires a careful monitoring of the mass balance of chloride entering and leaving the catalyst beds, to maintain the chloride content of the catalyst at about 1.0 wt%. A strict control of water in the feed is also required to minimize chloride loss.

6.6.1.2. Coke Formation

Hydrogenolysis reactions cause the formation of coke precursors on the active platinum surface, which can then diffuse to adjacent acid sites on the alumina support. A series of polymerization and dehydrogenation reactions form coke, which eventually covers the catalyst surface and reduces activity. Early experience showed that traces of sulfur in the catalyst during start-up reduced coking by inhibiting the excessive hydrogenolysis reaction and the interval before regeneration became necessary could be increased. Acid catalyzed cracking reactions were not affected, however, and some coke was still formed. No sulfur addition is required during normal operation because it shortens the cycle time.

The interval between regenerations can be extended by increasing hydrogen recycle and raising operating pressure from 15 to 35 atm. This improves the rate at which coke precursors are hydrogenated on the platinum surface, but these changes decrease yields and octane ratings (see Section 6.6.1.1).

6.6.2. Reforming Catalysts

Platinum was not only the most active and selective Group-VIII metal for the naphtha reforming process, but was also more readily available and relatively less expensive than iridium or rhodium.

Both γ- and η-aluminas were suitable as the acidic support. Although η-alumina is more acidic and has a higher isomerization activity, it is more expensive to produce and not much better than a stable γ-alumina to which chloride has been added during its preparation. The composition of typical reforming catalysts is shown in Table 6.19. Fluorided-alumina and silica/alumina supports were both used in some early reformers but were not really successful because of excess acidity, which promoted cracking reactions.

The success of platforming soon led to the rapid introduction of competitive processes as demand for higher-octane gasoline for high-compression automobile engines increased. Seven process licensors are still offering reforming processes differing only in the physical appearance of the unit and the catalysts used.[42] More than 50 different catalyst types are currently being supplied by at least ten catalyst producers. As usual with licensed processes, full details of catalyst compositions and operation are not described in detail.

Table 6.19. Typical Reforming Catalysts.

Catalyst	Properties	
Platinum	Early type, Pt 0.65%	Later type, Pt 0.35%
Platinum/rhenium	Balanced , Pt 0.3% / Re 0.3%	Skewed , Pt 0.3%/Re 0.6%
Platinum/tin	CCR units, Pt 0.3% / Sn 0.6%	No change
Platinum/iridium	Exxon units, Pt 0.3% / Ir 0.3%	No change
Support	γ-Al_2O_3	
Surface area (m^2 g^{-1})	200	
Pore volume (ml g^{-1})	0.7	
Pore diameter (nm)	6–7	
Reactor bulk density (kg liter^{-1})	0.5–0.6 (normal)	
	0.7–0.8 (dense packed)	
Chloride (wt%)	0.9–1.2	

6.6.2.1. *Bimetallic Catalysts*

During the 1950s and 1960s the performance of platinum reforming catalysts was considerably improved as operating procedures were optimized. More economic catalysts became available as the platinum content was reduced from more than 0.6 wt% to about 0.35 wt% and the sulfur content of feed naphtha was considerably reduced. Higher purity alumina supports, with better activity and lower levels of sodium and iron could be synthesized reproducibly. An optimum amount of chloride added to the support limited coke formation.

More significant changes, however, began in 1968 when Chevron introduced its bimetallic catalyst containing platinum and rhenium. The main advantage of the new catalyst was its relative stability, which extended operation before regeneration was necessary. The separation of small platinum crystallites by rhenium made the catalyst less sensitive to carbon deposition. In fact, bimetallic catalysts doubled the period before regeneration was necessary. This improvement allowed operation at lower pressures with increased yields. Improved dehydrocyclization and isomerization selectivity also increased the volume of aromatics produced and the octane number of the gasoline. The level of hydrocracking decreased and the production of light gases was reduced.

The success of the platinum/rhenium catalyst was evident, as more than 80% of all catalyst replacement charges were bimetallic by 1972. By then 30% of all reformers were using the rhenium catalyst.[43] Acceptance continued in all reforming processes so that more than two-thirds of installed catalyst capacity was bimetallic by 1979.[44] Catalyst improvements also continued during this period, and the original Chevron catalyst, Grade A, was replaced several times up to Grade F.[44]

Most of the changes were probably related to metals concentration but the original bimetallic catalyst contained equal weight percentages of platinum and rhenium and is now known as a *balanced* catalyst. Since the late 1980s, rhenium catalysts have contained about twice as much rhenium as platinum and are known as *skewed* catalysts. The skewed catalysts are even more stable, particularly when loaded to reactors at high bulk density, and have extended the operating cycle even further before regeneration is required. This is partly due to the ability of the catalyst to operate with increased coke content.

A disadvantage of platinum/rhenium catalysts is their sensitivity to sulfur poisoning. This requires that naphtha feed must be carefully hydrotreated before use.

The very few plants still using platinum catalysts (with no rhenium) are those without the necessary hydrotreating facilities.

It is necessary to pre-sulfide new or regenerated catalysts with a suitable sulfur compound prior to use when working with a low sulfur-content feed, to reduce excessive hydrogenolysis activity of the catalyst. Presulfiding causes deactivation of the *super-active edges* on the metal sites that are responsible for the bulk of the hydrogenolysis activity which would otherwise lead to premature coke formation and the hot spots that cause a reduction in reforming activity. The addition of 100–200 ppm of dimethyl sulfide to the recirculating gas raises the sulfur level of the catalyst to about 0.25% by weight. During subsequent operation, only the sulfur on the *superactive edges* remains in the catalyst, the rest being quickly removed. Subsequent operation of the catalyst is not affected and the presence of rhenium sulfides may limit the sintering of the platinum crystallites.

The information in Table 6.20 shows how the use of rhenium has improved operation and indicates the coke content before regeneration is necessary. The coke content shown is for the third bed in a semi-regenerative unit (the first and second beds contain considerably less coke.) It has been found that even 1 ppm of sulfur in the naphtha feed reduces the cycle time with *skewed* and *balanced* catalysts by about 25% and 35%, respectively.[45]

The first CCR unit in 1971 used a platinum/rhenium catalyst, but in 1973 UOP introduced a catalyst in which the platinum was promoted by tin.[39] The new catalyst not only resulted in lower coke levels at the low operating pressure but was easily regenerated throughout its long life. Tin reacts with the surface hydroxyl groups on the γ-alumina support. This gives better stability and increases the amount of coke deposited on the support rather than on the platinum. Satisfactory operation, however, depends on a good distribution of both platinum and tin. γ-Alumina is stabilized by the tin, which reacts with the surface hydroxyl groups. The first UOP catalytic reformer plant at 1000 tonnes per capacity is shown in Figure 6.6.

Iridium has also been used as a promoter in platinum reforming catalysts since it was introduced by Exxon in 1971–1972.[38] This was reported by *Chemical Week* in one of their *Catalyst Reviews*, when they noted a sharp increase in the reported sales of iridium.[46] Catalysts containing iridium have not been widely used except in plants designed by Exxon and Amoco. Extremely long periods between regenerations have been reported and may be due to the more efficient

Table 6.20. Improved Operation with Platinum/Rhenium Catalysts.

Catalyst	%Coke before regeneration	Relative cycle time
Pt/Al$_2$O$_3$	10	1
Pt/Re/Al$_2$O$_3$ (balanced)	15	x3
Pt/Re/Al$_2$O$_3$ (skewed) high density	20	x6

Sulfur in feed (ppm)	Relative cycle time	
	Balanced	Skewed
0	1	2
0.5	0.9	1.7
1.0	0.7	1.5

Figure 6.6. The first UOP catalytic reforming plant at 1000 tonne per day capacity. Reproduced with permission from UOP, LLC.

hydrogenation and removal of coke precursors. The active metals are claimed to be well distributed as bimetallic clusters and great care must be taken to limit the temperature during the oxidation and reduction of catalyst at start-up to prevent sintering.

6.6.2.2. Catalyst Preparation

Monometallic reforming catalysts are prepared by impregnating the γ-alumina support with chloroplatinic acid in a dilute solution of hydrochloric acid.[47] The solution usually contains the exact amount of platinum required. Hydrochloric acid plays an important role in the catalyst preparation by keeping the platinum in solution, by chloriding the alumina and ensuring good platinum distribution. The acid strength used is proportional to the alumina surface area but is not strong enough to dissolve the alumina. Chloroplatinate and chloride ions react with hydroxyl groups on the alumina surface. The replacement of hydroxyl groups with chloride gives neighboring hydroxyl groups increased acidity without increasing the total number of acid sites.

Impregnated alumina is carefully drained and dried at about 150°C to remove residual water and then oxidized in dry air at about 500°C. Drying and oxidation can replace some of the chlorine atoms from the adsorbed chloroplatinate ions on active sites with oxygen. At the same time, the chloride content of the alumina is decreased depending on the partial pressure of water vapor present in the air stream. Loss of chloride should, therefore, be carefully controlled by ensuring a low drying temperature and using dried air during oxidation to maintain the required chlorine content in the finished catalyst. The properties of the γ-alumina support are extremely important and the gradual improvements made to the catalytic reforming process up to about 1960 came mainly from an understanding of the catalyst properties and the use of purer, more stable aluminas.

Bimetallic catalysts have the same basic characteristics as the platinum/γ-alumina catalysts, with good dispersion of small platinum crystallites. The most widely used bimetallic catalysts are the platinum/rhenium formulations. The preparation of these catalysts is usually by impregnation of the γ-alumina support with a solution of chloroplatinic acid and perrhenic acid, or its ammonium salt in hydrochloric acid.[47] Successive impregnations with chloroplatinic and perrhenic acid solutions have also been used. The catalysts are then carefully dried, oxidized, and reduced. The volatility of the rhenium complexes or oxides requires careful control of temperature to avoid metal loss by vaporisation and to prevent alloy formation during reduction.

Platinum/iridium catalysts are made by impregnating γ-alumina with chloroplatinic and chloroiridic acids. Heating to temperatures about 375°C produces the highly dispersed bimetallic clusters that form the catalyst. Great care is required to achieve the necessary metal dispersion because iridium oxide is volatile, and it has even been suggested that a third metal such as chromium is required to promote support-metal interaction.

Platinum/tin catalysts are made by impregnating the support with tin chloride and calcining at 500°C, before impregnation with chloroplatinic acid. The catalyst is then dried, oxidized, and reduced. Temperature control is necessary to avoid reduction of the tin chloride to metallic tin, which would then alloy with the platinum. Tin oxide combines with acid sites on the support and increases resistance to deactivation. Typical catalyst formulations are shown in Table 6.19.

6.6.3. Catalyst Regeneration

When the yield from the reformer falls to an uneconomic level the catalyst must be regenerated.[48] This is the point where the C_5^+ yield falls from the initial 76–77% to about 73% and the purity of recovered hydrogen declines. Regeneration removes the coke deposited on the active sites, restores the catalyst acidity, and redisperses the active metal into smaller crystallites. Different quantities of coke are deposited in each bed of a semi-regenerative reformer. Beds 1, 2, and 3 contain about 1–3 wt%, 5–15 wt%, and 8–20 wt%, respectively. Great care must be taken to maintain a steady temperature profile in each bed during the regeneration procedure.

Catalyst is regenerated in the continuous catalytic reduction process using the same basic procedures. About 200 lb h^{-1} were reduced in the first CCR units in 1971 compared with 4000 lb h^{-1} in modern plants. These quantities are equivalent to regenerating the whole charge of catalyst in the reformer—a total of 12 and 150 times a year, respectively.

6.6.3.1. Carbon Burn

Coke is removed by burning in an air/nitrogen gas stream, containing up to 0.5 vol% of oxygen at a space velocity of about 5000 h^{-1}. Bed temperatures during the first stage should always be less than 350^{0}C. If necessary the oxygen content can be decreased to control hot spots. These conditions are maintained until the outlet temperature falls to the same level as the inlet temperature. At this point the oxygen content can be increased gradually and the bed temperature slowly raised to 500^{0}C. The final carbon content of the catalyst is normally less than 0.2 wt%.

6.6.3.2. Oxychlorination

After carbon has been removed from the catalyst, the chlorine content usually falls to 0.6–0.8 wt%. The loss of chlorine must, therefore, be compensated by passing dry air that contains a suitable chlorine compound such as carbon tetrachloride over the catalyst at 450^{0}C. Sufficient chlorine must be used to restore the original 0.9–1.3 wt% chlorine level. About 70% of the total chlorine in the airstream is retained by the catalyst. A high space velocity of up to 5000 h^{-1} can be used.

6.6.3.3. Platinum Re-Dispersal

Platinum that has sintered during plant operation must be redispersed into the small crystallites required for active catalysis. This is carried out by heating in a dry air flow at 500^{0}C. When the air leaving the reactor contains less than 5 ppm water the air should be purged from the catalyst bed until the oxygen content is less than 1 ppm. It is thought that the redispersion treatment breaks down the larger platinum particles by producing volatile oxide or oxychloride complexes. These then react with surface hydroxyl groups to form smaller crystallites and restore activity.

6.6.3.4. *Catalyst Reduction*

Before use the new or regenerated catalyst is reduced in dry hydrogen at 500°C at a space velocity of 400 h⁻¹ for about 2 h. This converts the platinum oxides into finely divided metal particles.

6.6.4. Catalyst Life

Most of the initial activity reforming catalysts can be recovered during regeneration, although it is inevitable that some activity is lost during the procedure and catalysts are eventually replaced. The catalyst in semi-regenerative reformers, which can be operated for 9 to 12 months before being regenerated, will not be changed for about 8 to 10 years. Catalyst in cyclic reformers, which are designed for regeneration after only a few days, has a shorter life of only 4 to 5 years. These estimates are only approximate and may be less depending on the feeds and operating conditions used.

Although catalyst life in the continuous catalytic regeneration reformers is probably up to seven years, it is usually progressively replaced gradually as catalyst activity declines. It may also be necessary from time to time to remove catalyst from the reactor to clear any accumulated dust. This is usually recognized by increased pressure drop or maldistribution of flow giving hot spots. The catalyst can be replaced for further use after sieving. Any *heel* catalyst, containing up to 50% carbon, which has not been properly regenerated in isolated *dead* spots of the bed, can be separated and discarded.

6.7. OCTANE BOOSTING

Shape-selective reactions have been extensively studied since zeolites were first used in catalytic crackers during 1967[49] and Mobil has introduced several octane-boosting processes since 1968.[50] Gasoline composition and octane number were reviewed when the Clean Air Act was passed in 1970. This mandated the phased removal of tetraethyl lead from gasoline as catalytic converters (see Chapter 11) were introduced to treat automobile exhaust gas. Octane boosting was the first in a series of measures that led to reformulated gasoline and improved exhaust emissions standards.

6.7.1. Selectoforming

Shape-selective zeolite catalysts were first used in the Selectoforming process by Mobil in 1968. Hydrogen-exchanged natural erionite, containing some nickel, was used to crack and isomerize low-octane normal paraffins in reformate to increase the concentration of isoparaffins and boost the octane number. The nickel minimized carbon formation but did not saturate aromatics. The erionite was later replaced by synthetic zeolite T, which has the twelve-membered offretite ring containing about 4% of eight-ring erionite and effectively prevents isoparaffins from entering the pores.

6.7.2. M-Forming

By about 1970 the benefits of ZSM-5 zeolite in cracking low-octane normal paraffins in FCC units were well known. By adding 1% of ZSM-5 as a promoter to reforming catalysts it was possible to increase the octane number of reformate at the expense of some liquid yield.

In the 1970's, the octane number of the product from the catalytic reforming of light naphthas was only RON 84. This was too low to satisfy the changing market, and the M-forming process was developed by Mobil to use ZSM-5 zeolites containing nickel as catalyst. Octane number was increased to RON 93 because the pores of ZSM-5 zeolite crack single-branched paraffins at about 315^0C. Higher-octane products are formed by oligomerization, aromatization, and alkylation reactions. There is an increase in yield and less gas formation compared with the Selectoforming process.

6.8. AROMATICS PRODUCTION

Reformers can also be operated to maximize the production of aromatics. When benzene is the required product, a narrow fraction boiling in the range 60–90^{0}C is used under severe operating conditions. If the feed is changed to a fraction boiling in the range 110–140^{0}C under similar high severity conditions, the main product is a mixture of toluene and mixed xylenes. Normally however, the main object of reforming is to maximize the octane rating of the product.

6.8.1. Aromatics Process

L-zeolite, potassium exchanged and also containing some platinum, is an active catalyst for the conversion of n-hexane to aromatic compounds. L-zeolite is a wide-pore zeolite, with a silica/alumina ratio of six and a unidimensional pore structure.

The commercial catalysts described and used in the Chevron Aromax process were partly exchanged with barium ions and contained platinum (0.6–0.8%) as small particles, corresponding to an approximate formula $Pt/Ba_2K_5Al_9Si_{27}O_{72}$.[61] Direct dehydrocyclization of paraffins is an alternative to isomerization and reforming to increase octane number of light straight-run naphthas. These zeolite catalysts are, however, more easily poisoned by sulfur compounds because they are based on platinum and feedstock must be desulfurized.

6.8.2. Cyclar Process

BP and UOP developed a process to convert liquid petroleum gas (LPG), a mixture of propane and butane, into aromatics.[52] A plant using this process began operating in 1989 at the BP refinery in Grangemouth, Scotland, and a second followed in 1999 operated by Saudi Basic Industries (SABIC) in Saudi Arabia.

The catalyst is based on a ZSM-5 zeolite containing gallium. The yield of aromatics is about 60%, with more than 90% as benzene, toluene, with some xylene and relatively small amounts of C_9 aromatics. Up to 2000 standard cubic feet of hydrogen per barrel of feed is produced. Reaction proceeds at 425^0C by dehydrogenation of the paraffins followed by oligomerization and cyclization. A four-bed UOP continuous catalytic regeneration reactor is used with interbed heating of the reaction mixture. Aromatics are separated from the remaining aliphatic hydrocarbons by distillation.

Japanese research on converting propane to aromatics in the presence of CO_2 on ZSM-5 catalyst loaded with zinc showed that equilibrium was improved as hydrogen was removed by the reverse water gas shift reaction:

$$H_2 + CO_2 \leftrightarrows H_2O + CO \tag{6.6}$$

6.8.3. M2-Forming Process

The M2-Forming process introduced by Mobil uses a ZSM-5 zeolite to convert both virgin naphtha (C_5–110^0C) and olefinic gasolines (C_6–110^0C) from FCC and thermal crackers into aromatic compounds. It operates at a higher temperature than the M-Forming process and at about 550^0C the selectivity to aromatics approaches 100%.[53]

6.9. CATALYTIC DEWAXING

The viscosity of high-boiling fuels and lube oils can be reduced by shape-selective cracking of long-chain paraffins in the presence of hydrogen. Several zeolites introduced during the 1990s have been used: H-mordenite containing platinum by BP; ferrierite containing palladium by Shell; ZSM-5 by Mobil. The exchange of zeolite with the precious metal promoted the hydrogenation of coke precursors at the high operating pressures,[54] and extended catalyst lifetimes.

Typical operating conditions are: reactor temperature, 300–350°C; pressure, 20–130 atm; hydrogen partial pressure, 15–100 atm.

6.10. ISOMERIZATION

The catalytic reforming of straight-run naphtha to gasoline is usually restricted to feeds in the boiling range 100–180°C. This is because the equilibrium conversion of the lower-boiling normal C_5–C_6 paraffins under typical operating conditions is too low to obtain a reasonable improvement in higher-octane-number branched isomers.

A separate process to isomerize the straight-run C_5–C_6 cut is used to complement catalytic reforming and produce gasoline with good octane qualities over the whole boiling range. The change in octane number as a range of C_5–C_6 hydrocarbons is isomerized, is shown in Table 6.21. To achieve the most suitable product, however, a high–activity catalyst must operate at the lowest possible temperature. This improves equilibrium conversion and increases the proportion of dimethyl isomers rather than single-branch isomers.

6.10.1. Isomerization Catalysts

Isomerization processes developed slowly because of the low demand for higher octane numbers and operational problems with aluminum chloride, the first catalyst to be developed. Despite high activity at 115–120°C, aluminum chloride has the major disadvantages of the formation of sludges and acid corrosion of equipment.

TABLE 6.21. Isomerization of C_5–C_6 Hydrocarbons.

Wt%	Feed	Product	Octane number
n-Pentane	33	15	62
i-Pentane	22	40	93
n-Hexane	20	6	31
2-Methyl pentane	12	14	74
3-Methyl pentane	10	7	74
2,2-Methyl butane	1	11	96
2,3-Methyl butane	2	5	105
C_3–C_4 Hydrocarbons		2	
RON	70	82	
After *n*-C_5H_{12} recycle	83–85		
After *n*-C_5H_{12},*n*-C_6H_{14} recycle	87–90		

Note: Extract *n*-hydrocarbons with a shape-selected zeolite.

The introduction of dual-function catalysts for naphtha reforming and the demand for high-octane gasoline led to further interest in isomerization. The platinum/alumina (chlorided) catalysts were a success despite the resulting lower conversion to high-octane products from the need to operate at higher temperatures. The Shell Hysomer process, which used a 5A-zeolite to separate low-octane paraffins from the product allowed operators to recycle unconverted feed and achieve almost 100% conversion.[55]

A further important development was the use of hydrogen-exchanged mordenite zeolite as the catalyst support. The zeolite was more stable and water tolerant than the chlorided alumina and did not need chloride addition during operation. The zeolite catalyst had a lower activity than that of alumina, but the need to operate at a slightly higher operating temperature was acceptable since the normal paraffins could be recycled. Other zeolites, such as β-zeolite, have also been used as the acid support.

Operating conditions for the different catalysts are given in Table 6.22. Hydrogen is usually added to minimize coke formation but is not used in the reactions taking place.

6.10.2. Reaction Mechanism

The isomerization of normal C_5–C_6 paraffins proceeds by dehydrogenation on the platinum surface followed by isomerization of the olefin formed on the acid support. The isomer is then hydrogenated on the platinum to give the higher-octane branched paraffin.

Platinum in the dual-function catalyst is important in reducing the steady-state olefin concentration, otherwise polymerization and cracking would lead to extensive coke formation. Some coke, however, does deposit slowly and the catalyst has to be regenerated at intervals. Mordenite supports are more stable than chlorided aluminas and can operate for longer at more than 60% conversion and about 97% selectivity. Catalysts are reported to operate for as long as seven years before being replaced.

TABLE 6.22. Isomerization Operating Conditions.

Conditions	$Pt/Cl-Al_2O_3$	Pt/H-mordenite
Temperature (^{0}C)	150	250
Pressure (atm)	15	30
Hydrogen/hydrocarbon	2:1	2:1
LHSV (h^{-1})	1.5	2.5
Yield (%)	> 98	>98
Feed (RON ~70)	(150^0C)	(250^0C)
No recycle	RON 84-85	RON 82-83
With recycle	RON 90-91	RON 89-90

Butane can also be isomerized with platinum on chlorided alumina catalysts, under the same conditions, to produce isobutane for use as feedstock in
alkylation processes and more recently for the production of MTBE .

REFERENCES

1. The Royal Dutch Shell Petroleum Company 1890–1950, *Diamond Jubilee Book*, , Nijgh and Van Ditmar NV, The Hague, 1950, p. 75.
2. The Royal Dutch Shell Petroleum Company 1890–1950, *Diamond Jubilee Book*, Nijgh and Van Ditmar NV, The Hague, 1950, p. 79.
3. The Royal Dutch Shell Petroleum Company 1890–1950, *Diamond Jubilee Book*, Nijgh and Van Ditmar NV, The Hague, 1950, p. 82.
4. S. H. McAllister, *Refinery Nat Gasoline Mfr*, (Nov 1937) 493.
5. K. G, Mackenzie, *Refinery Nat Gasoline Mfr*, (Nov 1939) 494.
6. The Royal Dutch Shell Petroleum Company 1890–1950, *Diamond Jubilee Book*, Nijgh and Van Ditmar NV, The Hague, 1950, p. 83.
7. J. H. Kunkel, *Petroleum Eng*, Sept, 80 (1944); Scott and Cooper, *Oil Gas J.*, (March 1946) 204.
8. P. Weinert and G. Egloff, *Petroleum Proc.*, (June 1948) 585; *Sud. Chemie/Cata.l and Chem. Europe Literature*, C84-3.
9. K. Gordon, *J. Inst. Fuel* **20** (1946) 42.
10. P. C. Templeton and B. H. King, *Western Petroleum Refiners Association*, June (1956) 21; *Chem. and Eng. News,* (Feb 1993) 15.
11. R. M. Kennedy, *Catalysis*, **6** (1958) 5.
12. P. Rao and S. R. Vatcha, *Oil Gas J.*, Sept. 9 (1996), p. 56.
13. K. Hovis, K. Hoover and L. Shoemaker, Revap. Process, *API Operating Practices Symposium* (May 1995).
14. *Oil Gas J.*, April (1996) 69; *Topsoe Topics,* June (1995).
15. G. M. Brooner and M. W. Conn, *Oil Gas J.*, Oct. (1946) 96; Martin and Carlson, *Oil Gas J.*, March 26 (1942)138.
16. A. M. McAffee and Home, *Petroleum Processing*, (April 1956) 47.
17. H. Hoog, H. G. Klinkert and A. Schaafsma, *Oil Gas J.*, (June 8 1955) 92.
18. A. M. McAffee, *Petroleum Ref.*, (May 1955) 156.
19. J. Sonnemans and P. Mars, *J. Catal.* **31** (1973) 209.
20. *Ketjen Catalyst Literature* (1981).
21. D. Chadwick, D. W. Aitchison, R. Badilla-Ohlbaum and L. Josefsson, *Preparation of Catalysts III*, Ed. by Poncelet, Elsevier, Amsterdam, 1982, p. 323.
22. H. Koyama, E. Nagai, H. Kumagai and H. Torii, *Oil Gas J.*, (Nov. 13 1995) 82.
23. H. Topsøe and B. S. Clausen, *Catal. Rev.* **22** (1984)401.
24. S. Blashka, G. Bond and D. Ward, *Oil Gas J.* (Jan 5 1998).

25. Danzinger, *Oil Gas J.*, (May 3 1999) 97; D. J. Podratz, Kleemeier, W. J. Turner, and B. M. Moyse, *Oil Gas J.*, (May 17 1999) 41.

26. Catalyst Recovery Inc, Rodange, Luxembourg; Eurecat, La Voulte sur Rhone, France; Tricat Inc, MacAlester, Oklahoma, 1995–1996.

27. R. F. Sullivan and J. W. Scott, *ACS Symposia Series 222*, (1983), p. 293; J. W. Ward, *Fuel Proc. Tech.* **35** (1993) 55; J. W. Ward, *Prep. Catalysts III*, Ed. by Poncelet, Elsevier, Amsterdam, 1983, p. 587.

28. D. H. Stormont, *Oil Gas J.*, **44** (1959) 48.

29. W. J. Baral and H. C. Huffman, *8th World Petroleum Congress* **4** (1971) 119.

30. J. Magee and G. Dolbear, *Petroleum Catalysis*, Pennwell, Tulsa, Okla., 1998, p. 141.

31. J. W. Ward, US Patent 4107031 (1978).

32. J. W. Ward, US Patent 3849293 (1974).

33. V. Haensel, US Patent 2479109; 2479110 (1949); V. Haensel and G. R. Donaldson, *Ind. Eng. Chem.* **43** (1951) 2102.

34. Jaggard and J. P. L. Johnson, *Petroleum Ref*, (Aug 1956) 157.

35. V. Haensel, *ACS Symposium Series 222*, 1983, p. 141; V. Haensel, *Oil Gas J.*, (March 30, 1950) 82.

36. V. Haensel, US Patent 2479109; 2479111; 2629683 (1949).

37. R. L. Jacobson, H. E. Kluksdahl and C. S. McCoy, US Patent 3415737 (1968); 3434966 (1969); R. L. Jacobson, H. E. Kluksdahl, C. S. McCoy and R. W. Davis, *Proc. Amer. Petrol. Inst. (Div Ref)* **49** (1969) 504.

38. J. H. Sinfeld, US Patent 3953368 (1976); J. H. Sinfeld and G. H. Via, *J. Catal.* **56** (1979) 1.

39. CFR, French Patent 2031984 (1969); R. E. Ransch, UOP US Patent 3745112 (1973).

40. N. L. Gilsdorf, A. E. Doornbos and T. J. Gevelinger, *NPRA Meeting*, San Antonio, AM-93-10 (1993).

41. *Oil Gas J. Special: Worldwide Refining*, (Dec 20, 1999) 41.

42. A. M. Aitani, *Catalytic Naphtha Reforming*, Ed. by G. J. Antos, A. M. Aitani and J. M. Parera, Dekker, New York, 1995, p. 413.

43. A. S. Al Kabbani, *Hydrocarbon Processing*, (July 1999) 61.

44. D. P. Burke, Catalysts, *Chem. Week*, (Mar 28, 1979) 53.

45. D. P. Burke, Catalysts, *Chem. Week*, (Nov 1, 1972) 30.

46. D. P. Burke, Catalysts, *Chem. Week*, (Nov 1, 1972) 27.

47. J. P. Boitiaux, J. M. Deves, B. Didillon, and C. R. Marcilly, *Catalytic Naphtha Reforming*, Ed. by G. J. Antos, A. M. Aitani, and J. M. Parera, Dekker, New York, 1995, p. 79.

48. In *Applied Heterogeneous Catalysis*, Ed. by J. F. Le Page, Editions Technip, Paris, 1987, p. 491.

49. H. E. Robson, G. P. Hammer and W. F. Arey, *Molecular Sieve Zeolites II*, in Advances in Chemistry Series, No. 102, Ed. by E. M. Flanigen and L. B. Sand, 1971, p. 417.
50. S. Sivasanker and P. Ratnasamy, *Catalytic Naphtha Reforming*, Ed. by G. J. Antos, A. M. Aitani, and J. M. Parera, Dekker, New York, 1995, p. 489.
51. P. W. Tamm, D. H. Mohr and C. R. Wilson, *Catalysis 1987*, Ed. by D. W. Ward, Elsevier, Amsterdam, 1987, p. 375.
52. P. C. Doolan and R. P. Pujado, *Hydrocarbon Processing*, (Sept 1989) 72.
53. S. Sivasanker and P. Ratnasamy, *Catalytic Naphtha Reforming*, Ed. by G. J. Antos, A. M. Aitani, and J. M. Parera, Dekker, New York, 1995, p. 492.
54. N. Y. Chen, J. Garwood, and F. G. Dyer, *Shape Selective Cracking in Industrial Applications*, Dekker, New York, 1989.
55. A. P. Bolton, *Zeolite Chemistry and Catalysis*, ACS Monograph, No. 171, Ed. by J. A. Rabo, 1976, p. 714.

7

PETROCHEMICAL CATALYSTS

7.1. THE DEVELOPMENT OF PETROCHEMICALS

Organic chemical compounds have been produced on an industrial scale for over a century. In the early days, the source of carbon was coal, and many of the compounds were produced via the intermediate manufacture of acetylene and benzene. The development of crude oil refining and the use of new processes to satisfy the increase in demand for gasoline has provided *refinery off-gases* and other hydrocarbons that are now used as the main source of industrial organic compounds, and the term *petrochemicals* arose to describe this evolution of the chemical industry.

For example, in the early years following the development of the Haber-Bosch process for the manufacture of ammonia, both ammonia and methanol were manufactured from synthesis gas derived from coal. This was eventually replaced by a naphtha-like fraction from the refining of crude oil, and much of the World capacity for methanol and ammonia is now derived from natural gas. In general, refinery feeds have now replaced coal for the production of synthesis gas almost entirely, and both ammonia and methanol can rightly be regarded as petrochemicals. Such is the scale of modern refining, together with the increased demand for petrochemical products such as plastics, fibers, and rubber, petrochemical complexes are now frequently closely linked to refineries. For example, of the 666 million tonnes of naphtha produced in 1997, up to 40% were used in the steam crackers and aromatics units that provided feeds for petrochemical production. Many chemicals are made using continuously operating refinery-like processes, often with several reaction steps in sequence, and the

L. Lloyd, *Handbook of Industrial Catalysts*, Fundamental and Applied Catalysis, DOI 10.1007/978-0-387-49962-8_7, © Springer Science+Business Media, LLC 2011

products must be produced to tight chemical specifications. Some of the oxidation, hydrogenation/dehydrogenation reactions developed more than 80 years ago are still being used, although efficiency and selectivity have been greatly improved by the use of better catalysts and equipment.

About 1914 the volume of gasoline sold exceeded the demand for kerosene. Consequently thermal cracking processes were introduced, since these resulted in an increase in the gasoline yield from crude oil. However, cracking produced large volumes of low-molecular-weight gases, which at first were used as fuel. The almost worthless gas did, however, contain significant volumes of olefins, which were eventually to become the aliphatic building blocks used in producing high-octane gasoline and, later, petrochemicals.

The Mellon Institute in Pittsburgh was founded in 1911 by a group including the Mellon brothers, who were the main shareholders in the Gulf Oil Corporation. George Curme of the Institute worked with both Prest-O-Lite, which supplied acetylene lamps, and Union Carbide, the largest supplier of calcium carbide, on alternative ways to produce acetylene and commercially useful organic derivatives of both acetylene and ethylene. In a series of experiments, it was demonstrated that ethylene produced by cracking had potential to be a cheaper chemical intermediate than acetylene.

Union Carbide and Linde then began to develop the production of ethylene from the ethane present in natural gas. This led to the purchase of a small gasoline plant at Clendenin, West Virginia, where ethane-rich natural gas was available from the Kanawha Valley. An ethane-cracking unit began operation in 1921 and provided ethylene as well as liquefied petroleum gas that was sold for domestic use.

Following further development work, ethylene glycol became the first *petrochemical* to be produced at the *ethylene plant* in South Charleston that began operating about 1925. Glycol antifreeze was widely used in automobile engines by 1927. Numerous other organic syntheses were being developed at the same time for commercial production in the small petrochemical complex. By 1926, as many as six ethylene and two propylene derivatives were available, and by 1934 the number had increased to 35-ethylene and 15-propylene products.[1]

During the same period Universal Oil Products developed olefin polymerization and hydrogenation processes and the Shell Chemical Company became interested in olefin chemistry and petrochemicals. Shell opened its Emeryville, California, research center in 1928 and developed methods for the production of alcohols, ketones, and esters from propylene and butylenes, as well as synthetic chlorohydrins, glycols, and glycerol. Many of these petroleum chemicals were produced in Martinez, California, from 1933 and the use of petrochemical catalysts became established.

By 1940, after the start of World War II, the production of petroleum chemicals, as they were still known, and the use of catalysts were common practice in the United States. Subsequently, the 1930s became known as the *miracle dec-*

ade, because the catalysts petrochemical processes that had been introduced began to provide artificial fibers, polymers, and synthetic rubbers. The use and development of these processes and the associated technology were accelerated as demand caused by the war required plants to be built to produce large quantities of polyvinyl chloride, butadiene–styrene rubber, polyethylene, and nylon.

Until 1945, most developments in the petrochemical industry arose in the United States, where refinery processes were advanced and feedstocks were readily available. There were no similar refineries in Europe and chemical production was still firmly based on coal. Petrochemicals could not be produced in Europe until the late 1950s, when the first large refineries were built and cheap naphtha was used in steam crackers and for ammonia production. Estimates of petrochemical production in both the United States and Europe from 1935 to 1965 demonstrate the difference as shown in Table 7.1.

More than 90% of today's petrochemicals are produced from refinery products. Most are based on the use of C_2–C_4 olefins and aromatics from hydrocarbon steam cracking units, which are even more closely linked to refineries. In North America, the feedstock for steam cracker units have generally been ethane, propane, or LPG. As a result, most of the propylene and aromatics have been provided by FCC units and catalytic reformers. In many other parts of the world where naphtha feed has been more readily available, supplies of propylene and aromatics have been produced directly by steam cracking. When necessary, the catalytic dehydrogenation of paraffins or dealkylation of toluene can balance the supply of olefins or benzene. In Table 7.2 some of the catalytic processes that convert olefins and benzene from a steam cracker into basic petrochemicals for the modern chemical industry are shown.

Before 1965, the production of petrochemicals as a new branch of the chemical industry does not seem to have been widely mentioned in textbooks and little general information was available.[2] Industry, however, well aware of the potential demand for the new products, was developing large new plants to take advantage of the economies possible from large-scale operation. ICI, for example, opened its Petrochemical and Polymer Laboratory—later to become its Corporate Laboratory—about this time. Since then many new processes and

TABLE 7.1. Petroleum Chemicals Production 1935–1965 (Million Tonnes)

	United States	Europe
1935	0.5	–
1940	1.1	–
1945	4.5	–
1950	7.2	< 0.1
1955	14.4	0.4
1960	24.7	1.7
1965	42.3	~ 5

petrochemical products have been introduced. Generally the processes have used catalysts and include development of the hydrogenation and oxidation procedures first used in the 1920s and 1930s. Many new reactions and syntheses have now evolved and the range of catalysts has undergone continuous improvement. The industry has moved a long way from the time when DuPont claimed it had made a fiber—Nylon—from *coal-air-water*! Fully detailed information is not usually provided, but the importance of petrochemical processes has now been recognized and is described in many reviews and books.[3] Annual meetings and conferences are organized by chemical societies, chemical engineering societies and catalyst companies to review the latest processes, catalyst science and catalyst developments.

TABLE 7.2. Petrochemicals From Naphtha Steam Cracking.[a]

Feed	Primary products	Main secondary derivatives
	Fuel gas 270,000 tonnes/year (16.9%)	Fuel, ammonia/methanol Propane cracking, etc.
	Ethylene 500,000 tonnes/year (31.2%)	Polyethylene (58%) Ethylene oxide (10%) Styrene (7%) Vinyl chloride (15%) Ethanol (2%)
Naphtha 1.6 million tes.year−1	Propylene 250,000 tonnes/year (15.6%)	Polypropylene (50%) Acrylonitrile (11%) Cumene (7%) Oxoalcohols (10%) Propylene oxide (10%) Isopropylalcohol (4%)
	Butenes 40,000 tonnes/year (2.5%)	Polymers MTBE
	Butadiene 50,000 tonnes/year	Rubber
	Pyrolysis gasoline 450,000 tonnes/year (28%)	Benzene/styrene (46%) Cyclohexane (13%) Toluene Benzene Solvents Xylenes-polyester Gasoline
	Fuel oil 40,000 tonnes/year (2.5%)	

[a]About 80% of petrochemicals are produced with the olefins and aromatics from steam crackers. A typical 500,000 tes/year ethylene plant would crack up to 1.6 million tes/year of naphtha, giving the fractions shown. Uses of individual products are listed.

7.1.1. Isopropyl Alcohol

Isopropanol was one of the first petrochemicals to be produced on a large scale and is still widely used as a solvent and chemical intermediate. Originally, production was by direct catalytic hydration of propylene in the liquid phase:

$$C_3H_6 + H_2O \rightarrow C_3H_7OH \tag{7.1}$$

Concentrated 70–98% sulfuric acid was used as the catalyst at a low pressure and temperature. The reaction mechanism involved the sulfate monoester intermediate, which was then hydrolyzed to isopropanol. The process is similar to the production of ethanol from ethylene, although the hydration step becomes easier as the carbon number of the olefin increases. The acid strength required for effective operation decreases from more than 90% for ethylene hydration to about 70% for propylene hydration.

The homogeneous sulfuric acid process has since been replaced by phosphoric acid–impregnated silica and tungstic oxide catalysts. The tungstic acid catalyst operated at pressures as high as 250 atm and at 250^0–290^0C, whereas phosphoric acid was active at 25–50 atm and 170^0–190^0C. A high water/propylene ratio of about 2–5 was required in order to avoid polymer formation. Although selectivity was as high as 97%, the propylene conversion was only about 6%.

A trickle phase process using strong acid-ion exchangers has now been developed by Deutsche Texaco. The process operates with a high water/propylene ratio of 12–15 at 80–100 atm and 130^0–160^0C to give 75% propylene conversion at up to 95% selectivity. The ion exchange resin loses activity after about 8 months as the acid groups on the resin are hydrolyzed. Other olefins, ranging from ethylene to butylenes can also be hydrated with acid-ion exchange resins.

7.1.1.1. *Acetone*

Until about 1970 most of the acetone used was produced in the gas phase from the azeotropic mixture of isopropanol and water by a dehydrogenation process:

$$C_3H_7OH \rightarrow CH_3COCH_3 + H_2 \tag{7.2}$$

A copper oxide/zinc oxide catalyst (1 CuO / 2 ZnO) was operated in the temperature range 220^0–340^0C to give a 95% yield at about 90% conversion. The reaction was endothermic and took place in 20-ft long, 2-in diameter tubes with the operating temperature limited by the maximum allowable flue gas temperature. At 325^0C, the equilibrium conversion was about 97% at 5 atm. At higher temperatures, however, increasing byproducts formation decreased selectivity to less than 90–95% more than offsetting any advantage from increased conversion. Activity was greater with low impurity levels in the catalyst, such as the residual sodium remaining from the precipitation stage of preparation, and high physical strength was needed. Spent catalyst was removed from the narrow tubes by drilling, as polymeric by-products tended to gum up the reactor tubes.

Another gas phase dehydrogenation process was used to prepare methyl ethyl ketone (MEK) from secondary butanol with a similar copper oxide/zinc oxide catalyst:

$$C_2H_5CH(OH)CH_3 \rightarrow C_2H_5COCH_3 + H_2 \tag{7.3}$$

However, a liquid phase dehydrogenation process, using Raney nickel or Raney copper catalysts in a high-boiling solvent at about 150^0C, gave 80–95% secondary butanol conversion to MEK at more than 95% selectivity.

With the introduction of the cumene–phenol process, which provides large volumes of acetone by-product, the older dehydrogenation process is now rarely used.

7.1.1.2. Bisphenol-A

Bisphenol-A is produced at 500^0C by the condensation of acetone with phenol in the presence of an acid catalyst. In early processes sulfuric acid or dry hydrochloric acid was promoted with methyl mercaptans. More recently acid ion exchange resins have been used:

$$2\ C_6H_5OH + CH_3COCH_3 \rightarrow C_6H_4(OH)C(CH_3)_2C_6H_4(OH) + H_2O \tag{7.4}$$

Bisphenol-A is used in the manufacture of epoxy resins, plastics such as polycarbonates, and polysulfones.

7.1.1.3. Cumene

It was realized during World War II that alkylated benzene molecules could be used to increase the octane level of aviation gasoline. The Petroleum Administration for War had asked for a high-octane synthetic aromatic component in 1941 and propyl benzene, also known as cumene, and butyl benzene were produced in the United States and the United Kingdom, respectively. In 1944, the Hock and Lang process,[4] which produced phenol and acetone from cumene was introduced, and this industrial process became the basis of the major source of phenol throughout the world:

$$C_6H_6 + C_3H_6 \rightarrow C_6H_5.C_3H_7 \tag{7.5}$$

$$C_6H_5.C_3H_7 + O_2 \rightarrow C_6H_5.C_3H_6(OOH) \tag{7.6}$$

$$C_6H_5.C_3H_6(OOH) \rightarrow C_6H_5OH + CH_3COCH_3 \tag{7.7}$$

More recently, new processes for phenol which do not produce acetone by products have been described.

As in all alkylation reactions the catalyst used to produce cumene from propylene and benzene is a strong acid and a number of processes have been introduced. The first were developed by Distillers and Hercules in 1953:

- Sulfuric acid or aluminum trichloride in the liquid phase at 35^0–40^0C and 7 atm.
- Hydrofluoric acid in the liquid phase at 50^0–70^0C and 7 atm.
- Phosphoric acid supported on kieselguhr promoted with HF in the gas phase at 200^0–300^0C and 20–40 atm.

Phosphoric acid/kieselguhr with added steam or aluminum trichloride have been the most widely used catalytic processes. Excess benzene is usually added to the reaction mixture to limit the formation of di- and tri-propylbenzene and to maximize the selectivity of propylene to cumene. Pure propylene must be used to avoid the production of ethyl benzene.

Typical acid catalysts are not regenerable and particularly in the case of aluminium trichloride can lead to disposal problems, so zeolites are now increasingly used as active catalysts. Early problems were encountered with relatively fast deactivation or the formation of ethyl benzene and butyl benzene from metathesis (disproportionation) reactions. These were overcome when Dow introduced a dealuminated mordenite catalyst (3DDM—three-dimensional dealuminated mordenite) that was both stable over 18 months of testing and regenerable.[5] Yields with the mordenite catalyst are greater than 99% and purity is almost 100%. ENI in Italy used a selective ß-zeolite catalyst and process improvements led to an increased plant capacity of 30–40%.

Cumene is converted by a two-step, noncatalytic process into phenol and acetone. It is first oxidized to hydroperoxide with air at 5–10 atm in an alkaline solution at pH 8.5–10 and 90^0–130^0C. The hydroperoxide is then converted to phenol and acetone in either a homogeneous 0.1–2% sulfuric acid solution at 60^0–65^0C or a two-phase liquid mixture with 40% sulfuric acid at 50^0C.

7.1.2. Vinyl Chloride

Vinyl chloride was first prepared by Regnier in 1835[6] when working with Liebig. He chlorinated ethylene, or *olefiant* gas, as it was then known, and converted the ethylene dichloride, or *oil of Dutch chemists* (1795), to vinyl chloride by reaction with caustic potash solution:

$$C_2H_4 + Cl_2 \rightarrow C_2H_4Cl_2 \tag{7.8}$$

$$C_2H_4Cl_2 + KOH \rightarrow CH_2{=}CHCl + KCl + H_2O \qquad (7.9)$$

The first reference to the compound as vinyl chloride was by Kolbe in 1854.[7] Baumann later described polyvinyl chloride in 1872 when he gave a good summary of its properties.[8]

Around 1900, when industry began to take an interest in acetylene chemistry, Fritz Klatte produced vinyl chloride at 180^0C from acetylene and hydrogen chloride:

$$CH{\equiv}CH + HCl \rightarrow CH_2{=}CHCl \qquad (7.10)$$

Patents for his process were issued to Griesheim–Electron in 1912/1913.[9,10] They described a mercuric chloride catalyst supported on coke or pumice. The early catalysts described in 1912 could only operate for short periods because mercury compounds sublime at reaction temperature. Although the 1913 patent described reaction in aqueous solution with the mercuric chloride catalyst, this process was never used industrially. A further patent in 1913 claimed that the process only took place in the presence of the mercury catalyst that accelerated the reaction.[11]

The volatility of mercury compounds led the Consortium für Electrochemische Industrie to patent a catalyst in 1928 based on mercury or bismuth compounds supported on active carbon or silica gel,[12] which retained its activity for 100 hours. B. F. Goodrich also claimed that complex salts, such as $HgCl_2.2KCl$ or $HgCl_2.BaCl_2$, were not only less volatile but more active.[13] None of the later catalysts was as efficient as mercuric chloride on charcoal in terms of the space time yield, despite their longer lives. The early catalysts gave better results at lower temperatures with good temperature control. Developments prior to 1940 were reviewed by Kaufmann[14] and showed that vinyl chloride could be produced from either ethylene or acetylene. Early processes, however, still used acetylene, which was more readily available, although at that time demand was not great.

In the 1920s, Union Carbide recovered by-product ethylene dichloride from its ethylene oxide plant for conversion to vinyl chloride.[14] Vinylite, a relatively soft copolymer of vinyl chloride and vinyl acetate, was used to produce gramophone records and other useful items. Plasticizers were then developed to provide the softer grades of polyvinyl chloride that were widely used during the World War II. By 1939, the acetylene-based vinyl chloride process had been developed commercially and was used to supply PVC for the war effort. During the period from 1935 to 1940 the only industrial producers were Germany, with about 110 tons a month, and the United States, where, in 1936, production reached 950 tons a year. The German catalysts and processes were described in various BIOS, FIAT, and CIOS reports and other publications after the end of

TABLE 7.3. Production of Vinyl Chloride 1939–1945.[a]

	Operating conditions
Feed	Acetylene (C_2H_2) and HCl
Reactor	Tubes 50–80 mm diameter × 3 m long (vessel 2 m diameter)
Temperature	120°C new to 200°C old
Pressure	~1 atm
Catalyst life	10 months
	Catalyst composition
A: Catalyst (I.G. Farben)	10% $HgCl_2$/charcoal—1/4-in. cubes
Yield (on C_2H_2)	96–98%
(on HCl)	80–90%
Conversion	100%
B: Catalyst (Wacker)	30% $BaCl_2$; 1% $HgCl_2$/charcoal

[a] HCl available from cracking carbon tetrachloride to trichloroethylene over $BaCl_2$ catalyst.

the war; operation from 1939 to 1945 is shown in Table 7.3.[14] No similar details were released for the US process. The same mercuric chloride catalyst supported on charcoal was still being used, acetylene was still the preferred feedstock, and the use of PVC was rapidly expanding.

The demand for larger vinyl chloride plants coincided with the development of hydrocarbon steam crackers. Ethylene reacts directly with chlorine forming ethylene dichloride, which was decomposed thermally to produce vinyl chloride. The byproduct hydrogen chloride could then be used to produce more vinyl chloride as a feedstock for the acetylene process. This was described as the balanced ethylene–acetylene process:[15]

$$CH_2{=}CH_2 + Cl_2 \rightarrow CH_2Cl.CH_2Cl \text{ (direct addition)} \qquad (7.11)$$

$$CH_2Cl.CH_2Cl \rightarrow CH_2{=}CHCl + HCl \text{ (thermal reaction and HCl recycle)} \quad (7.12)$$

$$CH{\equiv}CH + HCl \rightarrow CH_2{=}CHCl \text{ (mercuric chloride catalyst)} \qquad (7.13)$$

The ethylene–acetylene balanced process could, therefore, solve the problem of hydrogen chloride disposal, although only half of the vinyl chloride was produced from ethylene. The manufacture of acetylene is a very energy-intensive process, and compared to ethylene, acetylene is very expensive to produce. Ethylene would, therefore always be the preferred feedstock for the production of petrochemicals, provided that suitable technology was available.

Development of the oxychlorination process, which used a copper chloride catalyst, improved the situation because it only required ethylene as feedstock. The oxychlorination of ethylene for the manufacture of vinyl chloride is a three-stage process. The first stage is the addition of chlorine gas to ethylene to produce 1-2 dichloroethane. The second stage is the cracking of the dichloroethane to vinyl chloride and hydrogen chloride. The third stage is the oxychlorination reaction itself, in which hydrogen chloride from the cracking stage is reacted with air, or in later processes, oxygen, to produce more dichloroethane, which is then recycled to the second stage.

$$CH_2=CH_2 + Cl_2 \rightarrow CH_2Cl.CH_2Cl \text{ (direct addition)} \qquad (7.14)$$

$$CH_2Cl.CH_2Cl \rightarrow CH_2=CHCl + HCl \text{ (thermal reaction and HCl recycle)} \quad (7.15)$$

$$CH_2=CH_2 + \tfrac{1}{2} O_2 + 2 HCl \rightarrow CH_2Cl.CH_2Cl + H_2O \text{ (copper chloride catalyst)} \qquad (7.16)$$

$$CH_2Cl.CH_2Cl \rightarrow CH_2=CHCl + HCl \text{ (thermal reaction and HCl recycle)} \quad (7.17)$$

The cupric chloride catalyst was promoted with potassium chloride and supported on activated alumina. It was related to the Deacon catalyst, which was developed for chlorine production in 1887.

The first oxychlorination plant was built by B. F. Goodrich in 1964[16] and produced 30 million lb.year^{-1} of vinyl chloride in a fluid bed. At about the same time the Stauffer Chemical Company built a unit using three tube-cooled oxychlorination reactors in series. This was the beginning of the end for routes based on acetylene.

7.1.2.1. *The Oxychlorination Reaction*

The need to recover hydrogen chloride from the ethylene dichloride cracking reaction introduced during the early 1960s led to a renewed interest in the Deacon process, although the low equilibrium concentration of chlorine in hydrogen chloride/air mixtures at a reasonable temperature made economic recovery impossible.

It was found, however, that the Deacon catalysts could chlorinate ethylene with a mixture of hydrogen chloride and air. The Deacon reaction did not produce molecular chlorine but chlorinated ethylene directly as chlorine species formed on the catalyst surface. Low equilibrium conversion to chlorine and its slow removal from the catalyst surface were no longer limitations and complete chlorination of the ethylene was achieved at temperatures in the range 210^0–240^0C. Free chlorine was never found even with low ethylene concentrations.

7.1.2.2. *Oxychlorination Catalyst*

Suitable catalysts are similar to the original Deacon catalyst. The copper chloride is supported on medium-surface-area alumina with pore sizes in the range 80–600 Å.[17] While copper chloride is more active than other metals it is also volatile at reaction temperature. Potassium chloride is added to decrease the loss of copper and reduce the formation of by-product ethyl chloride, even though it also reduces the overall activity. The copper chloride content of the catalyst must be limited because an excess would lead to catalyst caking during reaction, since the complexes $KCuCl_3$ and K_2CuCl_4 have naturally low melting points. The eutectic mixture with copper chloride melts at 150°C. Catalysts may also contain other alkali and rare earth chlorides and promoters that help to control the reaction or inhibit side reactions.[17]

The reaction mechanism was originally thought to be as follows:

$$C_2H_4 + 2\ CuCl_2\ \rightarrow\ CH_2Cl.CH_2Cl + Cu_2Cl_2 \qquad (7.18)$$

$$Cu_2Cl_2 + \tfrac{1}{2}\ O_2\ \rightarrow\ CuO.CuCl_2 \qquad (7.19)$$

$$CuO.CuCl_2 + 2\ HCl\ \rightarrow\ 2\ CuCl_2 + H_2O \qquad (7.20)$$

The overall reaction is:

$$C_2H_4 + 2HCl + 1/2O_2 \rightarrow CH_2Cl.CH_2Cl + H_2O \qquad (7.21)$$

Catalysts are usually prepared on a large scale by impregnating a suitable alumina support with aqueous solutions of cupric chloride and potassium chloride. Other alkali metals or rare earth chlorides have been used as promoters or to inhibit by-product formation.

7.1.2.3. Catalyst Operation

The overall reaction is very exothermic, and temperature control is extremely important in vinyl chloride production to maintain a satisfactory selectivity. This is relatively easy in fluidized beds, which is probably the most widely used process, but more difficult when fixed bed tubular reactors are used. High bed temperatures result in the dehydrochlorination of the ethylene dichloride to give vinyl chloride, and this can lead to the formation of other chlorinated products. Increased oxidation of the ethylene feed to carbon dioxide also occurs as the temperature increases. Coke formation and copper loss by sublimation also increase with higher bed temperatures.

In tubular reactors the proportions and concentration of the reactants can be changed to control the reaction or the tube diameter design can be decreased. The most common ways to control the reaction are to dilute the catalyst with inert balls or to vary the copper chloride content of the catalyst. There is a rapid increase in catalyst temperature at the top of the tubes where most of the exothermic reaction takes place giving a significant hot spot. This can be controlled to a certain extent by splitting the total air addition between the three beds. Good control at the inlet to the first bed is necessary in any case to ensure that the explosive limit is not exceeded. The addition of fused alumina, graphite, or other inert diluents to the catalyst can limit the temperature rise, although there are practical problems in mixing catalyst with inert materials.

Stauffer[18] described a process consisting of a single reactor containing up to four different layers of a catalyst containing 8.5% cupric chloride. Each successive layer was diluted with an inert substance to control the exothermical reaction. The first layer only contained 7% of catalyst, the others containing 15%, then 40% and the final layer contained 100% catalyst. However, better control with diluted catalyst was obtained in a series of several separate reactors. A top layer of special catalyst to initiate or *strike* the reaction was used in the first reactor. An even better degree of control was obtained with three different catalyst compositions containing varying amounts of cupric chloride. Typically, catalysts containing 6%, 10%, and finally 18% cupric chloride, promoted with 3%, 3%, and then 2% potassium chloride respectively, each packed in layers, were the optimum. Beds 1 and 2 contained 40% of both the 6% copper chloride and 10% copper chloride, respectively, on top of the 18% copper chloride catalyst. Bed 3 was completely filled with the more concentrated catalyst.

Catalyst life in tube-cooled reactors depends on the operating conditions, but in reactor 1 the catalyst is usually changed after one year. Life increases to one and a half years and three years in reactors 2 and 3, respectively, where conditions are less severe and the intensity of the temperature hot spots decrease. There are usually hot spots at the position in tubes where the catalyst type changes. Up to 50% of the air can be added to the first reactor and must be carefully controlled to avoid the explosive limit. The balance of air is added to the second and third reactors to minimize the hot spot and to maintain conversion. Oxygen can replace air when convenient because this reduces the gas to be vented from the system, simplifies the incineration of noxious products and makes environmental control easier.

In fluidized bed reactors the preheated gas mixture enters the base of the vessel and temperature is controlled either by cooling coils or a bundle of cooling tubes in the fluid bed. The oxychlorination reaction takes place at a relatively low temperature, around 220–225^{0}C, and at a low pressure of about 2 bar. The catalyst support consists of microspheroidal γ-alumina particles, with a median particle size around 50–60 microns. The particles are impregnated with an aqueous mixture of cupric and potassium chlorides, to give a copper content around 12%. One of the main process problems with oxychlorination[19] is a possible tendency for the catalyst particles to *cake* or to stick to the cooling bundles within the reactor. This results in poor temperature control, local overheating, and a loss in selectivity. The surface area of the alumina is in the range 150–250 $m^2.g^{-1}$, with 94% of the particles greater than 6 μm and 24% greater than 80 μm.[19] Fresh catalyst is added to replace fines lost from the bed. An excess of air is added to maximize hydrochloric acid conversion and unreacted ethylene is recycled. Yields of 98% on hydrochloric acid and 96% on ethylene can be obtained. By-products are chloral, 1,1,2-trichloroethane, chloroform, 1,2-dichloroethylene, and ethyl chloride. In modern processes pure oxygen is used instead of air. This enables much easier recovery of product and catalyst fines, also enabling a reduction in the volume of vent gas by at least 95% and a reduction in both energy and capital costs.[20] Since the oxychlorination process was introduced plant capacities have increased from about 10,000 to more than 600,000 tonnes a year.

7.2. SYNTHETIC RUBBER FROM BUTADIENE AND STYRENE

In the 1930s, a wide range of hydrogenation and dehydrogenation processes was already developed by companies in the petroleum industry to increase gasoline yield. Olefins had been used to provide polygasoline and alkylates, as well as many other petrochemical intermediates.   A form of synthetic rubber had been made in Germany since the 1914–1918 war and in 1929 two I. G. Farben chemists from Leverkusen patented a variety they named Buna-S.[21] It was a copolymer of butadiene containing 25% styrene and, as a result of many improvements, is still being used. However, at the time it was difficult to process, thermally unstable during use, and was expensive to produce. Other synthetic rubbers soon followed. Neoprene, produced from chlorobutadiene and originally named *Duprene*, was developed by DuPont and used in several applications.[22] I. G. Farben also produced Buna-N, replacing the styrene in Buna-S with acrylonitrile. At the time Buna-N was unsuccessful, although it did eventually lead to the present large-scale production of acrylonitrile. By 1940, when the Reserve Company was formed in the USA, two of the most significant rubber intermediates were butadiene and styrene. The potential wartime shortage of natural rubber then led the Rubber Reserve Company to make plans for four plants to produce synthetic rubber, which was known as GR-S rubber (Government Rubber-Styrene), and by November 1943 a total of 15 plants were operating.[23] Production of GR-S rubber increased from about 4000 tonnes in 1942 to almost 1 million tonnes in 1946. Production fell as the war ended but increased again to more than 600,000 tonnes a year from 1951 to 1953, while the Korean War was in progress.

Production on such a large scale led to an increasing demand for the relatively new petrochemicals butadiene and styrene. When wartime rubber production was being planned most commercial butadiene was produced as a by-product from the high-temperature steam cracking of naphtha or catalytic cracking of oil fractions. The quantity, however, was very small. It was therefore apparent that, in addition to making more of the *quickie* butadiene, other methods would have to be developed. Two sources were potentially available. One possibility was a process that had been used in Russia to produce butadiene from alcohol and the second was to produce butadiene by the dehydrogenation of *n*-butane and *n*-butene by processes that were still being developed.

Nine dehydrogenation plants were eventually authorized by the Rubber Reserve Company to make 330,000 tonnes year^{-1} of butadiene.[24] Two were based on butane and seven on butene. During construction of these plants it was also decided to make about 220,000 tonnes.year^{-1} of butadiene by the Russian Lebedev process based on ethanol:

$$C_4H_{10} \rightarrow C_4H_6 + 2H_2 \tag{7.22}$$

$$C_4H_8 \rightarrow C_4H_6 + H_2 \tag{7.23}$$

Little is known about the Lebedev process, which was developed in Russia in 1927 during the investigation of synthetic rubber production. It was a one-step process that converted ethanol to butadiene by simultaneous dehydrogenation, dehydration, and condensation stages:[25]

$$2\ C_2H_5OH \rightarrow C_4H_6 + 2\ H_2O + H_2 \tag{7.24}$$

The catalyst used was nominally a mixture of 75% magnesia and 25% silica or siliceous earth, which are well known as dehydrogenation and dehydration catalysts respectively. Up to 3% chromium oxide was included to inhibit the formation of magnesium silicate.

Carbon formed on the catalyst during operation at 270°–300°C, so that regeneration was required every 12–16 h. The catalyst was soon deactivated and had to be changed after only 30 days. A maximum butadiene yield of 60% was achieved.[26] As with most catalysts, the original recipe must have been modified as revealed in BIOS reports on the German use of the catalyst.[27] Whereas the composition was reported to be 100 parts magnesium oxide with 15 parts kaolin and 1–3 parts chromia, the analysis of an authentic sample was somewhat different.[28] The catalyst was made by wet mixing the components, extruding the mixture, and calcining at 500°C for 2–3 h before use. Commercial dehydrogenation processes are usually limited to propane and butane feeds because they operate below the thermal steam cracking temperature. Nevertheless, a high reaction temperature is still required to achieve a reasonable conversion close to the thermodynamic equilibrium. Frey and Huppke indicated that chromia catalysts gave a good yield of olefins from the corresponding alkanes.[29] Later, Burgin and Groll patented chromia/alumina catalysts containing 6–40% Cr_2O_3 and it was found that the optimum chromium content to dehydrogenate propane was lower than that required to dehydrogenate *n*-butane. Promoters such as potash or magnesia were used to reduce carbon decomposition during operation.[30]

7.2.1. Butadiene from Butane

All butane dehydrogenation processes used chromium catalysts supported on γ-alumina. Catalysts were generally made by impregnating a preformed support with chromic acid solution that also contained the magnesium or potassium promoter. Processes operated at a high temperature and low butane partial pressure to compensate for the high hydrogen content of reaction gases and to improve the equilibrium conversion. The Houdry or Catadiene process was used in only two of the Rubber Reserve Company plants and was the only process able to produce butadiene directly from butane. The chromia/alumina catalyst, containing 15–20% chromia, was used in several adiabatic reactors operating in parallel. Despite being diluted with a dense inert material the catalyst was deactivated by carbon deposition after about 7–15 min of operation at 620°–650°C and a reduced pressure of 3 psig. Liquid butane space velocity was in the range 1–2 h^{-1}. The reactor containing deactivated catalyst was taken out of service and replaced by a spare reactor to give continuous operation. Deactivated catalyst was then regenerated in air and diluted with an inert gas at 500°C before reuse.

Overall selectivity to butadiene was only 55–60%, with conversions in the range 30–40%. The feed gas was preheated using the heat evolved from the combustion of carbon that had been laid down on the surface of the catalyst during operation. Thus, carbon lay-down and catalyst regeneration became an integral part of the process.

Butane was also dehydrogenated, usually to butene, by the Phillips and UOP processes. Both processes used the same sort of chromia/alumina catalysts, loaded in tubular reactors, but were able to operate at atmospheric pressure because less hydrogen was produced during the reaction. In some cases the butene produced was subsequently dehydrogenated to butadiene in a separate unit.

ICI operated a plant to produce isobutene from isobutane in a UOP plant.[31] Yields of 75–80% isobutene were obtained at 30–40% conversion in a tubular reactor and residual isobutane was recycled.

I. G. Farben also produced butene by butane dehydrogenation.[32] A moving catalyst bed system was used in a tubular reactor. The total catalyst charge was 1.5 tonnes with a residence time of 4 h in the tubes. Yields of 85% at 20–25% conversion were obtained at a liquid space velocity of 2 h^{-1} and 620°–650°C operating temperature. This was an impressive result for a new reactor design that has now been developed as the continuous catalyst regeneration process and is widely used in refineries.

7.2.2. Butadiene from Butenes

Butene dehydrogenation was studied extensively in Russia, Germany, and the United States in the 1930s, and several promising processes were selected for the Rubber Reserve Company program. In all cases a steam ratio of up to 20:1 was used during operation. The steam diluent provided several benefits by:

- Decreasing the partial pressure of hydrocarbon and increasing equilibrium conversion at atmospheric pressure.
- Acting as a heat carrier so that the feed was not directly heated to the reaction temperature.
- Decreasing coke deposition to extend the operating cycle.

For these reasons the catalyst could be used in adiabatic beds rather than in tubular reactors. The inlet temperature lay in the range 600°–675°C and, more practically from a process point of view, the process was operated at atmospheric pressure. The use of less water-sensitive catalysts also meant that regeneration was possible with steam rather than using air at high-temperature.

A catalyst developed by the Standard Oil Development Company and first used in 1940 was used in six of the Government butadiene plants.[33] The composition of catalyst 1707 is shown in Table 7.4.

It was shown that iron oxide promoted dehydrogenation activity while potash reduced the extent of carbon deposition during operation and also assisted in carbon removal during regeneration. The catalyst gave 85% selectivity at 20% conversion.

Phillips produced a bauxite catalyst containing 5% barium hydroxide that was later impregnated with potash. Natural bauxite always contains a significant amount of iron impurities, and there was probably an unexpected beneficial effect from the iron impurity. The Shell Development Company introduced an iron catalyst promoted with chromia and containing potash, which by this time was recognized as essential to reduce the effects of carbon formation.[34] The compositions of the Shell catalyst 105 and the later Shell 205, which gradually replaced 1707 in wartime plants, are also shown in Table 7.4. It seems likely that the potash content of Shell 105 was gradually increased to 7% K_2CO_3.[35] By

TABLE 7.4. Early Butene Dehydrogenation Catalysts.[a]

Wt%	Standard Oil 1707	Shell 105 (1940s)	Shell 205 (1950s)	Dow (1950s)
Magnesium oxide	72.4	—	—	
Ferric oxide	18.4	93	62.5	$Ca_8Ni(PO_4)_6$
Copper oxide	4.6			
Chromium oxide	—	5	2.2	2
Potassium	4.6 (K_2O)	2 KOH)	33.3 (K_2CO_3)	
Operating temperature (°C)		625/650	650	625/650
Steam/butane		14:1	18:1	20:1
Conversion %		28–35	37	30–50
Selectivity %		70–75	76	86–90

[a]Lebedev catalyst 40–50% MgO; 10–12% SiO_2; 2–3% Al_2O_3; ~3% Cr_2O_3; balance CO_2/H_2O.

1954 the Shell catalyst 205, containing 33% potassium carbonate, became the preferred catalyst for butadiene production.

The Shell catalysts were made by dry mixing powdered iron oxide, chromic acid, and potassium carbonate, mixing with water, extruding before drying, and then calcining at 800°–950°C. The high-temperature calcinations was critical and later work indicated that it optimized surface area and pore size as the iron oxide was converted to α-hematite.

Dow introduced a calcium/nickel phosphate catalyst (see Table 7.4), which was used in several butene dehydrogenation units.[36] It gave a better yield at a higher steam ratio than the other catalysts but needed to be regenerated with air at intervals of about 15 min to remove the carbon deposits which collected in the catalyst bed, not on the catalyst itself. The operating life was about seven months before replacement, which was much shorter than that of the other catalysts.

Butadiene production from these dehydrogenation processes did not continue beyond the 1960s, as butadiene from steam cracking of naphtha became available. Later, however, from about 1990, as butadiene and butene shortages developed, several commercial processes were revived.

7.2.2.1. *Oxidative Dehydrogenation*

The addition of oxygen to the butene and steam reaction mixture improves conversion and selectivity during the dehydrogenation reaction by removing the hydrogen as water:

$$C_4H_8 + \tfrac{1}{2} O_2 \rightarrow C_4H_6 + H_2O \tag{7.25}$$

Not only does this improve the equilibrium conversion but the exothermic oxidation reaction also supplies heat to balance the endothermic dehydrogenation reaction. At 550°–600°C more than 90% selectivity is possible at conversions in the range 65–80% with steam mole ratios up to 12. Catalysts in the Petrotex (Mobay) process are zinc/chromium and magnesium/chromium ferrites.[37]

7.2.3. Propylene from Propane

The dehydrogenation processes used for the production of butadiene from *n*-butane and *n*-butenes during the development of GR-S rubber were modified in the 1990s for the dehydrogenation of propane to propylene. This compensated for the short supply of steam cracked propylene used to produce polypropylene. The new processes can also be used for the dehydrogenation of other paraffins.

The old Houdry process has been developed by UCI/Lummus to become the Catofin process while Phillips and UOP have introduced their Star and Oleflex processes. Linde in Germany has also introduced a process. A range of different catalysts is used and high selectivity is achieved at fairly low conversion. The Phillips Star process used a promoted noble metal catalyst supported on zinc aluminate, which had been originally developed for light paraffin dehydrogenation with minimum isomerization.[38] The catalyst is contained in a tubular reformer, operating at temperatures in the range 480°–620°C and at a pressure in the range 1–2 bar. The space velocity ranges between 0.5–1.0 LHSV and the steam ratio is in the range 2–10 moles of steam per mole of hydrocarbon. The process suffers from the deposition of carbon, and the catalyst must be regenerated after an eight hour cycle. The average lifetime of the catalyst is about 1–2 years. The selectivity to propylene is around 80–90% at a propane conversion of 30–40%. In the case of butane, the selectivity to butene is 85–95% at a conversion of 45–50%. The UOP Oleflex process uses a platinum catalyst supported on alumina under similar conditions and selectivity. A chromium-based catalyst is used in the Catofin process. A more recent process, announced by Linde in 1992 and using a chromium/alumina catalyst developed by Engelhard, is now operating in Germany.[39] The platinum catalyst, promoted with tin and supported on $CuO/ZnO/Al_2O_3$ (derived from calcination of a mixed precursor with a hydrotalcite structure) is different from the early formulations. It is reduced in hydrogen and steam before use. High conversions and selectivities are claimed in operation at high temperatures with extended cycle times.

7.2.4. Styrene

The Naugatuck Chemical Company, Connecticut, a subsidiary of U.S. Rubber, was the first company to manufacture styrene. The process, however, was based on the hydrodechlorination of ethylbenzene and the product was not sufficiently pure for use in polymer applications. The company had received assistance from Igor Ostromisslenski, a well-known Russian emigrant, who held several patents in the field.[40]

Both I. G. Farben in Germany[41] and the Dow Chemical Company, Midland, Michigan,[42] were also working, independently, on styrene and polystyrene and had applied for patents. By 1931 I. G. Farben were operating a 60-tonnes.day^{-1} plant at Ludwigshafen and, soon afterward, Dow built a plant in the United States. Ethylbenzene was dehydrogenated directly to styrene in both processes, although different reactor designs were used. The catalysts used by I. G. Farben[43] and Dow[44] were quite different. At first neither used potash, which is now known limit the degree of carbon deposition. I. G. Farben used supported zinc oxide whereas Dow used bauxite. The iron oxide impurity in the bauxite, as in the Phillips butane dehydrogenation catalyst, would have acted as a fortuitous catalyst promoter and is still an essential component of modern catalysts. Catalyst compositions are shown in Table 7.5.

Early patent literature does not always give a true idea of the process. Much technical information on various petrochemical processes became available in the years following World War II, when experts inspected the German chemical engineering and technical centers producing chemicals. The results were published in a number of reports:

- British teams wrote British Intelligence Objectives Sub-Committee Reports (BIOS).
- American teams prepared Field Information Agency Technical Reports (FIAT).
- There were also Combined Intelligence Objectives Sub-Committee Reports (CIOS).

The German process used banks of gas-fired, copper-lined tubes, 30 ft long and 8 in in diameter, containing catalyst. These were heated to 610°C. The feed to the reactor tubes consisted of a mixture of 1.25 lb steam for every pound of ethylbenzene, preheated to 560°C. Up to 1941 the yield was 75% at 40% conversion, but later the yield improved to 90%, presumably as a result of using an improved catalyst. The US process used a fixed bed of catalyst with 2.6-lb steam for every pound of ethylbenzene. Operation at 600°C gave a catalyst life of one year with no regeneration. As with other dehydrogenation reactions, the partial pressure of ethylbenzene was reduced by the addition of steam or even carbon dioxide. Modern processes are based on the use of isothermal tubular reactors or fixed bed adiabatic reactors. Unconverted ethylbenzene is recycled after the styrene product is removed. When the Rubber Reserve Company authorized butadiene production in 1942, they also included six styrene plants, five of which used the Dow process. The sixth plant used a Union Carbide process. A total of 187,500 tonnes/year of styrene was planned, based on ethylene and coal-based benzene. When Standard Oil introduced catalyst 1707 for butadiene production, it was also used to dehydrogenate ethylbenzene in the new plants. The improved iron oxide/chromia/ potash catalysts and Shell 105 and 205 were eventually preferred when they became available, particularly after the war.

Many other iron catalyst compositions have been examined experimentally as the styrene process has developed.[45] Chromium was replaced or combined with a wide range of other oxide promoters, such as copper oxide, zinc oxide, vanadium pentoxide, thoria, tungsten oxide, molybdena, ceria, and alkaline earth oxides. At the time, none was used in any full-scale process but some were later found to give good performances in improved plants.

7.2.4.1. Ethylbenzene Production

The alkylation process is the addition of an alkene to benzene, usually over an acidic catalyst to give the alkyl benzene. The reaction is non-selective, and poly-alkyl benzenes are regular impurities in the crude product stream. The degree of polysubstitution is usually limited by controlling the ratio of reactants.

TABLE 7.5. Ethylbenzene Dehydrogenation Processes.

	Operating conditions	
Catalyst volume (m^3.100,000 tonnes^{-1} year^{-1})	~40	
Inlet temperature (°C)	580–600	
Outlet temperature (°C)	540–600	
Pressure (atm)	<1–2	
Liquid space velocity (LHSV)	0.3–0.7	
Steam ratio (weight)	1–2	
Steam ratio (molar)	6–12	
Conversion (%)	~60	
Selectivity (%)	93–96	
Catalyst life (years)		
Adiabatic	1–2	
Isothermal	4–5	
Catalyst^{a,b}	(a)	(b)
%Fe$_2$O$_3$	60	70–80
%K$_2$O	30	10
%Cr$_2$O$_3$	1–2	—
%Ce$_2$O$_3$		5
%MoO$_3$		2–3
%CaO		2
%MgO	7–8	2
%Loss at 540°C	Balance	Balance
Particle size (mm)	3–7	3–7
Bulk density (kg l^{-1})	1.0–1.8	1.0–1.3
Surface area (m^2 g^{-1})	2–3	2–3
Pore volume (cm^3 g^{-1})	0.2–0.3	0.2–0.3

^{a}Some catalysts contain binders (e.g., cement)
^{b}Early World War II catalysts compositions were: I.G. Farben (a) 72% ZnO; 18% Al$_2$O$_3$; 9% CaCrO$_4$; (b) 77.4% ZnO; 9.4% Al$_2$O$_3$; 2.8% CaO; 2.8% K$_2$CrO$_4$; 2.8% K$_2$SO$_4$. Dow: bauxite (12% Fe$_2$O$_3$) + 5% Ba(OH)$_2$ or KOH.

Dry benzene was alkylated with ethylene in either the liquid or gas phases using acidic catalysts:

- An early liquid phase process used an aluminum trichloride catalyst at 85°– 95°C at pressures just above atmospheric. A low ethylene/benzene ratio was used to limit the formation of diethylbenzene and other polyethylbenzenes. By-products could, however, be recycled with benzene and were recovered as ethylbenzene by transalkylation. Ethylbenzene selectivity was about 94% based on benzene and higher on ethylene. The catalyst that formed in solution was thought to be HAlCl$_4$.n-C$_6$H$_5$C$_2$H$_5$, which gradually deactivated and was replenished as required. Other acid catalysts such as boron trifluoride can be used in the liquid phase process, which is still widely used in older plants.

- Gas phase processes have been used and operate at 300°C and higher pressures up to 60 atm. The first catalysts were the typical phosphoric acid/kieselguhr types (UOP Alkar process) or silica/alumina (Koppers) operating with a low ethylene/benzene molar ratio (0.2). The low ratio was necessary to minimize by-products that could not be dealkylated by the catalyst. The Alkar process using boron trifluoride supported on activated alumina did, however, dealkylate the by-products, thus giving higher selectivity at 100% ethylene conversion. It was useful for the production of ethylbenzene from dilute ethylene streams.
- Since 1980 high-activity modified ZSM-5 catalysts that operate at 450°C and 20 atm have been used in most new plants. Capacities of up to 800,000 tonnes/year have been possible. The catalyst is extremely active and forms few by-products or coke-forming polymers. Capital costs for equipment are lower and the use of highly corrosive catalysts is avoided.

7.2.4.2. Styrene Production after 1950

Although the use of dehydrogenation processes to supply butadiene declined as the more economical supplies from steam cracking of naphtha were introduced, the production of styrene from ethylbenzene dehydrogenation has been continuously developed, since styrene is not available in sufficient quantities as a by-product.

From about 1950, Shell 205 and similar catalysts based on alkalized iron and chromium oxides were used exclusively for styrene production. As plant capacities were rapidly expanded, efforts were increased to improve the performance of the catalyst. Higher potash levels were introduced[46] and cement binders were used to increase strength and selectivity.[47] Ethylbenzene conversion, which was still about 30–50% in the 1950's, was increased to at least 60% by 1960. Better plant designs were developed and reactors with up to three beds were introduced. One of the first higher selectivity catalysts included vanadium pentoxide with the conventional chromium oxide and potash.[48] Improvements often led to different catalysts being used in a single reactor to optimize operation.

The energy crisis of the 1970s made it even more important to reduce costs and make the process more efficient. Selectivity was increased even further by a new catalyst, in which a combination of molybdenum oxide and cerium oxide together with a cement binder replaced the chromium oxide completely. It was still necessary to reduce fuel costs and operators began to conserve energy by lowering the steam ratio during operation. Under these conditions chromium promoters were still effective, and as the steam ratio decreased, chromium-promoted catalysts, particularly those with high potash content and cement binders, operated with a much lower level of deactivation for longer periods than the new chrome-free catalysts. High-potash chromia catalysts operated at a steam mole ratio as low as 3:1, while catalysts containing mixed chromium, ceria, and molybdenum promoters, together with potash, were more selective but had to be operated at steam ratios of about 7:1.[50]

Increased efforts during the 1970s led to a rapid increase in the availability of catalysts containing various promoters and different potash contents. One supplier estimated that some 12–15 catalysts were available for both isothermal and adiabatic reactors operating at both high and low steam ratios.[51] This was partly because so many different processes were used and every operator probably had a favorite type of catalyst! It was still possible, however, to choose catalysts for both isothermal and adiabatic plants that had either a high selectivity and good activity at a moderate steam ratio or a high activity and long life at a higher steam ratio.

7.2.4.3. Styrene Plant Operation

The dehydrogenation of ethyl benzene is endothermic so that heat must be supplied during operation. The two commercial styrene processes either incorporate several adiabatic beds with interbed heat exchange/steam addition or isothermal tubular reactors with a suitable heating medium in order to maintain operating temperature.[52]

Carbon deposition and dealkylation reactions are both inhibited by the addition of potash to the catalyst. The equilibrium of the forward reaction is favored by high temperature and low operating pressures. A high proportion of superheated steam is also added to the ethyl benzene feed for the following reasons:

- Provide heat for the process.
- Reduce the partial pressure of ethylbenzene.
- Inhibit the formation of undesirable carbon deposits.
- Prevent reduction of magnetite to metallic iron.

After charging to the reactor, the catalyst is carefully heated, first in nitrogen and then in steam, to about 540°C, before ethylbenzene is admitted to the reactor. A high steam ratio is maintained for several days to activate the catalyst. During the activation period, the α-haematite (α-Fe2O3) component of the catalyst is reduced to magnetite (Fe3O4) and selectivity increases to a maximum during a period of two to three days. The level of conversion is maintained by controlling the temperature of the catalyst or by increasing the steam ratio. The presence of chloride in the feedstock or in the steam leads to an irreversible poisoning of the catalyst. Sulfur compounds also act as poisons, but the activity of the catalyst does recover when the sulfur contamination is removed from the feedstock. Operating conditions are summarized in Table 7.5.

Some catalysts can catalyze the dealkylation of the ethyl benzene to ethylene and benzene, as well as the required dehydrogenation reaction to give styrene and hydrogen. This undesirable reaction can be suppressed by allowing a small amount of carbon or coke to deposit on the catalyst surface. However, an excessive level of carbon deposition also inhibits the required dehydrogenation reaction. When this occurs, some of the carbon can be removed, and activity for hydrogenation restored, by treating the catalyst with superheated steam. If the conditions under which the catalyst is steamed lead to conversion of the potassium carbonate to the hydroxide, the catalyst selectivity decreases and potassium hydroxide, which is volatile, is slowly lost. Catalyst life is usually less than two years.

7.2.4.4. *Ethylbenzene Dehydrogenation (Styrene) Catalysts*

Catalysts require a high thermal stability to maintain the required surface area and pore size at the high operating temperature and steam ratio.[52] The composition of the catalyst is quite similar to the high-temperature-shift catalyst used in the manufacturing of ammonia, which is operated at much lower temperatures and which often loses strength when exposed to potassium compounds which have migrated from the reforming stages of the process. Perhaps for this reason, hydraulic cements have often been used to bind styrene catalysts and improve physical strength. Catalysts have been made by wet mixing of the components to form a paste.

Early catalysts were produced from calcined ferric oxide, potassium carbonate, a binder when required, and usually chromium oxide. Subsequently a wide range of other oxides replaced the chromium oxide;[51,52] typical compositions are shown in Table 7.5. The paste was extruded or granulated to produce a suitable shape and then calcined at a high temperature in the range 900°–950°C. Solid solutions of α-hematite and chromium oxide (the active catalyst precursors) were formed and these also contained potassium carbonate to inhibit coke formation. Catalyst surface area and pore volume were controlled by calcination conditions. It has been confirmed by X-ray diffraction studies that α-hematite is reduced to magnetite and that there is some combination of potash and the chromium oxide stabilizer. There is little change in the physical properties of the catalyst during reduction and subsequent operation.

Catalyst activity and selectivity, at high or low steam ratios, can be controlled by selection of both the stabilizers and the calcination temperature.

7.3. SYNTHETIC FIBERS

The production of nylons and polyesters, which were first synthesized in the 1930s, has led to the development of *man-made* fibers. The complex sequence of catalytic reactions involved in both processes was originally based on 1930s technology but, as the reaction mechanisms came to be understood, catalysts were improved and, in some cases, new *petrochemical* feeds were introduced.

Acrylic fibers based on acrylonitrile were then developed from about the 1940's and, although not as widely used as polyesters and nylons, have many applications.

It is shown in Table 7.6 how synthetic man-made fibers are replacing the viscose and acetate rayons produced from natural cotton fibers or wood pulp. During 1970 the production of cellulose fibers was 43% of world fiber output, but by 2000 the proportion had fallen to only 8%. Cellulose fiber production fell from 3.6 to less than 2.5 million tonnes during the same period. Polyester fiber dominates the market and far exceeds the use of nylon 66, nylon 6, and acrylic fibers. In 1993 man-made synthetic fiber production was about 17.5 million tonnes while man-made cellulose fiber production was about 2.5 million tonnes. At the same time about 15.5 million tonnes of natural cotton was used.[53]

7.3.1. Nylon 66

Wallace Carothers began his work on polymerization in 1928 at the DuPont laboratories. By 1930, he had found that aliphatic polyesters, formed by the condensation of dihydric alcohols and dicarboxylic acids, could be drawn into threads when melted.

Most of the polyesters he investigated, however, had low melting points and were too soluble in water for use as textiles, although polyamides were more promising. To form useful fibers, the polymer needs to have a molecular weight in the range 12,000–20,000. When amino acids were briefly investigated, those with fewer than five carbon atoms condensed to form lactams, together with a low-molecular-weight polymer. Carothers then assumed that only those amino acids containing more than seven carbon atoms would be able to produce useful polymers. These were difficult to synthesize, so he concentrated on the range of polyamides produced from diamines and dicarboxylic acids. He selected adipic acid and hexamethylene diamines for his work because cyclohexane was readily available and could be used as the raw material for conversion to both intermediates. The polymer produced from the two C_6 molecules formed good fibers

TABLE 7.6. Production of Synthetic Fibers 1970–2000.

Fiber	1970	1993	2000
% Polyester	20	50–60	62
% Nylon[a]	24	15–20	12
% Acrylics	12	~10	9
% Cellulose	43	~10	8
% Others[b]	1	~ 5	9
Total (million tonnes)	8.5	20	30

[a]Nylon also used, e.g., for carpets, and polyester also used, e.g., for bottles.
[b]Polyethylene, polypropylene, polyvinyl alcohol, polyvinyl chloride.

and production of nylon 66 began in 1938.[54] The term nylon 66 is derived from the fact that the diamine and the dicarboxylic acid both contain 6 carbon atoms.

7.3.1.1. Production of Nylon Intermediates

During the early years of nylon 66 production, cyclohexane was obtained by distillation from some natural gasolines, together with isomerization of the small volumes of methyl cyclopentane also present. In other parts of the world and as the demand for nylon fibers and resins increased, benzene became the major source of cyclohexane.

Sulfur-free benzene can be converted to cyclohexane by hydrogenation in a two stage process. Approximately 95% of the benzene in a circulating liquid phase is converted in the first reactor using a simple Raney nickel catalyst at temperatures less than 230°C and at pressures up to 50 bar. The catalyst also acts as a sulfur guard. Temperature control is achieved largely by relying on the latent heat of evaporation of cyclohexane. It is important to avoid isomerization to methyl cyclopentane. A second reactor with a fixed catalyst bed of nickel supported on kieselguhr or alumina, operating in the vapor phase at 200°–225°C and 50 bar, completes the benzene conversion.

7.3.1.2. Adipic Acid

The first stage in the production of adipic acid is the oxidation of liquid cyclohexane with air using a cobalt naphthenate catalyst at temperatures in the range at 140°–160°C and pressures about 8–12 bar. A mixed ketone/alcohol oil containing cyclohexanol ($C_6H_{11}OH$) and cyclohexanone ($C_6H_{10}O$) (KA oil) is produced at up to 85% selectivity, with the ketone/alcohol ratio of about 1:1. The conversion is only about 10% and unconverted cyclohexane is recycled. The process was improved by Bashkirov by the additions up to 5% boric acid to the cyclohexane and by restricting the oxygen content of the air to about 4%. The overall yield was increased to more than 90% and conversion to more than 12%. The ketone/alcohol ratio was decreased to 1:9.

The ketone/alcohol oil was then oxidized further to adipic acid in a 60% nitric acid solution with an ammonium metavanadate/copper nitrate catalyst at 50°–80°C. Selectivity to adipic acid was more than 95%. A second air oxidation process at 80°–85°C and 6 bar has since been developed. By using a copper acetate/manganese acetate catalyst in acetic acid solution, the process has the advantage of avoiding the use of the strongly corrosive nitric acid:

$$C_6H_{11}OH + 4\,O(HNO_3) \rightarrow COOH(CH_2)_4COOH + H_2O \qquad (7.26)$$

$$C_6H_{10}O + 3\,O(HNO_3) \rightarrow COOH(CH_2)_4COOH \qquad (7.27)$$

Pure cyclohexanol can also be obtained by the hydrogenation of phenol with a palladium catalyst at 150°C and 10 atm although the process is not widely used. Attempts were made by BASF to synthesize adipic acid from butadiene via a two-step carbonylation process in the presence of methanol. The first step of the synthesis operated at 130°C and 600 bar while the second operated at 170°C and a lower pressure of 160 bar. A typical cobalt catalyst with organic ligands was used, but the process was never developed industrially.

Phenol produced by the cumene–phenol process is relatively expensive but Solutia has recently claimed a new process developed in Russia.[55] Benzene is oxidized directly to phenol using nitrous oxide. Phenol is then converted to adipic acid by oxidature procedures. Nitrous oxide from the final nitric acid oxidation to adipic acid is recycled to the first stage. This has been reported as the first new commercial technology since DuPont introduced the direct hydrocyanation of butadiene to adiponitrile in 1971.

7.3.1.3. Hexamethylenediamine

The manufacture of hexamethylenediamine is usually a two stage process. In the first stage, adiponitrile is produced, and in the second stage, adiponitrile is hydrogenated to hexamethylenediamine. Adiponitrile has been manufactured, in the main, from three feedstocks, and these will be described briefly in turn.

In the first process adipic acid was converted to adiponitrile by a high temperature reaction with ammonia over a boron phosphate catalyst at temperatures in the range 300–350°C. The process was thought to proceed via the formation of diammonium adipate, followed by dehydration to adiponitrile:

$$COOH\text{-}(CH_2)_4\text{-}COOH + 2\ NH_3 \rightarrow COONH_4(CH_2)_4\text{-}COONH_4 \quad (7.28)$$

$$COONH_4(CH_2)_4\text{-}COONH_4 \rightarrow NC(CH_2)_4CN + 4\ H_2O \quad (7.29)$$

The selectivity to adiponitrile was about 80% in the early reactors.

The main problem with this reaction is that adipic acid in the liquid phase is thermally unstable and degrades to a carbon-rich, hard, resinous product. The vapor is rather more thermally stable at a similar temperature, but at higher temperatures also degrades to similar products. The overall reaction producing adiponitrile from adipic acid is very endothermic, and the reaction mixture is subjected to a significant measure of cooling due to this process. Thus, while all of the reaction components were initially in the gas phase this cooling result in the condensation of some liquid adipic acid, which then proceeds to degrade to the polymeric resin described above. The reactor tubes regularly become blocked with catalyst and polymeric resin, thus preventing flow of reactants through the reactor. At this stage, the reactor must be taken out of service, so that the polymeric resins can be drilled out of the reactor, and the catalyst replenished. Selectivity could be increased to 90% by converting the adipic acid to adiponitrile in a fluid bed with a phosphoric acid/kieselguhr catalyst that was continuously regenerated during operation.

As the demand for nylon increased during the 1950s and 1960s, an alternative, more readily available feedstock was sought for the production of adiponitrile. In this way more adipic acid would be released for the production of nylon salt. Consequently, by the 1960's, an alternative route to the production of adiponitrile using butadiene as the raw material was sought, either by direct or indirect hydrocyanation. The indirect route proceeded via chlorination with no catalyst, at 200°–300°C, followed by the replacement of chlorine with cyanide to produce a mixture of dicyanobutene isomers. The mixture was isomerized to the required 1,4-dicyano-2-butene by a copper complex in a reaction mixture containing hydrogen cyanide. Adiponitrile was formed by gas phase hydrogenation at 300°C using a palladium catalyst.

The direct addition of chlorine to butadiene followed by reaction with sodium cyanide is a very unattractive process. Both chlorine and sodium cyanide are very expensive to produce in terms of energy required, and then to be left with the problem of disposal of contaminated sodium chloride residues only serves to compound the problem. The direct addition of hydrogen cyanide to butadiene is much more attractive, provided that the addition reactions can be tailored to produce the desired linear product. The problems were solved by some excellent work by scientists from du Pont.[56,57] The chemistry of the process can be envisaged in three stages. In the first stage, one molecule of hydrogen cyanide adds across one of the double bonds to give a mixture of linear and branched nitriles. In the second stage, the branched isomer is isomerized to the linear form, and in the third stage, the second molecule of hydrogen cyanide adds across the remaining double bond to give adiponitrile:

$$CH_2=CH.CH=CH_2 + HCN \rightarrow \qquad\qquad (7.30)$$
$$CH_2=CH.CH_2CH_2CN + CH_2=CH.CH(CN).CH_3$$

$$CH_2=CH.CH(CN)CH_3 \rightarrow CH_2=CH.CH_2CH_2CN \qquad (7.31)$$

$$CH_2=CH.CH_2CH_2CN + HCN \rightarrow NC(CH_2)_4CN \qquad (7.32)$$

The catalyst is similar for all three steps, and consists of a zero valent nickel phosphite complex, promoted with zinc or aluminium chlorides. The direct addition of hydrogen cyanide to butadiene is particularly attractive with the availability of by-product hydrogen cyanide form the manufacture of acrylonitrile by the ammoxidation of propylene.

The third of the potential feedstocks for the manufacture of adiponitrile is acrylonitrile, prepared by the ammoxidation of propylene. The only route to have been commercialized to date is the non-catalytic electrodimerization process, as first used by Monsanto, and now by BASF:

$$2 \, CH_2=CHCN + 2 \, H^+ + 2 \, e^- \rightarrow NC(CH_2)_4CN \qquad (7.33)$$

The reaction probably proceeds via the formation of a dianion by electron transfer from an acrylonitrile molecule followed by combination with a second molecule and proton addition from a water molecule.

Two rather different approaches to the catalytic dimerisation of acrylonitrile have also been explored. In the first approach, a ruthenium catalyst was used in the presence of hydrogen, at temperatures in the range 200–350°C and pressures of 1–3 bar. Adiponitrile[58] was produced directly, but unfortunately, was accompanied by an equimolar amount of propionitrile, formed by direct hydrogenation of the acrylonitrile:

$$2 \, CH_2=CH.CN + H_2 \rightarrow NC(CH_2)_4CN \qquad (7.34)$$

$$CH_2=CH.CN + H_2 \rightarrow CH_3CH_2CN \qquad (7.35)$$

It was not possible to eliminate the formation of the propionitrile and the cost of converting the propionitrile to useful products was such that work on this route was discontinued.

In the second approach, trivalent organophosphorus derivatives were used. The main reaction product was α-methyleneglutaronitrile, which was not a useful product, and which could not easily be converted to linear from. In subsequent work at ICI,[59] a range of arylphosphonite and diarylphosphinite catalysts were developed, which under rigorously anhydrous conditions, and in isopropanol solution, gave essentially linear dimer[60] at a good rate at temperatures as low as 60°C. This process was developed successfully through the pilot plant stages, but work was eventually terminated when ICI withdrew from the nylon business for strategic commercial reasons.

Adiponitrile can be hydrogenated in the liquid phase, at about 99% selectivity, to produce hexamethylene diamine at pressures as low as 30 atm using nickel catalysts containing iron or cobalt at about 75°C. New routes to HMD are being developed by major producers.[61] Adiponitrile was easily hydrogenated to HMD, at pressures up to 600 atm, with a cobalt oxide catalyst containing small amounts of silica and titania. Hydrogenation at 300–350 atm was also possible with iron catalysts.

7.3.1.4. Nylon Polymer

Adipic acid and hexamethylene diamine are mixed in either a water or methanol solution, where they form the salt, known as *nylon salt*, which crystallizes as hexamethylene diammonium adipamide. The nylon salt is then heated to temperatures of 250°–270°C to eliminate water and hence form the polymer. Care is needed to control the dehydration conditions, to produce the optimum chain length and molecular weight. Acetic acid can be added to control chain termination.

7.3.2. Nylon 6

Nylon was such a success that other companies already working on polymerization began to search for competitive polyamides not covered by the DuPont patents. Paul Schleck of I.G. Farben quickly developed a type of nylon based on ε–caprolactam.[62] The pure dry ε-caprolactam was difficult to polymerize, but the problem was solved by the addition of catalytic amounts of an acid and water. Production details were soon developed. Production of polycaprolactam (trademark Perlon L) began in 1940. Nylons based on ε-caprolactam were soon being produced by many other companies because after the war, details of the process were released as intelligence reports and were free of patent restrictions.[63]

The DuPont polymer then became known as nylon 66 because it was made from two C_6 monomers, while the I.G. Farben polymer, from a single C_6 monomer, was nylon 6.

7.3.2.1. Caprolactam

The preparation of ε-caprolactam via the Beckman rearrangement of cyclohexanone oxime was first described by Wallach in 1900.[64] In the I.G. Farben process, phenol was hydrogenated to cyclohexanol, at 140°–160°C and 15 bar, with a nickel oxide/silica catalyst. The cyclohexanol was dehydrogenated to cyclohexanone using a zinc/iron catalyst at 400°C, and the cyclohexanone oxime was formed by reaction with hydroxylamine monosulfonate. Conversion to cyclohexanone oxime was maximized at pH 7 by neutralizing the solution with ammonia. The organic layer of crude oxime was crystallized and then isomerized to ε-caprolactam with 20% oleum at 120°C. ε-Caprolactam was purified and the ammonium sulfate recovered.

The basic industrial process is still essentially the same except that cyclohexane is now used to provide more than 60% of the ε-caprolactam produced, with phenol used for the balance. BASF switched to cyclohexane in feedstock 1963. A big disadvantage of the process has been the need to sell about 2.3 lb of by-product ammonium sulfate for every pound of ε–caprolactam produced, that is, about 5 kg of ammonium sulfate for each kilogram of ε- caprolactam. However, there is a ready market for ammonium sulphate in the fertilizer business.

7.3.2.2. Cyclohexanone

Cyclohexanone was recovered from ketone/alcohol oil by distillation and then dehydrogenating the remaining cyclohexanol using copper oxide/zinc oxide catalysts at 400°–450°C, with about 90% conversion and 95% selectivity.

The original preparation of cyclohexanone from phenol by hydrogenation to cyclohexanol followed by dehydrogenation has since been improved. Selective hydrogenation of phenol to cyclohexanone is possible using a palladium catalyst at 140°–170°C and 1–2 atm. An Allied/Vickers–Zimmer catalyst contained 0.5–5% palladium, supported on a low-surface-area calcium aluminate which contained about 8–9% calcium oxide.[65]

7.3.2.3. Cyclohexanone Oxime

Cyclohexanone reacts with hydroxylamine sulfate in the presence of ammonia, to produce the oxime with an equivalent of by-product ammonium sulfate. The number of steps required to form hydroxylamine sulfate complicates the procedure, and also gave rise to some further by-product ammonium sulphate.

In the DSM HPO process, nitrate ions are reduced to hydroxylamine using a palladium/carbon catalyst in an aqueous phosphate buffer solution. It was possible to prepare cyclohexanone oxime by a process involving the extraction of cyclohexanone from solution in toluene, into an immiscible aqueous solution of hydroxylamine in a two phase liquid reactor. The spent solution was then recycled with added nitric acid to the reduction step and sulfate production was eliminated:

$$NO_3^- + H^+ + 3H_2 \rightarrow NH_2OH + 2H_2O \tag{7.36}$$

$$C_6H_{10}O + NH_2OH \rightarrow C_6H_{10}(OH)NHOH \tag{7.37}$$

Further efforts to decrease by-product sulfate led to investigation of oxime production from cyclohexanone by one step ammoxidation reactions. Ammoxidation catalysts are usually unstable (phosphotungstates) or relatively inactive (silica), but a more efficient catalyst was introduced by Montedipe (now Enichem) in 1986.[66] This was based on a titanium silicalite catalyst with a silica/titania molar ratio greater than 30:1. Reaction took place with a hydrogen peroxide oxidant, in tertiary butanol as solvent at 70°–90°C. Although the conversion was almost 100% giving selectivity greater than 98%, ammoxidation has not been used commercially. The photochemical reaction of benzene with nitrosyl chloride, NOCl, has also been investigated in Japan, as an alternative route to cyclohexane oxime.

7.3.2.4. *Snia-Viscosa Process*

Snia-Viscosa developed an ε–caprolactam process in 1958 based on the use of toluene as the raw material,[67] which was licensed in 1962. The oxidation of toluene has been studied since at least 1892,[68] when the first products obtained were benzaldehyde and benzoic acid. Catalysts based on platinum,[69] vanadium pentoxide, and various other oxides[70] have been described:

$$C_6H_5.CH_3 + 3/2\ O_2 \rightarrow C_6H_5.COOH + H_2O \qquad (7.38)$$

$$C_6H_5.COOH + 3\ H_2 \rightarrow C_6H_{11}.COOH \qquad (7.39)$$

Weiss and Downs[71] described the use of vanadium pentoxide for the catalytic oxidation of toluene and naphthalene. Subsequently, Craver[72] suggested a mixed oxide catalyst derived from uranium oxide and molybdenum trioxide in molar ratios ranging from 3-13:1. Copper oxide was also suggested as a possible promoter. When using the vanadium pentoxide catalyst, benzoic acid was the main product at temperatures below 400°C, with some benzaldehyde formed at higher temperatures. Selectivity was only about 50% at 5% conversion.

The Snia-Viscosa process now uses benzoic acid as an intermediate in caprolactam production. The oxidation of toluene in the liquid phase is still used in modern processes, with selectivity greater than 90%, at about 30% conversion. Typical operating conditions are at temperatures around 140°C, and pressures in the range 8–10 bar, using a cobalt acetate catalyst.

The benzoic acid is now hydrogenated to cyclohexane carboxylic acid with a palladium/carbon catalyst at 170°C and 10–19 bar, whereas originally a nickel catalyst was used. Reaction with nitrosyl sulfuric acid in oleum solution at 80°C then gives ε-caprolactam. In the original Snia-Viscosa process, about 25% more ammonium sulfate was formed than in the conventional process. However, solvent extraction of the caprolactam from the acid solution using alkyl phenols does avoid the production of the ammonium salt.

7.3.2.5. *Conversion of Cyclohexanone Oxime to Caprolactam*

Conversion of the oxime to caprolactam was achieved by mixing the oxime with an equal weight of 20% oleum at a temperature of 120°C. After 15 minutes, the yield of ε-caprolactam was about 90%. The solution was neutralized with ammonia and the crude lactam separated recovered and purified by distillation.

In the DSM process, ammonium sulfate formation is reduced from about 5 kg for each kilogram of caprolactam to 1.8 kg. Boric acid catalysts can be used for the rearrangement of the oxime to the lactam at 330°C in fixed or fluidized beds, but have not been used commercially. Strong acid exchange resins have also been tested in a modified Beckman rearrangement using dimethylsulfoxide solutions at 100°C.

7.3.2.6. Caprolactam from Butadiene

The DuPont direct hydrocyanation of butadiene process for the production of hexamethylene diamine can also be adapted for the production of caprolactam. Partial hydrogenation of the adiponitrile intermediate in which only one cyanide group is converted to the amine, followed by hydrolysis of the remaining cyanide group and ring closure can produce ε-caprolactam.[73] The process has not been developed commercially. It is also possible that a two-stage butadiene carbonylation process to produce caprolactam, developed by DSM and Shell, may be further developed.

Caprolactam is polymerized to produce nylon 6 by opening the ring with high temperature water at 250°–280°C to give ω-amino acid [$NH_2(CH_2)_4COOH$]. The amino acid can later react with either a second acid molecule or a caprolactam molecule. Temperature control and the water content of the mixture are important parameters during the polymerization process.

7.3.3. Polyesters

As a result of the DuPont patents for nylon, in 1937 Whinfield and Dickson, of the Calico Printers' Association in England, became interested in polyesters based on aromatic dicarboxylic acids and ethylene glycol. Carothers had, of course, abandoned his own early work following the disappointing results with aliphatic dicarboxylic acids. Whinfield and Dickson quickly produced a long-chain polyester and decided to patent the discovery, which was eventually licensed to ICI. The patent was actually granted in 1941 and ICI built the first commercial production unit in late 1954.[74]

By 1946 the ICI polyester technology had already been licensed to DuPont and interest in the new material was on the rise. However, there were severe practical difficulties in obtaining the almost insoluble terephthalic acid of sufficient purity from the oxidation of p-xylene, without first preparing and converting the p-toluic acid to the methyl ester. Direct oxidation of *p*-xylene with air was later developed, although complete oxidation remained a problem until Scientific Design and Amoco introduced the use of bromine as a cocatalyst in conjunction with manganese or cobalt salts. Ethylene glycol has remained the most widely used alcohol.

Polyethylene terephthalate, or polyester, has become the most widely used synthetic fiber, particularly when combined with cotton. It is also a very popular thermoplastic resin for bottle and film production and new polyesters are now being developed. Thermosetting unsaturated polyesters are based on maleic anhydride with a range of glycols. Polyethylene terephthalate is produced from ethylene glycol, obtained by the direct oxidation of ethylene, and terephthalic acid is obtained by oxidizing paraxylene. By the year 2000 almost 20 million tonnes were being produced annually throughout the world.[75]

7.3.3.1. Paraxylene

The main source of mixed xylenes is either the product from catalytic reformers or pyrolysis gasoline. The proportion of ortho-, meta-, and paraxylenes is in the ratio 1:2:1 in both reformate and pyrolysis gasoline. However, reformate aromatics contain less than 20% ethylbenzene, whereas pyrolysis gasoline aromatics contains up to 50%. Ethylbenzene and o-xylene can be separated from the mixture by fractional distillation, while p-xylene is separated by a zeolite adsorption process. The remaining fraction, which is largely the unwanted m-xylene, is usually isomerized with a catalyst to produce an equilibrium mixture of the three isomers. This corresponds, approximately, to the 1:2:1, o-, m-, p ratio in the original aromatic mixtures, depending on the isomerization temperature. In Table 7.7, typical catalysts and operating conditions for processes that have been used commercially are shown.

The early silica/alumina catalysts for the isomerization of xylene suffered from deactivation due to the deposition of carbon and the needed frequent regeneration. The process was improved by both the use of catalysts impregnated with platinum and the addition of hydrogen to the reactants, and thus led to a reduction in the need for frequent regeneration. These catalysts also converted ethyl benzene to xylenes. High-silica zeolites are now used to produce most of the p-xylene obtained by isomerization, because high selectivity can be achieved of equilibrium conversion. Mobil ZSM-5 is particularly useful because the pore size promotes *paraselectivity* and controls the unwanted disproportionation reac-

TABLE 7.7. Xylene Isomerization Catalysts.

Catalyst	Operation	Comments
Silica/alumina	400°–500°C, atmospheric pressure	Carbon deposition No ethylbenzene conversion
Platinum/silica alumina	430°–450°C, 10–25 atm	Hydrisomerization (+ H_2) No carbon—long life Ethylbenzene conversion
1% Platinum/Al_2O_3 + mordenite	425°C, 12 atm	Ammonia injection to control zeolite acidity/activity
Platinum exchanged ZSM-5	420°–460°C, 14–18 atm	Hydrisomerization (+H_2) Equilibrium shifted to increase conversion by 4%

tions that produce trimethyl benzenes. Control of p-xylene isomerization within the zeolite pores can lead to higher than equilibrium yields of the p-xylene. Zeolites are more resistant to poisons, and do not require frequent regeneration.

7.3.3.2. Terephthalic Acid

The range of catalysts and manner in which the air oxidation of p-xylene to terephthalic acid has developed since the first plant started operation during the 1950s are shown in Table 7.8. The original oxidation of p-xylene using nitric acid as oxidant, which led to nitrogen impurities in the product, soon became obsolete. Direct oxidation with air was then developed using homogeneous catalysts in the liquid phase. The oxidation of p-xylene produced p-toluic acid but was not able to produce terephthalic acid directly:

$$p\text{-}C_6H_4(CH_3)_2 + 3O_2(HNO_3) \rightarrow p\text{-}C_6H_4(COOH)_2 + 2H_2O \qquad (7.40)$$

$$p\text{-}C_6H_4(CH_3)_2 + 3/2O_2 \rightarrow p\text{-}CH_3.C_6H_4.COOH + H_2O \qquad (7.41)$$

$$\overset{Co^{3+}}{p\text{-}CH_3.C_6H_4.COOH} \rightarrow p\text{-}C^\cdot H_2.C_6H_4.COOH \overset{O_2}{\rightarrow} p\text{-}OOCH_2.C_6H_4.COOH \quad (7.42)$$

$$p\text{-}OOCH_2.C_6H_4.COOH \rightarrow p\text{-}C_6H_4(COOH)_2 \qquad (7.43)$$

At first, the p-toluic acid was esterified with methanol in a separate step before final oxidation to monomethyl terephthalic acid. Alternatively, if methanol was used as a solvent for the p-xylene, both methyl groups were esterified in a single step, if the residence time was longer than 20 h. The reaction mechanism involved the conversion of the methyl group to a benzyl radical by reaction with trivalent cobalt ions. The benzyl radical then reacts with oxygen and forms a peroxy species that decomposes to the oxidation product. Trivalent cobalt ions are regenerated as aldehydes from the peroxy or hydroperoxy species.

The purification of terephthalic acid is complicated because it does not melt and, as it was not soluble in either water or other solvents, it could not be crystallized. On the other hand, the dimethyl ester of terephthalic acid could be easily crystallized from methanol or xylene. When the Mid Century Process was introduced by Scientific Design and Amoco in 1956, it became possible to produce and purify terephthalic acid directly. This process used air oxidation conditions similar to those for previous processes, with a mixed trivalent cobalt and manganese acetate catalyst in glacial acetic acid, but introduced an ammonium bromide cocatalyst in conjunction with tetrabromomethane. Cobalt or molybdenum bromides or hydrobromic acid have also been used, and following reaction with the trivalent cobalt, provided a source of bromine free radicals. The free radicals activated the methyl groups of the p-xylene and led to the for-

TABLE 7.8. Development of *p*-Xylene Oxidation Processes.

Process	Catalyst	Operating conditions
(a) *p*-xylene to *p*-toluic acid. (b) *p*-toluic methyl ester to monomethyl terephthalic acid (Witton 1951)	Cobalt or manganese acetates promoted by *p*-toluene sulfonic acid.	140°–170°C, 4–8 atm 250°–280°C, 20–25 atm (*p*-xylene selectivity 85%)
p-xylene in methanol solvent to dimethyl terephthalate	Cobalt salts	100°–200°C, 5–20 atm (*p*-xylene selectivity 90%)
p-xylene to pure terephthalic acid (Amoco/Scientific Design, 1958)	Reaction: cobalt and manganese acetates in glacial acetic acid/cocatalyst NH$_4$BR plus tetra-bromomethane Purification: palladium supported on carbon	Reaction: 190°–205°C, 15–30 atm in stirred auto-claves (95% conversion; *p*-xylene selectivity 90%) Purification: 225°–275°C to hydrogenate 4-carboxy benzaldehyde to *p*-toluic acid in water solution.
p-xylene oxidized (with substance that provides hydrop-eroxides, e.g., paraldehyde) to pure terephthalic acid	Cobalt acetate in acetic acid Catalysts used represent about 0.1% of the TPA produced	100°–140°C, 30 atm (selectivity >97% but by-product acetic acid forms)

mation of aromatic aldehydes.[76] These were converted to the acids via the peroxy intermediates. A further novelty of the process was the purification of product by dissolving it at high pressure in water at 225°–275°C, and then hydrogenating any 4-carboxybenzaldehyde by-product to *p*-toluic acid with a palladium/carbon catalyst before allowing the terephthalic acid to crystallize. The same process was used to oxidize toluene to benzoic acid in the Snia-Viscosa caprolactam process. The Amoco process has been widely licensed and now produces the bulk of the terephthalic acid used in polyester manufacture. Terephthalic acid is preferable to dimethylterephthalate in polyester production because the reaction rate is faster and a better yield is obtained. The use of methanol is also avoided.

The polyester polymer is produced in two steps to avoid side reactions. The acid is esterified in an excess of ethylene glycol at 100°–150°C and 10–70 bar using a cobalt or zinc acetate catalyst. The monomer then condenses at 150°–270°C as a liquid melt under vacuum using an antimony oxide catalyst. Ethylene oxide is removed by distillation as the polymerization proceeds. The use of antimony oxide catalyst always led to a slightly grey-colored product, as a result of reduction of some of the antimony to the metallic state. Pure white polymer could be obtained by the use of germanium dioxide catalyst, but the problem was the very high cost of germanium relative to antimony. This problem was solved in ICI by W Hewertson with the addition of phosphine oxide ligands to the catalyst, which stabilized the antimony towards reduction. However, as is often the case in production, the more stable catalyst could be operated at a higher temperature than the conventional catalyst. This led to shorter polymerization times, and hence an increase in plant capacity, but at the expense of returning to the slightly grey polymer! The operator then had a choice between pure white polymer or an increase in total production.

A relatively low-molecular-weight polymer is produced and generally used to make fibers and other products. The higher-molecular-weight polymer needed for bottles or packaging is formed by further polymerization in the solid state.[77]

7.3.3.3. Alternative Routes for Terephthalic Acid Production

Other syntheses for terephthalic acid have been developed, although none has yet been used extensively. As with most chemical processes, the capital cost of new equipment usually means that modifications to existing processes are more easily adopted than completely different routes to the same product.

In the Henkel process, dipotassium phthalate can be isomerized to dipotassium terephthalate at about 430°C and 5–20 bar using carbon dioxide with a soluble organic zinc or cadmium salt. An alternative approach is the disproportionation of potassium benzoate to give dipotassium terephthalate and benzene using zinc or cadmium benzoates at up to 50 bar pressure of carbon dioxide. In both operations the potassium is recycled by reaction with either phthalic anhydride or benzoic acid.

In a process developed by Lummus, p-xylene and ammonia are converted to terephthalonitrile in a fluidized bed at 450°C. A supported vanadium pentoxide catalyst is used and gives about 50% conversion. The vanadium pentoxide is reduced during the reaction and can be reoxidized with air in a separate reactor. The dinitrile is then hydrolyzed to terephthalic acid. The process has not been used in polyester production, although a similar process used in Japan produces diamines from terephthalonitrile or isophthalonitriles.

7.3.3.4. Use of Polyesters

Polyesters are widely used as fibers, although the market share of fibers fell from more than 70% before 1995 to about 60% by 2000, as demand for polyester (PET) bottles increased. To cope with the increased demand for bottles since 1995 more isophthalic acid is being used as a copolymer with ethylene glycol. At the same time, naphthalene dicarboxylate (NDC) has been used to produce a new polyester known as polyethylene terenaphthalate (PEN).[78] The new polyester bottles, although originally very expensive, have lowered gas permeability and are stronger and more heat resistant. Nevertheless, as prices fall, the demand for all types of polyester will continue to rise.

7.4. HYDROFORMYLATION AND CARBONYLATION

Otto Roelen began studies on the OXO reaction in 1938, when he was working on the Fischer–Tropsch reaction with Ruhrchemie AG in Germany.[79] He discovered the presence of oxygenated organic compounds in the Fischer–Tropsch paraffin waxes and concluded that they were the result of carbonylation and hydrogenation of the unsaturated hydrocarbons present in the feed gases:

$$RCH=CH_2 + H_2 + CO \longrightarrow \begin{array}{l} RCH_2(CH_2)_2CHO \\ \\ RCH(CHO)CH_3 \end{array} \qquad (7.44)$$

This idea was confirmed by studies on the reaction of ethylene with synthesis gas, (H_2 + CO), at 150°–200°C and 150 bar using the standard Fischer–Tropsch cobalt catalyst. Reaction products included propanaldehyde and isopropanaldehyde. In 1945 Ruhrchemie opened a plant with a capacity of 10,000 tonnes/year at Holten, to produce C_{11}–C_{17} alcohols from Fischer–Tropsch olefins.[80] These were subsequently converted to detergent sulfates.

Other companies, including ICI in the United Kingdom, Shell in Holland, and the Enjay Chemical Company in the United States investigated the process from 1947 onwards and full-scale manufacture soon followed, often in redundant high-pressure coal hydrogenation equipment.[81] By 1963 this had led to a worldwide production of more than 500,000 tonnes/year, mainly as butanols and C_7–C_{10} alcohols. The process, also known as hydroformylation, gave a mixture of normal and isoalcohols, after hydrogenation of the initial *carbonylation* aldehydes. Some alcohols were formed during hydroformylation together with small amounts of paraffins and heavy oils. In the early plant operated by Roelen, the ratio of normal to isopropionaldehyde was about 2:1.

7.4.1. Cobalt Carbonyl Catalysts

During early experiments, Roelen used the standard Fischer–Tropsch catalyst ($30CoO/66SiO_2/2ThO_2/2MgO$) in both stages of alcohol production. He soon realized, however, that in the presence of carbon monoxide, cobalt octacarbonyl was formed and reacted reversibly with hydrogen to give a homogeneous catalyst. The reaction then proceeded in solution. For this reason, all further developments used a convenient and cheap soluble cobalt salt, such as cobalt naphthenate. Cobalt naphthenate was soluble in the organic reaction medium and was available commercially.

The partial pressures of hydrogen and carbon monoxide are critical in maintaining the concentration of the active species, cobalt hydridotetracarbonyl and cobalt hydridotricarbonyl, in the equilibrium reaction mixture with dicobalt octacarbonyl.

The usually accepted reaction mechanism with cobalt octacarbonyl can be simplified as follows:[82]

- An activation step gives an equilibrium mixture containing the catalyst, which coordinates with the olefin forming a π-complex:

$$Co_2(CO)_8 + H_2 \leftrightarrow 2\ HCo(CO)_4 \leftrightarrow 2\ HCo(CO)_3 + 2\ CO \qquad (7.45)$$

- The olefin is then inserted into the Co–H bond to form an alkyl species:

$$RCH=CH_2 + HCo(CO)_3 \leftrightarrow H(RCH=CH_2)Co(CO)_3 \qquad (7.46)$$

$$H(RH=CH_2)Co(CO)_3 \leftrightarrow RCH_2CH_2Co(CO)_3 \qquad (7.47)$$

- Molecular carbon monoxide is inserted into Co–C bond to form an acyl species:

$$RCH_2CH_2Co(CO)_3 + CO \leftrightarrow RCH_2CH_2COCo(CO)_3 \qquad (7.48)$$

- The final, irreversible step is hydrogenolysis of the acyl species to the aldehyde and catalyst regeneration:

$$RCH_2CH_2COCo(CO)_3 + H_2 \rightarrow RCH_2CHO + HCo(CO)_3 \qquad (7.49)$$

The reaction steps are reversible and isomerization of the olefin, alkyl, or acyl species can take place to allow the formation of isoaldehydes. The typical 4:1 product distribution of normal and isoaldehydes must be separated if the mixture cannot be used commercially. Efforts were therefore made to increase the proportion of useful normal aldehydes during operation. Partial success was achieved by operating at lower temperatures with higher carbon monoxide partial pressures, although this decreased conversion to aldehydes. A major problem with the cobalt catalyst was the tendency to decompose at high temperature and to deposit metal onto the reactor walls. This led to loss of activity and low catalyst recovery.

7.4.2. Phosphine Modified Catalysts

By the 1970s, the chemistry of the process was being investigated in an effort to improve performance. The main objectives were as follows:

- Reduce operating pressure and find a more stable and active catalyst.
- Increase the proportion of normal acetaldehyde produced.
- Reduce by-product formation.

Catalysts derived from rhodium were found to be much more active than those from cobalt, but rhodium was vastly more expensive than cobalt and gave a lower yield of the normal aldehyde. Around the same time, it was shown in work by Shell that thermally-stable complexes could be formed[83] by partial replacement of the carbonyl groups in the cobalt carbonyl catalyst by ligands such as organophosphines. This increase in stability allowed the reaction temperature to be increased to such an extent that the reaction pressure could be reduced with no loss in the rate of reaction. This resulted in a lowering of both operating and capital costs. However, a big disadvantage of the new catalyst was that the rate of reaction decreased by about 80% and hydrogenation activity increased. This led to lower yields, higher paraffin production, and more alcohol formation. The process is still being used to produce 2-ethyl hexanol directly from propylene and a range of higher-molecular weight detergent alcohols. Operating conditions for unmodified and modified cobalt catalysts are shown in Table 7.9.

TABLE 7.9. Hydroformylation Catalyst Operation.

Variable	Catalyst: Cobalt	
	(Used for higher-molecular-weight olefins: $>C_4$)	
Composition	$Co_2(CO)_8$	$Co_2(CO)_8$ $4(n$-butyl$)_3$P
Pressure (atm)	200–300	50–100
Temperature (°C)	110–160	160–200
CO/H_2	1:1	1:2
% Metal/olefin	0.1–1.0	0.6
Phosphine/metal	—	2:1
Normal/isoaldehyde	4:1	9:1
By-products (wt%)	5	1
Olefin hydrogenation (vol%)	<2	Up to 10
	Catalyst: Rhodium	
	(Used for low-molecular-weight olefins: up to C_4)	
Composition	$HRh(CO)(Ph_3P)_2$	$HRh(CO)(P(m$-sulfophenyl Na$)_3)_3$ [a]
Pressure (atm)	12–15	Up to 50
Temperature (°C)	~100	~120
CO/H_2	1:1	1:1
% Metal/olefin	0.01–0.1	
Phosphine/metal	2:1 (+ excess)	3:1 (+ excess)
Normal/isoaldehyde	Up to 16:1	15–20:1
By-products (wt%)	Low	0.2–0.8
Olefin hydrogenation (vol%)	5	—

[a]Operates in a single or two-phase system.

7.4.3. Low-Pressure Hydroformylation

Union Carbide Corporation was working on a low-pressure hydroformylation process as early as 1967, and used a rhodium carbonyl catalyst modified with a triphenyl phosphine ligand.[84] The ligand dramatically improved the normal/iso-aldehyde ratio of the product when compared to the free metal carbonyl. Reduced operating temperatures and pressures, low by-product formation, and the elimination of metallic rhodium deposition were also achieved. By 1975, a plant was operating using this catalyst, and the process was being licensed for the hydroformylation of propylene.

In the early version of the process, propylene and synthesis gas were circulated through the catalyst solution to minimize handling of the catalyst and thus to reduce catalyst losses. The product was recovered by using the reaction exotherm to vaporize the product, which was then removed from the reactor by the flow of the process gases. The life of the catalyst was relatively short in these early plants, despite careful purification of the propylene and the synthesis gas. The operating conditions are shown in Table 7.9.

The operating procedure was subsequently modified.[85] A single liquid phase containing a solution of the catalyst in the form of rhodium carbonyl with a modified ligand, together with excess ligand circulated through the reactor with the propylene and synthesis gas. Solubilizing agents such as N-methyl pyrrolidone or a polyalkylene glycol were added to ensure that both the catalyst and products were completely miscible. This gave better mixing of the catalyst with the propylene and synthesis gas thus promoting formation of a single liquid phase. Other ligands such as modified diorganophosphates could also be used.[86]

On leaving the reactor, the nonpolar products were separated from the catalyst by adding water or methanol to split the reaction mixture into two liquid phases. The two phases were separated by decantation, and the products thus recovered. The catalyst was dried and purified before being returned to the reactor. The lifetime of the catalyst was extended to several years, though make-up quantities had to be added to replenish any small losses. Increasing the solubility of the catalyst by use of more effective ligands was claimed to produce a catalyst that was active for the hydroformylation of octane, dodecene and even styrene, in addition to propylene and butene.

Ruhrchemie and Rhone Poulenc introduced a two-phase modification to the basic propylene hydroformylation process in 1984.[87] The aqueous phase contained the rhodium catalyst, and a phosphine derivative was circulated through the reactor with the propylene and synthesis gas. The specific phosphine used was triphenylphosphine in which the meta-position of the phenyl group had been substituted by the sodium salt of the sulphonic acid. The products formed an organic layer that was separated by decantation when the two immiscible liquid phases were removed from the reactor. The aqueous layer was returned to the reactor for further use. The catalyst is not affected by sulfur poisons, and can operate for as long as 12 years provided that additional ligand is regularly supplied to replace losses owing to degradation. The solubility of the propylene in the reaction medium can become a limitation in the process.

7.4.4. Commercial Operation

Hydroformylation processes are used extensively throughout the world and the single largest product is 2-ethyl hexanol, which is converted to dioctylphthalate, which is used mainly as a plasticizer for polyvinylchloride. Cobalt carbonyl catalysts are still important and produce large volumes of the higher-molecular-weight aldehydes, most of which are converted to detergent alcohols. When alcohols are required, cobalt carbonyl catalysts, modified with phosphine ligands are useful in that they provide a high proportion of the normal alcohol relative to the iso-alcohol.

Rhodium catalysts with phosphine ligands are used only for propylene hydroformylation, to produce butyraldehyde and then 2-ethyl hexanal, but are not used for the production of higher aldehydes. The available processes are complementary and do not compete. The main factor in choosing a process is the product required.

7.4.5. Acetic Acid

Acetic acid was originally produced by bacterial oxidation of ethanol, but from around 1914, synthetic acetic acid was produced by the oxidation of acetaldehyde. Hydrocarbon oxidation processes using butane or naphtha as feedstock were introduced in the 1950s. In a typical liquid phase oxidation process, a cobalt acetate catalyst operating at a temperature of 175°C, and a pressure of 54 bar was used, and by-products could be recycled. Conditions could be modified to produce methyl ethyl ketone.

The carbonylation of methanol based on Reppe chemistry was a major advance in acetic acid technology. BASF introduced this process in 1966, and used an iodide-promoted cobalt catalyst at high pressures in the range 500–700 bar.[88] The reaction was thought to proceed via cobalt hydridotetracarbonyl and hydrogen iodide as intermediates. The hydrogen iodide was believed to react with methanol to give methyl iodide, which underwent oxidative addition to the cobalt carbonyl intermediate, thus forming a methyl cobalt derivative. Carbon monoxide was inserted into the methyl cobalt bond to give an acyl cobalt species which on hydrolysis formed acetic acid and more hydrogen iodide, thus completing the catalytic cycle.

This process was rapidly rendered obsolete by the introduction of the low pressure process by Monsanto[89] in 1971. The catalyst was similar to the Union Carbide hydroformylation catalyst, and was based on rhodium carbonyl, promoted with iodine. The operating pressure was only around 30–40 bar, with a lower reaction temperature of about 180°C. These mild reaction conditions led to the formation of acetic acid at a selectivity of >99%, whereas the selectivity of the cobalt route was only 90%, and of course a major reduction in both capital and operating costs. Furthermore, the cobalt catalyst unlike the rhodium catalysts was very sensitive to traces of hydrogen in the feed gases.

While reaction rates for cobalt catalysts depend on the methanol concentration and partial pressure of carbon monoxide, the rates with rhodium catalysts are independent of both reactant and product concentrations. The reaction mechanism for cobalt catalyst in methanol carbonylation is similar to hydroformylation but is different for rhodium catalysts. The catalyst intermediate in the rhodium process is $Rh(CO)_2I_2$ and has been identified at 100°C and 6 atm by infrared spectroscopy.[90] Reaction with methyl iodide in the rate-determining step then forms a methyl rhodium complex that rapidly gives an acyl complex $(CH_3CORh(CO)I_3)_2^{2-}$ from reaction with the high carbon monoxide content in the reactor. The complex decomposes after reaction with water to produce acetic acid, and with simultaneous regeneration of the catalyst:

$$CH_3OH + HI \leftrightarrow CH_3I + H_2O \tag{7.50}$$

$$CH_3I + CO \rightarrow CH_3COI + H_2O \rightarrow CH_3COOH + HI \tag{7.51}$$

By 1999 more than two-thirds of worldwide acetic acid was made by the Monsanto process using the rhodium catalyst. BP Chemicals purchased the rights to the process in 1986 and in 1996 introduced a promoted iridium acetate catalyst that was claimed to be more efficient with easier purification.[91] A comparison of the BASF and Monsanto processes is shown in Table 7.10.

TABLE 7.10. Methanol Carbonylation Catalyst Operation

	Catalyst	
Variable	Cobalt	Rhodium
Composition	$HCo(CO)_4$	$(Rh(CO)_2I_2)^-$
Pressure (atm)	500–700	30–40
Temperature (°C)	230	180
Metal concentration	10^{-1}	10^{-3}
Promoter[a]	Iodide	Iodide
Selectivity (%)	~90	>99
By-products	Methane, acetaldehyde, ethanol, carbon dioxide	Virtually none

[a]Reaction via methyl iodide intermediate.

7.4.6. Acetaldehyde

An interesting process for the production of acetaldehyde was based on the work of F. C. Phillips, who showed that ethylene could be oxidized to acetaldehyde by an aqueous palladium chloride solution.[92] The palladium chloride was reduced to metallic palladium. During the late 1950's, Wacker Chemie introduced a new process for the manufacture of acetaldehyde by direct oxidation of ethylene with air. The palladium metal was converted back to $(PdCl_4)^{2-}$ by an acidic solution of cupric chloride, which was, itself, reduced to cuprous chloride. The cupric chloride was regenerated by reaction with air in hydrochloric acid solution. The reaction sequence is shown in the following equations:

$$C_2H_4 + Pd(Cl_4)^{2-} + H_2O \rightarrow CH_3CHO + Pd^{\circ} + 2\ HCl + 2\ Cl^- \qquad (7.52)$$

$$2\ Cl^- + Pd^{\circ} + 2\ CuCl_2 \rightarrow Pd(Cl_4)^{2-} + 2\ CuCl \qquad (7.53)$$

$$2\ CuCl + \tfrac{1}{2}\ O_2 + 2\ HCl \rightarrow 2\ CuCl_2 + H_2O \qquad (7.54)$$

Overall: $C_2H_4 + \tfrac{1}{2}\ O_2 \rightarrow CH_3CHO \qquad (7.55)$

In the Wacker two-stage process, the ethylene was completely oxidized with palladium chloride/cupric chloride solution at 110°C and 10 bar, with the stoichiometric volume of air in the reaction vessel. Acetaldehyde was recovered from the circulating solution and the palladium oxidized with air at 100°C and 100 bar pressure in a second vessel before the catalyst solution was returned to the reactor.

A one-stage process was then introduced by Hoechst. A mixture of almost completely pure ethylene and oxygen (above the upper flammability limit) was passed through the catalyst solution which contained about 200 g/liter cupric chloride and 4 g/liter palladium chloride. At a temperature of about 120°–130°C and a pressure of 3 bar, 35–45% of the ethylene was converted consuming all of the oxygen. The acetaldehyde product distilled out of the reaction mixture using the exotherm from the reaction, and the remaining ethylene was recycled. Selectivity in both processes is about 95%, and the by-products are acetic acid, crotonaldehyde, and methyl/ethyl chlorides, which must be removed periodically. The process was still used to make most of the acetaldehyde required in the early 1990s. In 1992, Catalytica suggested a possible replacement—phosphormolybdovanadate/polyoxoanions—for the cupric chloride used with palladium chloride in the two-stage process, to avoid corrosion problems from hydrochloric acid.

7.5. METATHESIS OF OLEFINS

Olefin metathesis is the disproportionation of two olefin molecules, via cleavage of the double bonds as shown in the equation:

$$2\ RCH{=}CH_2\ \rightarrow\ RCH{=}CHR + CH_2{=}CH_2 \tag{7.56}$$

and

$$RCH{=}CH_2 + R'CH{=}CH_2\ \rightarrow\ RCH{=}CHR' + CH_2{=}CH_2 \tag{7.57}$$

The reaction was announced by Phillips Petroleum in 1964 and became known as the *Triolefin* process because it was first used to balance the production of olefins from naphtha steam crackers by converting propylene into ethylene and butene-2.[94]

As the scope of the reaction increased, the name *Olefin Metathesis* was introduced by Goodyear, who also pioneered the use of homogeneous catalysts. The reaction had first been recognized during experiments on the development of a heterogeneous catalyst to replace mineral acids in alkylation reactions. Molybdenum hexacarbonyl catalyst supported on alumina produced 2-pentene (40%) from mixed *n*-butenes together with propylene (51%) and hexene (9%). Tungsten hexacarbonyl was less active than the molybdenum catalyst, and in total contrast, it was found that chromium hexacarbonyl acted as a polymerization catalyst.

Later experiments showed that metal oxides were more active than carbonyls and that the most active catalyst was actually a commercial cobalt/molybdate hydrodesulfurization catalyst containing 3% cobalt oxide and 8% molybdenum oxide supported on γ-alumina. Molybdenum and tungsten oxides supported on high-surface-area silica, similar to the familiar Phillips chromium/silica polyethylene catalyst, were also active in metathesis reactions and more resistant to poisons. The operating temperature range for a tungsten oxide/silica catalyst was, however, significantly higher at 300°–400°C than for the cobalt/molybdate catalyst which operated at 120°–210°C, although the catalyst volume required was reduced by a factor of ten. Operating pressure was in the range 25–30 bar.

BP has developed an even more active rhenium oxide catalyst supported on alumina that operates at 20°–50°C, with an equilibrium conversion higher than 60%. This follows from an observation by Phillips that alumina is a more suitable support than silica.

7.5.1. Process Development

The first commercial use of the process was in Canada in 1966, to convert a surplus of propylene into polymer-grade ethylene and 30,000 tonnes/year of butene-2, at about 40% propylene conversion. The Phillips neohexene process is used to convert di-isobutylene and ethylene to the product by a double-bond isomerization/metathesis mechanism. All metathesis catalysts are gradually deactivated by the formation of high-molecular-weight compounds and must be regenerated by heating with air to 500°–600°C. The cycle time before regeneration depends on the feed purity and the operating temperature.

The catalyst is installed in three reactors, two of which are on line at any given time, while the third is being regenerated ready for use. Continuous catalyst regeneration, similar to the method used in the platinum-catalyzed reforming of naphtha, could be used providing that the plant size is large enough to make the additional capital cost economic.[96] Metathesis is, of course, an equilibrium process, so that depending on the ratio of the reacting molecules and the temperature, the nature of the products can be selected.

Following experience at an Arco plant in the United States, it was shown that propylene production from a naphtha steam cracker can be increased by using a metathesis unit. This has the advantage of maintaining maximum ethylene production at high cracking severity. The largest naphtha steam cracker built by the year 2000 was designed to incorporate an isomerization unit to provide butene-2 from the mixed C_4 stream that could produce propylene when reacted with an equivalent proportion of ethylene. While the steam cracker on its own was designed to produce 950,000 tonnes.year^{-1} of ethylene and 540,000 metric tons/year of propylene, the use of the metathesis unit increased the level of production to 860,000 tonnes/year of both propylene and ethylene.[96]

The rhenium/alumina catalyst operating at 20°–50°C can produce 63% conversion at 35°C, whereas a tungsten oxide/silica catalyst operating at 300°–400°C gave 55% conversion at 330°C.[97]

It is important to remove all oxygen, dienes and acetylenes from the feed to the metathesis reactor. Furthermore, the C_4 stream needs to contain the minimum practical level of butene-1 and isobutene to minimize the metathesis reaction of the C_4 hydrocarbons, which results in the formation of C_5 and C_6 olefins. The higher olefins lead to polymer formation and catalyst deactivation. An excess of ethylene normally suppresses the C_4 reactions. A typical steam cracker C_4 stream, which has been subjected to selective hydrogenation to remove impurities and fractionation to provide a suitable butene-2 rich *raffinate-2*, or the butene-2 rich effluent from an MTBE unit, can provide a suitable feed for the metathesis reactor. The catalyst operating cycle with a rhenium catalyst is usually fairly long. Regular catalyst regeneration may be necessary and the catalyst can last for several years.

7.5.2. The Shell Higher-Olefins Process

The Shell Higher-Olefins Process (SHOP), which includes a commercial application of metathesis, was developed to convert ethylene into a range (C_{10}–C_{18})

of straight-chain detergent alkenes containing an even number of carbon atoms.[97] Three catalytic steps are involved.

In the first stage, ethylene is oligomerized to give a Schultz-Flory distribution of linear olefins in the range C_4 to C_{20}. The catalyst is a soluble nickel complex formed by the reaction of bis(cyclo-octadienyl) nickel with a phenyl or cyclohexyl-substituted phosphine derivative of the type $R_2PCH_2COO^-$, in butanediol solution. When the ethylene is removed, the products separate as a hydrocarbon phase, and the catalyst solution can then be recycled.

About 30% of the α-olefin products are in the C_{10}–C_{18} range. These are separated by distillation and the residual C_4–C_8 and C_{20+} fractions pass into an isomerization reactor containing a suitable heterogeneous catalyst. An almost equilibrium distribution of straight-chain internal olefins forms at 80°–140°C and 3.5–17 bar. Hardly any α-olefins remain. The bottom C_{20} fraction from the distillation column can be changed to any boiling range depending on the products required. The final metathesis step uses a typical molybdena/alumina catalyst to converted the isomerized internal olefins into a new range of mainly linear internal olefins from which about 10–15% of C_{11}–C_{14} olefins can be separated by distillation. The fraction of olefins with fewer than ten carbon atoms is recycled to the metathesis reactor and the fraction with more than fifteen carbon atoms is recycled first to the isomerization reactor. It is also possible to add ethylene to the internal olefin feed entering the metathesis reactor to increase the yield of α-olefins. The olefins produced are converted into detergent alcohols by hydroformylation to aldehydes, which are then hydrogenated to alcohols.

REFERENCES

1. R. F. Goldstein and A. L. Waddams, *The Petroleum Chemical Industry (3rd Edition)*, E. and F. N. Spon., London, 1967; P. H. Spitz, *Petrochemicals: The Rise of an Industry*, Wiley, New York, 1988.

2. A. L. Waddams, *Chemicals from Petroleum (An Introductory Survey)*, John Murray, London, 1963/1968; R. Long, *Production of Polymer/Plastic Intermediates from Petroleum*, Butterworths, London, 1967; H. Steiner, *Introduction to Petroleum Chemicals*, Pergamon, London/NewYork, 1961.

3. P. H. Spitz, *Petrochemicals: The Rise of an Industry*, Wiley-Interscience, New York, 1988; K. Weissermel and H-J. Arpe, *Industrial Organic Chemistry*, VCH, Weinheim, 1993; S. Matar and L. F. Hatch, *Chemistry of Petroleum Processes*, Gulf., Houston, 1994; C. N. Satterfield, *Heterogeneous Catalysis in Industrial Practice (2nd Edition)*, Krieger, Malabar, Florida, 1996; C. Masters, *Homogeneous Transition Metal Catalysts*, Chapman & Hall, London, 1981; R. Pearce and W. R. Patterson, *Catalysis and Chemical Processes*, Leonard Hill, Glasgow, 1981; *Catalysis Science Technology*, Ed. by

J. R. Anderson and M. Boudart, Vols. 1–8, Springer-Verlag, Berlin, 1981–1987; *Preparation of Catalysts—A Series*, Ed. by B. Delmon, G. Poncelet, P. A. Jacobs, and Grange, Elsevier, Amsterdam, 1976–present; G. M. Wells, *Handbook of Petrochemicals and Processes (2nd Edition)*, Ashgate, Oxford, 1999.

4. H. Hock and S. Lang, *Berichte* **77** (1944) 257.
5. *Chem. Eng. News*, April 4 (1994) 23.
6. H. V. Regnier, *Ann. Chim. (Phys.)* **58** (1835) 307.
7. H. Kolbe, *Ausfs Lehrb Organ Chem.* **1** (1854) 346.
8. E. Baumann, *Liebigs Ann.* **163** (1872) 649.
9. Griesheim Electron German Patent 278249 (1912) (Gas phase).
10. Griesheim Electron German Patent 281584 (1913) (Liquid phase).
11. Griesheim Electron German Patent 281887 (1913).
12. Consortium für Electrochemische Industrie, British Patent 339093 (1928).
13. B. F. Goodrich US Patent 2225635 (1938).
14. M Kaufman, *The History of PVC*, Ch. 3, MacLaren, London (1969).
15. D. P. K. Keane, R. E. S. Stobaugh, and P. L. T. Miller Townsend, *Hydrocarbon Processing*, Feb (1973) 100.
16. Burke, Miller, *Chemical Week*, (August 22, 1964) 93.
17. US Patent 4206180 (1980).
18. J. S. Naworski and E. S. Velez, *Oxychlorination of Ethylene in Applied Industrial Catalysts*, Vol. 1, Ch. 9, Ed. by B. E. Leach, Academic, New York, 1983, p. 239.
19. J. W. Harpring, A. E. Van Antwerp, R. F. Sterbenz, and T. L. Kang, US Patent 3488398 (1970).
20. N. V. EVC International, *Hydrocarbon Processing*, (March 1, 1997) 164.
21. German Patent 570980 (1929).
22. P. Morris, *The American Synthetic Rubber Research Program*, University of Pennsylvania Press, Philadelphia, 1989, p. 7.
23. P. Morris, *The American Synthetic Rubber Research Program*, University of Pennsylvania Press, Philadelphia, 1989, p. 13.
24. *Rubber Reserve Co. Report on Rubber Program 1940–45*, Schedule II, p. 61 (1945) Supplement No.1, p. 41/45.
25. S. V. Lebedev, British Patent 331402.
26. G. Egloff and G. Hulla, *Chem. Rev.* **36** (1945) 63.
27. *BIOS Report 1060.*
28. *BIOS Report 356.*
29. F. E. Frey and W. F. Huppke, *Ind. Eng. Chem.* **25** (1933) 54; US Patent 2098959 (1937).
30. *Report on Investigations of Fuels and Lubricants Teams: US Bureau of Mines Information Circulars (Potash)*, 7370, 7375 (1946); B. B. Corston, US Patent 2375405 (1942) (Magnesia).

31. K. Gordon, Progress in the Hydrogenation of Coal and Tar, Institute of Mechanical Engineers, Dec. 9, 1946.
32. R. Holroyd, *Report on Investigations of Fuels and Lubricant Teams: US Bureau of Mines Information Circulars*, 7370/7375 (1946).
33. K. K. Kearby, *Ind. Eng. Chem.* **42** (1950) 295; K. K. Kearby, Emmett, US Patent 2370797-8 (1945).
34. H. E. Drennan, US Patent 2371809 (1945); F. T. Eggertsen and H. H. Voge (Shell) US Patent 2414585 (1947).
35. F. T. Eggertsen and E. P. Davies, US Patent 2461147 (1949).
36. E. C. Britton, A. J. Dietzler, and C. R. Noddings, *Ind. Eng. Chem.* **43** (1951) 2871.
37. Massoth and Scarpiello, *J. Catal.* **21** (1971) 294;. Rennard and Kehl, *J. Catal.* **21** (1971) 282.
38. Phillips Petroleum Star Process, *Hydrocarbon Processing*, June (1976) 133.
39. Linde Process, *Chem. Eng. News,* March, 1992; *Hydrocarbon Processing,* July (2000).
40. Styrene, its polymers, copolymers and derivatives, Ed. by R. H. Boundy, *ACS Monograph 115*, Reinhold, New York (1952).
41. I. G. Farben, US Patents 1986241 (1935) and 2110833 (1938).
42. Dow, US Patents 1985844 (1934) and 2036410 (1936).
43. H. W. Ashton and T. W. Flavell, *BIOS Report 3057.*
44. K. K. Kearby, Catalytic dehydrogenation, in *Catalysis*, Vol. 3, Ed. by P. H. Emmett, Reinhold, New York, 1955, p. 481.
45. K. K. Kearby, Catalytic dehydrogenation, in *Catalysis*, Vol. 3, Ed. by P. H. Emmett, Reinhold, New York, 1955, p. 480.
46. E. W. Pitzer, US Patent 2866790.
47. Fleming and W. R. Gutmann, US Patent 2990432.
48. W. R. Gutmann, US Patent 3361683.
49. F. J. O'Harra, US Patent 3904552.
50. Sud. Chemie, US Patent 4467046.
51. D. L. Williams, Styrene Catalysts, *AICE Meeting*, New Orleans, March 6– 10, 1988.
52. (a) Sud Chemie Literature; (b) P. R. Courty and C. Marcilly, A Scientific Approach to the Preparation of Bulk Mixed Oxide Catalysts, in *Preparation of Catalysts III*, Eds. by G. Poncelet, P. Grange and P. A. Jacobs, p. 485 (c) P. R. Courty and J. F. LePage, Relationship between average pore diameter and selectivity in iron-chromium-potassium dehydrogenation catalysts in *Preparation of Catalysts II*, Ed. by G. Poncelet, P. Grange and P. A. Jacobs, Elsevier Science Publishers, Amsterdam, 1978, p 293.
53. R. W. Moncrieff, Heywood and Co. Ltd., London, 1963; *Chem. Eng. News,* (March 21, 1994) 11; (May 2, 1994) 10.
54. W. H. Carothers, US Patents 2130523, 2130947-8 and 2163636; British Patents 461236 and 461367 (1937).

55. M. McCoy, *Chem. Eng. News*, (Aug 28, 2000) 10; (Oct 2, 2000) 33.
56. *Chem. Eng. News*, (April 26, 1971) 30.
57. C. A. Tolman, W. C. Seidel, J. D. Druliner, and P. J. Domaille, *J. Organometallic Chem.* **3** (1984) 33.
58. E Billig, C B Strow and R L Pruett, *Chem. Comm.* (1968) 1307.
59. P. J. Hogan and J. R. Jennings, UK Patent 1,546,807 (to ICI plc).
60. R. J. Cozens and J. R. Jennings, *Applied Catal.* **19** (1995) 297.
61. P. Schlack, I. G. Farben Industrie, German Patent *Chem. Eng. News* (Oct 2, 2000) 32; US Patent 2241321 (1941); Smith, in *The Production of Polymer and Plastics Intermediates from Petroleum*, Ed. by Long, Butterworths, London (1967).
62. German Patent 748253.
63. O. Wallach, *Synthetic Fiber Development in Germany*, *CIOS Report XXX*, HMSO, London, 1945, p. 13.
64. Wallach, *Liebigs Ann.* **312** (1900) 171.
65. *Hydrocarbon Processing* **49** (1970)137; British Patents 1257609 (1971) and 1332211 (1973); US Patents 4092360 (1978) and 4203923 (1980).
66. M. A. Mantegazza, G. Paparatto, G. Petrini, G. Fornasari, and F. Trifiro (Enichem), Ammoxidation Reaction in Gas and Liquid Phase, p. 353, *Catalysis of Organic Reactions*, edited by M. G. Scaros and M. L. Prunier, Dekker, New York (1995).
67. W. Muench, *Chimica Ind. Milano* **44** (1962) 636; SNIA Viscosa Italian Patents 604795; Belgian Patents 582793, 603606.
68. Walter, *J. Pract. Chem.* **45** (1892) 107.
69. P. Woog, *Comp. Rend.*, **145** (1907) 124.
70. P. Woog, French Patent 379715 (1907).
71. Weiss and Downs, US Patent 1321959 (1919).
72. Craver, British Patent 189091 (1920).
73. M. McCoy, *Chem. Eng. News*, (Oct 2, 2000) 32.
74. R. W. Moncrieff, *Man Made Fibers*, Wiley, New York, 1971; J. R. Whinfield and J. T. Dickson, British Patent 578,079 (1941); US Patent 2,465,319 (Assigned to DuPont)
75. P. Short, *Chem. Eng. News*, (May 15, 2000) 25.
76. C. Masters, *Homogeneous Transition Metal Catalysts*, Chapman & Hall, London (1981).
77. A. Tullo, DuPont NG-3 Process, *Chem. Eng. News*, (March 6, 2000) 25.
78. P. M. Morse, New Polyester, *Chem. Eng. News*, (Nov 10, 1997) 8.
79. O. Roelen, US Patent 2415102 (1947).
80. *Combined Intelligence Sub-Committee Report (CIOS)*, G 2 RAPO 1.
81. G. U. Ferguson, in *The Production of Polymer and Plastics Intermediates from Petroleum*, Ed. R. Long, Butterworth, London (1967), p. 86.

82. R. Wyman, in Selected developments in catalysts, Ed. J. R. Jennings, *Critical Reports on Applied Chemistry*, Vol. 12, Blackwell, Oxford (1985), p. 128.

83. L. H. Slaugh and R. D. Mullineaux, *Organomet. Chem.* **13** (1968) 469; E. R. Tucci, *Ind. Eng. Chem. Prod. Res. Dev.* **9** (1970) 516.

84. R. Fowler, H. Connor, and R. A. Baehl, *Hydrocarbon Processing*, (Sept. 1976) 247; *Chem. Eng. News* (March 9, 1998).

85. *Chem. Eng. News* (Oct 10, 1994).

86. K. Weissermel and H-J. Arpe, *Industrial Organic Chemistry*, VCH, Weinheim, 1993.

87. E. Wiebus and B. Cornils, *Hydrocarbon Processing*, (March 1996) 63.

88. N. von Kutepow,W. Himmele, and H. Hohenschitz, *Chem. Ingr. Tech.* **37** (1965) 383; *Hydrocarbon Processing* **45** (1966) 141.

89. F. E. Paulic and J. F. Roth, *J. Amer. Chem. Soc. Chem. Comm.* (1968) 1578.

90. D. Forster, *J. Amer. Chem. Soc.* **98** (1976) 846.

91. Catiwa Process, *Chem. Eng. News*, (July 1, 1996) 8

92. F. C. Phillips, *J. Amer. Chem. Soc.* **16** (1894) 255.

93. J. Smit, W. Hafner, R. Jira, R. Sieber, J. Sedlmeier, and A. Sabel, *Angew. Chem. (Int Edn)* **1** (1962) 80.

94. R. Banks, *Discovery and Development of Olefin Disproportionation (Metathesis)*, American Chemical Society (1983).

95. *Hydrocarbon Processing*, (March, 1998) 61.

96. *Oil Gas J.*, (Sep. 20 1999) 62.

97. E. R. Freitas and C. R. Gum, *Chem. Eng. Prog.*, (Jan 1979) 73.

8

OLEFIN POLYMERIZATION CATALYSTS

Catalytic polymerization processes have become increasingly important as the use of plastics has escalated throughout the world. Many common plastics were discovered in the 1930s, and after the 1939–1945 war this created a demand for petrochemical intermediates derived from the refining industry. Ethylene and propylene are now common building blocks for plastics and by the year 2000, polyethylene, polypropylene and their copolymers were the most widely used plastic materials. The gradual development of polyolefin production is shown in Table 8.1.

Even before Ziegler discovered his catalysts for the polymerization of olefins in 1953,[1] it was well known that high-molecular-weight oils and insoluble waxes could be formed by the treatment of olefins and synthesis gas with metal alkyls and transition metal catalysts.[2] However, the first polyethylene process was the result of research into high-pressure reactions such as those undertaken by ICI in the 1920s and 1930s.[3] By March 1933, an experiment with high-purity ethylene and benzaldehyde at 1900 bar pressure and 170°C had produced a substance like paraffin wax when the pressure vessel had apparently developed a leak. After several more experiments, in which it was confirmed that polymer had formed, and that the reaction could be explosive, work stopped until 1935, at which time it appeared that there might be a commercial demand for the product. When relatively impure ethylene was heated at 170°C, the pressure could not be maintained at the target of 2000 bar and polyethylene, in the form of a white powder, was formed. It had been recognized, of course, that benzaldehyde played no part in the reaction. By 1936, it was found that traces of oxygen

L. Lloyd, *Handbook of Industrial Catalysts*, Fundamental and Applied Catalysis,
DOI 10.1007/978-0-387-49962-8_8, © Springer Science+Business Media, LLC 2011

were acting as a catalyst and process details were filed in a patent.[4] The role of the oxygen is not to act as a catalyst in the normal sense of the word, but to act as a source of free radicals. The free radical reacts with a molecule of ethylene, which then undergoes chain growth by a free radical mechanism. The oxygen is better referred to as an initiator. Details of both molecular weight and operating conditions had been accidentally disclosed at a conference held in 1935, although experts, including Hermann Staudinger, refused to believe that the polymerization of ethylene was possible. Fortunately, the minutes of the meeting did not include details of the disclosure and the ICI patent application was subsequently granted.[3] A full-scale plant with a capacity of about 100 tons of polyethylene per year started operation in September 1939.

The ICI patent took precedence over work by Union Carbide in the United States and by I. G. Farben in Germany. Experts from Germany had visited the ICI polyethylene plant in 1937, and according to an official from BASF, then began the work that led to the development of the Lupolen Process during the war.[5]

8.1. LOW-PRESSURE POLYETHYLENE

In 1953 a further step in the development of polyethylene was announced by Karl Ziegler at the Max Plank Institute in Mulheim.[6] Ziegler began by studying the reaction of lithium hydride and ethylene. This was not very promising, and he turned to the reaction of aluminium hydride and ethylene, which gave triethylaluminium at temperatures in the range 60–80°C. Triethylaluminium is, of course, spontaneously inflammable in air and reacts explosively with water. The handling of these dangerous reactive compounds needed the development of a whole range of new experimental and production techniques. Polymers were formed by addition of further molecules of ethylene at temperatures around 100–120°C and these were eventually developed as low molecular weight α-

TABLE 8.1. Worldwide Production of Polyolefins.

Year	Polyethylene (million tones year^{-1})	Polypropylene (million tones year^{-1})	Typical new plant capacity (tones year^{-1})
1965	3	< 0.05	10,000
1970	6		
1975	10	4	
1980	15	6	
1985	22	8	80,000
1990	30	13	100,000
1995	38	19	
2000	50	30	300,000

olefins. The reaction then continued. The reaction became known as the Aufbau reaction and produced a linear polymer, in marked contrast to the highly branched product from the high pressure processes.

The subsequent development of the process for high-density polyethylene was the result of careful systematic observation by the experimental team. Further work on the oligomerization of ethylene to the commercially-valuable C_6 and C_8 α-olefins led to the discovery that the displacement reaction was catalyzed by traces of nickel in the reactor. The displacement reaction became faster than the growth reaction as the concentration of nickel was increased, and at 120°C, only the dimer was produced. In subsequent investigations of other metals as co-catalysts in conjunction with triethylaluminium, it was shown that zirconium acetylacetonate inhibited the displacement reaction completely, and that polyethylene was the only product. Further work led to the development of the well-known titanium chloride/triethyl aluminum catalyst, which operated at low temperatures and atmospheric pressure. The reaction became known as the *Mulheim Atmospheric Polyethylene Process.*[6]

At about the same time that news of the Ziegler discovery was released, the Phillips Petroleum Company in the United States announced that it had developed a medium-pressure, catalytic process (500 psig) to produce a high-density, crystalline polyethylene.[7] The process was discovered when traces of ethylene in a flue gas had polymerized over conventional cracking catalysts. The Phillips catalyst contained chromic oxide supported on silica. The Standard Oil Company of Indiana (later Amoco) also introduced a medium pressure process using a catalyst comprising molybdenum oxide supported on carbon or alumina, but it did not enjoy the success of the Ziegler or Phillips processes and was only operated in three full-scale plants.[8]

8.1.1. Polyethylene Process Development

The Phillips process, used a different type of catalyst from Ziegler, was available for license almost immediately and became a commercial success by producing a linear, highly crystalline product with higher density than the high-pressure polymer discovered by ICI. The new polymer became known as high-density polyethylene (HDPE), whereas the original ICI polyethylene was thereafter known as low density polyethylene (LDPE). Phillips did not develop a catalyst for the production of polypropylene.

Ziegler, working in the academic environment of the Max Planck Institut, did not develop his own process for the polymerization of olefins, but did offer the rights to his catalysts for license. Licensees were required to develop their own versions of a polymerization process that was based on Ziegler's discovery. It is not surprising that several different versions of the *Ziegler* processes evolved to produce HDPE, each based on somewhat different catalyst formulations and operating procedures.

8.1.2. The Development of Polypropylene Catalysts

Ziegler restricted his studies to the polyethylene catalyst, and the next major discovery was made by one of his licensees. Professor Guilio Natta, who worked with Montedison in Italy, investigated the polymerization of propylene and other α-olefins to produce solid crystalline polymers. He found that polymerization did not proceed when using the brown titanium trichloride used by Ziegler. He showed that crystalline stereoregular polypropylene[9] could be produced when using violet titanium trichloride.

8.2. ZIEGLER–NATTA CATALYSTS

The versatile range of low-pressure catalysts, discovered by Ziegler and modified by Natta in 1953 and 1954 was a tremendous advance in catalysis. It led to the introduction of a wide range of commercially useful polyolefins, including polypropylene and many copolymers as well as polyethylene. Since the initial discoveries, there have been many developments in both the catalysts and the commercial polymerization processes that have led to increased activity and productivity.

8.2.1. Early Polyolefin Catalysts

The original Ziegler catalyst comprised a metal alkyl and a transition metal compound that could be reduced and alkylated in the presence of the metal alkyl co-catalyst to give active initiating and propagating centers. Since 1954, the original catalyst has been developed continuously and the active centers modified with coordinating ligands. Most of the commercial catalysts are heterogeneous and the active component is supported on an inert material.

The most widely used transition metal salts have been the relatively cheap titanium or vanadium halides, reduced with aluminum alkyls. In Ziegler's work, titanium tetrachloride was reduced to brown β-titanium trichloride, which was able to polymerize ethylene. However, when β-titanium trichloride catalyst was used in the polymerization of propylene, the product contained a high proportion of the gum like atactic polymer, which was not viable for commercial use. In contrast, Natta, in his work on the polymerization of propylene and other α-olefins, showed that violet α-titanium trichloride could polymerize propylene to the useful, crystalline, isotactic polymer. Nevertheless, a relatively large quantity of atactic polypropylene still had to be separated from the commercial isotactic product.

Most of the early catalyst development was related to the need for a more active catalyst to improve productivity. With low-activity catalysts, the polymer must be *de-ashed* to remove solid catalyst residues so as to meet product specifications. To achieve a high productivity, the catalyst particles should *fragment* during the process and be diluted with high-molecular-weight polymers. The most significant difference between olefin polymerization and other catalytic processes is that the catalyst remains in the product.

In crystalline Ziegler–Natta catalysts, the properties of the active transition metal centers with their vacant coordination sites are different, depending on their position on the fractured crystal surface. This leads to variations in both activity and reducibility and consequently to the wide variations in polymer molecular weight observed in commercial products. The continuing development of catalysts has therefore been related to the need for the formation of active more regular sites, which can then produce polymer chains with a more uniform molecular weight and, in the case of polypropylene, higher stereospecificity. This subsequently led to the use of a range of supported titanium catalysts for both polyethylene and polypropylene production and, more recently, to the use of single-site metallocene catalysts. These improvements have led to increases in the range of possible products even further.

8.2.2. Ziegler's Brown Titanium Trichloride

The early heterogeneous Ziegler catalysts were prepared in the reactor by the reaction in solution from a mixture with a molar ratio of titanium tetrachloride to triethyl aluminium of less than 2.5:1 to prevent excessive reduction of the titanium species. The organotitanium complexes that formed decomposed spontaneously at ambient temperature to a precipitate of the brown metastable β-titanium trichloride.[10] This is a very superficial description because the nature of the titanium product depends on many factors, such as the relative concentration of the components, the solvent used, the order of addition, and the temperature of the reaction. The aluminum alkyl acts as an alkylating agent and the resulting alkyl titanium species decomposes to a lower valence state. Triethyl aluminum is a stronger alkylating agent than diethyl aluminum chloride and was preferred for polyethylene catalyst production.

A controlled molar excess of the aluminum alkyl cocatalyst was added to the reduced titanium (~2:1) before the catalyst entered the reaction zone. All the components were carefully purified and handled in a dry, inert atmosphere. Despite these precautions, the catalyst had a low surface area and was not very active. Most of the titanium was inaccessible due to the formation of relatively large crystals and relatively few active sites were formed at the crystal edges. Furthermore, ß-titanium trichloride was not stereospecific for the polymerization of higher olefins.

Activity and stereospecificity depended, therefore, not only on the catalyst preparation but also on the size and crystallinity of the particles. The careful control of conditions and the optimum concentration of reactants were necessary to form particles that fragmented during use to expose fresh active sites and to increase productivity. Higher productivity, that would allow the use of the polymer without *de-ashing* to remove catalyst particles, remained an objective.

8.2.3. Natta's Violet Titanium Trichloride

The important contribution made by Natta was the modification of the Ziegler catalyst for the polymerization of propylene. Pasquon, one of Natta's colleagues, reported that,

> "he assumed stereospecificity to be associated with the regularity of the catalyst surface and used the crystalline violet *β*-titanium trichloride instead of liquid titanium tetrachloride to prepare his catalysts."[11]

In this way, Natta was able to develop a stereospecific catalyst based on *α*-titanium trichloride, produced by reducing titanium tetrachloride in hydrogen at 500°–800°C, before combining it with the triethyl aluminum co-catalyst.

The catalyst was further improved by producing a stereospecific form of *α*-titanium trichloride. The metastable *β*-titanium trichloride which also contained co-crystallized aluminum trichloride was heated to temperatures of 300°–400°C. The transformation proceeded via intermediate *γ*-titanium trichloride, which could also produce stereoregular polymer at temperatures of 100°–200°C. The *α*- and *γ*-crystalline modifications of titanium trichloride have layered hexagonal and cubic crystal structures, respectively, and are both active catalysts.

Stauffer produced titanium trichloride commercially by reducing titanium tetrachloride with aluminum metal in an aromatic solvent at about 250°C.[12] The aluminum could be activated by ball milling, with the addition of a small amount of aluminum trichloride. This procedure formed high-activity *γ*-titanium trichloride containing co-crystallized aluminum trichloride. It was soon realized, from experience during commercial production of polypropylene catalysts, that both the *α* and *γ* forms of titanium trichloride were more than ten times as active after the crystals had been ground or milled to produce smaller particles.[13] Milling converts *α*- and *γ*-titanium trichloride to the more active *δ*-titanium trichloride and increases the surface area from about 5 m^2 g^{-1} to 20 m^2 g^{-1}. The crystal structure of a range of titanium trichloride catalysts is described in Table 8.2.

Early commercial catalysts were supplied to a strict specification. The average particle size was usually in the range 10–100 μm with an activity exceeding 1000 g polypropylene per gram of catalyst and an isotactic index exceeding 88%. Continuous, rather than batch, operation was possible with the new, externally formed catalyst but the efficiency was still low and the polypropylene still contained unacceptably high levels of both catalyst residues and atactic polymer.

TABLE 8.2. First-Generation Ziegler–Natta Catalysts (1957–1960).

Modification	Crystal form[a]	%Polypropylene stereoregularity (AlEt$_2$ Cl co-catalyst)[b]	Preparation
β-TiCl$_3$ (brown)	Globular; fibrous chainlike structure	40–50	In situ
α-TiCl$_3$ (violet)	Platelets; hexagonal close packed	80–92	External: H$_2$ reduction of TiCl$_4$ at 300°–400°C
γ-TiCl$_3$ (violet)	Platelets; cubic close packed	80–92	External: H$_2$ reduction of TiCl$_4$ at >150°C
TiCl$_3$·0.33AlCl$_3$	Hexagonal globules (10–100 μm)	>88	External: Al reduction
δ-TiCl$_3$ (violet)	Larger secondary particles; random alternation of α- and γ- packed particles	80–92	External: milling of the α-and γ-forms, from 1960 onwards

[a] Charges in the close-packed chlorine layers of α-and γ-TiCl$_3$ balanced by hexagonal clusters of smaller titanium atoms in spaces between chlorine layers. To maintain electroneutrality there are rows of empty spaces in the titanium layer with lower charge density, which simplifies disintegration when milled. The small crystals agglomerate and are more active.
[b] Catalyst more active with AlEt$_3$ but less stereospecific.

Although polypropylene was only produced on a small scale from 1957, and despite the relatively greater importance of high-density polyethylene, work on improving polypropylene catalysts led to a better overall understanding of both catalysts and process operation.

8.2.4. Second-Generation Propylene Polymerization Catalysts

By 1960, the original preparation of brown β-titanium trichloride in situ was no longer used by most producers. It was replaced by commercial γ-titanium trichloride, which contained an isomorphous form of aluminum chloride in the molar ratio (3TiCl$_3$·AlCl$_3$). This was ball milled to form δ-titanium trichloride, which increased activity and led to the formation of a higher proportion of isotactic polymer. However, even after these improvements in catalyst performance, there was still a limitation to the catalyst activity because the aluminum trichloride in the catalyst reacted with the triethyl aluminum co-catalyst to form quantities of ethyl aluminum dichloride, which is detrimental to the polymerization reactions. Titanium trichloride crystals were also relatively large despite the milling treatment to reduce particle size.

A further improvement in performance was achieved by the addition of electron donors (Lewis bases), such as esters, ethers, or amines, to the catalyst, which can form complexes with aluminum alkyls. Electron donors could be added either with the co-catalyst, or during the γ-titanium trichloride milling stage, depending on the compound used or the result required. In addition to modifying the effects of poisons, electron donors also deactivated nonstereospecific sites and improved the stereochemical and kinetic behavior of the active centers by acting as a template to control insertion of the monomer into the growing chain. In many cases, electron donors had a somewhat adverse effect on productivity, due to the deactivation above. They were, however, still useful. By increasing the isotactic index, more efficient use of propylene was obtained, because a smaller proportion of the worthless atactic polymer was formed. Details are given in Table 8.3.

Refinements to the Ziegler–Natta catalysts continued, with several attempts to convert brown β-titanium trichloride to the active purple γ-titanium trichloride at a lower temperature. The objective of this work was an increase in the activity of the catalyst by avoiding the formation of the aluminium trichloride that was isomorphous with the γ-titanium trichloride. High temperatures were normally required to reduce the aluminium chloride content, and this led to a reduction in the surface area, and hence, activity. In a development by Solvay, titanium tetrachloride was reduced with aluminum alkyl at about 1°C, before extraction of the co-crystallized aluminum chloride with either dibutyl- or di-isoamyl ether. The porous β-titanium trichloride was then heated with excess titanium tetrachloride at 65–100°C to remove excess ether and to produce high-surface-area (<75 m^2g^{-1}) purple α-titanium trichloride under controlled conditions.[14]

A feature of these catalysts, which were more active and stereospecific in polypropylene production, was that they formed uniform 25 to 35 μm diameter spherical particles and the shape of the polymer replicated the catalyst morphology. Unfortunately, despite the favorable effect of these catalysts on the shape of the polymer particles produced, the particles were unstable during storage and de-ashing was still necessary. The multistage recipes were only used where on-site production facilities were available.

Worldwide demand for polypropylene was still only about 1–6 million tones year^{-1} during the period from 1970 to 1980. As demand began to increase, more efficient catalysts, based largely on a successful range of supported polyethylene catalysts, were developed.

TABLE 8.3. Second Generation Ziegler–Natta Catalyst (1970).

Electron donor promoter	Productivity (kg polymer per1g Ti)	Stereospecificity
None (first generation)	<5	80–92
Hexamethyl phosphoric triamide	5	95
Terpenic ketones	6	92
Di-isoamyl ether (Solvay)^a	20	93

^{a} High-surface-area (35 m^2g^{-1}) ether removes AlCl$_3$ with small spherical catalyst particles (25–35 μm) giving improved handling of polymer. Electron donors remove less stereospecific sites and activate remaining sites.

8.2.5. Supported Polyethylene Catalysts

The demand for HDPE was always greater than that for polypropylene (PP) and more efficient polyethylene catalysts were needed from about the mid-1960s. The brown β-titanium trichloride catalyst introduced by Ziegler had very low productivity because less than 1% of the total titanium content actually took part in the polymerization reaction.[15] De-ashing the polymers to remove the titanium and halogen residues represented a significant production cost. Moreover, as soon as the potential of the polymerization products was realized, the scope for new and improved polymers with specific molecular weight and molecular weight distribution became commercially attractive.

A catalyst with a higher proportion of uniformly active centers was therefore required, not only to achieve better control of polymer chain length but to allow the introduction of α-olefins into the polymer chain. Regular branching in the polymer chain could give a range of lower density products and could improve on the variable branching in LDPE produced at high pressure. Such polymers became known as linear low-density polyethylene (LLDPE).

The use of commercial $3TiCl_3.AlCl_3$ catalysts for HDPE production had already avoided the need for 'in situ' catalyst preparation and more active supported catalysts became available in about 1970. These had almost replaced the first- and second generation Ziegler catalysts by 1980. A wide variety of typical supports had been investigated, including silica and alumina, although despite the success of Phillips catalyst, these had low activity.[16] The first useful supports to be commercialized were based on a range of crystalline magnesium compounds with surface hydroxyl groups.[17] For example, high yields were obtained when using a support based on magnesium hydroxide, combined with titanium tetrachloride and a triethyl aluminum co-catalyst.

Magnesium supports continued to show the most promise and were studied intensively during the 1970s. Magnesium alkyls and Grignard reagents also react with titanium tetrachloride to produce species that, in conjunction with triethyl aluminum co-catalysts, are also very productive and provide a high proportion of active centers. The solid magnesium-based catalysts were found to be nodular and to contain magnesium chloride.[18] Other very active high-surface-area porous catalysts were produced from magnesium ethoxide and titanium tetrachloride using a triethyl aluminum cocatalyst.[19] The high pore volume of the nodules led to catalyst disintegration during operation as polymer chains were formed. Productivity was related to the ratio of co-catalyst and titanium used, while the molecular weight distribution could be controlled by the addition of relatively low hydrogen levels to the olefin feed.

Highly active catalysts containing 3–4% titanium were prepared by co-milling titanium tetrachloride with magnesium chloride and then using a triethyl aluminum co-catalyst. The crystal structure of magnesium chloride is similar to both α- and γ-titanium trichloride; during milling, a disordered structure, similar to δ- titanium trichloride, develops and the extremely small particles have a high surface area of up to 200 m^2 g^{-1}.[20,21] The magnesium ions at the edges of crystallites react with the titanium tetrachloride and provide active sites.

Catalytic activity can be considerably increased if the magnesium chloride is first combined with a suitable Lewis base, such as an ester, acid, alcohol, or amine, before reaction with an excess of titanium tetrachloride and subsequent activation with triethyl aluminum. Polyethylene produced with these catalysts has a lower molecular weight than that from the titanium trichloride catalysts mentioned earlier and less hydrogen is needed for molecular weight control.

8.2.6. Supported Polypropylene Catalysts

More efficient use of titanium was also possible when using the same catalysts to produce polypropylene. The supported catalysts were more active and there was no need to remove catalyst residues from the polymer. The preparation of both nodular polyethylene catalysts, containing magnesium chloride from magnesium alkyls or Grignard reagents, and fragmenting catalysts made from magnesium ethoxide led to the production of polypropylene with controlled physical shape and better processing characteristics.

8.2.6.1. Third-Generation Catalysts

The magnesium supports used for polyethylene catalysts could be modified for use in polypropylene production, and magnesium chloride proved to be the most suitable when used with a Lewis base electron donor. Milled magnesium chloride was known to have the same layer structure as α- and γ-titanium trichloride with the quadrivalent titanium ion (0.068 nm diameter), being about the same size as the divalent magnesium ion (0.066 nm diameter). In 1968, Montedison and Mitsui Petrochemical Industries both disclosed the production of a highly active, very stereospecific catalyst that contained about 3% titanium on a magnesium chloride support,[22] promoted with a Lewis base, such as ethyl benzoate.[23] The polymer produced contained less than 1 ppm titanium with an isotactic index of more than 90%, which was improvement on the product made with previous catalysts.

During catalyst preparation, the magnesium chloride was milled with ethyl benzoate (mole ratio 5:1) before the small particles were digested with titanium tetrachloride at 80–130°C. The solid was washed with a hydrocarbon and dried. The electron donor formed complexes with surface sites on the magnesium chloride particles and also prevented particles from sticking together during subsequent handling. Titanium tetrachloride replaced some of the attached ethyl benzoate and entered the support lattice, particularly at edge and corner sites. The catalyst contained between 1–4% titanium and 5–20% ethyl benzoate, depending on the titanium content. During use, the catalyst was activated by triethyl aluminum and a second electron donor, such as ethyl anisate or *p*-methyl toluate, was added to deactivate nonstereospecific sites. The high initial activity of the catalyst decayed fairly quickly during operation, but several modifications to the preparation procedure could be made to improve performance.

8.2.6.2. Fourth-Generation Catalysts

Important objectives of the later production methods were to control the size and shape of the catalyst particles during precipitation of the magnesium chloride and to improve stability. Catalysts with better-controlled size and shape were based on the reaction of a precipitated magnesium chloride with titanium tetrachloride in a high-boiling-point hydrocarbon diluent at 80°C, with di-isobutyl phthalate added as an internal electron donor.[24] After separation, the solid formed was reacted with more titanium tetrachloride at 120°C, before being washed and dried. The catalyst contained between 2–3% titanium and the phthalates used were limited to C_4–C_8 esters to avoid potential problems with colloid formation. The catalysts produced with phthalates as the internal donor had much higher surface area and pore volume than when ethyl benzoate was used. This method provided more active and stereospecific catalysts when used with the same triethyl aluminum co-catalyst and an external electron donor such as phenyl triethoxy silane.

Very high isotactic index polymer, up to about 99%, could be produced with the early magnesium ethoxide–based catalysts using dibutyl phthalate as an internal donor and an organosilane external donor.[25] Size and shape of the catalyst particles produced from magnesium alkyls or ethoxides could be controlled by using polar organic solvents, such as tetrahydrofuran, as in the preparation of the hydrocarbyl carbonate catalysts described by Amoco.[26]

It was pointed out by Tait that the rapid initial increase in the rate of polymerization, with both the spherical magnesium chloride and magnesium ethoxide catalyst, decayed rapidly in the same way as ball-milled magnesium chloride catalysts.[26] The loss of activity was presumably due to the small particles or fragments becoming blocked with polymer.

The phenomenon of catalyst fragmentation was first recognized by Natta and Pasquon in 1959.[27] The fragmentation continued until the co-catalyst could no longer reach any potential active centers and the polymer had replicated the shape of the catalyst fragments. The polymer-coated particles that formed were about seven times as big as the catalyst fragments. On the other hand, large, uniform, spherical catalyst particles prepared from magnesium alcoholates by careful sieving have a higher activity that is very stable provided that care is taken to ensure that the particles do not fragment.

Particles of catalysts for polypropylene, and the corresponding polypropylene particles can be examined using a scanning electron microscopy to show that the polymer particle is about 20 times as big as the original catalyst. Porous

TABLE 8.4. Third- and Fourth-Generation Ziegler–Natta Catalysts (1977/1983).

Catalyst	Productivity (kg polymer/g Ti)	Stereospecificity
Third generation: TiCl$_4$/MgCl$_2$/EB/AlEt$_3$/EA Ethyl benzoate (EB) complexes MgCl$_2$. Ethyl anisate (EA) or p-methyl toluate added with cocatalyst.	300	92
Fourth-generation: TiCl$_4$/MgCl$_2$/DIB/AlEt$_3$/PhSi(OEt)$_3$ Initially based on magnesium ethoxide, with di-isobutyl phthalate (DIB) and PhSi(OEt)$_3$. Better, more stable spherical catalysts obtained with magnesium alcoholate. Polymer particles (1500 μm diameter) replicate expanding catalyst particles (100 μm diameter).	600	98

catalysts, which do not fragment, allow the co-catalyst and monomer to continue reaction until all of the catalyst crystallites are covered in polymer. In practice, this means that the primary catalyst particles expand and are cemented with polymer as it forms. The reaction stops when the pore structure is saturated. The productivity of these catalysts can be related to the relative ability of particles to expand to expose more of the crystallites, and each catalyst particle may be regarded as a separate micro-reactor. The catalysts[28] were produced by Montedison for their *Spheripol* polypropylene process in 1981 and for the corresponding *Spherilene* polyethylene process in the early 1990s. Third- and fourth-generation Ziegler–Natta catalysts are summarized in Table 8.4.

8.3. PHILLIPS POLYETHYLENE CATALYSTS

In 1954, the Phillips Petroleum Company announced a polyethylene process that had been discovered by Hogan and Banks,[29] in which a chromium catalyst was used. By 1956, nine companies had become process licensees and Phillips was producing high-density polyethylene (HDPE) in its plant at Pasadena, Texas.[29] A branched form of HDPE was made in 1958 by the introduction of a co-monomer, 1-butene.[29] This modified form of HDPE is now known as linear low-density polyethylene (LLDPE). Since 1956 the Phillips process has been widely used throughout the world and various new forms of the chromium catalyst have continued the extremely successful development of the process.

The catalyst preparation appeared to be a very simple procedure and consisted of impregnation a silica support, made by precipitation of silica gel, with a soluble chromium salt. It was found that siloxyl chromium complexes were formed by the reaction of chromic oxide with the hydroxyl groups on the silica surface, as the catalyst was activated prior to operation.[29] The structure of the silica gel support determined both the catalyst activity and the properties of the polymer formed. This contrasted strongly with Ziegler catalysts which needed the use of a co-catalyst and the polymer chains contained methyl and vinyl groups at either end with virtually no branching or internal double bonds. Since it was introduced, the productivity and applications of the support have been modified by control of the surface area, pore volume, and pore size, as well as by the addition of other oxides that influence both surface structure and the complexes formed with chromium compounds.

Although it has not been easy to understand the mechanism of the polymerization process with the chromium catalyst, many formulations have been introduced that enable the catalyst to provide a wide range of polymers.[30] These have provided increased catalyst activity and products with controlled molecular weight and chain length. For example, it is possible to modify the catalyst to form branched α-olefin co-polymers in situ and thus to prepared LLDPE directly. Moreover, by complexing two forms of chromium with the support, two distinct grades of polymer are formed that give a bimodal molecular weight distribution to the HDPE produced.

8.3.1. Catalyst Production

Low-sodium silica gels, with high surface area and pore volume, are usually chosen as supports, as they can be synthesized to give a wide range of properties. Catalyst activity is proportional to surface area, while the pore size controls productivity and the molecular weight of the polymer. Small catalyst particles are used but the actual size depends on the polymerization process.[31] Slurry reactors require particles in the size range 50–180 μm, whereas gas phase processes use smaller particles in the range 20–90 μm.

Silica hydrosols are formed by mixing sodium silicate with sulfuric acid at pH < 7. The sol polymerizes spontaneously, is washed to remove sodium sulfate, and then aged before drying to form a xerogel. The concentration of the silica solution and the aging/drying conditions must be carefully chosen to produce a gel with the appropriate properties. Water is removed by heating in air or, if a very high porosity is required, with a suitable organic solvent. The surface area and pore volume of the gels produced are in the range 50–1000 $m^2\,g^{-1}$ and 0.4–3.0 $cm^3\,g^{-1}$, respectively, depending on the aging and drying procedures, but a typical free-flowing product has a surface area of about 350 m^2g^{-1} and a pore volume of 1.8 $cm^3\,g^{-1}$.

Silica gel is simply impregnated with an aqueous solution of chromium (III) acetate to give about 1% chromium on the catalyst.[32] This low chromium content gives the maximum yield per unit weight of catalyst used during operation, and avoids the clumping of uncomplexed chromium oxide on the surface. It has been estimated that at this stage only about one-third of the total chromium forms active centers. Some of the chromium is needed to absorb or react with gaseous poisons in the monomer.

The impregnated support must be fluidized and activated for up to 12 h by heating in air at temperatures up to 500°–950°C.[33] This removes adsorbed surface water and decomposes some of the surface hydroxyl groups. Chromium (III) is oxidized to chromium (VI) by heating in air at temperatures above 300°C. The low melting oxide is very mobile and reacts with surface hydroxyl groups to give siloxy- chromates. Siloxy chromates are stable at the higher temperatures that are required to decompose more of the surface hydroxyl groups, and develop full activity. Dry air must be used during the thermal activation process, to avoid hydrolysis of the siloxy chromates to free chromium (VI) oxide. Any free chromium (VI) is reduced to chromium (III) at temperatures exceeding 400°–500°C, and this must be avoided to achieve the best performance from the catalyst. More than 90% of the chromium is combined with silica following activation. Higher-temperature activation of the catalyst gives a lower range of molecular weight in polymers produced during use.

8.3.2. Catalyst Reduction

Subsequent to startup of the catalyst, there is an induction period during which active centers are formed as chromium (VI) is reduced to chromium (II) by the ethylene monomer.[33] Alternatively, chromium can also be activated by pre-reduction with carbon monoxide at temperatures above 600°C.[34] This procedure results in an enhanced removal of surface hydroxyl groups from the support, with only producing carbon dioxide and hydrogen as by-products, in a reaction analogous to the water gas shift reaction:

$$H_2O + CO \rightarrow CO_2 + H_2 \tag{8.1}$$

The absence of water thus avoids the two problems of hydrolysis of the surface chromates and water-induced sintering of the support, both of which lead to a catalyst with lower activity. Reoxidation of the chromium after carbon monoxide reduction gives a different active site precursor, which produces a catalyst which gives a broader molecular weight distribution in the resulting polymer. Catalysts pre-reduced with carbon monoxide are more active than those reduced with ethylene in the reactor.

8.3.3. Catalyst Operation

Phillips catalysts were originally operated in solution or particle form (slurry) processes to produce HDPE, although the particle form process was preferred in 1983.[29] In the 1960s, following a lead by Phillips, licensees began to develop gas phase processes using mechanical agitation (BASF)[35] or fluidized beds (Union Carbide),[35] both of which avoided agglomeration of the catalyst or the polymer.

The silica support of the chromium catalyst influences operation by controlling the formation of active centers. The activity of the catalyst increases at higher activation temperatures and the molecular weight of the polymer decreases, which suggests that polymer propagation and chain termination are related to the decreasing density of hydroxyl groups on the silica surface. This may be a consequence of the way growing polymer chains interact with the surface hydroxyl groups, to favor chain growth.[36]

The shape and pore size of the particles controlled the catalyst activity and molecular weight of the polymer chain in the early catalysts. The molecular weight distribution of the polymer could not be controlled very easily because of differences in the active sites. It is accepted that several different active centers can form with chromic oxide and silica and that the ease of reduction of each type depends on the operating temperature. Catalysts which have been activated at 850°C contain fewer active sites that seem to reduce easily. This supports the observation that these catalysts produce polymers with a narrow molecular weight distribution. In general, however, the molecular weight of the polymer decreases as the pore size of the catalyst increases. Pore size controls both performance and selectivity.

The most important practical property of the support is that it disintegrates as the polymer chains form within the catalyst pores. This is why preparation of the support to control particle size is so important. Fragmentation provides continuous access to the active sites inside the pores and must continue during operation. From examination of the fragments of catalyst in particles of polymer, it has been shown that the typical particle size is as small as 0.05–0.1 μm.[37] Despite the need to fragment as pressure from the governing polymer chain develops within the pores, all supports must be physically strong to withstand attrition during handling.

The ability to produce catalyst supports that could fragment led to the development of high-productivity catalysts and processes in which de-ashing of the polymer was unnecessary.

8.3.4. Catalyst Modifiers

The ability to control molecular weight and molecular weight distribution is important in the development of new polymer products. It soon became apparent that the silica gel support could be modified by the addition of other components to make polymerization catalysts even more versatile.

Oxides such as titania, zirconia, or alumina when co-gelled with silica can influence the pore structure and improve the stability of silica gel during the aging stage of preparation. It was also shown, however, that these oxides formed a variety of additional sites that provided extra catalyst activity and a wider range of polymer densities. Silica could also be impregnated with precursors of the modifying oxides to achieve the same results.[32] Residual surface hydroxyls groups on the support in activated catalysts could also be replaced by fluoride ions. This had an indirect effect on the active centers and led to the formation of a different range of polymers.

8.3.4.1. *Titanium*

Pre-dried silica gel was impregnated with a titanium ester that could be decomposed to give the necessary titanium content.[38] The support could then be impregnated with the appropriate chromium salt and activated by the normal procedure. The modified catalyst was more active and produced a bimodal, broader molecular weight distribution of polymers with a lower average molecular weight than that from an unmodified catalyst. A wide range of polymers with different molecular weight can be formed by variations in the titanium content. Catalysts containing more than about 8% titanium are, however, less thermally stable. Co-gelled titanium catalysts are more active than unmodified catalysts and produce lower-molecular-weight polymer.[39] These catalysts are different from both the untreated and impregnated catalysts, because they produce a much narrower molecular weight distribution. The lower thermal stability of the co-gelled catalyst requires that the titanium content is limited to about 4%.

The different molecular weight distributions of polymers produced by titanium-modified catalysts can be explained by the formation of different active sites[40]. The presence of the less-electronegative titanium atom may lead to easier reduction of the chromium sites. An increased electron density at the chromium atom could destabilize the chromium-carbon bond, giving higher catalytic activity and lower molecular weight of the polymers. There is a higher surface concentration of titanium in impregnated catalysts (90%) than in co-gelled catalysts (60%). The impregnated catalysts therefore probably contain titanyl chromate sites as well as the siloxyl chromate sites of unmodified catalysts. This explanation is supported the observation of two exotherms in the temperature-programmed reduction of impregnated catalyst. One at 440°C corresponds to the siloxyl chromate reduction peak of unmodified catalyst and a second at 420°C corresponds to titanyl chromate reduction.[32] The two different active centers would lead to the bimodal distribution of the molecular weight of the polymer.

On the other hand, the more homogeneous mixture of silica and titania in co-gelled catalysts may lead to the formation of a predominant third kind of site consisting of mixed silyl/titanyl chromates. These sites would then provide the narrow molecular weight distribution and other properties of the polymer formed.[30]

8.3.4.2. *Alumina and Zirconia*

Silica/chromia catalysts have also been modified by the incorporation of other oxides such as alumina[41] or zirconia,[42] by impregnating a silica support with zirconium acetylacetonate or aluminum sec-butoxide or co-gelling the silica gel with appropriate soluble salts.[32] Zirconia-modified catalysts are similar to those with added titanium in their effect but have not been widely reported. Aluminum-modified catalysts have increased activity and provide lower-molecular-weight polymers, but the procedure for their preparation is complicated. Neither type of catalyst is widely described in the literature but they are reported as containing different active sites.[43]

8.3.4.3. *Fluorides*

The properties of silica/chromium catalysts can be modified indirectly by replacing hydroxyl groups with fluorine atoms.[44] Impregnation of the catalyst with ammonium hexafluorosilicate before activation causes the surface hydroxyl groups from the siloxane groups to react with the fluorine atoms and form silicon–fluorine bonds. This has been claimed to decrease the electron density at the chromium atoms and alter the distribution of active centers. These catalysts produce polymers with a narrow molecular weight distribution. Better co-monomer incorporation results when titanium impregnated catalysts are modified with fluoride, and a higher-molecular-weight polymer is produced.[42]

8.3.5. Use of Co-Catalysts

The early Phillips catalysts only needed to be activated and reduced to generate active centers, but several patents since have described the use of co-catalysts to promote the production of LLDPE. For example, a typical catalyst that had been modified with titanium, activated in air and reduced in carbon monoxide was then further activated by addition of triethylboron prior to operation.[45] This procedure led to the production of linear low density polyethylene, LLDPE, which had a density of 0.9726 g cm^{-3}, directly, without the need for the addition of an α-olefin to the pure ethylene feed. Olefins were produced in situ and these were incorporated into the polyethylene. A second catalyst was made using silica with a very high pore volume, modified with titanium, activated in air, and finally reduced with carbon monoxide at 350°C. This catalyst was then treated with triethylboron before use. LLDPE polymers with densities in the range 0.890 to 0.915 were obtained from a feedstock of ethylene and hexene-1, but the addition of some hydrogen to the gas stream was required to limit the length of the polymer chain.

Higher-molecular-weight HDPE polymers, with broad molecular weight distributions, can be made using an air-activated titanium-modified catalyst with the addition of diethylboron ethoxide in the concentration range 2–10 ppm as co-catalyst. When a similar catalyst is used with the addition of triethylboron in the same range of concentrations, the density of the polymer decreases. This suggests that the presence of triethylboron leads to the formation of some α-olefins and that these are incorporated into the HDPE to produce a polymer with a lower molecular weight, more analogous to LLDPE.[47]

It has been possible to introduce many different polyethylene varieties with Phillips catalysts by modifying the silica surface or using co-catalysts, which will continue to allow a much wider use of chromium catalysts in the future.

8.3.6. Organo-Chromium Catalysts

Catalysts derived from organo-chromium compounds supported on silica gel are very different to the original Phillips chromium oxide catalyst. In the original catalysts one molecule of chromium oxide is bound to two silica units, to form the silyloxy chromate sites. In contrast, organo-chromium compounds are usually only bound to a single site on the silica support and the other coordination sites are still occupied by the organic ligands. For example, tris-π-allyl chromium reacts with a hydroxylated support, with the evolution of propylene and the formation of a bis-π-allyl chromium siloxy derivative.

The support for an organo-chromium catalyst must, therefore, be carefully calcined to lower the concentration of surface hydroxyl groups, leaving a sufficiently wide separation to avoid the possible linkage of the chromium to two units of silica.[33] Excessive calcination has to be avoided as it removes too much water, leaving too few hydroxyl groups to achieve the required loading of chromium.

A new catalyst in which chromocene was supported on silica gel was developed by Union Carbide[48] for use in gas-phase polymerization reactions. A typical catalyst contained about 2.5% by weight of chromium, and was activated by the addition of small amounts of alkylaluminium alkoxides, such as diethylaluminium ethoxide. Use of such a catalyst formulation gave rise to a polymer with much narrower molecular weight distributions. The average molecular weight could be controlled by the addition of hydrogen, which caused termination of the growing chain.

Other organo-chromium catalysts include those produced by the impregnation of silica supports with diarene chromium species, chromium allyls, and bis(triphenyl) silyl chromate. These catalysts give very broad molecular weight distributions of the polymer. Branching in the polymer chains is often extensive, because low-molecular weight olefins are formed and can act as co-monomers. Polymer chain termination is usually by β-hydrogen abstraction, leaving a terminal double bond in the polymer chain.

Mixed catalysts have been prepared by reaction of an activated chromium oxide silica catalyst with an organo-chromium compound. Mixed catalysts are very active and produce branched polymers with a low density. The type of organochromium compound used in the catalyst controls the formation and isomerization of the olefins incorporated into the polymer chains,[49] so influencing polymer density.

8.4. OTHER CATALYSTS

Standard Oil of Indiana used a molybdenum oxide/alumina catalyst that could be prepared either by co-precipitation or impregnation methods.[50] The catalyst was activated by calcination in air at 500°C followed by reduction in hydrogen (or carbon monoxide) at a temperature in the range 450°–500°C, to give a product in which the valency of the molybdenum was less than six. The catalyst needed to be activated using sodium alkyls, possibly because the catalyst was difficult to reduce. The original catalyst had a low productivity. Better results were obtained when the alumina support had been halogenated, and the process operated in the liquid phase at an operating temperature of 200°–275°C and a pressure of 35–100 bar. Although other modifications were made to the catalyst and the process, only three plants seem to have been built and operated. A different catalyst, based on vanadium pentoxide/alumina, was mentioned in some early publications but it was only slightly easier to activate than the molybdenum oxide/alumina catalysts. Both molybdenum and vanadium oxides melt at higher temperatures than chromium trioxide, so consequently they are not as mobile on a silica surface.[29]

8.5. POLYMERIZATION PROCESSES

The first commercial process for the manufacture of polyethylene was commissioned in 1939, and required pressures of 2000 bar and temperatures of 200°C. Free-radical initiators were used, and low-density polymer was formed as highly branched chains. In the 1950s, new processes which did not require the use of free-radial initiators, and which operated at low temperatures and low pressures ranging up to 40 bar, were introduced. Catalysts were used and a higher-density polyethylene with little or no chain branching was produced.

Low-density polymers produced at high pressure became known as low-density polyethylene (LDPE), while the higher-density, low-pressure polymer is now known as high-density polyethylene (HDPE). When α-olefin co-monomers were used with ethylene, the low-pressure polymer was less dense than HDPE as a result of regular branching in the polymer chain. The product is now known as linear low-density polyethylene (LLDPE). The approximate proportions of the different polyethylene types are shown in Table 8.5.

TABLE 8.5. Relative Proportion of Polyethylene Types.[a]

Year	LDPE (%)	HDPE (%)	LLDPE (%)
1985	55	35	10
1990	45	40	15
1995	40	40	20
2000	30	40	30

[a]Relative proportions of HDPE/LLDPE depend on demand.

Polyethylene is now produced by several processes that are more convenient and do not require the high capital investment and expertise associated with high-pressure operation. When catalysts were first introduced, the most widely used processes were based on *solution* or *slurry* operation, in which polymerization took place on catalyst suspended in an inert liquid hydrocarbon. Since 1968, many new plants have used a gas phase process in which the catalyst and polymer particles are suspended as a fluid bed, by the circulation of gaseous monomer. An approximate estimate of the market share of each process by 1999 is given in Table 8.6. A typical plant for the production of polyethylene is shown in Figure 8.1.

As demand for polyethylene has increased, there has been a continuous effort to improve catalyst activity. This was important to achieve high productivity at relatively short residence times. These properties have avoided the need to remove catalyst residues from the product. It was also important for polypropylene to have a high isotactic index to eliminate the need for equipment for the extraction of atactic polymer.

Two features distinguish the use of modern polymerization catalysts. First, there is no separation of the catalyst from the product; second, the catalyst must be able to disintegrate, or fragment, during operation, to expose new active centers. Use of the original low-activity catalysts led to unacceptably high residues of transition metal, and in the case of Ziegler catalysts, halogen in the product, which had to be removed before the polymer could be sold.

Reactor design is important and plug flow is essential to prevent both back mixing of reactants and to maintain a reasonable product quality. This requirement led to the use of long tubular reactors, or a number of small vessels operating in series. Newer gas phase processes are efficient in this respect and operation can be controlled to remove product of the appropriate size and molecular weight from the bottom of the fluid bed as it is formed. Another requirement of the processes is the need to control reactor temperature by removing the heat of reaction, so preventing the formation of low-molecular-weight polymers. High temperatures also lead to agglomeration of the product and deactivation of the catalyst. The reaction exotherm would lead to a rise in temperature of about 16C for every % of ethylene converted.

Figure 8.1. Unipol/Dow plant for the production of polyethylene. The key unit in the polyolefins plant is the Unipol reactor, which is the largest of its type in the world. Reproduced with permission from Univation Technologies, a Joint Venture between ExxonMobil Chemical and Dow Chemical.

A further essential feature of operation, particularly in gas phase processes, is the removal of the typical impurities found in olefins derived from steam crackers. The most common impurities are acetylenes, dienes, carbon monoxide, oxygen, water vapor, and sulfur compounds and these are removed either by hydrogenation or by absorption using conventional procedures before the olefins enter the polymerization unit.

TABLE 8.6. Relative Use (%) of Commercial Processes (1999).

Process	Polyethylene (%)	Polypropylene (%)
Gas phase	45–50	40–45
Slurry	45–50	45–50
Solution	5–15	5–10

8.5.1. Slurry Processes

In the slurry process,[51] a solution of about 2–6 wt% of the monomers in solution in a hydrocarbon such as isobutane, *n*-pentane or *n*-hexane is slurred with the catalyst, and the polymerization proceeds in a loop reactor achieving conversions 97%. Typical co-monomers with ethylene can be butene-1, hexene-1, 4-methylpentene-1, and octene-1. The catalyst is suspended in the solution, which is then pumped as turbulent slurry through relatively narrow tubes. The polymer is essentially insoluble in the liquid diluent and forms a slurry containing up to 30 wt% of small polymer particles, which are removed continuously.

The operating temperature of the process must be below the melting or softening point of polymers and, depending on the type of polymer produced and the diluents used, is generally less than 100°C. A unique loop reactor was designed by Phillips Petroleum Corporation to prevent the deposition of solids onto the reactor walls. It is reported that loop reactors have operated continuously for as long as three years without the need for cleaning.[51] As an alternative, autoclave slurry reactors with efficient stirrers have been operated in batch or cascade processes. Changes in the partial pressure of the hydrogen moderator from one stage to another allow a range of different average molecular weights to be obtained within the product. Loop processes using Phillips chromium catalysts operate at pressures in the range 20–40 bar, whereas autoclave processes operate with Ziegler catalysts at lower pressures.

Residence times of up to 1 h are typical, with the continuous addition of monomer, catalyst, and fresh diluent. Productivity can exceed more than 10,000 lb of polymer per pound of catalyst. The use of low concentrations of catalyst allows the grade of the product to be changed within a few hours. The average molecular weight of the product can be controlled by the type of catalyst used, the activation procedure, operating conditions, and by changing the partial pressure of the hydrogen moderator.[52]

8.5.2. Solution Processes

In the solution process, polymerization takes place in solution, in a liquid hydrocarbon such as cyclohexane or iso-octane, in which both the monomer and polymer can dissolve. About 0.2–0.6 wt% of a Ziegler catalyst is normally suspended in the solvent, by stirring efficiently in a suitable reactor. The operating temperature must be above the melting point of the polymer but lower than the decomposition point of the catalyst. Typical temperatures are in the range 125–175°C.

The polymerization is carried out at a typical pressure of about 30 bar, and the solution of monomer and co-monomer is admitted into the reactor with fresh catalyst. After reaction, the solution containing some 10–20% of polymer is removed from the top of the reactor. The spent catalyst taken from the reactor is largely deactivated and is separated from the solution. In some of the early versions of the process, the catalyst could be recycled, provided that it still possessed sufficient activity for this to be worthwhile. The viscosity of the solution needs to be sufficiently low to avoid circulation and pumping problems, and this requirement can limit the magnitude of the average molecular weight of polymer.

The residence time in the reactor is only a few minutes and the conversion of monomers usually exceeds 95%, depending on the concentrations of ethylene and polymer in the solution. Unused ethylene is recycled. This makes the process very adaptable if it is necessary to change the grades of polymer at regular intervals. It is claimed that more than 500 tonnes of polymer can be produced per kilogram of titanium in the catalyst.[53] The use of solution processes was reported to be increasing and by 1990, represented up to 15% of world polyethylene production.[54]

8.5.3. Gas Phase Process

In the most successful process, the catalyst is fluidized in a large reactor by the circulation of gaseous olefin monomers at high space velocity.[55] Polymerization takes place in the gas phase at pressures less than 20 bar and an operating temperature in the range 80°–100°C, to ensure that the polymer does not soften. Conversion is limited to about 3% per pass. This allows circulation of the monomers through external coolers to remove the heat of reaction while polymer and catalyst particles reside in the reactor for an average of 2–3 h. The largest polymer particles gradually fall to the bottom of the fluid bed and are removed when the appropriate physical properties are achieved. Fresh catalyst is injected, often as a slurry, to make up for losses. An immediate attraction of the process is that the concentration of catalyst in the polymer is very low and the product does not have to be de-ashed. Other gas phase processes operate with vertical reactors, which use stirrers,[56] and horizontal vessels which use weirs and paddles.[57]

Gas phase polymerization is flexible and can produce a wider range of polymer grades than the earlier processes. Capital and operating costs are relatively low and the process has been very successful. Modern process developments include the injection of liquid hydrocarbon directly into the fluid bed, rather than to the recycle gas.

This makes the process more economic by allowing better temperature control and increased conversion.[53] This has become known as the condensation mode and can considerably increase the capacity of existing units. All types of active catalysts can be used.

8.6. METALLOCENE/SINGLE-SITE CATALYSTS

In the early 1990s a new type of olefin polymerization catalyst was introduced. Catalyst derived from metallocenes give a polymer with a very narrow molecular weight distribution with regular inclusion of co-monomers in the growing chains. This was a significant improvement on the relatively wide molecular weight distribution and a variable co-monomer content of Ziegler–Natta products.[58] Development has been slower than forecast owing to operational problems and the high cost of both metallocenes and the first methyl aluminoxane *activator*. However, the high level of industrial interest was demonstrated by 350 patent applications by the end of 1993 increasing to 1500 by 1997. Competition led to litigation over patent rights, followed by a series of joint ventures between the important producers. By 2000, up to 2% of the total worldwide polyethylene production was based on the use of metallocene, or single-site catalysts, as they became more accurately described.[59] This corresponded to about 1 million tonnes of polymer.

Exxon Chemicals tested their range of *Exxpol* catalysts in a demonstration plant at Baton Rouge, Louisiana, in 1991.[60] The unit had a capacity of 15,000 tonnes of polyethylene per year using an autoclave type of reactor. The plant was operated in the temperature range 100°–300°C and a pressure in the range 1000–2000 bar, using the monomer as a supercritical fluid solvent. These conditions led to the inclusion of a co-monomer into the polyethylene, over a wide range of molecular weights. A second, gas phase plant, was then retrofitted at Mont Belvieu, Texas. It was found that plant capacities could be increased with the new catalysts.

The Exxon catalyst is a derivative of cyclopentadiene, indene, or fluorene which may be bridged, and which may also be substituted with up to five different atoms or radicals. This organic group is bonded to a metal atom from Group IV b of the Periodic Table, titanium, zirconium or hafnium, which has been activated by addition of a strong Lewis acid, to give a cationic structure.

In 1993, Dow Chemicals started operation using its *Insite* catalyst for a solution process, in Freeport, Texas[61], producing a copolymer of ethylene containing octene-1 co-monomers as well as plastomers. A second plant in Tarrogona, Spain, began operation in 1996. The Dow catalyst consists of a Group-IV transition metal (titanium, zirconium, hafnium) covalently bonded to a substituted monocyclopentadienyl group bridged with a heteroatom such as nitrogen. The components form a "constrained cyclic structure with the metal center." Other companies, including BASF, Hoechst, and Phillips, were also testing single-site catalysts by 1995.

8.6.1. Early Development

Although single-site catalysts were only used industrially beginning in 1991/1993, they were not really new. In the late 1960's and early 1970's, the Ballard group[62] at ICI Corporate lab studied a range of π-ally derivatives[63] of transition metals, such as zirconium and chromium, and benzyl derivatives of titanium and zirconium as potential candidates for olefin polymerization catalysts. Relatively high activities were found, particularly when the π-ally or benzyl derivative were reacted with a support, though all of the derivatives showed significant activity as pure compounds in hydrocarbon solutions. This work showed significant promise, but in the end, was never commercialized. Even earlier, about 1953, both Natta[64] (Montedison) and Breslow[65] (Hercules) had investigated the catalytic potential of metallocenes during mechanistic studies on ethylene insertion. At the time they used a typical activator such as diethyl aluminium chloride to produce the metallocene cation intermediate in two ion-forming steps:[66]

$$Cp_2TiCl_2 + AlR_2Cl \leftrightarrow [Cp_2TiR^+. \ AlRCl_3^-] \tag{8.2}$$

$$[Cp_2TiR^+. \ AlRCl_3^-] \leftrightarrow Cp_2TiR^+ + AlRCl_3^- \tag{8.3}$$

Unfortunately the forward reaction equilibrium was unfavorable and the catalyst had a low activity.[67]

In later experiments, it was shown that the metallocene catalysts had higher activity when traces of water were present during the preparation of the catalyst. This is in direct contrast with the use of the original Ziegler Natta catalysts, for which the rigorous exclusion of water was essential. This observation was linked by Breslow to the formation of aluminoxanes by the hydrolysis of the alky aluminium derivative.[69]

The activation of zirconacenes by high molecular weight aluminoxanes derivatives led to the development of active and stable catalysts for the polymerization of ethylene. High yields of polymers with molecular weights in the range 10^3 to 1.5×10^6, depending on operating temperature, are now produced.[70] The high-molecular weight methyl aluminoxane alkylates the zirconacenes and converts the product into a cationic complex:

$$Cp_2ZrRX + MAO + Olefin \rightarrow Cp_2ZrR(Olefin)^+.MAOX \tag{8.4}$$

A 200-fold excess of MAO (methyl aluminoxane) was originally needed to provide a reasonable active catalyst. By the late 1990s, this proportion had been reduced to about 20. It was later realized that the active species in the catalyst formulation was a cation. This led to the replacement of the aluminoxanes species by compounds that could provide a source of a suitable non-coordinating anion, such as tetrakispentafluorophenylborate:[71]

$$Cp_2ZrMe_2 + PhN^+Me_2HB(C_6F_5)_4 \rightarrow Cp_2ZrMe^+B(C_6F_5)_4 + PhNMe_2 + CH_4$$
(8.5)

The first metallocene catalyst (Cp_2MCl_2.MAO) was completely nonstereo-specific for the polymerization of propylene and the polymer was atactic. It was then shown that when the catalyst structure was modified by replacing the halogen atoms with phenyl groups (Cp_2MPh_2.MAO) the catalyst became stereospecific at a temperature of $-30°C$.[72]

Ansa metallocenes[*] were the first soluble, stereospecific, single-site catalysts to be produced and were based on indenyl groups[73] [racemic tetrahydroethylene (1-indenyl)$_2$ZrCl$_2$]. By variations in the substitution of the indene group it was possible, with a MAO activator, to match the performance of the third-generation magnesium chloride–supported Ziegler–Natta catalysts.[74] In quantitative terms, this was equivalent to the production of about 1 tonne of polypropylene per gram of transition metal, with an isotactic index greater than 99% and molecular weight up to at least 500,000 g.mol^{-1}.[75]

8.6.2. Early Development

A single-site catalyst is made up of three components:

- The catalyst framework is usually a pair of organic ring compounds or ligands, linked by a bridging group. The rings can be substituted with various organic groups to modify the electronic properties of the catalyst sites to produce the specific grades of polymer as required. Common ligands include radicals from compounds such as cyclopentadiene, indene, fluorene and benzidene. Other ligands include imido-compounds, and some organo-phosphine derivatives, with bridges formed from organosilicon groups and a wide range of organic compounds such as ethylene. The ligand pairs are held in the form of an asymmetric V-shape, described as the ansa form (ansa is Latin for bent handle)[76] and are able to form stereospecific catalyst sites.
- A transition metal halide from the sub-group titanium, zirconium, and hafnium is coordinated between the ligand pairs. This provides a rigid, potentially stereo-selective framework that controls the access of monomers to the active site.
- The use of a co-catalyst to generate the active catalytic cationic centers is required. Derivatives of methyl aluminoxanes, salts containing the

---

[*] An ansa metallocene is one in which ligand pairs are attached to the form of an asymmetric V-shape.

tetrakis-pentafluorophenylborate anion, other non-coordinating anions or trispentafluorophenylaluminium can be used.

Some processes are based upon soluble catalysts but gas phase and slurry processes use about 1–2 wt% catalyst impregnated onto a suitable form of silica. Changes to the catalyst structure can modify catalytic activity, the molecular weight of the product, and the incorporation of co-monomers in the polymerization of ethylene. In the production of polypropylene, structural changes within the catalyst also direct the formation of isotactic, syndiotactic, and non-crystalline atactic polymers.[77] Better control of the polymerization process was therefore possible, as shown in Table 8.7.

The following general observations have been made:

- The presence of silicon bridges within the catalyst leads to an increase in the molecular weight, and a small increase in stereospecificity of the polymer. Longer bridges decrease both molecular weight and stereospecificity.
- Phenyl substitution within the bridge gives a further increase in molecular weight.
- In the case of catalysts based on indene, substitution of the both the indenyl group and the bridge gives an increase in molecular weight.
- Substitution of the indene ring with phenyl or naphthyl groups, and substitution of the bridge with alkyl groups leads polymer with a high molecular weight and stereospecificity.

Asymmetric single-site catalysts have been produced from bridged fluorenylcyclopentadienyl rings with alkyl substitution in both the bridge and the cyclopentadiene ring. These modifications can lead to control over the alternation of isotactic and syndiatactic sequences in the polymer and, consequently,

TABLE 8.7. Stereospecific Metallocene Catalysts.

Catalyst[a]	Activity	Molecular weight	Tacticity
Ethylene (indenyl-1)$_2$ ZrCl$_2$	180	2.4×10^4	80
Dimethyl silylene (2-Me, indenyl-1)$_2$ZrCl$_2$	99	1.9×10^5	88
Dimethyl silylene (2-Me, 5-benzene, indenyl-1)$_2$ZrCl$_2$	550	7.3×10^5	95
Dimethyl silylene (2-Me, 5-naphthalene, indenyl-1)$_2$ZrCl$_2$	880	9.2×10^5	99

[a] Many other catalysts have been introduced for the polymerization of propylene to produce a range of different structures ranging from isotactic, syndiotactic, to hemi-isotactic and isotactic/syndiotactic sequences. These include catalysts based on a bridged cyclopentadiene/fluorene methyl-substituted ligand framework that can alternate the substitution of ethylene and propylene in copolymers.

properties of the polymer. An typical catalyst contains a bridged fluorenyl-cyclopentadienyl framework, with a methyl group substituted in the 3-position of the cyclopentadiene ring. The bridging group can consist of a -CMe$_2$- or -SiMe$_2$- group.[74]

8.6.3. Industrial Operation

The initial reluctance to introduce single-site catalysts was due to the high costs of the catalysts and the methyl aluminoxane activator, as well as difficulties in the processing of the new polymer with existing equipment. By 1995, the ratio of the aluminoxanes required to activate the catalyst had fallen from about 200 to 20, thus offering a significant reduction in costs, and new activators based on non-coordinating anions (NCA) were being developed.

From 1995 the new catalysts, now known as single-site catalysts, were developed and many new structures were introduced. This was particularly true with catalysts used to produce a range of polypropylenes with new and different properties and to introduce long-chain co-monomers. A feature of the new polymers was a narrow range in the molecular weight distribution. By combining different single-site catalysts, a wider range of molecular weight distribution could be provided. The synthesis of single-site catalysts has been simplified and is now the domain of the specialist chemical companies. Catalyst costs have been reduced from about two cents per pound of polymer in 1993/1994 to 0.4 cents in 1998, compared with 0.2 cents for the older Ziegler–Natta types.[78] This reduction in overall catalyst cost is a result of both a reduction in the costs of making the catalysts and increases in the productivities of these catalysts. By 1998 it was estimated that 2–4 tonnes of the organometallic component of single-site catalysts could supply about 500,000 tonnes of polymer.

8.6.4. Catalyst Activators

Methyl aluminoxanes activate metallocene catalysts by alkylation of the metallocene precursor, forming a cationic complex of the type, Cp$_2$ZrR(olefin)$^+$. When these catalysts were first used, a very large excess of the aluminoxanes was required, but the need for such a large excess has now been removed. MAO is formed by the controlled reaction of a limited quantity of water with trimethylaluminium. Normally, water reacts with explosive violence with Me$_3$Al, to give methane and a hydroxylated form of aluminium oxide or aluminium hydroxide when an excess of water is used. However, by careful control of the reaction conditions, an intermediate degree of hydrolysis can be achieved. The product of reaction is a polymeric molecule, with a spherical shape, and has the formula (Al$_4$O$_3$(Me)$_6$)$_4$.Al(Me)$_3$. The MAO molecule contains a single unit of trimethylaluminium, and further reaction with methyl groups produces the complex anion, $(Al_{17}O_{12}(Me_3)_{28})^-$. Any excess trimethylaluminium can be extracted with diethyl ether.[79]

There is a tendency for the equilibrium between the complexes to be reversed, in polymerizations which use MAO as the activator. So-called *ion equilibrium deactivation* led to a reduction in the productivity of the catalyst, and a large excess of activator was required to maintain performance. It is now known that the active form of the catalyst is a cation formed in an equilibrium reaction during the activation stage, and more stable activators have since been developed. These can react irreversibly with the metallocene by the abstraction of anionic ligands during the alkylation step and thus generate a large non-coordinating anion, (NCA), that stabilizes the cationic catalytic species formed. The NCA must be large, inert to the catalytic cation, and have its negative charge localized within the structure. A most important feature is that it is labile and easily displaced by ethylene or propylene during the polymerization process.

The first NCA precursor to be identified was tetrafluoroboric acid, although this was not sufficiently labile to generate an active catalyst.[80] More efficient NCA precursors were later developed and included tri-n-butylammonium tetrakispentafluorophenyl borate and N,N-dimethylanilinium tetrakis-pentafluoro-phenyl borate., each used in a 1:3 molar ratio with the metallocene. Compounds such as these were of critical importance in the optimization of the catalyst.[81]

As the demand for new polymers increased, catalysts with two active sites per molecule were developed. It was found that tris-pentafluorophenyl-aluminium, $(C_6F_5)_3Al$, could be used to replace the borate derivatives, and when used in conjunction with the typical transition metallocene catalyst precursors, a series of di-cationic derivatives of titanium and zirconium could be produced. Work in Dow showed that these di-cationic catalysts were up to 30 times more reactive than boron activated catalysts, while still producing polymers with the same molecular weight and melting point.[82]

8.6.5. Molecular Weight Control

An early objective in the development of single-site catalysts was an increase in the molecular weight of the polymers formed. The polymerization process can be thought of as happening in three discrete steps, namely initiation, propagation and termination. The final molecular weight of the polymer is determined by the relative rates of the propagation and termination stages. The rate of propagation could be increased by the substitution of appropriate organic groups into the 2-position of the ligand rings, and by the introduction of bridging atoms or groups between the ligands. Typical processes of chain termination include abstraction of hydrogen from the β-position of the growing chain to produce a polymer molecule with a double bond at the end of the chain, and chain transfer with monomer to start a new chain.

The original metallocene catalyst, Cp_2ZrCl_2/MAO, produced a polymer with a molecular weight of less than 5000 g mol^{-1}. At 50°C, by using a single-site catalyst, ethylene (1-indenyl)$_2ZrCl_2$/MAO, the molecular weight of the polymer could be increased to about 50,000 g mol^{-1}.[82]

The ability to produce a wide range of polymers with different molecular weights is useful and ethylene oligomers are, of course, produced by the Aufbau reaction and SHOP process. Single-site catalysts are now able to provide a range of stereospecific oligomers of propylene to replace or extend the use of current petrochemical processes (Chapter 7). They are used as α-olefins or as intermediates in the manufacture of surfactants.[83]

8.6.6. New Catalyst Developments

New catalysts have been introduced since 1995 as a result of cooperation between industry and academia. DuPont and Professor Brookhart from the University of North Carolina discovered active nickel and palladium cationic complexes based on bulky di-imine ligands that formed catalysts in conjugation with a suitable activator.[84] Chain migration can be encouraged or restricted to produce extensive chain branching or chain growth, by variations in the operating temperature and pressure. This can lead to a wide range of polymer densities. Nickel catalysts are produced more easily than single-site catalysts and provide the basis for DuPont's Versipol technology.

Later, in 1998, the same group and, independently, BP Chemicals working with Imperial College, London, discovered *tridentate* iron(II) and cobalt(II) complexes with 2,6-bis(imino)pyridyl ligands that could be activated with MAO. The activity of the catalysts was as high as that of the single site polyethylene catalysts available at that time. Polypropylene and a wide range of other α-olefin polymers could also be produced with appropriate catalyst modifications.[85]

In a flashback to the early Ziegler experiments, a group at the University of California at Santa Barbara prepared organo-nickel catalysts that, when activated with the tetrakis-pentafluorophenyl borate anion, can produce butene-1 by the dimerization of ethylene, selectively at low temperatures and pressures. When the same nickel complex was combined with a conventional Dow-type single-site catalyst and the same NCA activator, a range of branched polyethylenes could be produced which varied according to the nickel/titanium ratio.[86] Other developments introduced by Professor Grubbs from the California Institute of Technology include neutral, single component nickel catalysts that do not require co-catalysts. These were based on the Shell SHOP catalysts, in which the phosphine ligand was replaced by the bulky salicylaldiamine molecule. Polymers with an average molecular weight greater than 250,000, few chain branches, and a narrow molecular weight distribution could be produced.[86]

8.7. THE MOLECULAR STRUCTURE OF POLYOLEFINS

The polymerization of olefins proceeds by the regular insertion of successive monomer units into the growing *head to tail* chain, and the stereochemical configuration of the growth reaction can be controlled by asymmetric carbon centers. This is particularly important for polypropylene, because the crystalline isotactic form is required, and each tertiary carbon (methyl group) must have the same orientation along the chain. Each propylene molecule must be adsorbed by the catalyst in an identical position on the active site with respect to the growing chain. This requires the same alignment of each secondary carbon atom in the double bonds during the *cis* insertion. The presence of uniformly active catalyst sites is desirable for both ethylene and propylene polymerization reactions but is not easy to achieve. The quantity of active sites that formed on the lateral crystal faces of α- and γ-titanium trichloride in the early catalysts was increased by grinding to give δ-titanium trichloride, and later by supporting titanium tetrachloride on magnesium chloride together with a suitable electron donor. The use of electron donors led to fewer inactive sites and greater stereoregularity in the polymer. The active sites on the catalysts still displayed differences, but in general, the polymers showed much less variations in the molecular weights. The application of metallocene compounds has resulted in the development of single site catalysts which can achieve excellent control over the average molecular weight, the molecular weight distribution, and the stereoregularity of the polymer.

8.7.1. Formation of Polymer Chains

The Cossee–Arlman mechanism for the polymerization of olefins is the most widely accepted theory but as yet it is not complete.[87] Cossee developed his early ideas of polyethylene growth at a titanium–carbon bond and supported the theory by molecular orbital calculations.[88] The role of the alkyl aluminium cocatalyst was in the generation of the active species, via the alkylation of the titanium chloride bonds, and to remove impurities in both the gas stream and catalyst preparative procedure. There was also the suggestion that it might be involved in the insertion of each monomer molecule, and also in the regeneration of dormant sites or the formation of new active sites.

Titanium atoms on the lateral faces or edges of the α-TiCl$_3$ crystals may be adjacent to vacant chlorine sites within the crystal structure. The active site is then formed when the titanium chloride bond is alkylated by the reaction with the alkylaluminium co-catalyst. This results in the formation of a titanium-carbon bond, adjacent to the vacancy. Olefin molecules, such as ethylene and propylene, can then co-ordinate with the vacant site, forming a γ-complex, which leads to a weakening of both the double bond of the olefin and the titanium-carbon bond itself. The olefin can then insert into the titanium-carbon bond, creating a new alkyl group of higher molecular weight, and thus initiating the polymerization reaction, but also regenerating the vacancy at which the olefin had originally been coordinated. Successive olefin molecules could then complex in turn with the vacancy, and thus propagate the growth reaction.

It has been shown by experiments using deuterium labeling of the propylene, that stereospecific polymerization of α-olefins to the isotactic form requires that the α-olefin is complexed to the active center in a cis- configuration.[89] The orientation of the olefin monomer can thus be arranged in the same relative position as the previous monomers in the growing chain. It is possible that other exposed titanium atoms may be converted to give active isotactic sites by the migration of the polymer chain, caused by steric overcrowding of surrounding chlorine atoms.[90] The different structural forms of polypropylene are described in Table 8.8.

8.7.2. Polymer Chain Termination

The growth of the polymer chain can be terminated by various reactions, and some procedures have been developed to stop the chain growth as a means of controlling the molecular weight of the polymer.[91] Some of these reactions are shown below:

- ß-hydrogen abstraction:

$$[Cat]-CH_2-CH_2-P_n \rightarrow [Cat]-H \ + \ CH_2=CH-P_n \qquad (8.6)$$

TABLE 8.8. Forms of Polypropylene.

Polymer	Structure
Head to tail	Crystallinity controlled by branching from the asymmetric carbon. Crystalline forms have helical conformation but, in practice, always contain amorphous regions.
Isotactic (typical stereospecific form)	Crystalline with three monomers per turn of helix. All methyl branches on same side of helix. Ideal melting point 176°C.
Syndiotactic	Crystalline with four monomers per turn of helix. Methyl branches alternate on opposite side of helix. Ideal melting point 183°C.
Atactic	Amorphous, does not crystallize. Methyl branches randomly placed along the chain.

Spontaneous termination by ß-hydrogen abstraction is less typical with Ziegler–Natta catalysts than with Phillips catalysts, because of the low temperature of operation at 60°–80°C. It can be induced at a higher temperature, however, and results in the formation of vinyl and vinylidene end groups. The polymer chain can also exchange with the monomer to form a new active center, with the release of free polymer, in a process known as chain transfer.

- Reaction with molecular hydrogen:

$$[Cat]-P_n + H_2 \rightarrow [Cat]-H + H-P_n \qquad (8.7)$$

Both Hercules[92] and Montecatini[93] have shown that hydrogen can act as a transfer agent for Ziegler–Natta catalysts and it can restore the active site. It is necessary to be able to control the magnitude of the molecular weight so that the resulting polymers can be processed more easily. The use of molecular hydrogen can be applied to all types of catalyst. The molecular weight of the polymer is approximately proportional to the square root of the partial pressure of the added hydrogen.[94] A typical example of the effect of the hydrogen concentration on the molecular weight of polypropylene is shown in Table 8.9. The results are taken from a stirred bed gas phase process at 70°C using a $TiCl_4/MgCl_2$/di-isobutyl phthalate, silica-supported catalyst with an $AlEt_3$ and dialkyldimethoxysilane co-catalyst.[95]

- Metal alkyls:

$$[Cat]-P_n + ZnEt_2 \rightarrow [Cat]-Et + EtZnP_n \qquad (8.8)$$

This method was also introduces by Montecatini,[92] and the rate of chain termination was dependent upon the concentration of the alkyl. The alkylated catalyst was still active, but the polymer chain contained zinc residues and the addition of further qualities of diethyl zinc was constantly required. The same effect could be achieved by the addition of more co-catalyst but this was not as effective, possibly because triethyl aluminium exists as a dimer, and can form a complex with the catalyst. In the case of chain termination with the Phillips catalyst, the addition of cyclopentadiene was effective.

TABLE 8.9. Polypropylene Molecular Weight Control with Hydrogen.

Hydrogen (mol %)	Molecular weight (g mol^{-1})	Productivity (g g^{-1} catalyst)
0.1	6×10^5	1.5×10^4
1.0	4×10^5	2.5×10^4
10	2×10^5	1×10^4

- Proton donors: Chain transfer and chain termination are also possible using proton donors. However, the use of this method is limited by the high probability that any proton donor would deactivate the catalyst completely and that further chain initiation would not be possible.

The use of molecular hydrogen to control the molecular weight of the polymer has been the most important method to date, particularly in the polymerization of propylene. The polymer chains growing on catalysts derived from metallocenes, however, appear to be terminated by β-hydrogen abstraction and the molecular weight can be controlled by the design of the ligand framework.[83]

8.7.3. Molecular Weight

The variations in the chain length of the polymers produced by different types of catalysts arise because of several different reasons, as shown below:

- Different active sites on the catalyst operate at different rates of reaction.
- The sites on the catalyst are not necessarily all active at the same time.
- The ratio of the rate of propagation to the rate of termination changes as the length of the polymer chain increases.
- The chain growth reaction is retarded as the active center is covered by growing polymer.
- The chain termination reaction can be retarded if the hydrogen or metal alkyl used to control the molecular weight of the polymer is unable to access the active site for chain transfer.

To define a polymer structure completely, the following properties are important:

- Average molecular weight.
- Molecular weight distribution.
- Distribution and type of branching within each polymer chain.

A broad molecular weight distribution gives a high polymer density with better alignment of chains in the crystals, whereas a narrower molecular weight distribution gives lower density with more tangling of the chain. The insertion of co-monomers can provide some control over density by disrupting the crystallization of the polymer. When co-monomers are introduced, they may be more easily inserted at different sites. Insertion is retarded as the chain length of the polymer increases.

The average molecular weight of a given polymer has a different value according to the method by which the measurement was made. The weight average molecular weight (M_ω), is measured by light scattering, or by gel permeation chromatography, whereas the number average molecular weight (M_{an}), can be measured relatively easily by ebullioscopy, or by measurement of the osmotic

TABLE 8.10. Properties of Commercial Polyolefins.

Polymer	Structure (branches per 1000 carbons)	Density	Melting point	M_w/M_n [a]
LDPE	20–30	0.92	110–120	20–50
HDPE (Ziegler)	7	0.95	125–130	3–7
HDPE (Phillips)	Linear	0.96		7–15
PE (metallocene)	Linear			~2
PP (conventional)	Linear isotactic	0.91	160–165	5–12
PP (metallocene)	Linear isotactic		140–160	~2

[a]Ratio of weight average molecular weight and number (molecules) average molecular weight gives an approximate measure of polymer chain length variation.

pressure of the solution. The ratio, M_w/M_{an} is used to define the molecular weight distribution. The melt index, a measure of the ease of extruding heated polymer through a standard die under controlled conditions, also gives an indication of the range of the molecular weight.

In practical terms, the *useful* range of molecular weights for a commercial polymer is from 30,000 to 1,000,000 g mol^{-1}. Below 30,000, the softening point of the polymer tends to be too low, and the properties of polymers with a molecular weight above the higher limit are beyond the capacity of most powder extrusion equipment. The range of properties of some typical polyolefins is shown in Table 8.10.

Since the introduction of Ziegler–Natta and Phillips catalysts, polyolefin processes are now used to produce a broad range of commercial polymers with controlled molecular weight. The use of a variety of catalysts can control the density of the polyolefin and its processing properties. All typical operating modes—slurry, solution, and gas phase—are used to satisfy market demands while allowing rapid changeover from one polymer grade to another.

"Dr Ziegler himself explained that his discovery was not the direct outcome of attempts to solve a *set* problem. He set out simply to follow a broad course of study in which 'my only guide was initially just the desire to do something which gave me pleasure'. This course threw up many interesting conclusions, some of them of highly practical value, and one of these led ultimately to a method of making polyethylene."[96]

REFERENCES

1. N. J. Gaylord and H. F. Mark, *Linear and Stereoregular Addition Polymers*, Wiley-Interscience, NewYork, 1959.
2. J. Boor, Junior, Shell Development Company, *Ziegler-Natta Catalysts and Polymerization*, Academic, New York, 1979.

3. C. Kennedy, *ICI The Company That Changed Our Lives*, Hutchinson, London, 1986; W. J. Reader, *ICI History*, Vol. 2, Oxford University Press, 1975.
4. Imperial Chemical Industries, British Patent 471590 (1936).
5. P. H. Spitz, *Petrochemicals: The Rise of an Industry*, Wiley, New York, 1988.
6. K. Ziegler, E. Holzkamp, B. Breil and H. Martin, *Angew Chem* **67** (1955) 426; German Patent 973626 (1953); British Patents 799392, 799823 (1953).
7. J. P. Hogan and R. L. Banks, Belgium Patent 530617 (1955); US Patent 2825721 (1958); 2846425 (1958); 2951816 (1960); *Ind. Eng. Chem.* **48**, (1956) 1152.
8. Standard Oil Company of Indiana US Patent 2692258 (1951).
9. G. Natta, *Science* **147** (1965) 261; G. Natta , P. Pino, G. Mazzanti, and P. Longi, *Gazz. Chem. Italia.* **87** (1957) 570.
10. K. Ziegler and H. Martin, *Makromol Chem* **18/19** (1956) 186.
11. I. Pasquon and U. Giannini, *Catalytic Olefin Polymerization*, in Catalysts—Science and Technology, Vol. 6, Ed. by G. Anderson and M. Boudart, Springer-Verlag, Berlin, 1989; G. Natta, *Chim. Ind. (Milan)* **38** (1956) 751; G. Natta, I. Pasquon, and E. Giachetti, *Angew. Chem.* **69** (1957) 213.
12. K. B. Tripplett, *Evolution of Ziegler–Natta Catalysts for PropylenePolymerization*, in Applied Industrial Catalysts, Vol. 1, Ed. by B. E. Leach, Academic, New York, 1983; US Patents 3032510 (1962), 3128252 (1964).
13. E. Tornquist, *J Catal* **8** (1967) 189.
14. Solvay, US Patents 3769233 (1973), 4210738 (1980).
15. G. Natta and I. Pasquon, *Advances in Catalysis*, Vol. 11, Academic, New York, 1959, p. 1; G Natta, *J. Polym. Sci.* **34** (1959) 21.
16. P. J. T. Tait, in *Comprehensive Polymer Science*, Ed. by G. Allen and J. C. Berington, Vol. 4, Pergamon, Oxford, 1989, p.1.
17. P. J. T. Tait, *New Trends in Polyolefin Science and Technology*, Ed. by S. Hosada, p. 1; P. J. T. Tait in *Research Signposts,* Trivandrum, India (1996) p.1.
18. Stamicarbon, Belgian Patent 751315 (1969); Shell International Research, German Patent 2003075 (1970); British Patent 1299862 (1970); Belgian Patent 776301 (1970); Haward et al., *Polymer* **14** (1973) 365.
19. L. L. Bohm, *Polymer* **19** (1978) 553.
20. X. Youchang, G. Linlin, L. Wangi, B. Naiyo and T. Yougi, *Sci. Sin.* **22** (1979) 1045; P. Galli, P. C. Barbe, G. Guidetti, A. Zanbetti, A. Marigo, M. Bergozza and A. Fichera, *Eur. Polym. J.* **19** (1983) 19;
21. P. Galli, P. C. Barbe and L. Noristi, *Makromol. Chem.* **120** (1984) 73.
22. Montecatini Edison, British Patent 1286807 (1968); Mitsui Petrochemical Industries, Italian Patent 912345 (1968).
23. B. L. Goodall, in *Transition Metal Catalyzed Polymerizations*, Vol. 4 (Part A), Ed. by R. P. Quirk, Harwood, New York, 1981, p.355; Kashiwa, in

Transition Metal Catalyzed Polymerizations, Vol. 4 (Part A), Ed. by R. P. Quirk, Harwood, New York, 1981, p. 379.

24. P. J. T. Tait, G. H. Zohuri, A. M. Kells and I. D. Mckenzie, in *Ziegler Catalysts*, Ed. by G. Fink, R. Mulhaupt and H. H. Brintzinger, Springer-Verlag, Berlin, 1995, p. 343.

25. Toho Titanium Co., US Patent 4829037 (1989).

26. Amoco, US Patent 5081090 (1992).

27. G. Natta and I. Pasquon, *Adv. Catal.* **11** (1959) 1.

28. P. Galli and J. C. Haylock, *Macromol. Chem. Symp.* **63** (1992) 19; P. Galli and J. C. Haylock, *Prog. Polym. Sci.* **16** (1991) 443; P. C. Barbe, G. Cecchin, and L. Noristi, *Adv. Polym. Sci.* **81** (1987).

29. J. P. Hogan, The Phillips Petroleum Polyethylene Process in *Applied Industrial Catalysis*, Vol. 1, Ed. by B. E. Leach, Academic, New York, 1983, p. 149.

30. C. E. Marsden, Advances in Supported Chromium Catalysts, S4A/3/1 in *Plastics & Rubber Institute—Polyethylene: The 1990s and Beyond*, London, May 1992.

31. R. L. Banks, US Patent 3225023 (1965); J. P. Hogan, D. D. Norwood and C. A. Ayres, in *Applied Polymer Symposia Series*, Vol. 36, Ed. by Mark, Wiley-Interscience, New York, 1981, p. 49.

32. C. E. Marsden, The Influence of Silica Support on Polymerization Catalyst in *Preparation of Catalysts V*, Ed. by B. Delmon et al., Elsevier, Amsterdam,1991.

33. M. P. McDaniel, Supported chromium catalysts for ethylene polymerization, in *Adv. Catal.* **33** (1985) 47; M. R. Welch and M. P. McDaniel, *J. Catal.* **82** (1983) 110; M. P. McDaniel and Welch, US Patent 4182815 (1980).

34. M. P. McDaniel, *J. Polym. Sci. (Polym. Chem. Ed.)* **19** (1981) 1967.

35. K. Wisseroth (BASF), *Angew. Makromol. Chem.* **8**(88) (1969) 41; D. M. Rasmussen (Union Carbide), *Chem. Eng.* **79**(21) (1972) 104.

36. W. K. Jozwiak, I. G. Dalla Lana,W. Przyastaijko and R. Fiedorow in *Proc. 9th Int. Congress on Catalysts*, (1988) p. 1340.

37. S. Wang, P. J. T. Tait and C. E. Marsden, *J. Mol. Catal.* **65** (1991) 237; P. J. T. Tait, Advances in Ziegler and Related Catalysts, Paper S4A/2/1, *Plastics and Rubber Institute—Polyethylene: The 1990s and Beyond* (May 1992).

38. British Petroleum, British Patent 1429174 (1973).

39. Phillips Petroleum, US Patents 3887494 (1975), 3950316 (1976).

40. T. J. Pullucat, R. E. Hoff and M. Shida, *J. Polym. Sci. (Polym. Chem. Ed.)* **18** (1980) 2857; M. P. McDaniel, M. B. Welsh and M. J. Dreiling, *J. Catal.* **82** (1983) 118.

41. BASF German Patent 2604548 (1977); US Patents 4110522 (1978); 4128500 (1998).

42. Union Carbide, US Patent 4011382 (1978).

43. B. Rebensdorf and S. L. T. Anderson, *J. Chem. Soc. (Faraday Trans.)* **86** (1990) 3153.
44. Phillips, US Patent 3130188 (1964).
45. Phillips, US Patent 4820785 (1989).
46. Phillips, US Patent 5208309 (1993).
47. Phillips, US Patent 4818800 (1989).
48. F. J. Karol, G. L. Karapinka, A. W. Wu Ch Dow, R. N. Johnson and W. L. Carrick, *J. Polym. Sci. (Pt A-1)* **10** (1972) 2609, 2621.
49. E. A. Benham, P. D. Smith, E. T. Hsieh and M. P. McDaniel, *J. Macromol. Sci. Chem. A* **4**(25) (1988) 259.
50. H. N. Friedlander, *J. Polym. Sci.* **38** (1959) 91; Juveland, Peters, and J. W. Shephard, *Polym. Repr.* **10** (1969) 263; Tabokoro et al., *Kogyo Koga Kuaschi* **70** (1967) 144; Juveland and Peters, French Patent 1521017 (1968).
51. D. D. Norwood, US Patents 3248179, 3257362 (1966); J. P. Hogan, D. D. Norwood and C. A. Ayres, *Applied Polymer Symposia Series*, Vol. 36, Ed. by Mark, Wiley-Interscience, New York (1981); Phillips Linear Polyethylene (3A/2/1), *Plastics and Rubber Institute—Polyethylene: The 1990s and Beyond*, London (May 1992).
52. Solvay, Belgian Patent 570981 (1958).
53. D. Newton, J. C. Chinh and M. Power, *Hydrocarbon Processing*, (March 1998) 86.
54. DuPont Canada Ltd., Sclairtech Solution Process (53A/3/1), *Plastics and Rubber Institute—Polyethylene: The 1990s and Beyond*, London (May 1992).
55. D. M. Rasmussen, *Chem. Eng.* **79** (21) (1972) 104; US Patents 3642749, 3687920; L. P. McMaster, The Gas Phase Process (53A/1/1), *Plastics and Rubber Institute—Polyethylene: The 1990s and Beyond*, London, May 1992.
56. K. Wisseroth, *Angew Makromol. Chem.* **8** (1969) 41; Muller-Tamm, *Soc. Plast. Eng. Tech.* **15**, (1969) Paper 27; US Patents 3300457 (1966), 3634382 (1971), 3639377 (1971), 3652527 (1972), 4212847 (1976).
57. US Patents 3965083, 3971768 (1976).
58. P. M. Morse, *Chem. & Eng. News*, (Dec. 1998) 25; A. I. Tullo, *Chem. & Eng. News*, (Aug 7, 2000) 35.
59. A. A. Montagno and J. C. Floyd, *Hydrocarbon Processing*, (March 1994) 57.
60. J. L. Hemmer, High Pressure Exxpol Technology (S2A/3/1), *Plastics and Rubber Institute—Polyethylene: The 1990s and Beyond*, London (May 1992); Exxon Chemical US Patents 5198401 (1991), 5384299 (1993), 5470927 (1994), 5324800 (1994), 5599761 (1997).

61. J. Krieger, Inventor of the Year Award (Dow) *Chem. Eng. News*, May 23 (1994) 6; Dow Chemical US Patents 5272236 (1993), 5470993 (1995), 5278272 (1994).
62. D. G. H. Ballard, W. H. James and J. D. Seddon, British Patent 1099116 (1968); D. G. H. Ballard and T. Medinger, British Patent 1145958 (1969); D. G. H. Ballard, T. Medinger and W. G. Oakes, German Patent 1904878 (1969).
63. G. Wilke et al., *Angew Chem. Intern. Ed.* **5**(2) (1966) 151.
64. G. Natta, G. Pino,G. Mazzanti,U. Giannini, E. Mantica and M. Peraldo, *Chim. Ind. (Milan)* **39**, 19G (1957); G. Natta, G. Pino, G. Mazzanti, U. Giannini, *J. Amer. Chem. Soc.* **79** (1957) 2975.
65. D. S. Breslow and N. R. Newbury, *J. Amer. Chem. Soc.* **79** (1957) 5072; D. S. Breslow and N. R. Newbury, **81** (1959) 81; W. P. Long and D. S. Breslow, *J. Amer. Chem. Soc.* **82**, (1960) 1953.
66. Dyachkovskii, Shilova, and Shilov, *J. Polym. Sci., Part C*, (1967) 2333; Eisch et al., *J. Amer. Chem. Soc.* **107** (1985) 7219.
67. A. Schindler in *Crystalline Olefin Polymers*, Ed. by R. A. Ruff and K. W. Doak, Wiley-Interscience, New York, 1965, p. 163; Zavorokhin, *Trans. Inst. Khim. Nauk. Akad. Nauk. Kaz.*, SSR **23** (1969) 3.
68. K. H. Reichert and K. R. Meyer, *Makromol Chem* **169** (1973) 163.
69. W. R. Long and D. S. Breslow, *Liebigs. Ann. Chem.* **1975** (1979) 463.
70. W. Kaminsky, J. Kopf, H. Sinn and H-J. Vollmor, *Angew. Chem.* **88** (1976) 688; H. Sinn, W. Kaminsky, H. J. Vollmor and R. Woldt, *Angew. Chem.* **92** (1980) 396; W. Kaminsky in *History of Polyolefins*, Ed. by R. B. Seymour and T. Cheng, Reidel, Dordrecht, 1986, p. 257; W. Kaminsky, *Nachr. Chem. Tech. Lab.* **29** (1981) 373.
71. G. G. Hlatky, R. R. Eckmann and H. W. Turner, *Organometallics* **11** (1992) 1413; Exxon, US Patent 5599761 (1997).
72. J. A. Ewen, *J. Amer. Chem. Soc.* **106** (1984) 6355.
73. W. Kaminsky, H. H. Brintzinger, K. Kulper and F. R. W. P. Wild, *Angew. Chem. Int. Ed.* (English) **24** (1985) 507.
74. M. Antberg, V. Dolle, R. Klein, J. Rohrmann, W. Spaleck and A. Winter, *Studies Surf. Sci. Catal.* **56**, (1990) 501; W. Spaleck, M. Antberg, V. Dolle, R. Klein, J. Rohrmann and A. Winter, *New J. Chem.* **14** (1990) 499; W. Spaleck, A. Winter, W. A. Hermann, J. Rohrmann and E. Hertweck, *Angew. Chem.* **101** (1989) 1536; P. Burger, K. Hortmann and H. H. Brintzinger, *Makromol. Chem. Symp.* **66** (1993) 127.
75. R. Mülhaupt, Novel polyolefin materials and processes in ziegler catalysts, Ed. by G. Fink, R. M̈ulhaupt and H. H. Brintzinger, Springer-Verlag, Berlin, 1995, p.42.
76. *Chem. Eng. News* (June 28,1999) 23.
77. Spalec et al., in *Ziegler Catalysts*, Ed. by Fink, Mulhaupt, and Brintzinger, Springer-Verlag, Berlin, 1995, p. 83.

78. P. M. Morse, *Chem. Eng. News*, (Dec 7, 1998) 25.
79. J. Bleimeister, W. Hagendort, A. Harder, B. Heitmann, I. Schimmel, E. Schmedt, W. Schnuchel, H. Sinn, L. Tikwe, N. von Thienen, K. Urlass, H. Winter and O. Zarnke, The role of MAO-activators, in: *Ziegler Catalysts*, Ed. by G. Fink, R. Mülhaupt and H. H. Brintzinger, Springer-Verlag, Berlin, 1995, p. 57; US Patents 4544762, 5015749, 5041584-5, 5542199.
80. Bochman and Wilson, *JCS Chem. Comm.* (1986) 1610.
81. Exxon, US Patent 5599761 (1997).
82. W. J. Kruper, D. R. Wilson and E. Y-X. Chen, *J. Amer. Chem. Soc.* **123** (2001) 745; M. C. Jacoby, *Chem. Eng. News*, (Feb 19, 2001) 57.
83. R. M"ulhaupt, Novel Polyolefin Materials and Processes: Overview and Prospects, in: *Ziegler Catalysts*,Ed. by G. Fink, R. Mülhaupt, and H. H. Brintzinger, Springer-Verlag, Berlin, 1995, p.45.
84. J. Haggin, *Chem. Eng. News*, (Feb 5, 1996) 6.
85. M. Freemantle, *Chem. Eng. News*, (April 13, 1998) 11.
86. F. Wilson, *Chem. Eng. News*, (April 10, 2000) 8; (March 6, 2000) 11; (Jan 24, 2000) 15.
87. T. K. Woo, L. Fan, T. Ziegler, A Combined Density Functional and Molecular Mechanics Study on Olefin Polymerization by Metallocene Catalysts in *Ziegler Catalysts*, Ed. by G. Fink, R. Mülhaupt and H. H. Brintzinger, Springer-Verlag, Berlin, 1995, p. 291.
88. P. Cossee, *Rec. Trav. Chim. Pays. Bas.* **85**, No. 9–10 (1966) 1152; P. Cossee, The Mechanism of Ziegler-Natta Polymerization, in: *The Stereochemistry of Macromolecules*, Vol. 1, A D Ketley editor, Marcel Dekker, New York, 1967, p. 145.
89. J. Boor, *Ziegler–Natta Catalysts and Polymerizations*, Academic Press, New York, 1979, p. 389.
90. E. J. Arlmann, *J. Catalysis* **3** (1964) 89; E. J. Arlmann and P. Cossee, *J. Catalysis* **3** (1964) 99.
91. J. Boor, *Ziegler–Natta Catalysts and Polymerizations*, Ch. 10, Academic, New York, 1979, p. 244.
92. Hercules, US Patent 3051690 (1962—applied July 1955).
93. Montedison, Italian Patents 554013, 557013 (1957); British Patents 584794, 850585; G. Natta, *Chim. Ind. (Milan)* **41** (6) (1959) 519.
94. G. Natta and I. Pasquon, *Advances in Catalysis* **11** (1959) 1.
95. K. D. Hungenberg, J. Kerth, F. Langhanser, B. Marczinde and R. Schlund, *Gas Phase Polymerization of Olefins with Ziegler-Natta and Metallocene/Aluminoxane Catalysts. A Comparison in Ziegler-Natta Catalysts,* Ed. by G. Fink, R. Mülhaupt and H. H. Brintzinger, Springer-Verlag, Berlin, 1995, p. 363.
96. Jewkes, Sawers, and Stillerman, *The Sources of Invention*, Macmillan/St. Martin's, New York, 1962.

9

SYNTHESIS GAS

Synthesis gas is a general term used in the chemical industry to describe a given feedstock. Unfortunately the term means different gas compositions to different groups of people and this can lead to confusion. In this chapter, the term *synthesis gas* is used to describe the gaseous feed used specifically for the manufacture of ammonia.

The replacement of charcoal by coke in the smelting of iron ores was a key step in establishing coal as a primary raw material for the manufacture of chemicals. Coke was produced via the *destructive distillation* of coal. The by-products included *Town Gas*, and a wide range of organic and inorganic compounds which could be used in a variety of ways. Coal gas was used for domestic illumination and, until natural gas became widely available, gas works were important catalyst users mainly for gas purification. Coal tar contained organic chemicals which were developed as intermediates in catalytic processes by the emerging chemical industry. Ammonium sulfate was also recovered from the process, and this became an important source of nitrogenous fertilizer. It was also used in conjunction with phosphates that were extracted from *phosphate rock* with sulfuric acid, to produce compound fertilisers. Prior to the recovery of ammonia from coke ovens, it was only available from natural wastes. Consequently, up to about 1920, most nitric acid was made from nitrates imported from South America (Chile saltpetre), even though it was already known that nitrates could be prepared via the catalytic oxidation of ammonia. By this time, the coal-based water gas and producer gas processes for the manufacture of town gas had been adapted to provide the synthesis gas needed for the Haber process, and thus coal became the principal raw material for the manufacture of ammonia.

L. Lloyd, *Handbook of Industrial Catalysts*, Fundamental and Applied Catalysis,
DOI 10.1007/978-0-387-49962-8_9, © Springer Science+Business Media, LLC 2011

9.1 AMMONIA SYNTHESIS GAS

The importance of synthetic ammonia, when it was first produced by BASF in 1913, cannot be over emphasized.[1] Early development of the Haber process and the widespread use of ammonia throughout the world will be described in Chapter 10.

Since the successful commissioning of the first plant, the Haber-Bosch process for the manufacture of ammonia has required increasingly large volumes of synthesis gas as the demand, initially for explosives, and subsequently, fertilizers continued to grow. For about 20 years, synthesis gas was made mainly from coke in water gas and producer gas units, as shown empirically below:

$$C + H_2O \rightarrow CO + H_2 \qquad \text{endothermic} \qquad (9.1)$$

$$2\,C + O_2 + 4\,N_2 \rightarrow 2\,CO + 4\,N_2 \qquad \text{exothermic} \qquad (9.2)$$

The oxides of carbon are poisons to the iron-based ammonia synthesis catalyst and must be removed from the synthesis gas before use. An important reaction that affected the economics of the manufacture of synthesis gas was the conversion of carbon monoxide into equal volumes of hydrogen and carbon dioxide by reaction with steam. This procedure, now commonly referred to as the water-gas shift reaction required a catalyst based on iron and chromium oxides.

$$CO + H_2O \rightarrow CO_2 + H_2 \qquad (9.3)$$

The carbon dioxide was washed out of the process gas with water at high pressure and could eventually be recovered as a useful byproduct. Residual carbon monoxide was usually removed with an ammoniacal copper solution before the final stage of ammonia synthesis.

The discovery of a *shift* catalyst by BASF was an important part of the process development because it recovered the *potential hydrogen* from carbon monoxide.[2] Originally known as the BAMAG method (Berlin Anhaltische Maschinenbau Aktien Gesellschaft) the process has been used in every subsequent major ammonia plant. There have been few changes in the catalyst since the first application in 1914, other than more complete removal of impurities introduced during catalyst manufacture, such as sodium and sulfate ions, and the use of more stable formulations at increasingly high operating pressure. Operating conditions for the first shift unit at Oppau are summarized in Table 9.1 and must be amongst the first to be reported in the early literature.[3]

TABLE 9.1. The BAMAG Carbon Monoxide Conversion Process.

Feed vol%	Inlet Catalyst	Outlet Catalyst
CO_2	3–5	27–30
CO	35–40	2–4
H_2	33–36	50–52
N_2	22–23	16–18
Inerts	<1	<1

Feed gas composed of 2–3 volumes of water gas combined with 1–2 volumes producer gas to give required N_2/H_3 ratio.

Gas rate (m^3h^{-1})	70,000
Temperature (°C)	400–600
Square reactor (ft)	16x12(cross section)x10 deep
Catalyst beds (two) (ft)	Each 16x12(cross section)x3 deep
Catalyst volume (m^3)	30–35
Conversion (%)	>90
Life (years)	2

Green, *Industrial Catalysis*, Benn, London, 1928, p. 327.
Partington, *J. Soc. Chem. Ind.* **46** (1921) 99R.

The strategic importance of ammonia, particularly in the production of explosives, led to competitive processes being developed in other parts of Europe and North America. From the mid-1920s onward, it became clear that the worldwide demand for fertilizers was becoming stagnant and further large-scale expansion in capacity was not justified. This meant that process improvements were delayed and even as late as 1937 the huge German wartime plants were still providing about 70% of worldwide ammonia requirements.[4] Typical ammonia plant capacity during this period was only 50–150 tonnes per day.

9.1.1 Process Developments

As a result of the post-war slump in the 1920's, there was a momentum in both Germany and the United Kingdom to be able to use surplus high-pressure equipment in more effective ways. For example, the high-pressure catalytic process for the hydrogenation of coal, introduced by Bergius in 1913, was eventually commercialized. However, the coal hydrogenation process provided not only the anticipated liquid hydrocarbons to be used as gasoline but also large volumes of byproduct off-gas.

Efforts were then made to improve the overall efficiency of the process by using the off-gas as a feedstock for the manufacture of hydrogen, to reduce the use of coke in the ammonia process. These efforts led to work on hydrocarbon steam reforming. A practical steam reforming process, using a solid nickel catalyst in a simple tubular furnace, was soon developed by BASF. The process was used for the first time by Standard Oil of New Jersey, during 1931, at their Baton Rouge, Louisiana, and Bay Way, New Jersey,[5] refineries in early work on the hydrocracking of gas oil to produce gasoline. During full scale operation of the steam reforming process, it was observed that the nickel catalysts suffered from poisoning by sulfur compounds still present in the feed gas, and from the deposition of carbon caused by olefin impurities. An improved furnace design, containing tubes packed with a better nickel catalyst in the form of raschig rings, was introduced by ICI in 1936 to make hydrogen from their own petrol plant off-gas.[6] The catalyst was protected from poisons by desulfurizing the feed with zinc oxide granules. Later on a cobalt molybdate catalyst was used, together with the zinc oxide, to decompose the less reactive organic sulfur compounds. By the 1940s it was recognized that methane could also be an excellent feed for steam reformers in hydrogen or ammonia plants close to sources of natural gas.

Once again the onset of war led to an increased demand for ammonia and several ordnance plants, using the ICI methane steam reforming process, were built in North America.[7] These plants used the latest tubular reformer design, with zinc oxide for feed desulfurization, and the raschig ring nickel catalyst supported on a cement based refractory. It is probable that the plants also included a reforming vessel, with raschig ring nickel catalyst, to burn some of the hydrogen in reformed gas with air to add nitrogen and produce synthesis gas.[8] Since a *secondary* reformer became part of the process, the original steam reforming furnace has been called the *primary* reformer. The only other catalysts used in these early plants were the original HTS catalyst, for the water gas shift reaction, and an ammonia synthesis catalyst.

From 1950, the demand for nitrogen fertilizers in North America led to the construction of many more ammonia plants all based on the steam reforming process. Modifications to the primary reforming catalysts by the incorporation of potash to reduce the level of carbon deposition have enabled operators in those parts of the World with no readily available supply of natural gas to use naphtha or refinery off-gases as feed for the primary reformer, and this has increased the versatility of the process even further.[9]

9.1.2   Increased Ammonia Production by Steam Reforming

Since 1945 the capacity of ammonia plants has grown to more than 2000 tonnes per day and the operating pressure of the steam reformer has been increased from atmospheric to about 30–40 atmospheres.

As the production capacity of ammonia plants grew from 50-100 tonnes per day up to 2000 tonnes per day the process was considerably improved. A more efficient furnace design was required which incorporated improved steel alloys able to operate at higher pressure. At the increased operating pressure, silica is volatile in steam. This required that the original catalysts, which were made by precipitation and which were based on hydrous silica cements, needed to be replaced by a new range of nickel-impregnated, silica-free catalysts. By 1966 large capacity, high-pressure single stream plants producing up to 1000 tons per day of ammonia were being built all over the world. These plants, however, had evolved in stages as a range of new catalysts and equipment which could operate a higher pressure became available.

Different versions of primary reforming and HTS catalysts, together with several new catalysts able to withstand the severe operating conditions, were gradually introduced. The rapid development of these new catalysts, designed for complex plants, led to an increase in the number of chemical engineering contracting companies able to supply *off the peg* process design and an extensive catalyst technical service for operators. *Commodity* catalysts were soon available from a number of specialist catalyst companies, which were not always in the in the ammonia business themselves. These companies needed to carry out their own catalyst research and plant design, so that such activities were no longer the preserve of the large chemical companies, other than when processes were provided under a strict license agreement. Intense competition developed among the new suppliers for the provision of new catalysts to independent operators, in what became a large and profitable business sector. A consequence of the rapid catalyst development, which was needed to solve operational problems, was that most of the competing catalyst suppliers evolved somewhat different production procedures and catalyst formulations.

The more severe process conditions that were being developed imposed greater stresses, both physical and chemical, on the catalysts and the economic penalty of premature catalyst failure required that catalyst reliability was of paramount importance in successful plant operation. Improvements in the testing of catalysts was needed to provide a better understanding of the various modes of catalyst failure, to meet the exacting standards now required by the users.

The major process changes and the new catalyst types which have been introduced since 1920 are listed in Table 9.2. Relatively slow technical developments from 1920 up to about 1950 correspond to the period when coal was the source of synthesis gas. Rapid changes between 1950 and 1970 reflect the expansion of the fertilizer and chemical industries as crude oil, natural gas, and petrochemicals became so important. Increasing the level of fertilizer production from natural gas was just as crucial as the new range of petrochemical products supporting the tremendous increase in world population.

9.2 MODERN AMMONIA PLANTS

Up to eight different catalysts and absorbent materials are regularly used in a modern ammonia plant to produce synthesis gas, and the design of any given process is determined by the type of feedstock available. A wide range of different catalyst compositions, shapes and sizes is now available to make operation more reliable and economic. The selection of the best choice of catalysts for a given process is made more complicated by the fact that several major suppliers

TABLE 9.2. Ammonia Synthesis Gas Process Developments since 1920.

Period	Process	Catalyst Developments
Pre 1930	Coal gasification.	Desulfurization evolving (bog iron ore).
1930–40	Steam reforming developments for hydrogen production.	Zinc oxide sulfur absorbent; raschig ring reforming catalyst; secondary reforming (in electrolytic hydrogen plants).
1940–55	Steam reforming process used in ammonia plants. Natural gas feed.	Sulfur absorption on activated carbon; secondary reforming; high temperature shift catalysts used. Reformer-pressure increasing from atmospheric to 9bar. Plant capacity increasing from 150 tpy to 300 tpy.
1960–63	Reforming process improvements and alkalized catalysts introduced. Use of naphtha feed.	Two-stage HTS conversion lowered CO slippage to 1.0% and methanation included. Plant capacity increased from 300 tpy to 600/1000 tpy. LTS with methanation replaced need for old CO scrubbing processes. Operating pressure at least 30 atm.

offer different competitive products, each with its own set of advantages and disadvantages. The complete range of catalyst types used in ammonia production and typical operating lives are shown in Table 9.3. The overall process can be divided into four parts and has been described previously:[10]

- Feed purification. This can include hydrogenation catalysts to convert any organic sulfur, nitrogen or chlorine to hydrogen sulfide, ammonia or hydrogen chloride, respectively, prior to removal by a suitable absorbent. Early processes based on water gas/producer gas derived from coal had to

TABLE 9.3. Introduction of Ammonia Plant Catalysts and Typical Operating Life.

Catalyst	Introduction	Approximately operating life years
Desulfurization:		
Activated carbon	1940	6
Zinc Oxide	1930	1–4 (depending on S content)
Cobalt molybdate	1930	5–10
Primary reforming:		
Precipitated	1920s	2–10 (decreasing with severity)
Impregnated	1970s	2–5
Shapes	1980s	2–5
Alkalized	1959	2
Secondary reforming	1940+	4–8
Carbon monoxide removal:		
High temperature	1912	2–4
Low temperature	1963	2–4
Methanation	1960	5–10
Catalyst costs estimated at ~1.5% of total production costs.		

deal with relatively large amounts of sulfur. The sulfur was removed in very large *oxide boxes* which contained bog iron ore. Small, low-pressure ammonia plants built before the introduction of the single-stream processes often used activated carbon absorbents to remove the small amounts of impurities present in natural gas. Modern plants use cobalt/molybdenum hydrodesulfurisation catalysts to convert organo-sulfur compounds to hydrogen sulfide prior to absorption by zinc oxide.

- Reforming. Hydrocarbons are converted catalytically to a mixture of hydrogen, nitrogen and oxides of carbon in two separate stages, namely primary and secondary reforming.
- Oxygenate removal. Carbon monoxide is converted catalytically to hydrogen and carbon dioxide in two separate stages, the high- and low-temperature shift reactions. Carbon dioxide is removed by any of the available proprietary processes, and traces of residual carbon monoxide are converted to methane over a nickel methanation catalyst. Any water formed in this stage can be removed by molecular sieves, or by washing with product ammonia.
- Ammonia Synthesis. This gives few problems in modern plants because the synthesis gas is extremely pure. Although the iron synthesis catalyst has not changed significantly since it was first used by BASF, a variety of different converters, using pre-reduced catalyst, now make the process more economic. A new, more active, catalyst made by impregnating an active carbon support with ruthenium has been developed.

9.3 FEEDSTOCK PURIFICATION

Hydrocarbons that can be fed to ammonia plants include natural gas, associated gas, liquid petroleum gas, and naphthas boiling up to 220°C. Higher hydrocarbons are not used in primary steam reforming because it would lead to coke formation on the catalysts. Hydrocarbons are usually contaminated with variable quantities of different sulfur compounds and often contain chlorides. These catalyst poisons must be removed before the other catalysts in the plant can operate in a satisfactory manner.

Impurities in natural gas are usually simple sulfur compounds such as hydrogen sulfide and mercaptans and are easily removed. Where the gas supply is also used for domestic purposes a *stenching* agent such as a thioether or thiophene may have been added. These less-reactive compounds have to be subjected to a hydrogenolysis treatment before they can be removed. Naphthas contain several hundred parts per million of more complex organic sulfur compounds and may be pre-treated in a hydrofiner before use in ammonia or hydrogen production. Variation in the sulfur content of feed does not affect the performance of the hydrodesulfurization catalysts and absorbents although larger volumes of absorbent must be included in the plant design to achieve a convenient operating cycle before replacement.

9.3.1 Activated Carbon

High-surface area, activated carbon has an affinity for some organo-sulfur compounds and can be used as an absorbent for sulfur in chemical processes. For example, some of the early, low pressure ammonia plants operating in North America used two parallel beds of activated carbon to adsorb any sulfur impurities from natural gas. One bed operated while the second bed was being regenerated with steam to allow continuous ammonia production. However, mercaptans and carbonyl sulfide are not strongly bound to activated carbon and can be displaced by higher molecular weight hydrocarbons in the gas stream. To improve adsorption efficiency the carbon was impregnated with iron or copper oxides which combined with sulfur but could be easily regenerated. Even so, sulfur retention was not reliable enough for the activated carbons to be used in modern high-pressure ammonia plants. The high cost of regular regeneration, plus sulfur poisoning of downstream catalysts, was excessive, as was the need to replace downstream catalysts poisoned by sulfur, and carbon was soon replaced by beds of more reliable zinc oxide. Zinc oxide beds were well established after having been used in hundreds of town gas, ammonia and hydrogen plants based on the naphtha steam reforming process. Since it replaced activated carbon zinc oxide has been almost exclusively used to protect steam reforming units.

9.3.2 Hydrodesulfurization

Zinc oxide can absorb simple sulfur compounds such as hydrogen sulfide and mercaptans to form zinc sulfide with mercaptans however some carbon may also be deposited. Other organic-sulfur compounds are not absorbed completely, and thermal cracking of these compounds results in the deposition of carbon onto the zinc oxide, thus reducing its absorption capacity. For these reasons, particularly when *refractory* organic sulfur compounds such as thiophene or thioethers are present in the feed, it is usual to add hydrogen and to include a bed of cobalt molybdate catalyst to hydrogenolyse the sulfur compounds to hydrogen sulfide, which can then be absorbed in the bed of zinc oxide. The cobalt/molybdate component is sulfided during commissioning to the active form and then operates in the same way as in refinery hydrotreating. Although cobalt/molybdate catalyst is the usual choice, nickel molybdate can also be used. This is important when the hydrogen stream contains a high proportion of carbon oxides. Unsulfided cobalt/molybdate can catalyse the methanation of carbon oxides at temperatures higher than about 300°C, and this may lead to excessive bed temperature while commissioning a new catalyst charge. This is less likely with the nickel catalyst. With liquid feeds, the catalyst can be presulfided at an appropriate low temperature.

It is not usually necessary, however, to presulfide cobalt molybdate catalysts before treating natural gas. Natural gas contains relatively small amounts of simple sulfur compounds that are hydrogenolyzed at a low temperature. Catalysts are sulfided during operation during operation and the eventual sulfur content of the catalyst depends on the sulfur content of the natural gas. Catalyst will normally operate for several years with no loss of activity. The sulfur content of typical feeds is shown in Table 9.4.

At the end of life in a typical ammonia plant, the cobalt/molybdate catalyst will contain 0.5–3.0wt% sulfur and 2-10wt% carbon. The carbon content may sometimes be higher than this but has no significant effect on performance. Hydrodesulfurization catalysts are usually pyrophoric after use because of adsorbed hydrogen and carbon deposits. Before catalysts are discharged the reactor must, therefore, be flushed with an inert gas such as nitrogen until the catalyst temperature has fallen to less than 30°C. Large quantities of dust or carbon deposits on the top of the catalyst layer can be sucked off and any broken or contaminated pellets replaced with fresh catalyst. If the catalyst bed is to be completely removed then a small volume of air can be added to the circulating nitrogen as the catalyst cools. Adsorbed hydrogen or hydrocarbons are oxidized. This will lead to a temperature increase and this can be controlled by limiting the volume of air added. The catalyst can then be discharged at ambient temperature after cooling.

TABLE 9.4. Sulfur Content of Typical Steam Reformer Feedstocks.

Natural Gas	North Sea	Holland	Associated Gas
Hydrocarbon/sulfur			
CH_4 (vol%)	81–94	81–82	70–75
C_2H_6 (vol%)	4–5	3	12–15
C_3H_8 (vol%)	0.1–0.5	0.5	5–8
C_4+ (vol%)	0.08	0.2	5–9
CO_2 (vol%)	0.2–0.5	1	Traces
N_2 (vol%)	0.8	14.3	–
Sulfur	5–30 ppm	1 ppm	1–3%
Naphtha			
Specific gravity	0.71		0.74
Initial boiling point (°C)	33		43
Final boiling point (°C)	172		214
Total sulfur ppm (w/w)	230–260		1500
RSH	70–80		200–300
R_2S_2	10–20		100–150
R_2S	毂00–130		700–800
Average molecular weight	~100		~110

It is not economic to regenerate discharged catalysts for further use following a typical life of several years.

9.3.3 Chlorine Removal

The presence of chlorides in the feed gas can cause the zinc oxide to sinter as they are absorbed to form zinc chloride. Any hydrogen chloride formed in the hydrodesulfurizer can be removed from the gas by installing a bed of alkalized alumina in front of the zinc oxide. After this treatment the natural gas usually contains less than 0.1mg chlorine per cubic meter. An average of about 10–12 wt% chloride can be absorbed before the alkalized alumina is saturated. Platformer hydrogen containing up to 2 ppm HCl is commonly treated in this way. Only a relatively small bed of absorbent is required at a space velocity of 10,000–15,000 h^{-1}.

9.3.4 Sulfur Absorption

Zinc oxide has been used to remove sulfur compounds from hydrocarbons since the 1930s, when the steam reforming process was first introduced. When a zinc oxide composition is specially prepared to have a high surface area and high degree of porosity, it can absorb more than 20 wt% sulfur in a single bed. It was the preferred choice in the early days, because the partial pressure of hydrogen sulfide at equilibrium under reaction conditions, particularly in the presence of water vapour, is very small compared to that of the bog iron ore used previously to purify water gas/producer gas.

This observation was confirmed during the 1960s when *luxmasse*, a cheap form of iron oxide, was investigated as an absorbent to desulfurize naphtha feed in the British Gas Council Catalytic Rich Gas (CRG) process. Under typical operating conditions of 370-400°C and with a vapour phase water concentration of 0.2–0.3%, the concentration of hydrogen sulfide in the vapour at equilibrium is less than 3×10^{-3} ppm when using zinc oxide, compared with 0.2–0.3 ppm when using luxmasse. The data presented in Table 9.5 shows an increasing differential as the concentration of water vapour increases. A further disadvantage of luxmasse is that it is also reduced to magnetite before conversion to iron sulfide by reaction with hydrogen sulfide. Thus more water vapour is produced by the conversion of luxmasse to iron sulfide, than is the case in the conversion of zinc oxide to zinc sulfide:

$$3 \text{ Fe}_2\text{O}_3 + \text{H}_2 \rightarrow 2 \text{ Fe}_3\text{O}_4 + \text{H}_2\text{O} \tag{9.4}$$

$$2 \text{ Fe}_3\text{O}_4 + 8 \text{ H}_2\text{S} \rightarrow 2 \text{ Fe}_3\text{S}_4 + 8 \text{ H}_2\text{O} \tag{9.5}$$

TABLE 9.5. Equilibrium Hydrogen Sulfide Concentrations.

Water	Equilibrium Concentration H_2S (ppm)					
(vol%)	200°C		300°C		400°C	
	ZnO	Fe_2O_3	ZnO	Fe_2O_3	ZnO	Fe_2O_3
0.17	1.3×10^{-5}	1.57×10^{-3}	3.0×10^{-4}	9.3×10^{-2}	3.3×10^{-3}	1.96×10^{-1}
0.33	2.6×10^{-5}	8.5×10^{-3}	7.0×10^{-4}	2.0×10^{-1}	6.5×10^{-3}	4.9×10^{-1}
1.7	1.3×10^{-4}	7.0×10^{-1}	3.0×10^{-3}	2.0	3.2×10^{-2}	4.23

Overall:

$$3\ Fe_2O_3 + H_2 + 8\ H_2S \rightarrow 2\ Fe_3S_4 + 9\ H_2O \qquad (9.6)$$

Whereas:

$$6\ ZnO + 6\ H_2S \rightarrow 6\ ZnS + 6\ H_2O \qquad (9.7)$$

Thus six atoms of iron will produce nine molecules of water compared with six atoms of zinc, which only produce six. Furthermore iron sulfide can release hydrogen sulfide by reaction with hydrogen and steam during plant start-up or shut-down conditions. In practice, therefore, when iron oxide is used it is always followed by a guard bed of zinc oxide.

9.3.4.1 Operation with Zinc Oxide

In general, more sulfur is absorbed by zinc oxide at higher operating temperatures, and when the porosity of the zinc oxide is increased. Consequently, operating temperatures are usually greater than 300°C, and porous particles are preferred. However, the techniques needed to produce porous zinc oxide also lead to a decrease in particle strength, which results in fracture of the particles during loading and service, which, in turn leads to an increase in pressure drop across the bed. Furthermore, there is also a decrease in the bulk density of the bed, meaning that a lower mass of zinc oxide is charged to the reactor, and thus the maximum amount of sulfur that can be absorbed by the bed also decreases proportionally. This means that it is possible for a fixed bed size of highly porous zinc oxide to absorb less sulfur than an equivalent bed of less porous, but more densely packed zinc oxide. The porosity of granules must therefore be optimized to give maximum absorbing efficiency and particle strength in any application within the temperature range from ambient to 400°C. Typical operating conditions are given in Table 9.6.

Up to about 10 ppm of low molecular weight mercaptans can be removed from natural gas without hydrogen addition, provided that the reaction tempera-

TABLE 9.6. Sulfur Absorbent Operation.

Operating Conditions		Zinc Oxide Bed	
Temperature (°C)		370–400	
Pressure (atm)		Up to 40	
Space velocity (h^{-1})		500–1000	
	400°C	300°C	200°C
Sulfur capacity wt%S (before slip observed)	25–30	20–25	8–10

By increasing the porosity of the absorbent the sulfur absorption can be increased by 200% in the range ambient to 100°C and almost 100% at 200°C.

---

ture is high enough to crack the sulfur compound. Carbon deposition does not immediately affect the sulfur absorption and a reasonable life can be achieved. For a longer life, or with increasing concentrations of organic sulfur compounds, a few percent of hydrogen is added to the gas stream being treated. This is routinely done in steam reforming plants which also nowadays include a bed of cobalt molybdate catalyst. Hydrogen addition, in this case, is also beneficial because it helps to reduce the reforming catalyst at the top of the steam reformer tubes.

Carbonyl sulfide may be formed during natural gas treatment by the reaction of hydrogen sulfide and carbon dioxide and this can be difficult to remove completely using zinc oxide alone. This is not a problem if a bed of cobalt molybdate catalyst is included in the desulfurizer. It has been shown that lead oxide is an efficient absorbent of carbonyl sulfide so the natural lead oxide impurity in some zinc oxides may promote absorption, especially if traces of water vapour are present to hydrolyze the carbonyl sulfide.

The operating temperature and space velocity through the beds of cobalt molybdate and zinc oxide are chosen to give optimum performance for the system as a whole. The maximum inlet temperature to the bed containing the cobalt/molybdate is usually about 400°C and the volume of zinc oxide is calculated to give an acceptable period of time on line prior to discharge, for a hydrocarbon feed with given sulfur content. This usually results in a space velocity of about 500-1000 hr^{-1} in both beds and an operating life for the zinc oxide of at least one year.

It can be more economical to have two beds of zinc oxide in series, with a by-pass line, so that one bed can be changed with the other still operating. This procedure allows an average of 25–30wt% sulfur pick-up during the period of operation. With one bed the average pick-up will probably be about 20%. Zinc oxide can desulfurize hydrocarbons at ambient temperature although the total sulfur pick-up will only be about 5% before saturation.

9.3.4.2 *Preparation of Zinc Oxide*

Zinc oxide absorbents are supplied in a variety of shapes such as spheres, extrudates, or pellets. These are generally made by granulation or extrusion of the zinc oxide with small amounts of a suitable binding agent such as a cement. The final zinc oxide content ranges from 90-100% and the bulk density from about 0.9–1.5 kg liter^{-1}. High density catalysts do not necessarily have the largest capacity for sulfur because the pore volume can also be low and this limits absorption. An optimum balance between bulk density and pore volume is, therefore, essential to ensure that the maximum amount of the zinc oxide in a particle is fully used. This not only extends the useful operating life but also minimizes operating costs. The importance of proprietary catalyst *recipes* and raw materials to improve catalyst performance have been explained in Chapter 1. Typical compositions are given in Table 9.7.

9.3.4.3 *Desulfurization of Other Gases*

Carbon dioxide may be desulfurized with zinc oxide at the same operating conditions as for hydrocarbons. Below about 150°C, however, zinc carbonate is also formed at temperatures below about 150°C but this is subsequently converted to zinc sulfide until saturation is reached.

9.4 STEAM REFORMING

Hydrocarbons are converted into a mixture of hydrogen and oxides of carbon by reaction with steam over steam reforming catalysts. The reforming reaction is endothermic and the catalysts are packed into narrow tubes, which are heated in a furnace. The reforming furnace is commonly known as a reformer. An efficient methane steam reforming process was developed by 1936[6,7] and was first used on a large scale in North America during World War Two as shown in Ta-

TABLE 9.7. Sulfur Absorbent Composition.

Property	A	B
Zinc oxide (wt%)	90	75–85
Calcium oxide (wt%)	3–4	–
Alumina (wt%)	Balance	4–5
Silica (wt%)	–	5–10
Ferric oxide (wt%)	1–2	–
Bulk density (kg liter^{-1})	1.0–1.1	1.0–1.1
Surface area (m^{2}g^{-1})	25–35	20–25
Shape	Granules	Extrudates

ble 9.8. Early reformers operated at slightly higher than atmospheric pressure. As demand for fertilizer nitrogen increased, however, and better alloys for reformer tubes were developed higher reforming pressures could be used in larger ammonia plants.

The original raschig ring catalyst, which contained precipitated nickel oxide, kaolin and a silica cement, was more or less unchanged until high-pressure reformers were introduced. Some of the early catalyst charges to be used lasted for about 20 years.[11] Eventually, improved catalysts were required in modern plants because silica was found to be volatile in high-pressure steam and the catalyst rings part way down the tubes became weakened. At first, silica free catalysts were made by exactly the same procedure, simply excluding silica from the ingredients and using alumina cements. However, all modern, high activity, reforming catalysts are now based on preformed, thermally stable, silica-free supports impregnated with nickel oxide. Ring catalysts are still available but the same formulation can be provided as special, high geometric surface area shapes. Shapes require pressure drop through the tubes and are able to transfer heat rapidly from the tube wall to the reacting gases. As a result of better heat transfer and higher activity, lower gas temperatures are effective in the tubes and the furnace.

Until 1960, steam reformers were only used in areas with readily available supplies of natural gas. As the cost of handling large quantities of coal increased during the 1950s and large volumes of cheap naphtha became available from refineries, the range of feedstock that could be reformed was extended by the introduction of a new alkalized reforming catalyst based on nickel. This began a revolution in synthesis gas production throughout the world. This new catalyst enabled the use of cheap naphtha with a boiling point up to 220C to be used, without the deposition of carbon on the catalyst and furthermore, there were no compression costs in the operation of the reformers at higher pressures. The de-

TABLE 9.8. The First US Steam Reforming Ammonia Plants.[a,b]

Location	Contractor
Ozark Ordnance Works, Eldorado, Arkansas.	Chemico (four furnaces)
Jayhawk Ordnance Works, Baxter Springs, Kansas.	Chemico (two furnaces)
Cactus Ordnance Works, Etter, Texas.	Chemico (three furnaces)
Ammonia Plant, Sterlington, Louisiana.	M W Kellogg (two furnaces)
Alberta Nitrogen Products Ltd, Calgary, Alberta.	M W Kellogg (two furnaces)

[a]All except Sterlington used ICI catalyst.
[b]Each furnace contained 66 tubes
Chem Week Report (Ammonia), Sept. 11 (1965) 11. Other plants listed in article are based on coal feed.
N Gard, Thirty Years of Steam Reforming – A Review of ICI Developments & Experience, *Nitrogen*, Jan/Feb 1966.

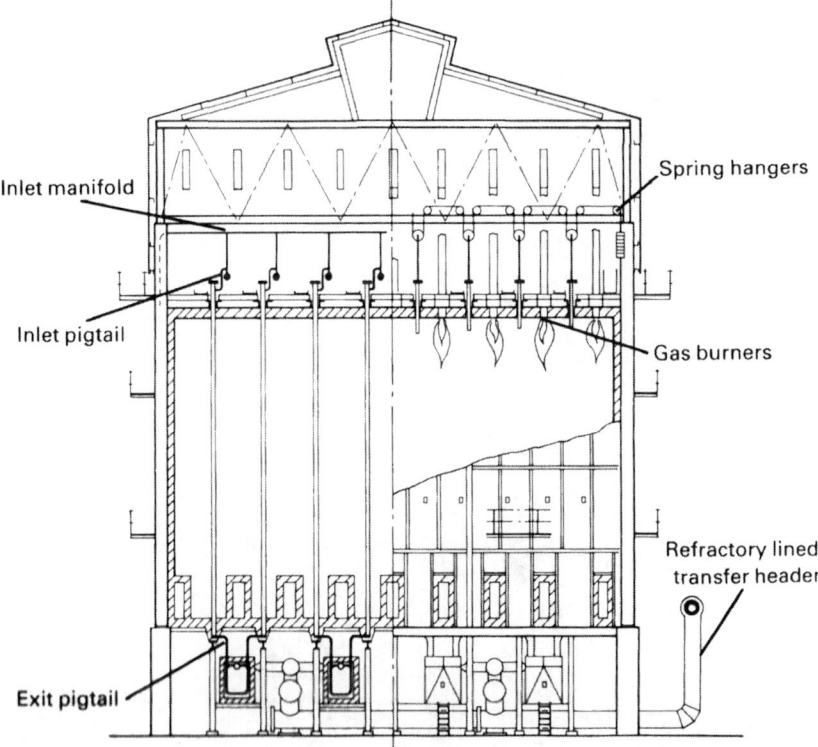

Inlet manifold

Spring hangers

Inlet pigtail

Gas burners

Refractory lined
transfer header

Exit pigtail

Figure 9.1. Schematic arrangement of a top-fired furnace for the primary reformer in synthesis gas production. Reprinted from *Catalyst Handbook*, 2nd ed., by kind permission of M. Twigg.

velopment of naphtha reforming eventually led to the use of the large single stream ammonia plants which are, of course, mainly based on natural gas feeds.

9.4.1 Reformer Design

The continual development of the steam reforming process ensured that this became the most practicable way to produce synthesis gas and ammonia on the large scale. A mixture of hydrocarbons and super-heated steam is passed through the reactor tubes that are packed with the nickel catalyst, and suspended in a furnace that operates at temperatures around 1000°C (Figs. 9.1 and 9.2). The steam reforming reaction is extremely endothermic and the heat of reaction must be supplied continually at a very high operating temperature. The catalyst must

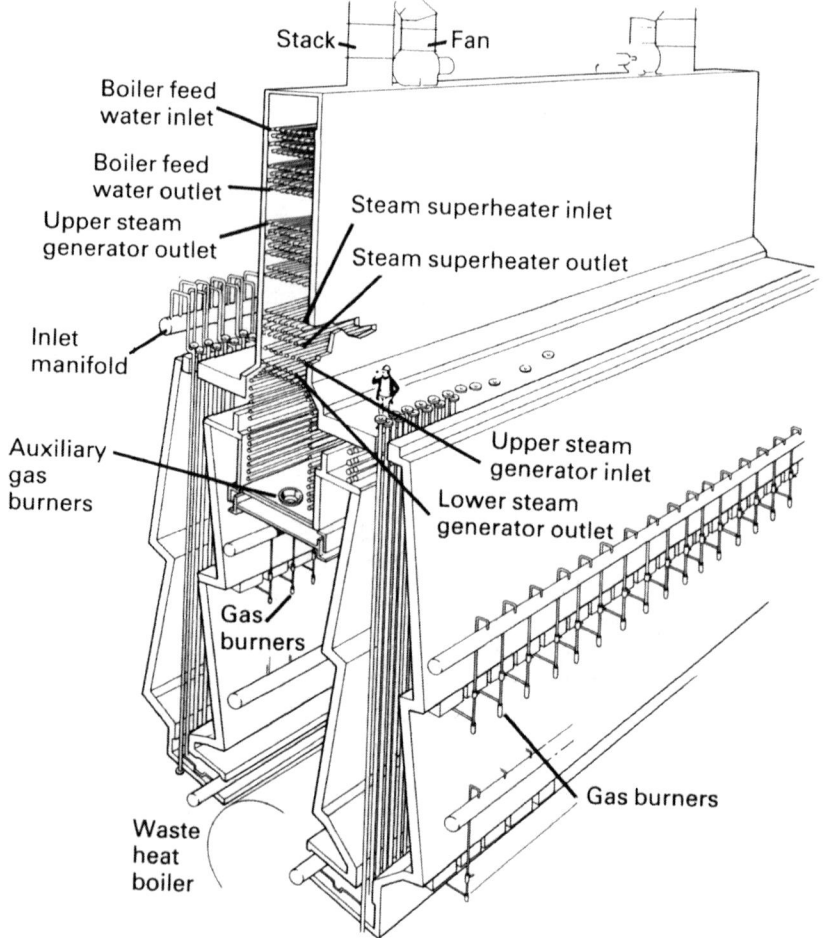

Figure 9.2. Schematic arrangement of a terrace wall fired furnace for the primary reformer in synthesis gas production. Reprinted from *Catalyst Handbook*, 2nd ed., by kind permission of M. Twigg.

therefore have exceptional chemical and physical stability to retain activity and strength during such arduous operating conditions.

    Low-pressure reformers were used in the early ammonia plants, partly because of the simple design and low production capacity and mainly because of the limited physical properties of the stainless steel that was available at the time. The operating pressure could not be increased because the reactor tubes would deform or creep at the high operating temperatures required for the steam reforming process. The lifetime of the reactor tubes before failure was the main limitation to progress, until more thermally resistant alloys became available. When designing a reformer the effects of stress at the high operating temperature and pressure are taken into account to select the most economic tube material for a life of about 100,000 hours.

For effective operation, heat must be transferred rapidly from the furnace itself to the surface of the catalyst, particularly at the top of the tubes. This is to ensure that the hydrocarbon feed reacts rapidly with the steam, as this helps to avoid the cracking reactions that lead to deactivation of the catalyst, and which in turn, helps to avoid overheating of the reactor tubes. This is achieved by the production of a catalyst with a high stable activity and by shaping the catalyst in such a way that allows rapid mixing of the gases in the tube, to allow effective heat transfer from the wall of the tube to the catalyst. This also results in a uniform temperature gradient throughout the tube. Careful packing of the catalyst rings or shapes into the narrow, 3-4 inch diameter tubes is necessary to equalize the gas flow and pressure drop through each tube. This also helps to avoid catalyst breakage and voids within the catalyst bed. Any of these problems can lead to maldistribution of gas flow, which causes variable tube temperatures and hot spots.

Reformers built during the 1940s, which operated at up to 50 psig, consisted of about 66 tubes made of rolled and welded tubes made from type 310 stainless steel. The early, high-pressure reformers built during 1965–1970 operated at up to 500psig and contained up to 400 cast tubes made from HK40. Eventually, when even better alloys like Manaurite 36X were introduced the HK40 tubes were also replaced. Reforming capacity could be increased by up to 10–15% at the same heat flux but with a lower wall temperature by putting a larger volume of catalyst into Mauranite tubes of the same diameter as usual, but with thinner walls. This was particularly useful when better catalysts, giving a lower pressure drop and even better heat transfer, became available.

The hydrocarbon reforming reaction gives a mixture of carbon monoxide and water which is close to equilibrium at the usual operating conditions. The reaction between steam and the hydrocarbon gives a mixture of carbon monoxide and hydrogen that is close to thermodynamic equilibrium at the temperature and pressure of the reactor.

$$CH_4 + H_2O \leftrightarrow CO + 3\,H_2 \qquad (9.8)$$

The operating steam to hydrocarbon ratio, or steam ratio, must, however, be higher than the stoichiometric level to avoid carbon formation on the catalyst, by cracking reactions, and to provide enough steam to operate the water gas shift reaction later in the process.

The presence of excess steam in the process gas to the reformer results in the formation of carbon dioxide by the water gas shift reaction. Thus the gas leaving the steam reformer also contains between 7 and 15% carbon dioxide:

$$CO + H_2O \leftrightarrow CO_2 + H_2 \tag{9.9}$$

Typical operating conditions for a modern steam reformer are given in Table 9.9 although these can be slightly different depending on the overall plant design. The table also includes typical conditions for the steam reformers in methanol and hydrogen plants which use the same catalysts. These examples illustrate the wide variations in gas composition which can be achieved by changing the operating steam ratio, temperature and pressure, to increase methane conversion.

The long catalyst tubes are suspended vertically in rows within the furnace (Figure 9.3). The hydrocarbon/steam mixture, preheated to about 500°C, passes into the reformer through a *header* pipe connected to the tops of each tube by *pigtails* or flexible joints. Following reaction, the products leave the reformer via a *manifold* pipe to the secondary reformer. Heat can be supplied to the furnace from burners at the top, sides or bottom of the insulated box. The intensity of the heat, or heat flux, in the furnace can be varied depending on the demands of the reaction but is generally highest at about one third from the top of the tube, where most of the endothermic reaction takes place.

Furnace design is complex and is fully described in specialized publications.[12] Catalyst operation at extremely high temperatures is critical in the production of ammonia and full instructions on handling and operation are issued by all catalyst suppliers.

TABLE 9.9. Steam Reforming Operating Conditions with Natural Gas.

Process	Ammonia 1000 t/day	Methanol 1000 t/day	Hydrogen 30 mm cfd
Catalyst volume (m³)	17–20	20–25	20
Steam/carbon	3–5	5	6
Inlet pressure (atm)	35	20	20
Temperature inlet (°C)	525	515	520
Temperature outlet (°C)	790	850	780
Outlet gas composition (vol%):			
Nitrogen	<2	Trace	Trace
Hydrogen	70	73	75[b]
Carbon monoxide	8	14[a]	10
Carbon dioxide	12	8	12
Methane	10	4–5	2–3

[a]Addition of CO_2 required before reforming or synthesis.
[b]CO removed by catalysts up to methanation or pressure swing absorption after HTS converter.

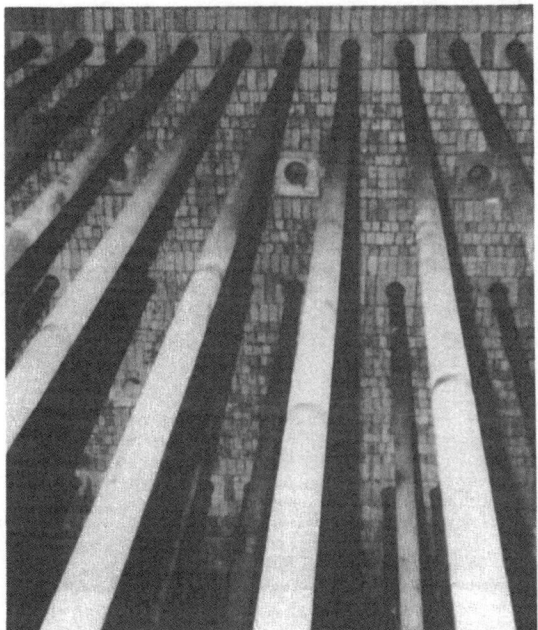

Figure 9.3. Arrangement of burner and tubes in a typical top-fired furnace for the primary reformer in synthesis gas production. Reprinted from *Catalyst Handbook*, 2nd ed., by kind permission of M. Twigg.

9.4.2 Reforming Catalysts

Early reforming catalysts were made by mixing precipitated nickel oxides with magnesia and kaolin. These were active but easily poisoned by sulfur compounds and olefins in the feed gas obtained from coal hydrogenation processes. Ciment fondu was then found to be a useful additive by increasing the catalyst strength and giving better resistance to sulfur poisoning. The best catalyst was a mixture of precipitated nickel oxide with magnesia, kaolin and ciment fondu. The composition of a number of early catalysts is shown in Table 9.10. The first I. G. Farben catalyst to be used in the large-scale plants operated by Standard Oil in 1932 was probably produced as three-quarter inch cubes. Raschig ring reforming catalysts were found by ICI to give better performance and lower pressure drop. During the 1930s, ICI also began to desulfurize the feed gas to the reforming process. This led to the modern cobalt molybdate hydrodesulfurization and zinc oxide absorption procedures.

TABLE 9.10. Examples of Early Reforming Catalyst Composition.

| Wt% | First Catalysts | | | Raschig Ring Catalyst |
	A	B	C	D
Nickel	20	10–16	20	15
Magnesia	20	0–5	15	12
Kaolin	60	–	35	40
Alumina	–	10–16	–	–
Ciment Fondu	–	68–75	30	35

Early catalysts were slowly developed to improve resistance to carbon formation.
A Operated well with pure methane but when poisoned by sulfur and unsaturated hydrocarbons led to carbon deposition.
B Ciment fondu acted as a filler and partly inhibited carbon formation.
C Gave activity as high as A and less carbon than B but still sensitive to unsaturated hydrocarbons.
D Could be used with most natural gases and off gas, providing that poisons were removed. Some early charges operated for 10–20 years with occasional steaming to clean catalyst.

The early catalyst formulations, with only minor changes, were still being used until about 1965 when methane was first reformed at pressures above about 15atm. Silica is volatile in steam at the reforming temperature in high-pressure plants and it is then deposited at a lower temperature in the waste heat boilers, or on to the high-temperature shift catalyst. This led to the development of silica-free reforming catalysts made from alumina, rather than kaolin or magnesia, and alumina cement, rather than ciment fondu. Although alumina based catalysts were active, it was found that they lost strength at high temperature. This led to some breakage and caused hot or patchy tubes. Attempts were made to improve strength by incorporating titanium dioxide into the catalyst recipe rather than alumina although these were not really successful. The best solution was found to be the use of silica free preformed rings, impregnated with nickel nitrate, which could be decomposed to give nickel oxide. Problems in the early days in the selection of a support that was stable in the severe operating conditions led to the introduction of several different catalysts. The supports ranged from pure alpha alumina to calcium aluminate and magnesium aluminate. Operational problems were experienced in the early days, arising from poor thermal stability and loss of activity when using α–alumina supports. Carbon was deposited about one third of the way down from the top of the tube and hot bands formed in high-heat-flux converters.

A partial solution to the low activity problem was to use shorter rings to improve heat transfer although this did, of course, this did increase the pressure drop through the tubes. Hot bands could also be avoided by using alkalized catalysts in the top part of tubes to control carbon deposition. It was clear that improved reforming catalysts were required, not only with a higher, stable activity but also, as it turned out, with a different shape.

By the time most ammonia plants were being operated beyond the limits of their design capacity, new catalyst shapes were eventually developed. The original intention was to increase the geometric surface area of the ring and thus to increase activity. A further benefit was, however, found to be a reduction of the pressure drop through the tubes and improved heat transfer from the tube wall to the reacting gases. Once again a variety of different shapes were introduced ranging from rings, spoked wagon wheels and cylinders with either four or seven holes. Part of the range of catalyst compositions and shapes available is shown in Table 9.11.

9.4.3. Reformer Operation

The catalyst should be loaded carefully into the reformer tubes to avoid breaking the rings. The normal procedure is to fill socks or soft tubes with a bottom flap which can be lowered into each of the tubes before use. The sock is then withdrawn from the tube and the flap opens to leave the catalyst in place. To ensure even packing, the same weight of catalyst is placed in each tube and the pressure drop measured. Gentle tapping, at the bottom of the tubes, may be needed to pack the catalyst properly. In common with all other tubular reactors charging catalyst into the tubes of a steam reformer is a laborious process.

The catalyst must be activated by reduction with a hydrogen-rich gas added to the steam passing through the tubes. A steam to hydrogen ratio of about six is used during reduction at operating temperature. The procedure can be completed in about six to eight hours. Feed may then be gradually introduced and operation begins. Catalyst breakage can result from shortage of steam or thermal shock during operation. This leads to an uneven gas flow with hot tubes and possible

TABLE 9.11. Reforming Catalysts: Composition and Shapes.

	Calcium aluminate	α-aluminate	Magnesium aluminate
Composition (wt%):			
NiO	15–20	15–20	15–20
Al$_2$O$_3$	65–70	80–85	60–65
CaO	10–15	–	–
MgO	–	–	15–20
Bulk density (kg liter^{-1})	~1.0	~0.8–0.9	~1.0
Surface area (m^{2}g^{-1})	10–20	2–5	10–20

Shapes	Size: diameter x length x wall
Raschig Rings mm	16x16x6
	16x16x6
Rings 16mm	7 or 9 ribbed rings
Rings 16mm	4 to 9 holes

carbon deposition. In badly affected tubes the catalyst must be replaced. The leakage of sulfur or chlorine from the purification stage also results in poisoning and deactivation of the catalyst. Any decrease in activity leads to a high methane content in the reformed gas from the reformer tubes which results in hot spots in the tube and the deposition of more carbon. A mild occurrence of sulfur or chlorine poisoning can generally be reversed by increasing the steam ratio for a short period. Arsenic is a permanent poison to the catalyst. If the arsenic concentration in the catalyst at the top of the tubes exceeds about 150 ppm, then the catalyst must be changed, and the tubes cleaned very carefully. The arsenic usually found its way into the reformers in the early days either as a component of the feedstock, or through some maloperation of the Vetrocoke system, used for the absorption of carbon dioxide. A typical early warning of catalyst poisoning is a gradual increase in the methane content of gas leaving the reformer and, in serious cases, an increase in pressure drop.

The introduction of the first 1000 tons per day high-pressure steam reformers in single stream plants led to some operating difficulties. For example, the new high heat flux reformers developed overheated bands, about one third of the way down the reformer tubes. Hot bands, as they were called, resulted from the deposition of carbon from the thermal cracking reaction as the catalyst lost activity. Carbon can most easily form from methane cracking:

$$CH_4 \leftrightarrow C + 2\,H_2 \tag{9.10}$$

This normally takes place when flow of steam stops for any reason but is also thermodynamically possible during full-scale operation of a reformer before the gas temperature reaches about 650°C.

Fortunately, two other reactions lead to the removal of carbon under normal reforming conditions:

$$C + CO_2 \leftrightarrow 2\,CO \qquad \text{Boudouard reaction} \tag{9.11}$$

$$C + H_2O \leftrightarrow CO + H_2 \qquad \text{Water-gas reaction} \tag{9.12}$$

Thus, any carbon formation arising from the cracking reaction at the tube inlet is controlled to some extent by the two carbon removal reactions, which tend to be rather faster than the cracking reaction. Unfortunately, the cracking reaction becomes the faster reaction at temperatures above about 650°C and this temperature corresponds to the position in the reactor tube where the hot bands tend to form. The rate of carbon deposition is greater than its rate of removal, unless the activity of the catalyst is sufficient to provide sufficient hydrogen by the reforming reaction to prevent the deposition of carbon in the first place.

It was clear that a more active and thermally stable catalyst was needed. For some time the situation was controlled by the use of shorter rings at the top of the tube. This provided more geometric surface area and increased reforming activity. Even more successful operation was possible if the shorter rings were alkalized. The addition of potash to catalysts under development for the reforming of naphtha was introduced by ICI during the 1960s and was a major factor in the control of the formation of hot bands. The potash had the effect of accelerating the steam carbon reaction and inhibiting cracking and polymerization on the catalyst surface.

During the 1970s and 1980s the cost of oil had increased substantially and production of ammonia in the older plants was becoming uneconomical. This led to the revamping of many units and the use of more energy efficient processes. A steam reformer uses about one third of the total energy required by a large ammonia plant. It was therefore suggested that the steam ratio could be decreased, from 3.5–4.0 to less than 3.0, even though this could mean that more carbon would deposit on the catalyst and that higher tube wall temperature would lead to shorter tube life. A penalty of operating a reformer with hot tubes, for any reason, is that a temperature only 10°C above the design level can reduce the tube life by up to 50%. The cost of tube failures resulting from the use of low activity catalysts or decreasing the steam ratio was very high.

The new catalyst shapes that were under development gave some compensation for these disadvantages. Higher activity, particularly at lower temperature, could help to avoid carbon formation. Carefully optimized shapes also gave a lower pressure drop and provided better heat transfer within the tubes to decrease the temperature of the tube wall. The use of manaurite tubes also helped to improve operation.

Some operators decided to use a small pre-reformer able to ease the load on the primary reformer and to give further energy savings. The small adiabatic reactor was loaded with a stable high activity catalyst operating at 530°C. The alkalized nickel alumina catalyst, introduced by British Gas in their CRG Process during the 1960s, can operate continuously for about two years. All traces of high hydrocarbons were removed in the prereformer and methane was partially reformed to give about 20% hydrogen in the feed gas entering the primary reformer. This allowed the minimum practicable steam ratio to be used.

An overall fuel saving of 7–10% was possible, and the tube wall temperature of the reformer tubes can be reduced by about 20°C by decreasing the firing rate and heat flux in the reformer. This leads to a longer tube life. Alternatively, up to 5–10% extra ammonia could be produced by increasing the feed rate. When mixed feeds were used, as in some hydrogen plants, steadier operation was possible. Brief details of the CRG pre-reforming process operation are given in Table 9.12.

TABLE 9.12. Prereforming of Natural Gas or Refinery Feeds

Process	Composition (dry outlet gas)		
Feed	Natural Gas	Butane	Naphtha
Steam ratio	0.3	1	1.5
H_2	22–24	30–34	20–22
CH_4	66–68	50–55	60
C_2H_6+	<10 ppm	<10 ppm	–
Balance	CO/CO_2	CO/CO_2	CO/CO_2
Temperature inlet (°C)	530	530	450–530
ΔT (°C)	–64	+11	+40
Pressure		Reformer design	

9.4.4. Secondary Reforming

When coke was the basic raw material for the production of ammonia synthesis gas, the hydrogen and nitrogen required was supplied by mixing water gas and producer gas. Several small ammonia plants were, in complete contrast, built in the United States during the 1920s to use electrolytic hydrogen.[8] The operators found it relatively easy to introduce the necessary nitrogen by burning some of the hydrogen in air. When the war time methane steam reformers were built by the US Government, the same procedure was modified by burning primary reformer outlet gas in a high temperature adiabatic reactor containing more primary reforming catalyst. The additional reactor became known as the secondary reformer. Eventually a catalyst with lower nickel content, and an even higher thermal stability, was developed to withstand the very high temperature reached at the top of the bed.

In modern ammonia plants, air containing the design volume of nitrogen needed for ammonia synthesis is added to the secondary reformer. Two important reactions can take place over the heat-resistant nickel catalyst. Two important reactions take place over the heat resistant nickel catalyst in the secondary reformer. Firstly, the oxygen component of the air is consumed by combustion with hydrogen and possibly carbon monoxide. Secondly, some of the residual methane from the primary reformer undergoes further reforming. Both reactions reach equilibrium at a bed outlet temperature of about 1000°C. The methane content of gas leaving the secondary reformer is usually in the range 0.2–0.5% while oxygen is completely removed.

The main problems in the operation of a secondary reformer are associated with the high temperatures generated by the combustion reactions. The gas distributor must be well designed so that the process gas and air are mixed as rapidly and thoroughly as possible. The catalyst bed itself is protected from the very high temperatures generated by the homogeneous combustion of air, by a layer of refractory material that is placed on top of the large, temperature-resistant catalyst particles. The rest of the bed is filled with secondary reforming catalyst.

TABLE 9.13. Secondary Reforming Catalyst Operation.

Gas Composition	Secondary Reformer	
	Inlet	Outlet
CH$_4$ (dry vol%)	9.4	0.2
CO$_2$ (vol%)	11.6	8.8
CO (vol%)	8.3	11.5
N$_2$ (vol%)	0.5	22.1
H$_2$ (vol%)	70.2	57.1
Air (vol% of dry reformed gas)	~38	
Pressure (atm)	30	29
Temperature (°C)	790	971
Catalyst volume (m^3)(1000 tonnes/day plant)	30	

The catalyst can operate for several years provided that the reacting gases are well mixed and that no hot spots (which in extreme cases could reach 1500°C) are allowed to develop in the catalyst bed. The gas distributor plays an important part in protecting the catalyst. The temperature of the catalyst bed in the secondary reformer falls as gas passes through the bed. At the inlet to the bed, the temperature is extremely high due to the highly exothermic combustion reaction in which all of the oxygen is consumed. The temperature of the bed towards the outlet then falls as the endothermic reforming reaction takes place. It is interesting to note that when the temperature of the primary reformer outlet falls and the slippage of methane increases, the temperature of the outlet of the secondary reformer also falls. Typical operating conditions are shown in Table 9.13.

Secondary reforming catalyst contains about 7% nickel oxide, supported on temperature-resistant α–alumina or calcium aluminate, in the form of raschig rings. Additional catalyst used at the top of the bed as a heat guard is in the form of large solid cylinders of α-alumina containing 5% nickel oxide.

9.5. CARBON MONOXIDE REMOVAL

Carbon monoxide leaving the secondary reformer is converted to useful hydrogen by the water gas shift reaction in a two-stage recovery section. The original high temperature shift reactor is now combined with a low temperature reactor filled with a copper catalyst. The removal of carbon monoxide is shown in Figure 9.4. Since copper catalysts are extremely prone to poisoning by sulfur and chlorine compounds, it is therefore essential that the concentration of these contaminants is reduced to an absolute minimum.

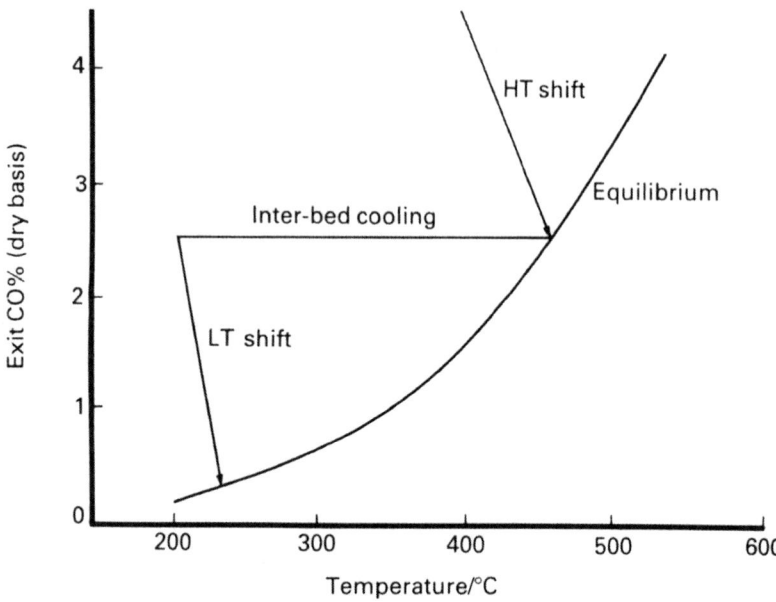

Figure 9.4. Typical variations of carbon monoxide levels in HT and LT shift catalytic beds. Reprinted from *Catalyst Handbook*, 2nd ed., by kind permission of M. Twigg.

9.5.1. High Temperature Carbon Monoxide Conversion

In 1915, the water-gas shift reaction has been used to remove carbon monoxide, by reaction with steam and to increase the *potential* hydrogen production since the first commercial ammonia plant began operation in 1913. The process was based on catalysts discovered by Wild of BASF in 1912[2] while following up work dating from the late 1800s.

Since then the basic formulation of high temperature shift catalyst (HTS) has been iron oxide, stabilized with chromium oxide, although the methods of stabilizing the catalysts and producing it in large quantities have been refined as operating conditions have changed

The catalyst as supplied consists mainly of hematite and chromium oxide, together with a variable amount of the hexavalent chromium trioxide. Before use, the catalyst must be activated by reduction to magnetite using process gas, and at the same time, any chromium trioxide is also reduced to the trivalent state. Under the typical conditions for the shift reaction, magnetite is the thermodynamically-stable state for the oxides of, and there is no further reduction to the metallic state. Some HTS catalysts may contain traces of sulfur, particularly if the catalyst has been manufactured from ferrous sulfate. When commissioning such a catalyst, most of the sulfur is reduced to hydrogen sulfide by the process gas and this amount would be sufficient to destroy the LTS catalyst if it were allowed to pass into the catalyst bed. It is normal practise, therefore, to flare the gas product until it is free from sulfur, before allowing the low temperature shift to come on line.

The forward shift reaction that is the conversion of carbon monoxide into hydrogen and carbon dioxide is quite strongly exothermic, and in common with all exothermic reactions, the level of conversion of carbon monoxide to products at equilibrium is greater at lower temperatures. The equilibrium constant is independent of pressure, but higher conversions are also obtained at higher steam ratios. Unfortunately, the preferred iron catalyst is not active at low temperature and plants operate in the temperature range 350° - 500°C. Furthermore, the more active catalysts based on copper that are active at low temperatures are not sufficiently stable to be able to withstand the exotherm associated with the shift reaction when the feed gas contains high levels of carbon monoxide. The exact conditions for operation of the high temperature shift converter are therefore determined by the carbon monoxide content of the gas entering the reactor, and the steam ratio used in the primary reformer.

At the time when natural gas was introduced as feed to a steam reformer from about 1940 onwards, most ammonia plants were designed to use single beds of HTS catalyst and the concentration of carbon monoxide in the outlet gas was about 2%. The synthesis gas from the older plants, which used coal as feedstock, contained higher concentrations of carbon monoxide, and it was often necessary to control the temperature by splitting the reactor into two or more separate beds, with inter-bed cooling or by the incorporation of a quench system. More recently, when the first ammonia plants based on natural gas were first operated at the higher pressures, a two bed/inter-cooled reactor design was also used and the concentration of carbon monoxide in the exit gas was lowered to about 1% It was then practicable to use a methanator to hydrogenate the residual carbon monoxide to methane, rather than operating with the copper liquor scrubbing stage which had previously been used for the final removal of carbon monoxide.

9.5.2 High Temperature Conversion Catalysts

The first HTS catalysts were reported to operate for about two years before replacement was required. As production techniques were developed, however, catalyst lives improved so that by 1940, lives of more than 14 years were regularly achieved. There were few poisons which affected the catalyst performance although sulfur, which was the most common impurity in early plants, did sulfide the magnetite. This reaction was, nevertheless, reversible. If hydrogen sulfide levels exceeded about 300 ppm, sulfided catalysts could not be regenerated and about twice the normal volume of catalyst had to be used to compensate for the lower activity of iron sulfide.

The catalysts were originally produced from readily available materials such as copperas, the ferrous sulfate waste product from steel works, and chrome tan, an industrial form of chromic acid. The process involved precipitation from ferrous sulfate solution with sodium carbonate. The precipitate was carefully washed, to remove soluble impurities but early catalysts still contained up to 1-2% insoluble ferric and chromic hydroxyl sulfates, which formed hydrogen sulfide during catalyst reduction. It was possible for the concentration of hydrogen sulfide in the process gas to reach more than 250 ppm, gradually falling over a period of several days, before the concentration fell to a an equilibrium level. Of course, on most occasions hydrogen sulfide levels were lower than this. When catalysts were used in naphtha based town gas plants during the 1960s and, later, in single stream ammonia plants with copper based LTS catalyst, the HTS catalyst production method had to be modified. With *low sulfur* catalysts it was possible to produce sulfur free gas following reduction in less than twelve hours.

A further undesirable feature of catalysts containing chromium is that during the calcination stage of preparation, a proportion of the chromium oxide is oxidized to hexavalent chromium at temperatures in the range 250°–350°C. The concentration of hexavalent chromium falls to an acceptable level of less than 1% when the calcination temperature is increased to about 430°C. High levels of hexavalent chromium in the catalyst lead to a significant exotherm during reduction, which can lower the activity of the catalyst. Furthermore, handling of the catalyst during manufacture can be a very dusty procedure, and contact with the dust containing hexavalent chromium is hazardous to the workforce, since it believed to have carcinogenic properties.

9.5.2.1 *Operating Conditions*

In modern, single stream ammonia plants there is little scope in the design to make significant changes to the operating conditions in any of the individual catalyst reactors. Operating conditions for the carbon monoxide conversion reaction are shown in Table 9.14. The only practical variable is operating temperature which can be slowly increased as catalyst loses activity.

The composition of the catalyst can affect its performance in many ways. For example, changes in the formulation to lower the amount of hydrogen sulfide evolved during reduction led to a significant physical weakening of the catalyst, which became much more prone to damage caused by the condensation of water or contamination by potash. Any water that condensed during plant start-up could also wash out any soluble chromium and lead to loss of stability. Most problems led to an increase in pressure drop or maldistribution of gas. Cat

TABLE 9.14. Carbon Monoxide Conversion Catalysts Operation.

	Inlet HTS	Outlet HTS	Inlet LTS	Outlet LTS
Gas rate (m^3/h)	133,000		146,000	
Steam ratio	0.6		0.46	
Pressure (atm)	29		28	
Catalyst volume (m^3)	60		64	
Temperature (°C)	360±20	430±20	210±10	230±10
Gas composition (%):				
CO	13		3.5	<0.5
CO$_2$	8		16	18
H$_2$	56		60	61
N$_2$	22		20	20
CH$_4$/A	~0.3		0.5	0.5

alysts were usually, however, able to operate for a period ranging from two to four years.

With the drive to more energy-efficient ammonia plants in the 1980s, the steam ratio fed to the primary reformer was significantly decreased. Apart from the effects in the reformer which have already been described, a number of changes were detected in the operation of the HTS catalyst. The lower steam ratio was still greater than the level calculated from the thermodynamics at which magnetite could be reduced to the metallic state (Figure 9.5). However, the lower concentration of steam resulted in a higher level of carbon monoxide in the process gas, as described earlier in this chapter, and this increase in the carbon monoxide/carbon dioxide ratio gave rise to a more reducing atmosphere. This, in turn, led to the formation of more carbon by the disproportionation of carbon monoxide, also known as the Boudouard reaction. The end result was an increase in pressure drop, and further disintegration of the catalyst.

Carbon monoxide also reacted with magnetite forming carbides which catalyzed the production of hydrocarbons by Fischer-Tropsch reactions. It has since been shown that both reactions can be suppressed by the addition of copper to the existing iron-chromium catalyst. This has allowed operation of HTS catalysts in existing plants down to a steam ratio as low as 0.4 compared with about 0.6 in the early single stream plants.

In more recent ammonia plant designs, in which the large steam reforming furnace has been replaced with gas-heated reformers or combined autothermal reformers, the HTS catalyst can be replaced by a temperature resistant copper catalyst that can operate at temperature as low as 260°–270°C.

9.5.3. Low Temperature Carbon Monoxide Conversion

As the industry moved towards cleaner feedstocks, such as naphtha and naturals gas, and the use of increasingly efficient gas purification systems, the synthesis

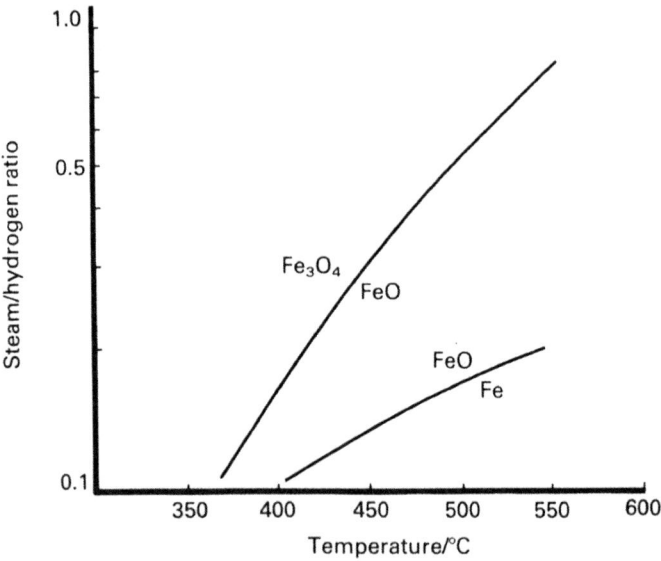

Figure 9.5. Minimum ratio of steam to hydrogen for reduction of conventional HT shift catalysts. Fe_2O_3 becomes stable with respect to Fe_3O_4 at $H_2O/H_2 > 2 \times 10^3$ at $\sim 550^0C$ and about half this value at $\sim 400^0C$. Reprinted from *Catalyst Handbook*, 2nd ed., by kind permission of M. Twigg.

gas from the reformers contained only trace amounts of catalyst poisons. This made it possible to consider the use of copper catalysts for carbon monoxide conversion and to simplify plant design even further. The introduction of a copper oxide-zinc oxide catalyst in 1963, operating at temperatures in the range 200°–250°C, lowered the concentration of carbon monoxide in the synthesis gas to 0.2–0.3%. Conditions were only limited by the need to prevent damage to the catalyst from the condensation of water at low temperature. Unfortunately, operation of the new copper catalyst was not reliable. Once the manufacturing procedure was improved however and the effect of catalyst poisons, particularly chlorine, was understood the low temperature shift (LTS) catalyst became an important part of the ammonia synthesis process.

It is important to understand the commercial significance of an active catalyst which can operate at maximum conversion for long, predictable periods in large ammonia plants. In a typical 1000 tonnes per day ammonia plant the most obvious penalty is the daily cost of an unexpected plant closure. Anther less obvious cost is the overall effect of an increase in the volume of carbon monoxide leaving the low temperature reactor. One volume of hydrogen is lost for every volume of carbon monoxide which is not converted by the LTS catalyst and again an extra three volumes of hydrogen are lost as the carbon monoxide is removed in the methanator.

$$CO + H_2O \rightarrow H_2 + CO_2 \qquad \text{1 vol. } H_2 \text{ not made} \qquad (9.13)$$

$$CO + 3 H_2 \rightarrow CH_4 + H_2O \qquad \text{3 vol. } H_2 \text{ consumed} \qquad (9.14)$$

The additional methane formed must then be purged from the synthesis loop which leads to even more hydrogen loss as significant hydrogen is also removed during the purge. Overall, the equivalent of failing to convert just 0.1% of carbon monoxide is more than 4000 tons of ammonia every year, or about 1% of the design production.

Every catalyst in an ammonia plant is important but, because low-temperature carbon monoxide conversion catalyst is likely to be the most sensitive to poisons, good operation of this catalyst is probably the most significant. Apart from using the best available catalyst, it is usual to make an allowance for catalyst poisoning by increasing the catalyst volume in the reactor.

9.5.3.1 *Operation*

An LTS catalyst should be sufficiently active to give a high conversion for a given volume of catalyst at the minimum practicable temperature. It should also be thermally stable and operate for the design period with maximum carbon monoxide conversion. With proper design and good upstream poisons removal, a typical catalyst lifetime is about three years.

The catalyst must be activated by careful reduction to convert the copper oxide component to metallic copper before use (Figure 9.6). The reduction reaction is exothermic and the reduction process must be carried out using an inert carrier gas to which a low concentration of hydrogen has been added, so that the temperature of the bed does not exceed 250°C. After reduction and when a new charge of catalyst is commissioned, the bed should be heated to a temperature greater than the dew point of the feed gas so that liquid water does not condense onto the catalyst when the feed is first admitted. The temperature is then increased gradually until the reaction begins. At this stage, the temperature can be adjusted until the required degree of conversion is achieved and a satisfactory temperature profile is seen in the bed.

As the catalyst ages, the peak temperature moves slowly down the bed (Figure 9.7). The rate of movement depends both on the thermal stability of the catalyst and on the total amounts of different poisons that accumulate on the bed. The temperature at the inlet to the bed can be increased gradually to compensate for deactivation of the catalyst. Eventually the catalyst must be changed because the concentration of carbon monoxide leaving the bed at equilibrium

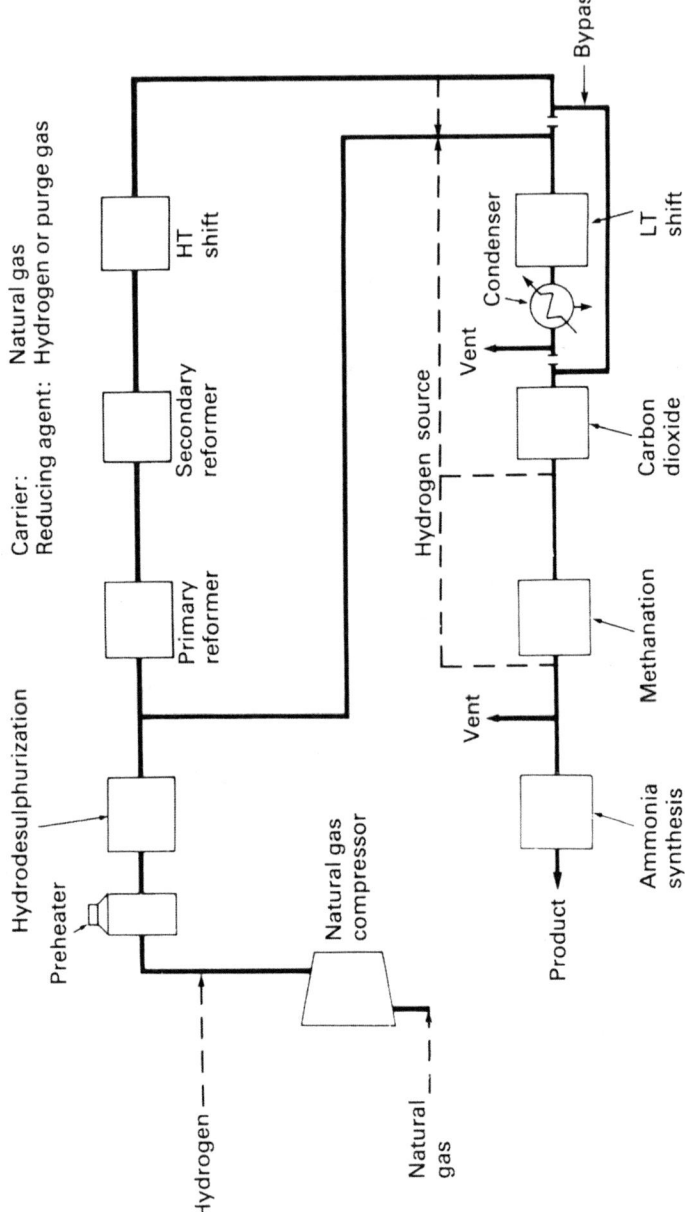

Figure 9.6. Schematic arrangement for the reduction of LT shift catalyst using a typical once-through system. Reprinted from *Catalyst Handbook*, 2nd ed., by kind permission of M. Twigg.

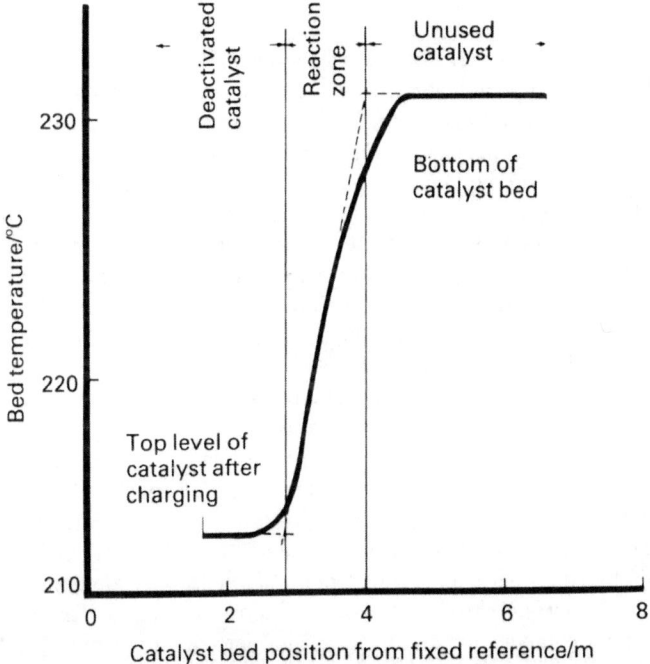

Figure 9.7. Typical temperature profile through a bed of ICI Catalyst 53-1 LT shift catalyst. Reprinted from *Catalyst Handbook*, 2nd ed., by kind permission of M. Twigg.

increases as the temperature is increased, and a point is eventually reached when the degree of slippage of carbon monoxide is not economically viable. The temperature profile in the bed should be measured regularly so that the remaining life can be estimated or any maloperation of the plant detected. For example, any condensation of water in the catalyst bed might wash soluble poisons down into the active layer.

It is possible to minimize the effects of poisons in the process gas if there is no way of removing them and a short catalyst life is unacceptable. A small guard bed can be installed in a separate reactor before the main bed so that any poisons can be removed. This bed need to be changed regularly. Originally, zinc oxide was used as a guard to remove sulfur compounds but, because the most serious poison for LTS catalyst is chlorine, ordinary LTS catalyst is preferable. The use of catalyst as guard has the advantage of always ensuring maximum carbon monoxide conversion.

9.5.3.2 *Catalyst*

The first LTS catalysts were based on copper and zinc oxides, (CuO)/2ZnO), and were prepared by calcination of the precipitated hydroxycarbonates of zinc and copper. By 1965, it was found that incorporation of some alumina into the formulation led to an increase in thermal stability and improved resistance to poisons. Depending on the plant conditions, the operating life of the catalyst was increased from about six months to more than two years. The most successful catalysts were produced by simultaneous precipitation of the copper, zinc and aluminum hydroxycarbonates, at about pH = 7, to optimize the particle size and distribution of the oxides. They were similar in composition to the successful low-pressure methanol catalysts produced in the same way although the copper content was significantly lower. Other catalysts were introduced which had high copper content in an effort to achieve a high activity. High copper catalysts were more susceptible to poisoning in some plants. Another catalyst, based on copper and zinc oxides but with high chromium oxide content, was expected to be re-generable, to be able to compensate for the effects of poisons. However, it was very susceptible to poisoning by sulfur, needed a long period off line for regeneration, and its use was not successful. This catalyst has now proved to be an acceptable chlorine guard for a much more stable copper-zinc-aluminum catalyst. Details of approximate catalyst compositions are given in Table 9.15.

Most spent catalysts contain varying amounts of sulfur, chlorine and silica poisons depending on the age of the catalyst and the purity of synthesis gas. The effects of each poison can therefore be considered separately in relation to the differing operating conditions. In some cases, up to 3–4% sulfur has been meas-

TABLE 9.15. Carbon Monoxide Conversion Catalyst Compositions.

Catalyst	Composition		
Low Temperature Shift:			
Composition (wt%) loss free	Low Copper	High Copper	Chromium
Copper oxide	32	42	20
Zinc oxide	55	47	35
Alumina	13	10	
Chromia	–	–	45
Ignition loss 900°C	<20	<10	<10
Bulk density (kg liter^{-1})	1	1.4	1.2
High Temperature Shift:			
Composition (wt%) loss free	High Steam Ratio	Low Steam Ratio	
Ferric oxide	89–91	85–89	
Chromia	9–10	7–11	
Water soluble CrO₃	<2		
Copper oxide	–	2	
Ignition loss 900°C	<10	<10	
Bulk density (kg liter^{-1})	1.3	1.3	

ured in samples from the top of catalyst beds when the life has been as long as seven years. The level of chloride in the same samples has been less than 0.02-0.04%, and it can be concluded that sulfur is held firmly by the catalyst and may not be a serious problem. A possible explanation for this observation is that although sulfur compounds are initially adsorbed by the copper crystallites, they are rapidly transferred to the small zinc oxide crystals. Zinc sulfide, which is thermodynamically more stable, is easily formed. This mechanism is dependent upon the manufacturing process to provide small crystallites in the catalyst structure. Similarly, silica deposits of up to 1.5wt% at the top of the LTS catalyst bed have been detected after satisfactory lives of two to four years.

On the other hand, while chlorides accumulate near the top of the catalyst, they are more mobile and can be detected in significant concentrations, up to 0.05%, at all levels in a deactivated bed. Although reasonable lives of at least two years can often be achieved in the presence of chloride there is more rapid movement of the peak in temperature profile, and the concentration of carbon monoxide in the outlet gas increases more rapidly. Surface chlorides, which are formed by reaction with zinc oxide, are mobile and sinter the catalyst surface. Chlorides are also soluble in condensed steam and can be washed down onto lower, more active catalyst layers.

Sulfur and chlorine are both present in the hydrocarbon feed, the process steam and the lubricating oils used while sulfur may also come from the high temperature shift catalyst. A major source of chlorine is from the air used in the secondary reformer. Silica is present in process steam but also comes from the refractory linings or support materials used in the reforming section.

9.6 METHANATION

Carbon monoxide and carbon dioxide cause the temporary deactivation of ammonia catalysts. Carbon dioxide can also lead to further problems because it forms ammonium carbonate in the make-up gas compressor and the synthesis loop. The removal of these impurities is, therefore, a vital step in the purification of synthesis gas. Removal of carbon dioxide has generally been via absorption in some suitable solvent, whereas at the present time, the concentration of carbon monoxide is reduced to a low level by reaction with steam in the water gas shift reaction, prior to almost complete removal by an additional procedure.

Methanation, the conversion of carbon monoxide and dioxide to methane, was described by Sabatier in 1905, during his work on hydrogenation catalysts, and since then it has been used in several industrial processes. Since about 1955, as the use of large, single-stream\m ammonia plants became more widespread, the methanation reaction has become the preferred way to remove the final traces of carbon oxides from the process gas.

Figure 9.8. Photograph of a methanator in a typical ammonia synthesis plant. This vessel contained about 25 tonne catalyst. Reprinted from *Catalyst Handbook*, 2nd ed., by kind permission of M. Twigg.

9.6.1 Operation

When the plant is operating normally, the synthesis gas contains around 0.3–0.7% oxides of carbon, which need to be removed during the methanation stage. A simple, single-bed reactor can be used at an inlet temperature in the range 250°–320°C depending on the catalyst activity and the concentration of the oxides of carbon (Figure 9.8). There is an adiabatic temperature rise of about 7.4°C for every 0.1% carbon monoxide and 6°C for every 0.1% carbon dioxide converted. Although the catalyst should not normally be operated at temperatures exceeding 450°C it is frequently allowed to overheat to temperatures as high as 600°C during periods of maloperation, but seems to incur little deactivation.. There is probably more chance of damage to the reactor in these situations. At a space velocity in the range 5000-7000 hours^{-1} methanation is virtually complete at a space velocity in the range 5000-7000h-1 and the outlet carbon oxides level is usually less than 5ppm. Good methanation catalysts have a normal life of about ten years.

The most common catalyst poisons are derived from traces of solvent carried over from carbon dioxide removal process. These simply block the catalyst pores and, providing no sulfur or arsenic compounds are used, can easily be removed by washing the catalyst with water to restore activity. If sulfur enters the catalyst bed from any other source it will rapidly poison the methanation catalyst. The low temperature shift catalyst, up-stream of the methanator, usually acts as a very efficient sulfur guard!

9.6.2. Catalyst

Different suppliers have produced methanation catalyst in a wide range of shapes and sizes, ranging from spheres and pellets to extrusions and granules, containing between 15–30% nickel oxide. Some typical support materials react with nickel oxide to form relatively stable spinels, which cannot readily be reduced to give an active catalyst, and these materials are not used. It is better if the nickel oxide can be precipitated in solid solution with a thermally stable matrix similar to those supports used in reforming catalysts. Typical supports are alumina, silica and magnesia, sometimes bound by alumina cement. The most important catalyst property is that it can be reduced easily at a temperature at or below 300°C to give an active and stable catalyst. The nickel can be combined with the support by either impregnation or precipitation, provided that the final catalyst is thermally stable.

The nickel oxide component of the catalyst is reduced to the metal with process gas, by gradually increasing the temperature of the bed to about 300°C. If necessary, the reduction procedure can be assisted by allowing a small proportion of the process gas to bypass the low temperature shift converter. This results in a temporary increase in the concentration of the oxides of carbon in the reducing gas to around 1–2%, which are then hydrogenated into methane over the active part of the catalyst. The resulting exotherm from the reaction increases the temperature of the bed to about 400°C, so that any remaining nickel oxide can be reduced fully. During the reduction period the carbon monoxide and carbon dioxide content of outlet gas is analyzed at regular intervals. Full operation can start when the level falls to less than 20 ppm. This normally takes 10–15 hours. If the LTS reactor cannot be by-passed the catalyst reduction should be completed, more slowly, at the highest inlet temperature possible up to a maximum of 450°C.

9.6.3. Other Methanation Processes

Methanation catalysts or, alternatively, pressure-swing absorption are used to purify synthesis gas and refinery hydrogen throughout the world. The methanation reaction has also been applied in other interesting and useful processes although not yet on a very significant scale.

When natural gas was felt to be running short in the US the British Gas Council CRG (Catalytic Rich Gas) process was converted to produce a substitute natural gas process simply by adding two methanation reactors to the original town gas process. The rich gas, including steam, was cooled and then methanated at 300°C limiting the outlet temperature to less than 400°C. The gas was cooled again and the remaining oxides of carbon were converted to methane in a second reactor, with a rise in temperature of only about 40°C, as shown in Table 9.16. Unalkalized CRG catalysts were used in the methanation reactors. Other processes were introduced by Lurgi and Japan Gasoline but, in the end, there no need for additional supplies of natural gas and the plants were only operated for a short time. It was also shown that the product from coal gasification, which contained oxides of carbon, was an alternative feed for conversion to methane.

Pure hydrogen can be recovered by a methanation procedure from olefin plant tail gas that contains methane, ethylene and carbon monoxide. Typical methanation catalysts will remove up to 0.3% carbon monoxide and 0.3% ethylene at 270°C, 30 atm pressure and about 6000 h^{-1} space velocity. A catalyst which does not crack ethylene to form carbon but which can produce ethane or

TABLE 9.16. Catalytic Rich Gas (CRG) Reforming and Substitute Natural Gas (SNG).

Process	Composition (dry outlet gas)		
Naphtha Feed	CRG	SNG	
		Two-stage methanation/CO_2 removal	
CO_2	20–22		<1
CO	1		–
H_2	16–18		<1
CH_4	64		98.5
Temperature inlet (°C)	450	300	300
Temperature outlet (°C)	510	375	340
Pressure (atm)	>10	>10	>10
Steam ratio weight	2 steam / 1 naphtha		
Hydrogen recycle	0.06 m^3 H_2 / kg naphtha		
	0.01 m^3 H_2 / kg propane		

methane by hydrogenation, should be chosen for the duty. Carbon monoxide is completely removed by this procedure.

The EVA-ADAM process was proposed as a means whereby energy could be transferred over fairly long distance. The process was based on a combination of the steam reforming and methanation reactions. The energy required to fuel the endothermic steam reforming reaction could be obtained from any remote source, such as a nuclear reactor, in contrast to the normal process in which gas is burnt to supply the heat. The products from the steam reforming reaction contained both hydrogen and the oxides of carbon. Energy, in the form of superheated steam, was then recovered from the exothermic heat of methanation at a second location. Methane was also available as fuel if necessary. The high temperature of the operation in the methanators, at 700°C, 500°C and 300°C, inlet the three reactors, required that special active, yet heat resistant, catalysts were necessary. The catalysts were deactivated by a combination of sulfur poisoning and carbon formation, through partial regeneration was possible by treatment with hydrogen at 500°C. The reaction product contained about 80% methane. Adiabatic beds could be replaced by internally cooled, tubular reactors (IRMA). The maximum temperature in tubes reached more than 600°C. The process has an exit temperature of 320°C giving up to 83% methane. Once again, the process has not yet been commercialized.

9.7 OTHER APPLICATIONS OF STEAM REFORMING

The steam reforming process with typical catalysts has been widely used to produce a range of hydrogen rich gases. The examples in Table 9.17 show how the operating conditions have been modified to obtain the appropriate gas composition.

9.7.1 Methanol Synthesis Gas

The gas produced by steam reforming any hydrocarbon using normal catalysts can be used directly as methanol synthesis gas with no further treatment. When methane is used as feedstock for the steam reformer, the gas produced contains 50% more hydrogen than that which is required for the methanol synthesis reaction, and this gas must be recovered at some stage in the process. However, as the molecular weight of the hydrocarbon increases, the amount of excess hydrogen produced falls with increasing molecular weight. When naphtha is used as the feedstock, the resulting synthesis gas is virtually stoichiometric for methanol synthesis. This is illustrated in the following equations:

TABLE 9.17. Other Applications of Steam Reforming Process.

Process	Lean town gas	OXO alcohol synthesis gas	Direct iron ore reducing gas
Feed	Naphtha	Natural Gas	Methane & Top Gas
Steam/carbon	3	<2	2
Inlet pressure (atm)	27	15	5
Temperature inlet (°C)	~450	540	400–450
Temperature outlet (°C)	750	870	950
Outlet gas composition (vol%):			
Hydrogen	60	64	74
Carbon monoxide	10	22	24
Carbon dioxide	16	11	1
Methane	12	3	1-2

$$CH_4 + H_2O \rightarrow CO + 3\,H_2 \tag{9.15}$$

$$C_nH_{2n+2} + n\,H_2O \rightarrow n\,CO + (2n + 2)\,H_2 \tag{9.16}$$

$$n\,CO + 2n\,H_2 \rightarrow n\,CH_3OH \tag{9.17}$$

Synthesis gas from natural gas feeds has, however, an excess of hydrogen but may be used directly if a high purge rate from the synthesis loop is acceptable.

It is possible to adjust the ratio of hydrogen to carbon oxides by the addition of carbon dioxide to the synthesis gas, either just before the methanol synthesis loop, or directly into the feed for the steam reformer. It is also possible to use a *secondary* reformer using oxygen rather than air, after the primary reformer. The amount of oxygen used is adjusted to provide the stoichiometric ratio of hydrogen and carbon oxides and, at the same time, reduce the amount of inert methane in synthesis gas. (Table 9.9)

9.7.2 OXO Synthesis Gas

Typical steam reformers are used to produce synthesis gas for OXO reactions. It is usual to operate with a steam/carbon ratio of less than two with a recycle of sufficient carbon dioxide to provide a carbon dioxide/methane ratio of about two so that an appropriate composition of the gas product is obtained.

9.7.3 Hydrogen Production

For the production of hydrogen, the steam reformer is operated at high severity to obtain maximum methane conversion and early hydrogen plants used exactly the same catalysts as ammonia plants, except for the secondary reformer. The hydrogen product is purified by catalytic conversion of the carbon monoxide, removal of carbon dioxide followed by methanation. This product is then acceptable for use in most refinery and chemical processes (see Table 9.9).

More recently pressure swing absorption units using molecular sieves have replaced the carbon dioxide removal and methanation steps and, often, the LTS stage, to give very pure hydrogen. Only the HTS catalyst is retained to remove most of the carbon monoxide. Hydrogen of even greater purity can be obtained by operating at a low steam ratio in the steam reformer.

9.7.4 Reducing Gas

The direct reduction of iron ore in blast furnaces decreases the amount of high quality coke required and can also lead to an increase in the production capacity of the furnace. Gas with a high content of hydrogen and carbon monoxide is produced at low pressure when natural gas is reformed using an almost stoichiometric steam to carbon ratio. However, the low steam ratio results in a tendency for the deposition of carbon. Some of the top gas from the blast furnace, which has a carbon dioxide content of about 20%, is recycled to the steam reformer to compensate for the low steam-to-carbon ratio and to provide the necessary volumes of hydrogen and carbon monoxide. The gas flow is upwards through the tubes, which are filled with large-diameter rings of catalyst to reduce pressure drop. The feed gas is pre-heated over inert rings, or even low activity catalyst, before passing into the reformer stage in which high-activity, thermally-stable catalysts are used (see Table 9.17).

9.7.5 Town Gas Production

Town gas was first made on a large scale by the Chartered Gas Company in London during 1812. Production then spread throughout Europe and the USA. The discovery of natural gas in California during 1928 and then in other states led to its domestic use in the US particularly after long-distance pipelines were installed. By 1934 almost 40% of the US domestic supply came from natural gas and this increased rapidly to more than 95% by 1960. During the 1970s, when it was thought that US natural gas supplies were declining, about ten substitute natural gas (SNG) plants were built in the eastern states using the UK Gas Council's catalytic rich gas (CRG) process. (Table 9.16)

The burners used in domestic appliances at that time were not suitable for use with natural gas, because of its high calorific value. For this reason, natural gas was not used directly until the 1960s. Small cyclic reforming processes were introduced to meet peak demand when the supply of coal gas was insufficient to meet the requirements. Larger plants were installed to meet the needs of larger areas, and the relatively low calorific value of the reformer gas was increased to the required level by the addition of low-molecular-weight hydrocarbons.[14]

A wide range of catalysts produced by impregnating α-alumina or magnesia with low concentrations of nickel oxide were used in cyclic reformers. These catalysts had to be extremely heat resistant because the process generated carbon which was removed at intervals by burning in air.

By 1960 the ICI naphtha steam reforming process was being used to make ammonia synthesis gas from the very cheap naphtha then available from European refineries. This process was easily adapted to make town gas simply by operating at lower reforming temperature. The reformed gas then contained a higher methane concentration. The calorific value of the gas was increased from 385Btu per cubic foot to 500Btu per cubic foot by enrichment with hydrocarbons.[14] More than 200 of these *lean gas* plants were built throughout the world before natural gas was used directly. (Table 9.17)

At about the same time the British Gas Council developed its adiabatic process in which naphtha or lighter hydrocarbons were converted into a high methane content gas with a calorific value of almost 700 Btu per cubic foot. This gas could then be converted into town gas with a calorific value of 500 Btu per cubic foot by continuous steam reforming in a lean gas plant and appropriate blending. Since then the CRG process has also been widely used throughout the world to produce methane rich gas from naphtha and LPG.

The CRG process operates adiabatically in a small single bed reactor. The catalyst has high nickel oxide content and is precipitated under carefully controlled conditions with alumina to give a high activity and thermally stable structure. The finished catalyst is alkalized to prevent carbon formation and is reduced before operation. Steam to carbon ratio can be as low as 1.5 when reforming naphtha to produce town gas.

When operating with naphtha as feed, the endothermic reforming reaction is dominant at the top of the bed. As the concentrations of both hydrogen and carbon monoxide increase, the exothermic methanation reaction becomes the dominant reaction so that the overall reaction is exothermic. Gradual catalyst deactivation leads to the temperature profile moving down the bed and catalyst life can be as long as two years. With natural gas feed any hydrocarbons heavier than methane are reformed, the methane steam equilibrium is established, and the overall reaction is endothermic. Operating details are shown in Table 9.16.

9.7.6 Substitute Natural Gas

Substitute natural gas (SNG) may be obtained by methanation of CRG process gas with a stable nickel-alumina catalyst after removal of the carbon dioxide formed in the conventional process. Gas containing more than 98% methane can be produced. The CRG process is operated at low steam ratio to maximize methane production. Two stages of methanation follow the CRG reactor and intercooling is required to control the temperature. Standard alkalized, high nickel

Figure 9.9. Combined steam reformer/autothermal reformer.

CRG catalyst is used in the reforming stage with a non-alkalized version of the catalyst in the two methanation stages.

9.7.7 Autothermal Reforming

In an attempt to improve energy efficiency, the steam reformer has been replaced in some plant designs by an autothermal reformer (Figure 9.9). In two relatively small ICI ammonia plants up to 50% excess air is added to the secondary reformer and the hot process gas used to provide heat for a small, pressurized primary reformer. Improved versions of the conventional reforming catalysts are used. A single bed of a low-temperature shift catalyst is used to convert carbon monoxide to hydrogen at about 2700–2800°C. Excess nitrogen, carbon dioxide, residual carbon monoxide and inerts are removed in a pressure swing adsorption system. Both plants were rapidly commissioned and operate successfully.

In another process, tested by Uhde, reforming of hydrocarbons in catalyst-filled tubes was combined with the partial oxidation of the reformed gas at 1300°C within a single vessel (Figure 9.10). The *primary* reforming reaction was similar to that of the CRG process and operated endothermically at a low steam/ratio. The composition of the product gas and control of the reaction was

Figure 9.10. Model of the Uhde ammonia plant. Reprinted with permission from Uhde GmbH.

achieved by variation in the steam and feed rates. The methane content was dependent upon the amount of oxygen used in the enriched air, or the flow of oxygen to the partial oxidation section of the process. The process could be adjusted to produce synthesis gas for methanol or OXO reactions by adding part of the hydrocarbon feed directly to the oxidation section. The appropriate carbon monoxide/carbon dioxide and hydrogen/carbon monoxide ratios in the gas composition were obtained by adjusting the volume of secondary feed and the steam/carbon ratio of the primary feed, respectively.

REFERENCES

1. Alwin Mittasch, Early Studies of Multicomponent Catalysts, *Advances in Catalysis,* Vol. 2, Academic Press, N. Y., 1950, p. 81.
2. German patents 292615 (1913), 279582 (1913), and 293585 (1914); US Patent 1330772 (1918); British Patents 27955 (1912), 8864 (1913), and 27963 (1913).
3. Green, *Industrial Catalysis*, Ernest Benn, London, 1928, p. 327; Partington, *J. Soc. Chem. Ind.* **40** (1921) 99R.
4. M. Appl, Brief History of ammonia production, *Nitrogen,* No. 100, British Sulphur Publishing (March-April 1976), p. 49.
5. Byrne, Gohr, Haslam, *Ind. Eng. Chem.,* **24**(10) (1932) 1129.
6. Gard, *Nitrogen*, No 39 (Jan-Feb 1966) 25.
7. *Chemical Week Report (Ammonia)* (Sept 11 1965) 11.
8. Lewis B. Nelson, *History of the US fertilizer industry,* Tennessee Valley Authority, 1990, p. 227.
9. British patents 953877 and 1003702 (1960s); Bridger, Safety in ammonia and related facilities, *AIChE Symposium,* Boston (Sept 7-10 1975).
10. *Catalyst Handbook,* Ed. byM. V. Twigg, Wolfe Publishing Co, London, 1989; M. Appl, Modern ammonia technology, *Nitrogen,* No. 200, British Sulphur Publishing, 1992.
11. Private communication, Alberta Nitrogen Products Ltd, Alberta, Canada.
12. *Materials Technology in Steam Reforming Processes,* Edeleanu, Pergamon Press, London, 1966; J. K. Rostrup-Nielson, *Catalysis, Science & Technology,* Vol. 5, Ed. By Anderson and Boudart, Springer Verlag, Berlin, 1984, p. 1; Howe-Baker, *Hydrocarbon Processing* (Nov 1995) 25 and (May 2000) 39.
13. *Oil Gas Journal Special,* (March 16 1998) 64; Cromarty and Crewdson, *ICI Symposium Thai Cat* (1990).
14. *Gas Making and Natural Gas,* BP Trading Ltd, London, 1972.

AMMONIA AND METHANOL SYNTHESIS

10.1 AMMONIA SYNTHESIS

In the early days, the composition of catalysts discovered and developed by Mittasch that were used by BASF in the plants at Oppau and Leuna was kept secret. Apart from the fact that iron was used and that the process operated at high pressure, little other technical information was available until after the war had ended. When the new synthesis process was investigated in other European countries and the United States, many different catalyst promoters were listed in a large number of patent applications and in the technical literature.[1] The wide variety of processes and catalysts developed during the 1920s reflects the difficulties experienced in establishing suitable operating conditions. These technical difficulties, together with a static demand for ammonia, meant that only a few plants were actually built before 1940 and BASF retained a virtual monopoly on the manufacture of ammonia.

Not all of the potential competitors used the catalyst promoters first introduced by BASF and then later examined by Larson of the US Fixed Nitrogen Research laboratory.[2] Other fused magnetite catalysts, containing promoters such as magnesia, silica and chromium, as well as those with the more efficient alumina, with potash and, perhaps calcium oxide, were used for a further twenty to thirty years. These gave reasonable results in the converters then in use[3] as shown in Table 10.1. Poor catalysts could operate adequately when the synthesis gas was not purified efficiently and a short operating life was the norm.

L. Lloyd, *Handbook of Industrial Catalysts*, Fundamental and Applied Catalysis, DOI 10.1007/978-0-387-49962-8_10, © Springer Science+Business Media, LLC 2011

TABLE 10.1. Typical Ammonia Synthesis Catalysts before 1950.

Composition wt%	A	B	C
Fe_3O_4	98.80	98.2	76.8
Al_2O_3	0.85	–	11.5
MgO	–	1.2	11.5
SiO_2	–	0.2	–
K_2O	0.35	0.4	0.2

Note: Nielson referred to catalysts promoted with Al_2O_3, CaO, K_2O in 1950/52: Nielson, Promoted Iron Catalysts for Ammonia Synthesis, Copenhagen 1950. *JACS*, 74, 963 (1952).

By the time that the large capacity, single-stream ammonia plants were introduced during the 1960s, all of the producers of ammonia had moved to the use of the iron oxide catalyst containing alumina, calcium oxide and potash. Magnetite was no longer produced by burning iron but was obtained naturally, probably from the source discovered by BASF in 1913-15. Magnesia, silica, and the other trace oxides present in the catalyst were, therefore, impurities from the natural magnetite and not added deliberately as promoters. Analytical details for a typical sample of magnetite are given in Table 10.2.

The large new plants contained huge quantities of catalyst and the time taken for reduction led to undesirable delays in starting production. A pre-reduced catalyst that could be activated quickly became available between 1955–60 and was widely accepted. The pre-reduced catalyst was based on the existing oxide catalysts and did not improve operation but could be commissioned more quickly.

The first chemically different catalyst, containing cobalt oxide, was developed by ICI and used in several large energy efficient plants during the 1980s.

TABLE 10.2. Analysis of a Pure Swedish Magnetite (Malmberget A).

Analysis	Content wt%
Fe_3O_4	95.94
Fe_2O_3	2.75
FeO	–
Al_2O_3	0.33
CaO	0.12
K_2O	0.05
SiO_2	0.43
MgO	0.28
TiO_2	0.20
V_2O_5	0.18
S	0.013
Balance	P_2O_5, MnO, Na_2O, CuO, CO_2
Bulk density kg liter^{-1}	2.6–3.3
Surface area m^{2}g^{-1}	0.95

The new catalyst was, however, still based on promoted magnetite containing up to about 20% cobalt oxide.[4] The reaction kinetics were the same and, although the catalyst improved operation in several respects, large catalyst volumes were still required and the actual process did not really change. A new version of the cobalt-containing catalyst, prepared by precipitation from nitrate solution, was later evaluated by ICI.[5] It showed extremely high activity, but the pellets were quite fragile after reduction and of lower density than fused catalysts. The catalyst was not developed further.

Commercial catalysts containing cobalt oxide could be reduced faster and operated at temperatures and pressures considerably lower than iron catalysts. Normal operation was at about 80 atmospheres compared with the usual 150 atmospheres. While catalyst activity was the same, the catalyst was apparently more stable than the iron catalyst over the whole range of pressure and operating conditions.

A significant advance was made in 1979 when BP demonstrated a new catalyst based on ruthenium supported on carbon and promoted with barium and potassium.[5]

10.1.1 Process Development from 1920

The Haber-Bosch ammonia synthesis process led to similar developments in other European countries and the US. As a result, many other commercial processes were being operated during the 1920s. Production rates were very small compared with modern plants but an extremely wide range of operating conditions was introduced together with a number of different catalysts. Despite these initial differences, most plants built in recent years still operate with more or less the same conditions as those chosen for the first BASF process. Some of the early ammonia processes are listed in Table 10.3.

10.1.1.1 *Haber-Bosch Process*

The first reactors at Oppau and Leuna operated at about 200 bar pressure with temperatures in the range 500°-600°C. The system was not quite autothermal, that is, the exotherm from the synthesis reaction was not sufficient to sustain the process given the design of the reactor at that time. The catalyst was charged to a single reactor, somewhat like a tube, equipped with various heating devices used during start-up and operation. A modified Haber-Bosch reactor using chrome/vanadium steel that could operate at 300–350 bar pressure and lower temperatures led to an increase in the conversion of synthesis gas to ammonia because higher pressures and lower temperatures favour the concentration of ammonia at equilibrium. This also made the condensation of product ammonia easier and more efficient. The basic design operated without the use of *sophisticated* injector circulators and high pressure steam was not recovered. It has

TABLE 10.3. Ammonia Synthesis Processes 1920-1950.

Process	Characteristics
Casale (1920)	Developed in Italy as a simple catalyst tube with heat exchange. High pressure up to 750atm.
Haber-Bosch (1920)	Modified Haber converter. Pressures up to 350atm.
ICI (1923)	Tube cooled converter operating up to 250atm.
NEC-Chemico (1928)	Based on the Fixed Nitrogen Research Laboratory work. Developed by NEC and Chemico with co-current cooling for tubes.
Claude (1921)	Pressure up to 1000atm in tube cooled converter with no need for gas circulation.
Fauser (1920s)	Catalyst beds. Inter-bed heat exchange cooling.
Mont Cenis/Uhde (1925)	Low pressure with novel cyanide intermediate for catalyst.
T V A (1940s)	Modern version of tube cooled converter.
Kellogg (1940s)	Developed inter-bed quench cooling with synthesis gas and low-pressure operation.
Topsoe (1940s)	First converters tube cooled. Later quench cooling and radial flow.

been reported that a converter at Oppau, in which 20 tonnes ammonia per day was produced, weighed 70 tonnes, was 12 meters long and had an internal diameter of 0.8 meters.[6] The time taken to change the catalyst and restart operation after a shut-down period was three days. There was a great deal of secrecy about the catalyst used by BASF at Oppau and patents during the 1920s give a confusing impression of the actual catalysts being developed by other companies.[7] This is perfectly reasonable when it is remembered that experimental work was being reported. Most processes eventually used very similar catalyst types.

10.1.1.2 *Claude Process*

Georges Claude began work in France on an ammonia synthesis process during 1917 and, by 1919, l'Air Liquide had formed the Society Chimique de la Grande Paroisse and built a three tonnes per day Claude plant near Bethune. Initially, the pure synthesis gas was derived from hydrogen recovered from various sources such as coke-oven gas, and the liquid nitrogen was obtained from the production of oxygen. L'Air Liquide later helped to form Societé Belge de l'Azote (SBA) who produced ammonia from 1923.

The operating conditions for the process were a pressure of 1000 bar and a temperature of 550°C. Under these conditions, a high conversion of around 40%, close to the theoretical equilibrium conversion, was achieved despite impurities that must have been present in the synthesis gas. The process was autothermal, and even required some form of cooling. The converter was small and weighed only 11 tonnes for a production of 20 tonnes ammonia per day.

Claude ammonia plants did not need to include recirculation of synthesis gas because of the high conversion to ammonia. Instead, four small converters, or 'contacts', were used. Two of the converters were operated in parallel, with the other two in series[8]. The gas was cooled and the ammonia removed between the "contacts". This two-stage procedure combined with intermediate product removal probably accounts for such a high level of conversion, at such an early stage in the evolution of the industry. It was also claimed that the catalyst packed in an iron tube inside the pressure vessel and weighing some 750 kilograms could be changed in only 10 minutes! Since discharged ammonia synthesis catalyst is pyrophoric, one can only wonder about the procedure.

The catalyst developed by Claude was made by oxidizing iron in a magnesia crucible under a strong jet of oxygen and contained 5-10% calcium oxide with "a little alkaline oxide" that acted as promoters. Some magnesium oxide dissolved in the magnetite as it melted.[8] The catalyst was reported to operate for several hundred hours. Most metallic iron at that time contained considerable amount of sulphur, which is a well-known poison for the ammonia synthesis catalyst. The removal of sulphur by burning the iron in air to form oxides of iron was developed by BASF. Although sulphur-free magnetite produced in this way led to the production of a satisfactory catalyst when the appropriate promoters had been added, the use of Swedish magnetite was, and still is, preferred.

A useful procedure adopted by both Claude and BASF was the inclusion of a guard bed to absorb poisons, and to convert any residual carbon monoxide to methane. The guard bed usually contained spent catalyst that had been discharged from the main converters, and operated at a temperature of 400°C. A similar procedure was later used by many other operators to extend the life of the catalyst.

The Claude Process was not very popular, and was developed further by Grande Paroisse to operate at a pressure of 600 bar, and to use pure make-up gas from a nitrogen wash system. The catalyst used in this process appears to be one of the first reported to use calcium oxide as a catalyst promoter.

10.1.1.3   *Casale Process*

Luigi Casale developed a process in Italy that usually operated at pressures up to 600 bar and gave a high conversion. The process operated more or less isothermally, and the high concentration of ammonia in the recycle gas was sufficient to allow product ammonia to be recovered by condensation with cooling water[9] at a temperature of 30°C. In several early patents from Casale, it was shown that magnetite, free from volatile impurities, could be prepared by burning iron metal with oxygen in a magnesia crucible.[10] Various promoters[11] such as magnesia, lime and alumina were added to produce the fused catalyst. Casale was one of the earliest contractors to the chemical industry, and is still active in licensing a modified synthesis loop which is often used to re-vamp existing units.

10.1.1.4   *United States of America*

Construction of the Oppau plant was announced in New York at the 8th International Congress of Applied Chemistry during September 1912 by Dr H A Bernthsen of BASF.[12] At the time, however, demand for ammonia was not sufficient to interest US companies in using the process. The General Chemical Company, who had discussed the sulfuric acid contact process with BASF in Germany, began pilot plant work in about 1913 to investigate ammonia synthesis and to develop catalysts. The Bureau of Soils also established a small fixed nitrogen laboratory at the Arlington Farm Research Station in Virginia.

The National Defense Act of June 1916 incorporated a bill providing for large-scale production of ammonia. By the time that the US had declared war in 1917, several nitrate plants were being considered by a Nitrates Supply Committee. These included an ammonia plant, with three units to make up to 30 tonnes per day of ammonia, at Sheffield, Alabama, designed by the General Chemical Company. Although one small unit, with a capacity of about seven tonnes per day, had been tested by the end of the war, there was difficulty in producing a reliable catalyst and the plant was closed down during January 1919.

The US Fixed Nitrogen Committee visited Oppau during 1919 and was able to discover the main differences between the BASF and General Chemical Company processes.[13] The main difference lay in the process conditions. The synthesis loop of the process in Germany was operated at a pressure of 200 bar, with a reaction temperature in the range 500–600°C. This contrasts strongly with the process in Sheffield, which was operated at a pressure of only 100 bar and a much lower temperature of 400–450°C. Such moderate reaction conditions would have required a significantly more active catalyst to have been successful. However, the catalyst used by BASF in Germany was derived from iron, promoted with alumina and was more active than the catalyst used by the General Chemical Company in America which was made from "spongy" iron, promoted with sodamide. The Sheffield catalyst, which was not very efficient. It was made by impregnating pumice with iron nitrate before heating it at 550°C and then reducing the oxide. Sodamide was deposited on the 'spongy' metal by a treatment with ammonia and melted sodium at 450°C. The catalyst was deactivated by traces of water.

Despite the problems at Sheffield, the Atmospheric Nitrogen Corporation, a subsidiary of the General Chemical Company, had built and operated a second plant with the same design at Syracuse, New York by 1921.[14] The original capacity was 15 tonnes per day ammonia but it was later increased to 40 tonnes per day. This plant eventually used a fused iron oxide catalyst, promoted with alumina and potash, developed at the Fixed Nitrogen Research Laboratory by A T Larson. It had first tested a catalyst developed by de Jahn of the General Chemical Company.

The Fixed Nitrogen Research Laboratory, at the American University in Washington DC, had been set up in March 1919 to continue work on nitrogen fixation. Larson had determined that fused iron oxide containing 3% alumina and 1% potash was the most satisfactory catalyst. A guard bed of catalyst was also recommended to remove impurities from the synthesis gas before it entered the main reactor. Eventually the process was known as the American Process. Although it was originally based on the use of electrolytic hydrogen, many plants were built by the Nitrogen Engineering Corporation, later the Chemical Construction Corporation (Chemico), using other sources of synthesis gas.[15] The Chemico ammonia process with a tube-cooled converter operating at 350 bar and 500°C maximum catalyst temperature was used in about 25% of US ammonia plants up to the 1960s.[16] The relatively low catalyst temperature was probably a result of the use of a refrigerated ammonia cooler/condenser which gave a lower ammonia content in the converter make-up gas and removed catalyst poisons.

10.1.1.5 *Mont Cenis/Uhde Process*

Friedrich Uhde signed a contract to build an ammonia plant for Gewerkschaft Mont Cenis of Sodingen during 1925. Uhde's ambition had been to design an ammonia plant which did not infringe the BASF patents. He was able to do this after meeting Ivar Cederberg from Sweden who had been active in ammonia process design for many years. Cederberg, with the Norsk Hydro-Elektrisk Company had developed catalysts based on complex iron cyanides.[17] Other metals such as ruthenium or osmium could be used instead of iron but were more expensive and unobtainable in large quantities. Following reduction to alpha-iron at 300° the catalysts derived from the complex cyanides were active at temperatures as low as 400°C, but suffered from poor thermal stability.[18]

The Mont Cenis Process was operated at a pressure in the range 100-120 bar, and at temperatures below 450°C because the catalyst was unstable at higher temperatures. Several of these low-pressure plants were built before 1930. Since then Uhde has designed many conventional ammonia plants and the associated ammonia converters (Figure 10.1). One of his innovations was the use of tube-cooled converters to improve heat exchange and to preheat inlet gas to maintain a more uniform temperature through the catalyst.[19]

10.1.1.6 *United Kingdom*

A Nitrogen Products Committee was set up in 1916 by the Ministry of Munitions to consider the production of nitrogen compounds.[20] H. C. Greenwood,

Figure 10.1. A modern Uhde plant for ammonia synthesis. Reprinted with permission from Uhde GmbH.

who had worked with Haber, then constructed a small-scale unit which obtained its synthesis gas by the thermal cracking of ammonia and which operated at 120 atmospheres. A fused iron catalyst promoted with molybdenum oxide was used. Tests led to the recommendation that a 120,000 tonnes per year ammonia plant should be built and a site was purchased at Billingham, England, during 1918.

The proposed plant was cancelled at the end of the war. An official chemical commission did, however, visit Oppau during April/June 1919 and, in April 1921, Brunner Mond Ltd purchased the British Patents taken out by BASF. Brunner Mond also bought the Government site at Billingham and began work on the development of ammonia process. During early tests they used a fused magnetite catalyst promoted with alumina, similar to those used at Oppau and in the US. The catalyst was produced by fusion with an electric arc in a water-cooled iron pot. The magnetite was probably produced by burning iron in the form of horseshoe nails. A small unit producing about a tonne of ammonia per day was successfully operated at Runcorn in Cheshire from May 1921. A guard bed of synthesis catalyst was incorporated to remove carbon monoxide and other catalyst poisons from synthesis gas. The main plant was then built at Billingham and began operation on 22 December 1923. It operated at 200 bar and produced 30 tonnes per day ammonia. Capacity was increased and several larger tube-cooled plants (Figure 10.2), operating at 250 bar, had been commissioned by 1930. Total production then exceeded 270,000 tonnes per year.

Figure 10.2. Reactors in an early process for the manufacture of ammonia.

Synthesis gas was produced from coke via water gas/producer gas. The iron-chromium, high-temperature shift catalyst used was probably the first to be made in the form of pellets.

10.1.2 Ammonia Synthesis Catalysts

10.1.2.1 *Catalyst Production*

Natural magnetites with low impurity levels are now used to produce ammonia synthesis catalysts. A typical catalyst analysis is shown in Table 10.4. The significance of promoters in the catalyst was only understood by Mittasch when he tested catalysts made from a Swedish magnetite.[21] Later on, other companies

TABLE 10.4. Typical Composition of Modern Ammonia Catalyst.

Property	Value
Composition (-wt%)	
Magnetite	Bulk
Alumina	2.5–3.0
Calcium oxide	2.0–2.5
Potash	0.4–0.6
Magnesium oxide	Less than 0.5
Silica	Less than 0.5
Size (mm)	6–10; 1.5–3.0; other sizes available
Surface area (m^2g^{-1})	
Oxidic	~1
Pre-reduced	9–14

made catalysts from a variety of magnetites. These included burnt horseshoe nails and drum scrap, or any other source of cheap iron provided that volatile poisons could be removed during oxidation at high temperature.

The catalyst is prepared by fusion of a mixture of magnetite and the promoters, which normally includes alumina and potash, in an electric arc furnace at a temperature of approximately 1700°C. The melt should be thoroughly homogeneous, before being chill-cast in a shallow tray, and then broken into small particles. The particles are then sieved into several size fractions, to be used according to the requirements of the converter. Any oversized material is normally crushed again, sieved, and all of the combined undersized particles are then recycled back to the furnace. The catalyst mix crystallises rapidly when cooled below a temperature of 570°C to form a solid solution of wustite and alumina in the magnetite crystal lattice. Promoters are usually added as carbonates, oxides or hydroxides, although potassium is often added as a nitrate. A visual examination of the particles will show any inadequate mixing of promoters as white specks.

The final product is essentially a mixture of magnetite (Fe_3O_4) and wustite (FeO) with a surface area in the range 1–2 m^2/g. Magnetite has a spinel structure of cubic packed oxygen atoms with one third ferrous and two thirds ferric ions occupying the 24 tetrahedral and octahedral positions.

Alumina is incorporated as a solid solution of the iron aluminate spinel, hercynite, in the crystal lattice. The alumina concentration should be less than the solubility of alumina in magnetite. This corresponds to a maximum content of about 3% alumina. Any excess of alumina does not go into solid solution, and leads to a reduction in catalytic activity, particularly when using catalysts promoted with alumina. The presence of alumina as a structural promoter also leads to the formation of wustite and stabilizes the reduced catalyst. Small amounts of magnesia can also dissolve into magnetite and act as a promoter. The calcium component exists in the form of ferrites or aluminates by neutralizing acidic components—such as silica—and protects the potash that activates the catalyst.

Calcium compounds can also help in stabilizing the reduced catalyst but do have a potentially adverse effect as high levels make catalyst reduction more difficult.

The potassium ferrites form during catalyst production increase catalyst activity up to about 0.8% potash but above this level activity falls. Ammonia catalysts are said to be *doubly-promoted*. This is because alumina—a *structural* promoter—helps control and stabilize both the porosity and surface area of the reduced catalyst; on the other hand, potash—an *electronic* promoter—increases the catalyst activity. Other promoters may have similar effects but usually interact with poisons or impurities to protect the catalyst.

10.1.2.2 Pre-reduced Catalysts

Ammonia synthesis catalysts are prepared in the form of magnetite which must be activated by reduction to metallic iron before use. In plants using large volumes of catalyst, this can take several days. The reduction process must be carried out slowly to avoid the accumulation of high levels of water in the reducing gas. Reduced catalysts lose activity if they are subjected to a high partial pressure of water vapour and since the linear gas velocity is relatively low in large diameter converters, the reduced catalyst can suffer some re-oxidation due to back mixing of the reducing gas. As a rule, the water content of the gas exiting the converter has been limited to a maximum of 3000 ppm, so that a reasonable compromise between maximum activity and the shortest realistic reduction time can be achieved.

For this reason, it is now common practice for catalyst producers to manufacture ammonia synthesis catalyst in the pre-reduced form. This is a routine procedure that takes in specially designed units where the process can be carefully controlled (Figure 10.3). After reduction, the catalyst is carefully stabilized by a controlled flow of a mixture of nitrogen gas and air so that less than 10% of the iron is re-oxidized with a thin film of oxide forming around the iron particles. This allows the catalyst to be loaded into commercial reactors and commissioned in a fraction of the time taken with standard catalyst. Reduction begins at about 330°C compared with 400°C for standard catalyst. One of the main advantages in using pre-reduced catalyst is that a smaller capacity start-up heater can be used, and that the reduction period can be much shorter. The rapid reduction of pre-reduced catalyst in the top bed of a multi-bed converter allows the exothermic ammonia synthesis reaction to begin much sooner. The heat from this exotherm is sufficient to supplement the heat from the smaller start-up heater so that ordinary catalyst can be reduced in the lower beds, and overall, the production of ammonia can begin within one to two days, compared with the usual four to seven days.

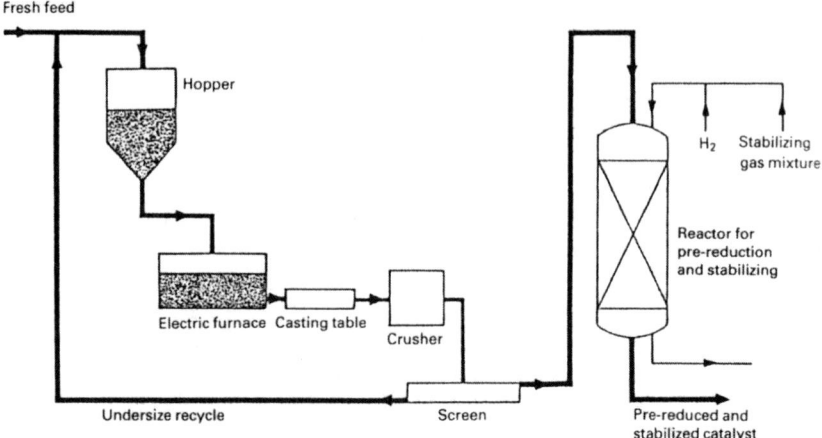

Figure 10.3. Flow sheet for a plant to make a pre-reduced ammonia synthesis catalyst by the fusion route. Reprinted from *Catalyst Handbook*, 2nd ed., Ed. by M. Twigg, Copyright (1989) by kind permission of M. Twigg.

10.1.2.3   *Loading Catalyst to Converter*

Catalyst can be loaded from a small hopper or a large bag fitted with a suitable flexible chute. The chute should be moved around during charging to maintain a regular level in the converter. Tube-cooled converters can be filled completely by pouring catalyst into the top of the opening around the tubes. The tubes must first be blocked to prevent catalyst entering during the loading procedure. When loading a quench converter it is important to maintain the design bulk density in all parts of the bed. This requires fixed levels to be marked around the converter diameter and a predetermined weight of catalyst to be loaded as uniformly as possible. Vibrators are often used to settle the bed carefully, which is necessary to achieve maximum activity.

Pre-reduced catalyst can be re-oxidized if heated in air or not handled properly in hot climates. Loading should be completed as quickly as possible while making sure that friction or vibration is limited. The temperature of the catalyst bed should be checked and access of air restricted. Nitrogen can be used to quench any hot spots.

10.1.2.4 *Catalyst Discharge from the Converter*

Spent ammonia synthesis catalyst contains very small crystallites of iron, which are extremely reactive, much more so than bulk iron metal. These residues are potentially pyrophoric in air and can react with water to produce hydrogen. Extreme care is therefore required when they are discharged from a converter.

The catalysts are not usually oxidized before being discharged from a converter but, after cooling to ambient temperature, it is important that the converter should not be exposed to air. The converter is usually purged with nitrogen. Catalyst suppliers and contractors provide special instructions for the procedure. Water should be sprayed onto catalyst as it is discharged from the converter.

10.1.3   Catalyst Reduction

Ammonia synthesis catalysts must be carefully reduced before use. Either hydrogen or, more usually, synthesis gas can be used, at pressures in the range 70–100 bar. Gas leaving the bed being reduced should be cooled until reduction is complete. Providing that there are no unexpected problems, the complete procedure within a given bed for up to four beds can take about four to seven days.

10.1.3.1 *Reduction of Oxidized Catalyst*

A typical converter, in a plant with a dosage capacity 1000 tonnes per day ammonia, will hold 100–200 tonnes of catalyst, which generates as much as 35–70 tonnes of water during reduction. Ammonia synthesis begins before all of the catalyst has been reduced, the aqueous ammonia produced must be used in the most convenient way. The reduction procedure is deliberately carried out at a very low rate, to ensure optimum activity of the final catalyst. The presence of high partial pressures of water vapour in contact with the freshly reduced catalyst results in the growth of the small crystallites within the catalyst, via a re-oxidation mechanism. This growth in crystallite size is inevitably accompanied by a reduction of the metal surface area, and hence an irreversible loss of activity. The concentration of water vapour in the process gas is therefore limited to a maximum of about 3000 ppm. Concentrations higher than these have been allowed but this requires good gas mixing in the wide diameter beds and rapid disengagement of water from the reduced particles. Back mixing of reducing gas, containing water, can lead to re-oxidation and loss of activity. Gas flow should therefore be as high as possible, the temperature of the bed carefully controlled and water content of reducing gas at the converter inlet should also be as low as possible. The cumulative volume of water collected during reduction can be used as a practical indication of how reduction is proceeding and should be carefully measured.

During the early stages of catalyst reduction, the maximum rate of gas flow is limited by the thermal capacity of the start-up heater to maintain the temperature of the bed. As more catalyst becomes reduced, the exothermic synthesis reaction begins. The heat from this exotherm supplements the output from the start-up heater, and the gas rate can be increased. The rate of increase of a bed inlet temperature should be limited to about 5°C per hour until a maximum outlet temperature of 500°C is reached. During this procedure, the temperature of the remaining unreduced catalyst beds should be just below the reduction temperature that is about 350°C for typical catalysts. As more beds are reduced, gas rates can be increased and the reduction rate is faster despite the larger catalyst volumes in the lower beds.

10.1.3.2 *Reduction of Pre-reduced Catalyst*

Pre-reduced catalysts can be reduced much more quickly than oxidized catalyst because only about 10% of the catalyst is re-oxidized during stabilization. The procedure, therefore, only takes about one day. Early difficulties when pre-reduced catalysts were unstable and could overheat on exposure to air have been overcome and the catalyst is now very stable. It is wise, however, to take sensible precautions while loading a converter because of the large quantities of catalyst involved.

The proportion of pre-reduced catalyst used depends on the time allowed for reduction and how easily the by-product aqueous ammonia can be used. Operating costs are balanced with the higher price of the catalyst. Using pre-reduced catalyst in only the top section of a reactor simplifies the early stages of reduction and speeds up commissioning. A full charge of pre-reduced catalyst allows much earlier production of ammonia but is more expensive. Although water evolution from pre-reduced catalyst is normally low, it is still usual to reduce only one bed at a time.

10.1.3.3 *Mechanism of Catalyst Reduction*

α-Iron is produced when ammonia synthesis catalyst is reduced.[22] The catalyst is initially non-porous, with a surface area of only $1–2$ m^2 g^{-1}, but on reduction, the catalyst becomes porous and develops a surface area around 20 m^2 g^{-1}. The reduction mechanism is unusual and is controlled by the wustite in the catalyst. Large pores form at the surface as wustite surrounding the magnetite crystals is reduced to form iron at a lower temperature than that normally required for bulk catalyst reduction. This provides iron nuclei for further reduction. Ferrous ions then diffuse through the magnetite structure to the reaction centers where they combine with magnetite to form wustite, which is also reduced, and the cycle continues, gradually forming the pore structure needed. There is an optimum wustite concentration so that diffusion of ferrous ions is not restricted.

Figure 10.4. Scanning electron micrographs (SEM) of ammonia synthesis catalyst: (a) unreduced catalyst showing well-formed crystals of magnetite 230X; (b)-(d) reduced catalyst at different magnifications showing porous iron pseudomorphs at magnifications of 2300, 7700 and 23,000. Reprinted from *Catalyst Handbook*, 2nd ed., Ed. by M. Twigg, Copyright (1989) by kind permission of M. Twigg.

After reduction, the catalyst consists of porous magnetite pseudomorphs in which the original structure has been replaced by small, plate-like iron crystals separated by the newly created pores (Figure 10.4). The structure is stabilized by alumina that has come out of solid solution to be re-deposited in the pores where it prevents crystal growth as more magnetite is reduced. Up to 90% of the iron surface is covered by a film of potash that promotes the adsorption of molecular nitrogen and weakens the nitrogen-nitrogen bond prior to reaction with adsorbed hydrogen. Ammonia is less strongly bound to the catalyst surface and rapidly desorbs.[22,23]

The formation of ammonia on the reduced iron surface is extremely structure sensitive and the 111 and 211 planes are by far the most reactive of the five possible crystal surfaces. The close-packed 111 plane, for example, at the base of each plate-like crystal can expose three layers of iron atoms. Potential active sites may, therefore, have seven near neighbour iron atoms that are more active and less easily poisoned than sites with fewer near neighbour atoms. The most active sites are known as C7 sites.[24]

10.1.4   The Ammonia Synthesis Process

Since the Haber process was first introduced, there has been a gradual evolution of both the process and converter designs as the production capacity of the plants has increased. Until the 1950s, there were very few changes to either the converter, synthesis loop or the catalyst formulations used by individual companies shown earlier in the chapter, in Table 10.3. This was partly because of inertia and because most plants operated with several ammonia converters each holding a few tonnes of a relatively cheap catalyst that could be conveniently changed during shutdown periods. This situation altered during the 1960s as larger capacity, single stream plants, using steam reforming and with a single ammonia converter, were built in all parts of the world.[25]

10.1.4.1   *The Ammonia Synthesis Loop*

The ammonia synthesis reaction occurs in a loop where synthesis gas can be circulated continuously through a converter containing the catalyst (Figure 10.5). Ammonia is condensed and removed from the loop while synthesis gas, known as make-up gas, is added to maintain the design space velocity through the catalyst. In most ammonia plants, the inert gases, such as methane and argon, accumulating in the loop are continuously removed in a purge gas stream, so that they do not reach an unacceptable level. There is no longer any need in modern plants to install a guard bed to protect the catalyst now that steam-reformed synthesis gas is poison free. Fresh make-up gas is usually added before ammonia is condensed. This means that poisons such as traces of compressor oil or oxygen compounds are removed with the liquid ammonia. Typical compositions of circulating gas, converter outlet gas and purge gas for a low-pressure ammonia loop are show in Table 10.5.

A *liquid nitrogen wash* used to remove inerts from the make-up gas which then contained less than 240 ppm methane was incorporated into some of the early plants. However, it was found that the inerts were preferentially dissolved in the product ammonia and they did not accumulate in the recycle gas in

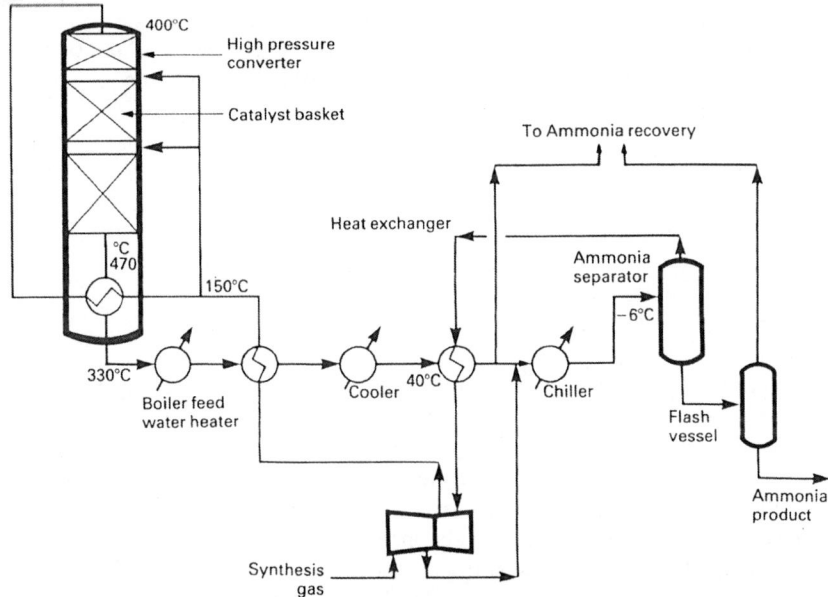

Figure 10.5. Ammonia synthesis loop. This layout is typical of a loop for a large capacity (1000 tonne day^{-1}) plant. Preheat of the converter feed gas to 150^{0}C allows high-grade heat recovery from the converter exit gas. Reprinted from *Catalyst Handbook*, 2nd ed., Ed. by M. Twigg, Copyright (1989) by kind permission of M. Twigg.

those plants which were operated at high pressure. In both of these cases, a purge was not required, and any apparent purge was entirely involuntary.

As the synthesis loops became simplified, and were operated at lower pressures, it became necessary to cool the converter exit gas with air or water, and then to incorporate a refrigeration stage using liquid ammonia. The system operated with two separators and more product ammonia was removed from the

TABLE 10.5. Gas Compositions in a Low Pressure Ammonia Loop.

Composition vol%	Circulating Gas	Converter Outlet	Purge Gas
Nitrogen	21.0	18.2	21.0
Argon	3.1	3.4	3.6
Helium	0.4	0.5	0.5
Hydrogen	63.4	54.9	63.0
Methane	10.0	11.0	12.0
Ammonia	2.0	12.0	–

Notes: 1. Make up gas equivalent to methanator outlet gas containing <5 ppm carbon oxides.
2. Purge gas is used for hydrodesulfurization and primary reformer fuel.

loop, resulting in a lower concentration of ammonia in the recycle gas. Since the maximum conversion of the synthesis gas to ammonia is limited by the thermodynamics of the reaction, the lower concentration of ammonia in the recycle gas led to an overall increase in the conversion of synthesis gas to ammonia. It also became possible to operate with a smaller converter, and still maintain the same level of production. By the 1960s, refrigeration had become essential when operating the low-pressure ammonia converters, as the equilibrium concentration of ammonia at 50-bar pressure is only about 12%, and the concentration of ammonia in the recycle gas was reduced to about 2%.

10.1.4.2 Converter Design

The majority of early ammonia plants used adiabatic beds of catalyst or tube-cooled converters that acted like a heat exchanger with the cold synthesis gas passing through the tubes to cool the catalyst. Tube-cooled reactors, such as those introduced by the Tennessee Valley Authority (TVA), did not operate isothermally and the exothermic reaction led to both axial and radial temperature profiles developing. The temperature difference depended on the number of tubes passing through the catalyst bed.[26] The converter design took this into account by maximizing the number of tubes although the temperature profile could only be controlled by changing the inlet gas temperature. There were problems in loading catalyst into the space between tubes to achieve the right packing density as well as in discharging catalyst when it was deactivated. Large tube-cooled converters were also expensive.

Modern converters are designed with several catalyst beds in which the hot gas can be cooled at each bed exit either by heat exchange or by the addition of cold synthesis gas, often referred to as quench cooling (Figures 10.6 and 10.7). Quench cooling has usually been preferred in plants using multi-bed converters despite the disadvantage of using a larger catalyst volume and having to by-pass some of the catalyst with a significant volume of the synthesis gas. The same process is detailed for a three or four bed operation is shown in Table 10.6.

Problems were experienced in wide, multi-bed converters, with gas flowing axially through the beds, because big catalyst particles were required to limit the pressure drop through the catalyst. Since activity is inversely proportional to particle size, increased volumes of catalyst were needed and the large reactors increased the capital cost of a plant. By designing converters in such a way that gas flowed radially through the catalyst bed, it was possible to decrease the overall pressure drop and to use smaller catalyst particles that did not suffer from the limitations of pore diffusion to the same extent, and thus showed greater activity per unit volume.[27] Gas distribution problems in the top bed—resulting from a low-pressure drop—were overcome by the use of an improved converter design. In one case, an axial-radial flow system was used. In another,

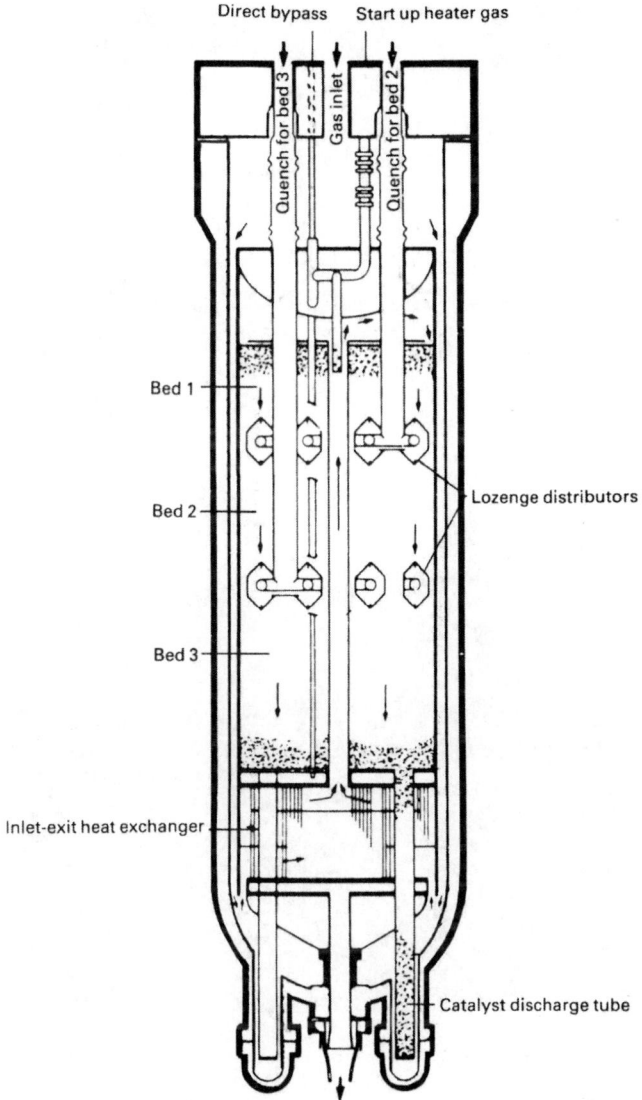

Figure 10.6. ICI quench converter for ammonia synthesis. Quench gas is injected and mixed by means of lozenges which also allow gravity discharge of the catalyst bed. Reprinted from *Catalyst Handbook*, 2nd ed., Ed. by M. Twigg, Copyright (1989) by kind permission of M. Twigg.

Figure 10.7. Topsøe converters for ammonia synthesis. Reproduced with permission from Haldor Topsøe A/S.

which was particularly useful in large capacity ammonia plants, transverse gas flow through the catalyst beds, in a horizontal converter was used.[28] Some recent ammonia converter designs are outlined in Table 10.7.

TABLE 10.6. Operation of Low-Pressure Ammonia Converters at 140–150 Bar.

Operation	Bed 1	Bed 2	Bed 3	Bed 4
A:				
Catalyst volume (m³)	12	22	32	—
Temperature inlet (°C)	410	405	400	—
Temperature outlet (°C))	500	470	450	—
Ammonia concentration (vol %)			12	
Cooling	Quench	Quench		
B:				
Catalyst volume (m³)	10	14	18	22
Temperature inlet (°C)	410	400	400	400
Temperature outlet (°C)	480	480	480	450
Ammonia concentration (vol %)				12
Cooling	Quench	Quench	Quench	

Notes: 1. Total synthesis gas volume for 1000 tes/day production is >600,00 m³/h with 33—42% being used as quench at 140°C.
2. At 220 atm pressure, 3 beds containing 7 m³, 12 m³, 22 m³ (ΔT 95°C, 45°C, 40°C) gave a conversion from 3% $\rightarrow$ 15% ammonia.

10.1.5 New Catalyst Developments

At the stage of development of the ammonia synthesis reaction as described in the previous sections, the main problem was the low level of conversion of synthesis gas to ammonia. This could be increased by operation at yet higher pressure, but this would not be cost effective. The other option would be to operate at lower temperature, under which conditions, the equilibrium concentration of ammonia would be appreciably higher. Unfortunately, the catalysts available at the time were not sufficiently active to operate at the lower temperature required to make a meaningful difference to the concentration of ammonia in the product

TABLE 10.7. Ammonia Converter Design.

Contractor	Converter Design
Kellogg	a. Quench cooled, axial flow reactor with 3–4 beds using 6–10 mm catalyst.
	b. Horizontal reactor, with down flow through three shallow beds and quench cooling. Small 1.5–3.0 mm catalyst.
Topsoe	Two radial flow beds with intermediate heat exchange and possibly quench cooling in a single vessel. Small 1.5–3.0 mm catalyst.
Ammonia Casale	Axial-radial flow through three catalyst beds with inter-bed heat exchange or quench cooling. A second converter holding the third bed has been used.
Uhde	Three catalyst beds; two in first converter with inlet/outlet heat exchange and waste heat boiler and a third in a second converter with a waste heat boiler.
C F Braun	Three beds in separate converters with an inlet/outlet heat exchanger after the first, and waste heat boilers after the second and third.

gas, and hence a significant difference to the economics of the process. Any real improvement to the process was therefore dependent upon the development of a new catalyst with significantly improved activity. The clear target would be a catalyst with sufficient activity to give a satisfactory level of conversion at the same pressure as that within the reformer.

10.1.5.1 *Magnetite Catalyst Containing Cobalt*

The surface area of the ammonia synthesis catalyst is only about $1-2$ m^2g^{-1}. It was known from experience with other catalysts that precipitation from aqueous solution always led to a product with a much higher surface area than one pre-pared by fusion, and therefore potentially at least, more active sites per unit weight of catalyst. This process route was studied in ICI, initially by Topham. Subsequent developments led to the preparation and testing of a precipitated catalyst containing cobalt which showed levels of activity approximately 3 fold higher that of the best conventional catalyst available at the time.[4a] Ammonia could be synthesised at temperatures below 350°C, but the catalyst was not commercialised for several reasons. The major reason was that the pelleted ox-ide was significantly weakened during the reduction procedure, and became too weak to withstand the rigours of an axial flow converter. It also suffered from shrinkage during reduction, leading to settling of the bed, and the likely devel-opment of hot-spots.

The ICI cobalt catalyst, which is reported to contain 10-20% cobalt oxide, was developed for use at 80 bar synthesis pressure and maximum temperatures as low as 460°C. These conditions were significantly different from the usual single-stream ammonia plants which operated at 150 bar or higher and tempera-tures up to 500°C.

From experimental work with precipitated cobalt-iron catalysts it appeared that the cobalt reacted with alumina giving cobalt spinels, which helped to form smaller iron and cobalt crystallites during reduction and to increase the surface area to make the catalyst more stable and less sensitive to oxide forming poi-sons. In the presence of potash, cobalt was able to decrease the reduction tem-perature.

Development work was successful and, following a typical fusion produc-tion procedure, the new catalyst could be reduced and operated at lower temper-ature under normal conditions in conventional plants. Reduction began at 380°C in the new AMV and LCA processes introduced by ICI in 1985 and 1988 re-spectively. The AMV plants have operated at more than 100% capacity of a pressure of only 60 bar. The life of the new catalyst has been very satisfactory, and could be operated of pressures much closer to those of the reformers, albeit with large converters.

10.1.5.2   *Ruthenium Catalyst*

The announcement in 1979 by BP of a ruthenium catalyst was the first real advance in improving the process.[29] The catalyst was more active and operated at a lower temperature to produce a higher equilibrium ammonia concentration. A relatively high operating pressure was still needed, however, when using a bed charged with Ru catalyst in conjunction with other beds containing magnetite catalysts.[30] The catalyst was reversibly poisoned by some of the impurities present in typical synthesis gas.[31]

10.1.5.3   *Catalyst Preparation*

Little information is available on the commercial production of the ruthenium synthesis catalyst. A typical catalyst prepared according to a Canadian patent has, however, been compared with typical iron catalysts. Many other patent references have also been included in a review of non-iron catalysts.[32] Descriptions stress the importance of the thermal treatment of the carbon support to control its structure.

Carbon is heated up to at least 1000°C in an inert or reducing atmosphere. The acid surface of the carbon then becomes basic with no effect on the pore structure. The heat treatment deactivates the carbon against the reaction with hydrogen to form methane during subsequent operation. Commercial coconut charcoals, or those produced from hard woods or cellulose, can be used. It is important to use a support with a relatively low surface area of less than 700 m^2/g.

The carbon can be impregnated with the active metal and two promoters, in the following sequence, being dried and heated to 250°C between each stage:

- 2% of barium as nitrate;
- 4% of ruthenium as chloride which is more stable than other salts and gives a better metal distribution;
- 12% of potassium or cesium as hydroxide.

A black, shiny catalyst is produced. Vacuum impregnation can give a uniform distribution of the metals in the carbon although this procedure may not be used for large-scale production. Addition of ruthenium as a volatile carbonyl was also considered during early development. Before use, the catalyst is activated by heating in synthesis gas for twelve hours, at a temperature of 400°C and pressures up to 25 bar, until more than 95% of the chloride content has been removed.

Two important points are significant with catalysts prepared in this way. The order of impregnation influences the catalyst activity and the catalyst is sensitive to hydrogen concentration. A typical 3:1 hydrogen-nitrogen ratio inhibits the activity of ruthenium. Much better operation is possible at temperatures as low as 400°C with 1:1 hydrogen-nitrogen ratio.

Further work has shown that cesium is a better promoter than potassium and that the alkali hydroxides are, surprisingly, reduced to a metallic state during catalyst activation. This seems thermodynamically improbable, but may result from the high heat of adsorption of the alkali metal hydroxides on carbon, which leads to charge transfer to the carbon, and could drive the reduction of the hydroxide to the metal. The alkali promoter may neutralize residual chloride ions and develop catalyst activity during reduction as ruthenium ions diffuse from the lattice to form crystallites on the edges of graphite crystals.

Ruthenium, promoted by alkali metals and barium and then supported on carbon, has provided a significant increase in activity compared with iron catalysts. The high cost of ruthenium when compared with iron must have led to it being used with the cheap support material. In retrospect, the complex interaction between carbon, alkali metal and barium with the active ruthenium led to the development of a successful catalyst. A magnesium oxide support has now been claimed to give a longer life than carbon.[33]

10.1.5.4   *Full-scale Operation with Ruthenium Catalyst*

The new ruthenium catalyst has been used in the Pacific Ammonia Inc plant at Kitimat in British Columbia since 1992.[31] Ammonia had been made there, from the methanol plant purge gas, since 1986 using a conventional iron synthesis catalyst. The new ruthenium catalyst converter was in series with the old converter. Although in 1992 there was no additional synthesis gas to increase production capacity, the ruthenium catalyst operated well in a radial flow reactor and reduced both the steam and electricity used by 30-40% and 5-10% respectively. The new catalyst was said to be twenty times as active as the iron catalyst, and the effluent gas contained about 20% ammonia.

The catalyst was reduced at 300°C, and took less than one day, with the evolution of only small amounts of water. The catalyst was operated at lower pressures and temperatures than the iron catalyst, and in marked contrast, could also tolerate the presence of 4000 ppm of carbon monoxide, with little or no effect on activity.

Other large ammonia plants are now using single beds of the ruthenium catalyst in conjunction with the magnetite catalyst. They confirm the higher activity of ruthenium, with the benefits of a lower operating pressure and temperature, while maintaining a high concentration of ammonia in the exit gas. For future process designs, the need for less gas compression and less high-pressure equipment will lead to lower operating and capital costs. Two plants in Trinidad, using one bed of iron catalyst and three beds of ruthenium catalyst, have operated for several years.[32] It is reported that the capacity of typical 1100 tonnes per day plant designs could be increased to 1850 tonnes per day, provided there is sufficient reformer capacity.

10.2   METHANOL SYNTHESIS

Experiments during the nineteenth century had shown that certain oxides could be used as dehydrogenation catalysts. For example, Jahn dehydrogenated methanol by passing the vapour over finely divided zinc or zinc oxide to produce a stoichiometric mixture of hydrogen and carbon monoxide.[35]

Subsequently, patents covering the conversion of synthesis gas to complex mixtures of organic oxygen compounds, including methanol, were issued to BASF during 1913.[36] This followed work by Mittasch and Schneider. Full-scale production of methanol was not attempted, however, until 1923. By that time high-pressure equipment had been in operation for several years in the new ammonia process. The methanol process was developed by Piers and the plant, built at Leuna, used mixed zinc oxide-chromic oxide catalyst. The use of metallic iron for the internal parts of the reactor was avoided to prevent the formation of the volatile iron pentacarbonyl. The would have decomposed on the surface of the catalyst, to deposit finely divided iron metal, which in turn would have promoted the exothermic formation of methane.

10.2.1 High-Pressure Synthesis

Although coal based water gas was used as feed to the methanol plant the catalyst was probably developed in conjunction with work on coal hydrogenation and the Fischer Tropsch process. The aim of the process had been the production of a range of liquid hydrocarbons but it was found that the high-pressure reaction with water gas formed very pure methanol. The methanol was tested as a fuel for automobiles, but this was unsuccessful because of *knocking* in the low compression engines then being used. Other applications for methanol were soon developed as, for example, bakelite was introduced.

10.2.1.1   *Zinc Oxide-Chromium Oxide Catalysts*

The BASF catalysts were made by mixing solid chromic acid with an aqueous suspension of zinc hydroxide, drying the paste and forming granules. Later an even simpler catalyst was made by mixing zinc oxide powder with chromic oxide, making a paste with water and then granulating.[38] The catalysts were not particularly active and heat evolution had to be carefully controlled during reduction.

During the 1920s the methanol process and various catalysts were widely studied both in Europe and in the US. It has been assumed that Patart, in France, investigated zinc oxide catalysts, containing other metals, as a result of Jahn's observations. Patart operated a semi-technical scale reactor with a zinc oxide catalyst supported on asbestos. The patent, issued in 1921,[39] and other published work[40] gave many details of methanol production. Later on Patart worked with zinc oxide catalysts containing copper oxide although best results were obtained with a 3:1 molar mixture of zinc oxide and chromic acid.[41]

Active and stable catalysts containing an excess of zinc oxide were eventually made by co-precipitation of soluble zinc and chromium salts. Natta did, however, show that the natural zinc carbonate, Smithsonite, which contained impurities such as cadmium, magnesium and copper, had a high stable activity.[40] While copper oxide in combination with zinc oxide or other supports was an active catalyst, it was unstable at high temperature and easily poisoned by the impurities in water gas.

From about 1930 onwards, all catalysts used industrially, for example by BASF, Du Pont, Montedison and ICI, were based on zinc oxide stabilized by the high melting chromium oxide. The actual composition depended largely on whether the catalyst was precipitated or was just a simple mixture of oxides.

The catalysts were prepared either by mixing or precipitation of the components, drying, followed by calcination in the temperature range, 400–450°C. The catalyst powder was then compacted and formed into granules or pellets. When the temperature of the catalyst rose above 200°C, it was found that the chromium oxide component, particularly in the case of precipitated catalysts, was partially oxidised to the hexavalent state. However, as the temperature rose higher, the hexavalent chromium decomposed back to the trivalent state. Consequently, the high chrome catalysts prepared by mixing the oxides must have been partially pre-reduced before use, because of the redox cycles. Nevertheless, they often cracked or lost particle strength. All precipitated catalysts had to be calcined at a high enough temperature to limit the presence of hexavalent chromium, which evolved heat during reduction in the reactor, leading to thermal processing and sintering of the catalyst. Catalyst activity was shown to be proportional to the surface area of the zinc oxide[43] and that these properties could be controlled by the precipitation and calcination conditions.

Zinc oxide-chromium oxide catalysts are often referred to as zinc chromite but there is always a considerable excess of zinc oxide. The catalysts are relatively stable at temperatures when zinc oxide alone would lose activity, the chromium, perhaps present as a spinel,[44] is believed to stabilize the zinc oxide. Maximum activity was claimed for precipitated catalysts containing 20-30% chromium oxide.[45] However, these catalysts were rather unstable, due to shrinkage in use, and longer life could be achieved with a lower chromium contents. From the limited information available, precipitated industrial catalysts generally contained less than 20% chromium oxide as shown in Table 10.8.

TABLE 10.8. High-Pressure Methanol Catalysts.

Composition	Catalyst Type			
	Mixed Oxides		Low Chromium	High Chromium
Zinc oxide	59–64	69	80	67–68
Chromium oxide	30–26 (CrO_3)	21 (CrO_3)	11–13	20–21
Loss at 900°C	9–10	10	Balance	9–10
Typical impurities:				
Sulfate		up to 1%		
Sodium		<0.5%		
Carbonate		<0.5%		
Water		up to 10%		
Cr_2O_3		<3%-calcined 430°C		
		<4-5%-calcined 400°C		

It was suggested that the relatively unstable mixed oxide catalysts could be supplied in the pre-reduced form that was, effectively, pre-shrunk. While these catalysts were offered for use commercially, there is no evidence that operation was actually improved.[46] In any case, the maximum operating temperature in a high-pressure methanol plant was limited to 390°C by the onset of methanation so that a stable activity was more important than high activity in the multibed reactors used.

Lazier of Du Pont published many patents covering the preparation of metal chromites.[47] These were formed by precipitation from a solution of zinc nitrate and chromic acid with ammonia at pH 6.8. The zinc ammine complex obtained was decomposed at about 400°C to give the mixed oxides. As Adkins noted, his copper chromite equivalent was extracted with dilute acetic acid solution to adjust the copper content. It is not clear whether the same treatment was ever used in producing methanol catalysts or even whether the zinc ammine intermediate was produced commercially. One problem with the Lazier preparation was the difficulty in controlling the exothermic decomposition of the ammine that could affect the catalyst activity.

Other companies preferred to precipitate the catalyst from solutions of the cheaper zinc sulfate and chromium sulfate (chrome tan) with sodium carbonate. An active catalyst, with a long life, could be produced although only about 90% of the metals were recovered. This resulted from the variable pH of the solution and the partial solubility of the precipitates at about pH 7 during the precipitation procedure. Significant amounts of sulfate and sodium, in the form of chromium complexes, also remained in the final catalyst although this did not appear to result in loss of activity or to cause operating problems.

10.2.1.2 High-Pressure Operation

Most methanol plants operating before 1965 were small, producing about 150 tonnes per day at 250–260 bar pressure. The small catalyst volume of up to 4.5–

TABLE 10.9. High-Pressure Methanol Process (150 Tonne Day^{-1}, 260 Bar).

Variable	Catalyst/Temperature Distribution				
Bed number	1	2	3	4	5
Catalyst volume (m^3)	1.2	0.6	0.8	0.9	1.0
Temperature inlet (°C)	340	360	360	360	360
Temperature outlet (°C)	364	375	380	380	375
	Quench cooling between beds				
Methanol exit reactor (vol%)				~3	

Synthesis Gas Composition	Volume %
Carbon monoxide	9
Hydrogen	78
Carbon dioxide	0.5
Methane	10
Nitrogen	2.5

5.0 cubic meters was divided into five beds, with inter-bed cooling to limit the temperature rise, in a one-meter diameter reactor. The recycle loop, similar to that in ammonia synthesis, had a high circulation rate of 150,000 m^{3}hr^{-1} so that space velocity in the catalyst was up to 30,000 hr^{-1}. A typical set of operating conditions is shown in Table 10.9 for a plant of this kind, which also shows the temperature increase in each bed, the methanol concentration in effluent leaving the final bed and the composition of synthesis gas entering the reactor.

About 2–5wt% of organic byproducts including about 1% of dimethyl ether, apart from water, formed during the reaction and the amount usually increased as the catalyst aged (a list of by-product formation is shown in Table 10.14.) A significant amount of methane was also formed, particularly at temperatures in excess of 390°C. Much of this was formed on metallic iron, deposited on the catalyst by the thermal decomposition of iron pentacarbonyl. Thus, the maximum operating temperature of the catalyst was severely limited. And the temperature of the reactor was controlled by inter-bed quench cooling with cold synthesis gas.

The loss of activity and selectivity from catalyst sintering and deactivation from the deposition of iron and solids in the bed were partly compensated by increasing the bed temperatures. The main operating problem was usually an increase in pressure drop. For this reason, after about two years' use, the catalyst was carefully removed from the beds in layers and sieved to remove dust. About 70% of the catalyst could often be replaced providing that it had adequate physical strength and an acceptable surface area, exceeding about 35 m^2/gm. A typical catalyst life, after two or three sieving cycles, was 4–5 years. Some surface area measurements on discharged samples from a five-bed unit are given in Table 10.10. The surface area of new, precipitated catalyst was about 50m^2/gm although, following reduction, the surface area increased to 65–70m^2/gm.

TABLE 10.10. Surface Area of Samples Used for Two Years.

Position Bed	1	2	Surface Area (m^2g^{-1}) 3	4	5
Top of bed	65	50	55	45	50
Middle	50	50	35	40	40
Bottom	40	50	40	35	35
Re-use catalyst		>40 m^2g^{-1}	Recovery up to 70% of volume		
Overall life		4–5 years			

Micromeritic measurements on the catalysts shown in Table 10.8 are given in Table 10.11. Before their use, the catalysts—particularly those prepared from mixed oxides— had to be carefully reduced in process gas to convert hexavalent chromic oxide to trivalent chromium oxide.

10.2.2   Low-pressure Synthesis

The first industrial methanol catalysts, used from 1923, were based on zinc oxide-chromium oxide mixtures. Experimental work during development of the process had, however, demonstrated the high activity of zinc oxide catalysts containing copper. This was particularly true if a third refractory oxide such as chromia or alumina was also added.[48] Catalysts containing copper are particularly susceptible to poisoning. At that time the synthesis gas was produced from water gas, and contained poisons such as sulfur and chlorine compounds so that copper catalysts were unsuitable.

When the hydrocarbon steam reforming process could provide poison free synthesis gas, the benefits of the more active copper catalysts were quickly reviewed. It was soon shown that the high activity of copper oxide-zinc oxide catalysts, compared with the zinc oxide-chromium oxide types, particularly when alumina or chromia promoters were added, could revolutionize methanol production. New, more efficient processes were urgently needed in view of the increasing demand for methanol and the economies of high capacity units.

TABLE 10.11. Micromeritics of High Pressure Methanol Catalysts.

Measurement	Mixed Oxide	Catalyst Type Mixed Oxide (reduced)	Low Cr	High Cr
Surface area (m^2g^{-1})	1.6	17.5	51	51
Pore volume (mlg^{-1})	0.04	0.23	0.38	0.17
Mean pore radius (nm)	45	26	13	6.5

TABLE 10.12. Low Pressure Methanol Catalysts.

Licensor	Commercial Catalysts	Reference
ICI	CuO.ZnO.Al$_2$O$_3$	DOS 2302658 London (1973)
Lurgi	CuO.ZnO.Cr$_2$O$_3$	German Pat 1300917 (1969)
CCI	CuO.ZnO.Al$_2$O$_3$	DOS 1956007 Louisville (1969)
Mitsubishi	CuO.ZnO.Cr$_2$O$_3$	DOS 2165378 Tokyo (1971)
Topsoe	CuO.ZnO.Cr$_2$O$_3$	
BASF	CuO.ZnO.Al$_2$O$_3$	DOS 2056612 Frankfurt (1970)
	CuO.ZnO.MnO$_2$.Cr$_2$O$_3$.	DOS 1930003 (1969)
	CuO.ZnO.MnO$_2$.Al$_2$O$_3$.Cr$_2$O$_3$	DOS 2026165 (1970)
	CuO.ZnO.MnO$_2$.Al$_2$O$_3$	DOS 2026182 (1970)

10.2.2.1   Copper Oxide Catalysts

The first full-scale trial of a copper catalyst to be reported took place in the Polish Chemical works at Oswiecim during 1963 but was not successful.[49] Tests showed that the precipitated copper oxide-zinc oxide catalysts available at the time were unstable and either sintered or were poisoned very quickly. Better formulations were investigated to develop large-scale production procedures and increase catalyst stability. Improved catalyst testing procedures provided detailed information on the catalyst structure and activity. This demonstrated that the unstable catalysts contained large oxide crystals and that the metals were distributed unevenly within the structure. Better catalysts required very small crystals which were uniformly distributed.[50] It was soon found that catalysts containing copper oxide-zinc oxide-alumina could be made with the appropriate properties and a large, new plant was operating successfully by 1966.[51] Only one new high-pressure methanol plant has been built since the low-pressure process was introduced.

Soon after the introduction of the catalyst stabilized with alumina, others, containing chromia, were also being produced. Processes based on both of the new catalysts were operating from 1971[52] and many patents were published as shown in Table 10.12.

Synthesis gas was obtained from a variety of sources including the steam reforming of natural gas or naphtha and oil gasification. A flowsheet for a typical low-pressure methanol synthesis plant is given in Figure 10.8. The different reactor designs used in methanol synthesis are listed in Table 10.13 and some are shown in Figures 10.9 and 10.10. The correct carbon dioxide, as well as carbon monoxide, for the synthesis reaction to proceed.

10.2.2.2   Copper Catalyst Production

Until 1960, copper oxide-zinc oxide catalysts were usually precipitated in batches. As with the zinc oxide-chromium oxide catalysts, an alkaline solution, such as sodium carbonate, was added to the relatively acidic solution of the metal

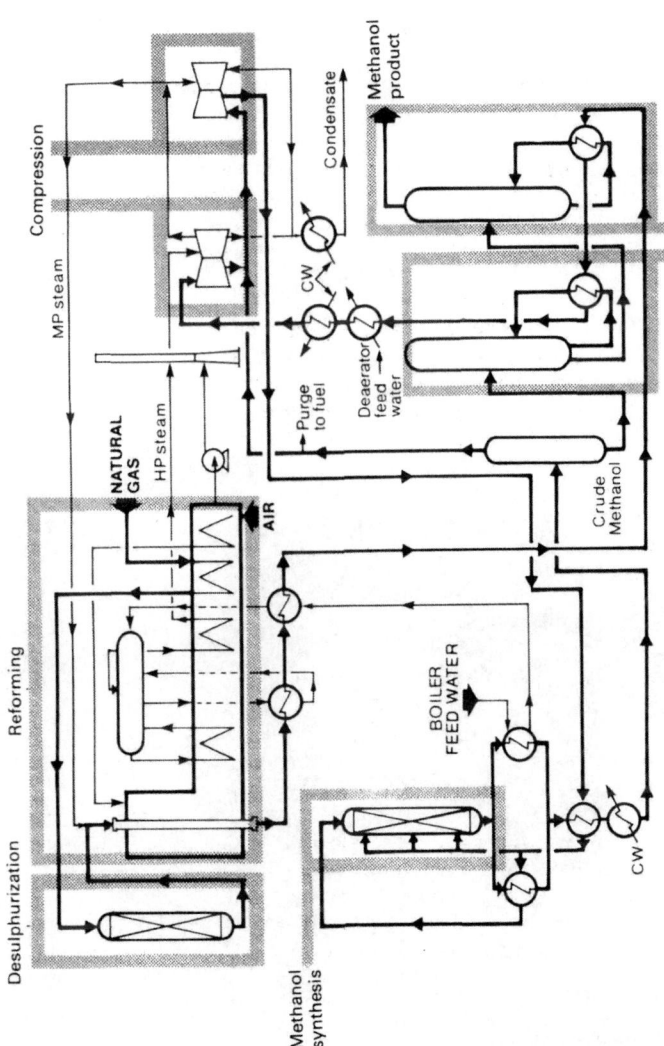

Figure 10.8. Flow sheet of typical low-pressure methanol synthesis plant based on the steam reforming of natural gas. Reprinted from *Catalyst Handbook*, 2nd ed., Ed. by M. Twigg, Copyright (1989) by kind permission of M. Twigg.

TABLE 10.13. Low Pressure Methanol Converter Types.

Process	Reactor	Pressure (bar)
ICI[a]	a. Inter-bed quench lozenges.	50–150
	b. Tubular.	
	c. Steam raising (advanced radial concept).	
Lurgi[b]	Water-cooled tubular.	40–100
Topsoe[c]	Three radial flow reactors with water coolingat outlets..	50–90
Mitsubishi[d]	Multibed with inter-bed steam generation.	50–200

[a]ICI, Pinto and Rogerson, *Chem. Eng. Prog.* (July 1977) 95; *Nitrogen* (July-Aug 1995) 32.
[b]Lurgi, E Supp, *Nitrogen* **36-40** (1977) 109.
[c]Topsoe, Dybkyaer, *Chem. Econ. Eng. Review* **13**(1981) 149.
[d]Mitsubishi, Takahashi and Tado, *Chem. Econ. Eng. Review*, **6** (1974) 79.

nitrates. With the new three component catalysts this meant that the first carbonate particles, precipitated in the acid solution, were large, rich in aluminum and deficient in zinc. The final stages of precipitation, under alkaline conditions, produced large particles of carbonate which were deficient in copper. This explained why the catalysts had a variable composition, contained crystals of different sizes and deactivated rapidly during operation.

Figure 10.9. Lurgi reactor for a low-pressure methanol synthesis plant. Reproduced with permission from Lurgi Aktiengesellschraft.

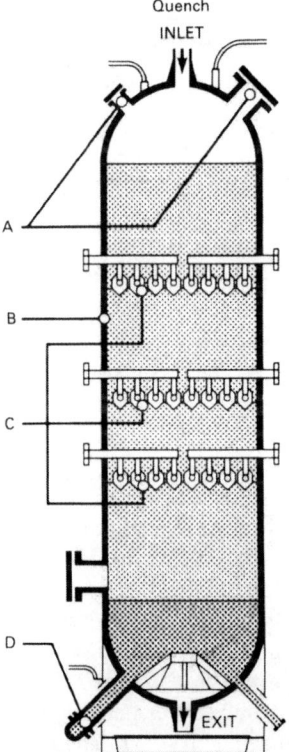

Figure 10.10. A modern low-pressure methanol synthesis converter, showing quench distributors for catalyst temperature control. A: catalyst is charged and inspected through these man-holes; B: the pressure vessel is of a simple design – no internal catalyst basket is required; C: the ICI lozenge quench distributors ensure good gas distribution and allow the free passage of catalyst for charging and discharging; D: gravity discharge of catalyst permits rapid preparation for maintenance or recharging. Reprinted from *Catalyst Handbook*, 2nd ed., Ed. by M. Twigg, Copyright (1989) by kind permission of M. Twigg.

TABLE 10.14. Byproduct Formation in Methanol Production.

Composition ppm (vol)	Process	
	High Pressure	Low Pressure
Methanol wt%	90–94	
Water wt%	~10	~20
Dimethyl ether	7800	<300
Ethanol	2500	150–650
n propanol	1200	<100
i-propanol	2700	<50
C_5+ alcohols	<400	<150
Acetone	<30	<10
Methyl formate	900	<1000
Other esters	<200	<20
Other ketones	<250	<20
Hydrocarbons		20–80
Methane	< 20 m³/te product	
Total (excluding H_2O)wt%	2–5	<0.5

A novel way to precipitate small particles was introduced to overcome the problem.[51] By continuous mixing of reactants in a jet, the carbonates could be precipitated at pH 7 so that the precipitate contained uniformly small, well-mixed crystals and provided catalysts with a long, stable life. Continuous operation at 50-bar pressure was possible for three years in a 300 tonnes per day plant. By-product formation was significantly decreased in comparison with the zinc oxide-chromia catalysts as outlined in Table 10.14.

As the demand for large, 1000 tonnes per day capacity plants increased, the operating pressure was optimized at 100 bar. The original catalyst was less stable at this pressure. However, by combining the alumina with a proportion of the zinc oxide, possibly in the form of a spinel, before precipitation of the copper component, an even more stable catalyst was produced.[50] Catalyst activity was proportional to the surface area of the metallic copper reduction, which also controlled the stability of the catalyst during use. Continuous improvement of production techniques and a better understanding of the catalyst precursors could increase catalyst life increased to about five years.

10.2.2.3 *Precipitates Forming During Production*

A mixture of basic copper and zinc carbonates is formed during the precipitation of the metal nitrates. These have general formulae similar to malachite—$Cu_2(OH)_2(CO_3)$—in which some of the copper is replaced by varying proportions of zinc. The amorphous mineral georgeite, with the formula $Cu_2(OH)_2CO_3$, has been identified in the precipitate. As the precipitate ages other minerals, such as aurichalcite and hydrozincite, form depending on the proportions of copper and zinc. Free aluminium ions in the solution can be incorporated in a complex basic carbonate $M_6Al_2(OH)_{16}CO_3.4H_2O$, hydrotalcite, which is related to the Feitknecht compounds. The mineral azurite can also be formed by ageing the precipitate at slightly elevated pressures.[52]

A vital stage of the catalyst preparation is the aging period, in which the structure of the precipitate changes, to produce the appropriate copper zinc and aluminium distribution in the final catalyst. During aging, it is significant that the initial blue precipitate changes color to blue-green. The sequence of changes taking place has been summarized as:[53]

- Amorphous copper/zinc georgeite $(Cu,Zn)_2CO_3(OH)_2$ forms during precipitation.
- Aging gives low zinc (15%) malachite $(Cu/Zn)_2CO_3(OH)_2$ and high zinc (36%) aurichalcite $(Cu/Zn)_5(CO_3)_2(OH)_6$.
- Further aging gives finely divided copper enriched (70+%) malachite $(Cu,Zn)_2CO_3(OH)_2$.

The very small copper crystals that form after careful drying, calcination and reduction are very stable and the catalyst remains active for several years of plant operation. It has been suggested that the reduced copper is stabilized as clusters of copper atoms at the oxygen rich (001) surface of the zinc oxide. In turn, the zinc oxide is stabilized by small zinc aluminate particles or other refractory oxides. The addition of small amounts of magnesium oxide enhances the stabilizing effect of zinc oxide.[54] Zinc oxide can also absorb any traces of sulfur or chlorine in synthesis gas.

10.2.2.4 *Operation with Copper Catalysts*

It is clear that the mixed copper oxide/zinc oxide/alumina formulation is an excellent catalyst in the low-pressure methanol process. There has been much debate in explaining the role of copper in the reaction and how methanol is produced. The group that invented the catalyst examined a number of possibilities:[53]

- Was methanol synthesized from carbon monoxide or carbon dioxide?
- Was the copper oxide reduced or oxidized during the synthesis process and what active sites were formed?
- What reactions were taking place on the active surface and how did this affect the process?

It has been concluded from experimental work with catalysts containing alumina that methanol forms from carbon dioxide and that the catalyst activity is proportional to the copper metal surface area. The presence of carbon dioxide in the gas increases the synthesis rate. The zinc oxide and alumina play little part in the actual reaction apart from stabilizing the reduced copper and protecting it from the effect of any poisons. On the other hand, with catalysts containing chromia, the carbon dioxide leads to a decrease in the reaction rate.[56]

10.2.2.5   *Reaction Mechanism with Copper Catalysts*

During operation, the reduced copper metal surface is partly covered by mobile oxygen atoms and carbon dioxide is adsorbed on the 'oxidized' copper. Carbon dioxide is then able to react with hydrogen adsorbed on the reduced copper surface. A formate intermediate is reduced first to a methoxy group and then to methanol. Oxygen remaining on the copper surface can either react with adsorbed carbon monoxide to form carbon dioxide or to react with hydrogen to form water which also provides more carbon dioxide by the water gas shift reaction with carbon monoxide. Depending on the ratio of carbon dioxide to carbon monoxide in synthesis gas, up to 30% of the reduced copper can be covered with adsorbed oxygen. The proportion of adsorbed oxygen also depends on the oxides in the support. These reactions are summarized as follows:

$$H_2O \leftrightarrow H_2 + O(ads) \ (H_2O + CO \leftrightarrow CO_2 + H_2) \tag{10.1}$$

$$H_2 \rightarrow 2\ H(ads) \tag{10.2}$$

$$CO_2 \rightarrow CO_2(ads) \tag{10.3}$$

$$CO_2(ads) + H(ads) \rightarrow HCOO(ads) \tag{10.4}$$

$$HCOO(ads) + 2\ H(ads) \rightarrow CH_3O(ads) + O(ads) \tag{10.5}$$

$$H(ads) + CH_3O(ads) \rightarrow CH_3OH \tag{10.6}$$

$$O(ads) + CO \leftrightarrow CO_2 \tag{10.7}$$

It is important to emphasize that the mobile, active sites are formed from adjacent copper atoms and surface oxide. Much of this work has been the subject of a review by the ICI group.[57]

10.2.2.6 *Selectivity*

Mixed oxide methanol synthesis catalysts containing copper are very selective and the selectivity of the conversion of carbon oxides to methanol exceeds 99%. This is simply explained because there is no splitting of carbon-oxygen bonds when carbon oxides are adsorbed on the copper surface.

Consequently, the methane and higher alcohols formed with zinc oxide-chromium oxide catalysts are no longer major by-products. Formaldehyde and methyl formate form only in small, equilibrium amounts.

Although dimethyl ether could be produced with an acid catalyst by a simple dehydration reaction, the alumina component of the catalyst has low acidity and less than 200 ppm ether is found in crude methanol. This is most probably due to the neutralizing effect of the basic zinc oxide.

10.2.2.7 *Low-pressure Methanol Reactor Types*

The first low-pressure methanol process was designed to produce 300-600 tonnes per day of methanol, at about 50-bar pressure with a copper oxide-zinc oxide-alumina catalyst. This was the maximum size of plants being used at the time. By using a simple process design, based on a centrifugal compressor/circulator, a novel quench converter with heat exchanger cooler and catchpot, the ICI process eliminated the costly high-pressure equipment and reciprocating compressors of the old process. The cost of making methanol was cut to about 25-30% at that current level of the time.[51]

The process was further optimized to operate at 100 bar pressure as demand for methanol increased and plants producing 1000 tonnes per day were developed. A more stable and efficient version of the original catalyst was also introduced. As methanol demand continued to increase in the 1970s, other similar processes were developed using more energy-efficient designs. Apart from the first, quench cooled converters, tube cooled, radial flow and steam-raising converters have all been used.

Catalyst life and activity have also been improved. For various reasons the catalysts used in some of the later processes were based on copper oxide-zinc oxide-chromia catalysts.

10.2.2.8 *Catalyst Reduction*

The copper oxide component of low-pressure methanol catalysts must be reduced to copper metal, with a dilute stream of hydrogen in nitrogen, before use. Careful temperature control is essential during this procedure in which a large volume of water is evolved, depending on the copper content of the catalyst used.

Reducing gas contains up to 1% hydrogen at about 10-bar pressure and is circulated through the converter. The catalyst temperature is increased slowly and reduction should begin between 150°-160°C. The reduction of the catalyst is extremely exothermic and the progress of the reduction can be monitored by observation of a high-temperature zone passing down the bed as reduction proceeds. To prevent damage to the catalyst the inlet temperature is restricted to 180°C with a maximum bed temperature of 230°C. If necessary, the maximum temperature can be controlled by changes to the hydrogen concentration as the hot spot passes down the catalyst bed. After condensing water from the outlet gas, the nitrogen is simply re-circulated after the addition of more hydrogen.

Any carbon dioxide, formed from the decomposition of residual carbonate in the catalyst, can be purged from the system if the concentration exceeds 2-3%.

When the peak catalyst temperature reaches the bottom of the reactor, the inlet temperature can be increased to 220°–240°C and the hydrogen concentration to 10-20%. It can take about two to five days to reduce the 100–200 tonnes of catalyst required for large plants.

The catalyst is most stable at low temperatures, and the purity of the product methanol decreases as the temperature of the converter is increased. For these reasons, the process is usually operated at the lowest possible temperature and pressure, consistent with achieving the design production rate. As the catalyst becomes deactivated, it usually becomes necessary to increase the reaction temperature to maintain production capacity. The inlet temperature to the bed is usually about 220°C in the initial stages, gradually increasing to 250°C over a three to five year period.

10.2.3 Novel Catalysts

The possible use of palladium as a potential catalyst for the synthesis of methanol received much of attention during the 1980s.[56,57,58] These catalysts were never commercialised because the rate of reaction was relatively low, a significant amount of methane was always produced and the high cost of palladium made this approach too unattractive.

A much more radical approach was first proposed by Baglin and coworkers,[61] who showed that careful oxidation of a range of copper/thorium intermetallic compounds gave rise to a new generation of extremely active catalysts. Thorium, however, is weakly radio-active and would not be acceptable for large scale use on safety grounds. A similar range of catalysts based on intermetallic compounds derived from copper and rare earth elements was then developed by workers from ICI, who also discovered that the pure rare earth metal could be substituted by the much cheaper mixed alloy, Mischmetal, with no loss in activity. Addition of a little aluminium metal to the molten copper/Mischmetal mixture improved the stability of the catalyst towards deactivation.[62] Methanol could be prepared over this generation of catalyst at temperatures as low as 70°C and pressures in the range 5–10 bar. The mechanism of methanol synthesis over this catalyst is quite different to that of the conventional copper/zinc oxide/alumina catalyst. It was demonstrated by radioactive labelling experiments that the methanol was produced directly from carbon monoxide, rather than carbon dioxide, in strong contrast with the conventional catalyst.[63] Indeed, carbon dioxide was even shown to be a poison for this catalyst. It is believed that the active component of this catalyst is extremely small particles of copper metal, perhaps even in the atomic state, supported on an oxide matrix, on the basis of in situ x-ray diffraction analysis of the evolution of the catalyst from the intermetallic precursor.[64] This approach has not been commercialised at the present time. The mains reasons were that poisoning by carbon dioxide and water, meant that purification of the synthesis gas was very difficult and costly. Furthermore, the catalyst expanded in volume greatly during the in situ activation procedure, and this led to poor gas distribution through the converter. These problems have yet to be overcome.

REFERENCES

1. S. J. Green, *Industrial Catalysis,* Benn, London, 1928, p. 334.
2. Larson and Richardson, *J. Ind. Eng. Chem.,* **17**, (1925)971; Larson and Brooks, *J. Ind. Eng. Chem.,* **18** (1926) 1305.
3. C. Ernst, *The Fixation of Atmospheric Nitrogen,* Chapman and Hall, London, 1928; J. B. Allen, *Chem. Eng. Progress* (Sept 1965) 1; S. Strelzoff and L. C. Pan, *Synthetic Ammonia,* Chemico, 1962.
4. ICI European Patent Application 78302769 (March 7 1979); *Nitrogen* (Nov-Dec 1979) 122; S. A. Topham, S. A. Hall, and D. G. Heath, Development of ICI Low Pressure Ammonia Synthesis Catalyst, ICI/CFDC Technical Symposium, Shanghai (April 1989*).
5. J. R. Jennings, ICI, US Patents 4654320 and 4668657.
6. British Petroleum US Patents 4142993 (1979) & 4453775 (1979).
7. S. J. Green, *Industrial Catalysis,* Ernest Benn, London, 1928, p. 343.
8. S. J. Green, *Industrial Catalysis,* Ernest Benn, London, 1928, p. 349-352; British Patent 153254 (1921); Thompson et al., *Chem. Eng. Prog.* **48**, (1952) 468.
9. *Chem. & Met. Eng.* **30** (1924) 198; USP 1408987 (1922); 1478549-50 (1923).
10. British Patent 197199 (1922); 218237 (1923); Canadian Patent 229485 (1923).
11. L. Casale, *J. Ind. Chem.* **17** (1925) 971; Bitish Pat. 227491 (1923).
12. L. B. Nelson, *History of the US Fertilizer Industry,* Tennessee Valley Authority, 1990, p. 201.
13. R. S. Tour, *Chemical & Mech. Eng.,* Parts 1-5, (June–Oct 1922).
14. L. B. Nelson, *History of the US Fertilizer Industry,* Tennessee Valley Authority, 1990, p. 227.
15. L. B. Nelson, *History of th US Fertilizer Industry,* Tennessee Valley Authority, 1990, p. 229.
16. Chemico US Pat. 3041151 (1962).
17. I. Cedersberg, British Patents 153290 (1920); 168902 (1921); USP 1452027 (1923); Can Pat 219825; 223453 (1922).
18. F. Uhde, German Patent 493793 (1925); Br Pat 273735; 253122; 247225 (1925-6).
19. F. Uhde British Pat. 259230 (1926).
20. V. E. Parke, *Billingham – The First Ten Years,* ICI Ltd, Billingham, 1957.

21. A. Mittasch, *Advances in Catalysis,* Vol. 2, Academic Press, NY, 1950, p. 82.
22. J. R. Jennings and S. A. Ward, *Catalyst Handbook,* 2nd edition, Ed. by M. V. Twigg, Wolfe, London, 1989, p. 397; R. Schlögl, Catalytic Ammonia Synthesis, Ed. by J. R. Jennings, Plenum, New York, 1991, p. 39.
23. G. Ertl, *Catalytic Ammonia Synthesis*, Ed. by J. R. Jennings, Plenum, New York, 1991, p. 39.
24. D. R. Strongin and G. A. Somorjai, *Catalytic Ammonia Synthesis*, Ed. by J. R. Jennings, Plenum, New York, 1991, p. 39.
25. W. Turner, Kellogg International Corporation: Ten Years of Single Train Ammonia Plants, Paper 1, *AMPO 74, ICI Operating Symposium,* Amsterdam (1974).
26. Slack, Allgood, Manne, *Chem. Eng. Prog.,* **49** (1953) 393; Bridger, Pole, Beinlich, Thompson, *Chem Eng Prog,* **43** (1947) 291.
27. Topsoe's New Ammonia Process, *Topsoe Topics* (March 1976); I. Dybkjaer, Fertilizer Nitrogen, *British Sulfur International Conference,* London (January 1981).
28. Quartulli & Wagner, Why Horizontal Ammonia Converters?, *Hydrocarbon Processing*, (December 1978); Le Blank, Retrofit Ammonia Plants to Save Energy, *Hydrocarbon Processing,* (Aug 1986) 39.
29. S. R. Tennison, US Patent 4163775 (1979).
30. Kellogg US Patent 4568531-2.
31. A. K. Rhodes, *Oil Gas Journal,* (Nov 18 1996) 37.
32. Canadian Pat. 1094532; S. R. Tennison, *Catalytic Ammonia Synthesis*, Ed. by J. R. Jennings, Plenum, New York, 1991, p. 39. *Angew Chem. Int. Ed.,* **40** (2001) 1061.
33. S. Stalica, *Hydrocarbon Processing,* (July 1999) 85.
34. Jahn, *Berichte,* **13** (1880) 983.
35. German Pat. 293379 (1913); British Pat. 20488 (1913); US Patent 1201850 (1914).
36. German Pat. 415686; 441433; 462837 (1923); US Patent 1558559; 1569775 (1923).
37. *Fiat Reports No 888,* (Aug 1946) p. 7-8; British Pat. 227147.
38. French Pat 540343 (1921).
39. Patart, *Comptes Rendus,* **179** (1924) 1330.
40. Patart, *Chimie et Ind.* (Nov 1926).
41. Italian Pat. 267698 (1928); French Pat. 670763 (1930); G Natta, *Catalysis,* Vol. 3, Ed. byEmmett, Reinhold Pub Corpn, N.Y., 1955, p. 349.
42. Natta and Corradini, *Proceedings Int. Symposium on the Reactivity of Solids,* Gothenberg, 1952, p. 619.
43. Lazier and Vaughan, *J. Am. Chem. Soc.,* (1932) 3080.
44. Molstead and Dodge, *Ind. Eng. Chem.,* **27** (1935) 134.

45. C. L. Thomas, *Catalytic Processes & Proven Catalysts,* Academic Press, N. Y., 1970, p. 150.
46. US Patent 1746783 (1930).
47. Fenske and Frohlich, *Ind. Eng. Chem.,* **21** (1929) 1052; Frohlich and Lewis, *Ind. Eng. Chem.,* **20** (1928) 285.
48. W. Katowski, *Chem. Tech.,* **15** (1963) 204.
49. G. W. Bridger, *MacRobert Award Lecture,* Manufacture of High Activity Catalysts (November 25, 1975).
50. M. V. Twigg and M. S. Spencer, *Topics Catal.* **22** (2003) 191.
51. M. S. Spencer, *Catal. Letters* **66** (2000) 255; A. M. Pollard, M. S. Spencer, R. G. Thomas, P. A. Williams, J. R. Jennings and J. Holt, *Appl. Catal. A* **85** (1992) 1.
52. Methanol the ICI Way, *Chemical Week,* (Jan 6 1968) 34.
53. Marschner and Moeller, *Applied Industrial Catalysis,* Vol. 2, Ed. by B. E. Leach, Academic Press, N.Y., 1983, p. 215.
54. M. S. Spencer, *Applied Catalysis A General,* **85** (1992) 1; M. S. Spencer, *Topics Catal.* **8** (1999) 259.
55. Doesburg, Hoppener, de Koning, Xiaoding, and Scholten, in *Preparation of Catalysts,* Vol. 4, Ed. By Delmon et al., Elsevier, Amsterdam, 1987, p. 767.
56. Chinchen et al., *ACS Div. Fuel Chem.,* **29**(5) (1984) 178.
57. Blaziak and Kotowski, *Prez. Chem.* **43** (1964) 657.
58. G. C. Chinchen, P. J. Denny, J. R. Jennings, M. S. Spencer and K. C. Waugh, *Appl. Catal.* **36** (1988) 1.
59. M. L. Poutsma, L. F. Elek, P. A. Ibarbia, A. P. Risch, and J. A. Rabo, *J. Catal.* **52** (1978) 157.
60. E. K. Poels, P. J. Mangnus, J. van Welzen, and V. Ponec, *Proc 8th Int Congress Catal* **2** (1984) 59.
61. R. F. Hicks and A. T. Bell, *J. Catal.* **91** (1985) 104.
62. E. G. Baglin, G. B. Atkinson, and L. J. Nicks, *Ind. Eng. Chem., Prod. Res. Dev.* **20** (1981) 87.
63. G. D. Short and J. R. Jennings, ICI, European Patent 117944 (1984).
64. J. R. Jennings, R. M. Lambert, R. M. Nix, G. Owen, and D. G. Parker, *Appl. Catal.* **50** (1989) 157.
65. R. M. Lambert, R. M. Nix, T. Rayment, J. R. Jennings, and G. Owen, *J. Catal.* **106** (1987) 216.

11

ENVIRONMENTAL CATALYSTS

The effects of atmospheric pollution from the combustion of fossil fuel have been recognized for many years. Acid rain from industrial areas of Europe and North America devastated forests. The *pea-souper* fogs in Donora, Pennsylvania, London, UK and Belgium caused by emissions from power plants, steelworks and metal smelters as well as domestic coal fires affected public health and led to thousands of deaths. All this, together with the photochemical smogs in Los Angeles, California, brought a growing demand for environmental protection.[1] Early measures were almost all restricted to imposing limits on particulate emission and in the UK, the first Clean Air Act of 1956 resulted in the introduction of smokeless zones. Since then further legislation has led to the development of catalytic processes that also reduce the concentration of nitrogen oxides, hydrocarbons and carbon monoxide in the various exhaust emissions to specified levels.

Legislation was passed in the 1970s to limit the release of sulfur and nitrogen oxides from power plants in Japan and later in Germany. This led to the investigation of new catalytic processes. Selective Catalytic Reduction (SCR) was developed to enable power companies to reduce NOX emissions from coal and oil based power plants. Since then another procedure has been developed to remove NOX from the gaseous effluent produced by several other combustion and chemical processes.

The Clean Air Act, introduced by the US Government during 1970, set specific targets for the control of emissions from automobiles to be achieved by 1975. Catalytic converters are now fitted to automobiles in many parts of the world to remove nitrogen oxides, hydrocarbons and carbon monoxide in accord-

L. Lloyd, *Handbook of Industrial Catalysts*, Fundamental and Applied Catalysis, DOI 10.1007/978-0-387-49962-8_11, © Springer Science+Business Media, LLC 2011

ance with local regulations. An additional environmental benefit from the use of catalytic converters has been the need to stop the addition of tetraethyl lead to gasoline. Tetraethyl lead was used originally to increase the octane rating of the gasoline, but lead is a powerful catalyst poison, and would cause a rapid deactivation of the catalytic converter. This led to a need for major changes to be made to the composition of the US gasoline pool and new Environmental Protection Agency (EPA) regulations covering the reformulated gasoline were developed. The use of low-boiling, high-octane hydrocarbons and toxic aromatic components which may be released to the atmosphere is now restricted.

A third major group of emissions controlled by environmental legislation comprises volatile organic compounds (VOCs) from all industrial and refinery effluents. The US Clean Air Act amendments of 1990 called for the reduction in the concentration of no fewer than 188 toxic chemical air pollutants before 2000. Many of these chemicals are VOCs and significant reductions have been achieved by the introduction of a range of new catalytic oxidation processes. The magnitude of the problems involved in environmental control and the potential demand for catalysts is demonstrated by Table 11.1 that shows the estimated volume of some effluents in the USA during 1995.[2]

Reports show that about 1.27 million tons of major chemicals were released in the US during 1993, some of which are listed in Table 11.2.[3] Sulfuric acid made up about 73,000 tons of this total. During 1999 up to 22 million tons of sulfur was recovered from crude oil fractions and hydrocarbon gases by the catalytic Claus Process.[4] Although sulfur from power plant emissions has usually been removed chemically as gypsum, it can be converted to useful sulfuric acid by a modified form of the contact process.[5]

TABLE 11.1. US Gaseous Emissions During 1995.

Emission	Source	%
NOX	Transport	49
	Combustion	46
	Industrial Processes	3
	Chemical Processes	1
	Miscellaneous	1
	Total 21.8 million tons	
VOCs	Non Chemical Processes	51
	Transport	37
	Chemical Processes	7
	Fuel Combustion	3
	Miscellaneous	2
	Total 22.9 million tons	
SOX (1986)	Fuels	85
	Transport	7
	Industry	8
	Total 18.5 million tons	

TABLE 11.2. Major Chemicals Released to Atmosphere in 1993 in the United States.

Chemical	Percentage of total
Ammonia	12.6
Hydrochloric Acid	8.0
Phosphoric Acid	7.6
Methanol	7.5
Toluene	6.4
Sulfuric Acid	5.7
Acetone	4.6
Xylenes	4.0
Carbon Disulfide	3.3
Methyl Ethyl Ketone	3.0
Chlorine	2.7
Zinc compounds	2.6
Dichloromethane	2.3
1,1,3 Trichloromethane	2.2
Total released	1.27 million tons

The catalytic processes used for environmental protection demand an enormous capital investment but are relatively uncomplicated and more efficient than physical procedures. Demands imposed by environmental legislation have created a significant and rapidly expanding sector of the worldwide catalyst business.

11.1 STATIONARY SOURCES

High temperature combustion of coal and oil to provide heat and generate electricity resulted in huge volumes of nitrogen oxides (NOX), sulfur dioxide and carbon oxides being vented to the atmosphere. The most widely used fixed, or *stationary*, sources which produce nitrogen oxides are steam generating boilers, combustion turbines, process heaters, waste incinerators and reciprocating engines. A number of chemical processes such as nitric acid, adipic acid, caprolactam and acrylonitrile also produce effluents containing large volumes of nitrogen oxides.

It is now generally necessary to treat the flue or exhaust gases from new stationary sources to remove NOX to a statutory level by the most convenient and economic procedure available (Figure 11.1). This usually requires a reduction process using added ammonia. Although some power plants have been retrofitted with SCR equipment, many older installations have not been covered by the legislation. This has been described as being 'grandfathered' but, as a typical power plant has a life of at least 30-40 years, means that considerable NOX would still release continue for several years.

Figure 11.1. Coal-fired power plant fitted with SCR (Selective Catalytic Reduction) unit to decrease NOX content of the flue gas. Reproduced with permission from Haldor Topsøe A/S.

Coal and fuel oil can contain up to 5 wt% sulfur and during combustion most is converted to gaseous sulfur oxides. This amounts to about 1800 volume ppm sulfur dioxide in flue gas for 2.5 wt% sulfur in coal. Up to 95% of the sulfur in the flue gas can be removed by scrubbing with an aqueous slurry of limestone to produce calcium sulphate, or gypsum. This procedure, however, required facilities for the disposal of the large amount of gypsum thus produced. Lower sulfur dioxide emissions can be obtained where possible by using low sulfur fuel. Alternatively, a mini sulfuric acid unit may now be installed to recover several tonnes of acid per day from a single power plant unit.[5]

Nitrogen oxides are formed at the high boiler temperatures, which can exceed 1500°C, from either the nitrogen compounds present in the fuel or by the direct oxidation of elemental nitrogen in air as shown in Table 11.3.[6] The NOX content of flue gas in modern power plants is usually in the range 800–1800 ppm as shown in Table 11.4.[7]

TABLE 11.3. Formation of Nitrogen Monoxide.

Mechanism	Reaction
Thermal (>1500°C)	$O + N_2 \rightarrow NO + N$
Oxidation of molecular nitrogen	$N + O_2 \rightarrow NO + O$
	$N + OH \rightarrow NO + H$
Prompt (low temperature, fuel rich)	$CH + N_2 \rightarrow HCN + N$
Initiation by hydrocarbon radicals	N + HCN converted to NO by reaction with oxygen and hydrogen atoms in the flame.

This can be controlled to a certain extent by the use of low nitrogen fuels or by modifying the combustion conditions. Low excess air addition, flue gas recirculation, low NOX burners or water/steam injection can provide 50-60% reductions in NOX concentration. The scale of the operation is enormous and a typical US power plant burning 9500 tons of selected coal per day will have to remove about 135,000 tons of sulfur dioxide (95% conversion) and 45,000 tons of NOX per year. During 2000 about 50% of US power was generated from coal but by 2020 60% is expected to come from natural gas.[8]

TABLE 11.4. NOX Concentration in Flue Gas.

A: Power Plant	
Plant Type	NOX Concentration (mg/m³)
Coal (Pre 1980)	1200–2800
Coal (Post 1984)	800–1800
Lignite	600–800
Oil	500–800
Gas	500–1200

B: Gas Turbines				
Fuel	NOX Concentrations (mg/m³)			
	Normal	Wet Injection	Low NOX Burner	SCR
Natural gas	100–400	25–40	15–25	5–10
Distillate	150–700	40–65	65	15–18

11.1.1 Selective Catalytic Reduction

Selective Catalytic Reduction (SCR) of nitrogen oxides with ammonia using specially developed catalysts was investigated in Japan from 1974 and then introduced commercially from about 1980.[9] It is now the most important process for removing nitrogen oxides (NOX) from the effluent gas of power plants and many other stationary sources. German power plants installed SCR units based on Japanese licenses from about 1984 following the introduction of legislation limiting the NOX content of effluent gases in 1983.[10]

Japanese power plants generally used low sulfur coal with dry fired boilers and accepted relatively low NOX conversion with the SCR catalysts. This also limited the catalytic oxidation of sulfur dioxide to sulfur trioxide and avoided excessive deposition of sulfates in downstream equipment. German power plants, with higher sulfur coals and slagging boilers, had difficulties resulting from high sulfur dioxide conversion to sulfur trioxide as well as catalyst poisoning from volatile arsenic oxides. These problems were gradually overcome through catalyst development and process modifications that included NOX removal following sulfur oxide removal.

In the SCR process, NOX impurities are reduced with added ammonia in the presence of some residual oxygen from the furnace. The main NOX reduction reactions are shown in Table 11.5 together with some of the undesirable oxidation reactions, which can both produce sulfur trioxide and waste some of the added ammonia. Between 0.6–0.9 moles of ammonia per mole of NOX are added to limit the ammonia slip to downstream equipment where it would deposit as sulfates. NOX conversion is therefore limited to between 60-90%. At low NOX levels, there is little conversion to nitrous oxide. Nitrous oxide formation is also inhibited by water. Gas leaving the boiler is usually at a temperature in the range 300–430°C and contains dust. Dust is removed in an electrostatic precipitator with little heat loss before sulfur dioxide is removed as gypsum by reaction with lime. Alternatively, sulfur dioxide can also be converted to sulfuric acid.[5] The effluent gas is then vented to atmosphere. In the first power plants to be retrofitted with SCR units there were three possible locations for the catalyst bed:

- At the boiler exit. In this position the temperature was suitable for the NOX reaction and allowed some conversion of sulfur dioxide to trioxide. Dust in the gas stream led to catalyst problems.
- Following the electrostatic precipitator. The gas temperature was still high enough for both NOX removal and some sulfur dioxide oxidation. Dust problems were, however, avoided.
- At the tail end following flue gas desulfurization. Dust problems and sulfur dioxide oxidation were avoided but the gas had to be reheated for the catalytic NOX reduction to be effective.

TABLE 11.5. Reduction/Oxidation Reactions with SCR Catalysts.

Process	Reaction
DENOX	$4 NO + 4 NH_3 + O_2 \rightarrow 4 N_2 + 6 H_2O$
	$6 NO_2 + 8 NH_3 \rightarrow 7 N_2 + 12 H_2O$
OXIDATION	$2 SO_2 + O_2 \rightarrow 2 SO_3$
	$4 NH_3 + 5 O_2 \rightarrow 4 NO + 6 H_2O$
	$4 NH_3 + 3 O_2 \rightarrow 2 N_2 + 6 H_2O$

Figure 11.2. DeNOX catalyst for use in SCR (Selective Catalytic Reduction) plants. Reproduced with permission from Haldor Topsøe A/S.

In the first processes, the choice was to use the catalyst in the most convenient high dust location at the boiler exit where dust blocked the catalyst bed and eroded the catalyst surface in a relatively short period of time. This led Japanese operators to prefer the low dust location after the electrostatic precipitator. German operators, however, decided to remove NOX after the sulfur dioxide removal stage despite the need for heat exchange to reheat the gas when using typical catalysts.

From 1980, a variety of catalyst shapes and compositions was developed to overcome operating problems and provide better catalyst selectivity. Following the US Clean Air Act Amendments in 1990, it became necessary to consider SCR flue gas treatment from some coal based plants and to install suitable NOX removal procedures with new boilers, turbines and furnaces.

11.1.2 Selective Catalytic Reduction Catalysts

The most important reactions taking place over SCR catalysts are shown in Table 11.5. In power plants using coal or hydrocarbon fuels there is typically a larger excess of nitric monoxide compared with nitrogen dioxide (~19:1) than in, say, nitric acid plant tail gas (~1:1). A number of catalysts have been introduced (Figure 11.2) that are able to operate in the available temperature range from about 200°–430°C. The development, preparation and structure of the catalysts, the supports used and operating performance are described in several reviews.[11]

11.1.2.1 Catalyst Composition

SCR catalysts are prepared from vanadium pentoxide promoted with either molybdenum trioxide or tungsten trioxide and supported on the anatase form of titanium dioxide. The kinetics of the reaction are diffusion limited, and consequently, a thin layer of catalyst is usually deposited on a suitable high geometric area framework or honeycomb. When the flue gas has high dust content, at the boiler outlet, the honeycomb is formed from thin, notched, metal plates. These can be stacked as layers in modules that provide appropriate channel openings in the reactor. Alternatively, an extruded ceramic honeycomb with smaller channels and a higher surface area has been used when the gas has low dust content, after having passed through the electrostatic precipitator or desulfurization unit. Small honeycomb blocks are stacked together and loaded as beds in the reactor. Details of catalyst composition and dimensions, together with operating conditions, are summarized in Table 11.6.[10]

Catalysts based on zeolites have also been used successfully to remove NOX from the emissions of large and small-scale stationary sources. The zeolite can be extruded directly as a monolith or applied as a washcoat on preformed cordierite supports. The use of several zeolites—including wide-pore mordenite—for this reaction has been patented.[12] Up to 95% NOX conversion can be achieved with mordenite, with no promoters, and some zeolites are stable at temperatures up to 600°C. Long catalyst lives have been achieved in retrofitted coal based SCR units with low sulfur trioxide formation and good poisons resistance. Spent zeolite catalysts can be disposed of in approved landfill sites because they have negligible heavy-metal content.[12]

Promoted vanadium pentoxide/titania and zeolite based catalysts have been used to reduce NOX emissions at low temperature from nitric acid plants, gas turbines, industrial heaters, incinerators and boilers.[13]

TABLE 11.6. SCR Catalyst and Operating Conditions.

	Exit Boiler	Exit deduster or low dust fuels[a]
Gas Temperature (°C)	320–430	280–430
Dust content (g/SCF)	> 6.5	0.02–6.5
Honeycomb	Metal	Titania/ceramic
Channel opening (mm)	4–6	3–4
Surface area (m^2/g)	280–350	400–800
Honeycomb block size (cm)	65 x 46 x 46	10 x 15 x 15
Catalyst volume (m^3 per megawatt)	~0.6–1.2	
Reactor volume (m^3 per megawatt)	~1.7–3.4	
Catalyst composition	V_2O_5/MoO_3 or WO_3/TiO_2	

[a]Low dust fuels are hard coal, oil or natural gas.

11.1.2.2 *Catalyst Operation*

Most catalysts supported on titanium dioxide reach an optimum NOX reduction temperature that depends on the catalyst composition and the treated gas. Activity then declines as the secondary reactions compete for the ammonia reductant and sulfur dioxide oxidation becomes excessive. Typical operation is in the range 300°–425°C although zeolite catalysts operate from 300°–600°C.[12,13]

Catalysts may therefore be designed for use in specific duties. For power plant, the design must balance the reaction rates of NOX reduction and sulfur dioxide oxidation in the restricted range of temperature of flue gas leaving the boiler, or at the dust and sulfur dioxide removal stages. A low activity catalyst that reaches maximum NOX reduction between, say 380°–400°C, can be more efficient than a catalyst that is more active between 300°–350°C because, overall, it produces less sulfur trioxide at the fixed operating temperature.[14]

Vanadium pentoxide/titania catalysts, promoted with molybdenum trioxide or tungsten trioxide, have lower sulfur dioxide oxidation activity under power plant conditions than unpromoted catalysts. However, while catalysts containing tungsten trioxide are initially more active, any arsenic oxides in flue gas more easily poison them than similar catalysts containing molybdenum trioxide. Arsenic is a typical impurity in coal. Flue gas from slagging boilers contains up to twenty times more than flue gas from dry ash boilers. Catalysts promoted with molybdenum trioxide have been more widely used in Europe for this reason. The vanadium pentoxide content of NOX reduction catalysts is probably less than 1 wt% which maximizes the rate of nitrogen oxide reduction while limiting the rate of sulfur dioxide oxidation. The total oxide content including promoter would be less than 10 wt%.[15]

11.1.2.3 *Reaction Mechanism*

Selective Catalytic Reduction catalysts are similar to the vanadium pentoxide-anatase catalysts introduced by BASF and von Hayden in the 1960s for the oxidation of methyl groups in ortho-xylene. They were also coated onto cordierite supports.[15] Vanadium pentoxide reacts with surface hydroxyl groups on the titania to form active surface sites. In the case of oxidation catalysts, the mono-vanadyl species are active. However, for NOX reduction, at least two vanadyl groups, linked by an oxygen atom, form the selective site. These sites must be maximized.[15]

When the vanadium content of the catalyst is low, the monovanadyl surface species predominates. At a concentration around 5%, the monovanadyl species begin to polymerize, forming V-O-V bridges by dehydration mechanism, and at concentrations above about 10%, crystalline vanadium pentoxide is deposited. Preparation of the necessary active sites for maximum activity and selectivity has been achieved by promoting the catalyst with an excess of molybdenum trioxide or tungsten trioxide. Both oxides are known to form a wide range of mixed oxides with vanadium pentoxide and can presumable spread active sites over the catalyst surface.[15]

The reaction is through to proceed via the following mechanism. Ammonia is adsorbed on one of the V=O groups of an active site and then reacts with nitric oxide:

$$NH_3 + NO + V=O \rightarrow N_2 + H_2O + V–OH \qquad (11.1)$$

The V-OH group is then reoxidized by lattice or molecular oxygen leaving the active site in its original state:

$$2 \ V–OH + O \rightarrow 2 \ V=O + H_2O \qquad (11.2)$$

It has been suggested that the V=O and V–OH groups are in equilibrium on the catalyst surface and that the hydrogen atoms of the hydroxyl groups are mobile.[15] It is possible that at low vanadia loading on unpromoted titania, with no promoter, that part of the vanadium is reduced to low activity V^{4+} and there is a lower proportion of Bronsted acid sites.[15] The presence of a promoter may avoid this and provide the active catalyst. Mixed oxides equivalent to the compound $V_9Mo_6O_{40}$ (β-bronze, see Table 4.7), containing no V_2O_5 also undergo a phase change to $V_6Mo_4O_{25}$, with some reduction of V^5 to V^{4+}, in the presence of arsenic oxide.[15]

The following substances are known to poison the reaction:

- Arsenic oxide which reduces the vanadium component of the active sites.
- Alkalis which neutralize the acid sites.
- Calcium sulfate or silica deposition on the catalyst surface , which covers the active sites.

11.1.2.4 *Removal of Sulfur Dioxide as Sulfuric Acid*

The SNOX process developed by Haldor Topsøe avoids problems associated with undesirable sulfur dioxide oxidation in the SCR process.[5] Sulfur dioxide is removed from the flue gas in a small contact process unit after the nitrogen oxides have been removed by the usual SCR procedure. Nitrogen oxide conversion is as high as 95% and sulfur oxides are almost completely recovered as commercially useful sulfuric acid. Apart from the environmental benefits, such a design allows the use of higher sulfur content coals in power generation and avoids the need for gypsum disposal.

Conventional SCR and sulfuric acid catalysts are used. Nitrogen oxides are removed with maximum conversion at about 390°C, following dust removal, with a typical promoted vanadium pentoxide/titania catalyst. Sulfur dioxide is then converted to sulfuric acid, using a vanadium pentoxide/kieselguhr catalyst, by the wet gas sulfuric acid process.[17] Residual ammonia is also removed during the oxidation.

This process was first employed in 1991 in a 300 MW coal-based power plant in Denmark. It produced about 5 tonne h-1 of sulfuric acid from a total gas flow of up to 900,000 m^3/hr at 100% load using 0.38 tonnes per hour of ammonia for the SCR reaction.[5]

11.1.3 Gas Turbine Exhausts

Up to one million tonnes of NOX emissions per year were produced by gas turbines fired with natural gas and distillates in 1992/3.[18] Typical volumes of NOX in exhaust gases for different operating conditions involving a number of mechanical or catalytic processes are shown in Table 11.4. In future, to cope with the rapid growth of power generation, the cheaper and cleaner gas turbines will replace coal-fired boilers as a major source of energy.[8] At the same time, more stringent regulations will probably limit NOX emissions to less than 5 ppm.

11.1.3.1   Low Temperature Vanadium Pentoxide Catalysts

It has been convenient to use a low temperature selective catalytic reduction catalyst operating in the range 160°–190°C after the heat recovery steam generator in a gas turbine. A promoted vanadium pentoxide catalyst supported on titania extrudates has been used in these duties, and is stable between 160°–360°C, with no sintering.[19] Nitrogen oxides are reduced by up to 95% and turbines of up to 20MW capacity have been retrofitted. If sulfur dioxide is present in the exhaust gas then a small conversion to sulfur trioxide may occur depending on the temperature.

11.1.3.2   Catalytic Combustion Processes

Catalytic combustion processes are now being developed in which the high temperature thermal formation of NOX is avoided by operating at a lower temperature. This could avoid the current use of SCR processes to remove thermally generated NOX from the effluent of conventional gas turbines. Reactions that generate NOX during the typical homogeneous combustion hydrocarbons with air take place at temperatures as high as 1500°C and are listed in Table 11.3. The NOX concentration depends on the maximum temperature and contact time. While water/steam injection or low fuel/air ratios do improve performance they increase operating costs or affect reliability.

Catalytic combustion can produce a stable surface flame at low fuel/air ratio and temperatures as low as 1300°C, and this avoids the formation of thermal NOX. Operating costs are also significantly lower.[20] Compression up to an operating level of 12–16 bar can increase the gas temperature to 350°–410°C. This may not be high enough to initiate catalytic combustion and a gas burner may be required to trigger the reaction. Catalytic combustion units must be designed to reach an outlet temperature of up to 1300°C in modern turbines so a very stable catalyst has to be used. Typical space velocities in the small turbine combustion chamber are up to 300,000 h^{-1} which demands a high catalyst activity. A problem during operation, however, is that both thermal and catalytic combustion occur at temperatures exceeding 1000°C.

So far, it has not been possible to develop a thermally stable catalyst that can operate at temperatures of up to 1300°C when burning the fuel/air mixture in an economic combustor design. This has led to several modified catalytic procedures. Two-stage burning procedures have been developed. One burns only part of the fuel in the first catalytic stage. This limits the catalyst temperature to less than 1000°C and forms less than 3 ppm NOX. Remaining fuel is added to the second stage and good mixing with the residual air is essential to achieve a maximum temperature below 1300°C and avoid temperature hot spots that can then give rise to additional NOX.[21]

In a second process, air is added in two stages. A high fuel ratio in the first stage limits the temperature rise in the catalyst and produces a carbon monoxide/hydrogen mixture with 2–14 ppm NOX. Once again, good mixing of fuel with the remaining air is needed in the second thermal stage to avoid the formation of more NOX.[22]

In a more interesting process, all of the fuel/air mixture is added to a combustor with three sections. The first section contains an active palladium oxide catalyst that can operate up to about 800°C before being reduced to palladium metal which is less active. The palladium oxide catalyst is regenerated by reoxidation of the metal as temperature falls. A more stable catalyst in the second section continues the catalytic combustion. In the third section, combustion is completed by thermal reaction and the gas temperature increases to 1300°–1400°C. Overall, less than 1 ppm NOX is formed. Palladium oxide is supported on a monolith coated with temperature resistant barium hexaaluminate.[23]

11.1.4 Nitric Acid Plant Exhaust Gas

Nitrogen oxides are removed from the exhaust gas of nitric acid plants by various procedures. A typical exhaust gas contains about 4000 ppm NOX (NO:NO$_2$ = 1) with 3% oxygen and the balance nitrogen. Non-selective removal is possible using a conventional supported palladium catalyst, with added hydrogen, carbon monoxide or a hydrocarbon, at a temperature in the range 200°–400°C. Both oxygen and NOX are removed with a temperature rise depending on the oxygen content.[24]

Investigations to develop base metal catalysts have led to the use of selective catalytic reduction processes with supported vanadium pentoxide catalysts and ammonia as the reductant. More than 90% NOX conversion can be achieved, with acceptable levels of ammonia in effluent, by operating at temperatures in the range 170°–250°C and a space velocity up to 40,000 h⁻¹.

Vanadium pentoxide catalysts, similar to those used to remove nitrogen oxides from gas turbine effluent, have also been used to treat the effluents from a nitric acid plant.[25] The catalyst is based on extruded titania, impregnated with vanadium pentoxide and any promoters, or cordierite monoliths covered with a washcoat of the active material.

An active catalyst based on an extruded large-pore mordenite support has also been used.[12]

11.1.5 Ion-exchanged ZSM-5 Zeolites

Direct decomposition of nitric oxide could be more attractive than selective catalytic reduction but, although the reaction is thermodynamically favourable up to 1000°C, this has not so far been achieved. It is possible, however, that catalysts may be developed in the future to make the process attractive at commercial temperatures. Nitrogen oxides decompose in the ten membered rings of the zeolite structure where metal has been exchanged. Copper exchanged ZSM-5 zeolite has a stable activity in the range 400°-500°C with conversion up to 60% when using ammonia as reductant, particularly when promoted with a second metal such as magnesium.[26] The presence of oxygen in the flue gas does, however, reduce activity.

Nitric oxide can be reduced by low molecular weight hydrocarbons in the presence of oxygen. This avoids the use of ammonia and, potentially, makes the metal exchanged ZSM-5 suitable for NOX removal from some exhaust emissions. Palladium exchanged ZSM-5 was found to be more active than the copper exchanged catalysts, at significantly lower temperature.[26]

So far, the practical problems of commercial operation and the effects of poisons limit the efficiency of metal exchanged ZSM-5 catalysts. No large-scale operations have been reported for the treatment of effluents from either stationary or mobile sources.

In general, the use of most zeolites has been limited both by the significantly higher costs of these materials and by problems with poisons and deactivation. The efficiency and longer life of cheaper alternatives are still more attractive. As the energy business changes from coal to natural gas as the preferred source of energy during the next 10-20 years, retrofits and any changes to the type of catalysts used will be delayed.

11.2   MOBILE SOURCES

The rapid increase in the use of automobiles throughout the world has generated a high proportion of the pollution responsible for low level ozone and smog formation since the 1950s. The significance of ozone in photochemical smog was discovered following work by Arie J Haagen Smit on the components of pineapple fragrance in 1952! He realized that the ozone concentration during his experiments was higher on foggy days than on clear days.[27] The paper on 'The Chemistry and Physics of Los Angeles Smog' described the importance of sunlight in smog formation.[28] It has been reported that Los Angeles, with mountains on three sides, has up to 320 days per year with inversion conditions.

The term *inversion* in this context refers to the atmospheric condition whereby there is an increase in temperature with altitude, in contrast with the norm, where there would be a decrease in temperature with altitude. Such an inversion can lead to pollution and smog being trapped close to the ground. It is believed that the exhaust emissions from automobiles contribute strongly to the genesis of this type of inversion.

The scale of the problem led to the introduction of legislation. In 1966 the California state government specified state limits for hydrocarbon and carbon monoxide emissions from automobiles. These were followed, in 1970, by the US Federal Clean Air Act. Since then the emission limits from automobiles have become increasingly rigorous, and a zero emission vehicle limit (ZEVL) in 1998 requiring that, in certain cases, hydrocarbons, carbon monoxide and nitrogen oxides be completely removed was imposed in California. This must obviously refer to electric powered vehicles but does indicate how tight the limits may become in future.

The trends in the automobile emission standards for the USA and Europe since 1966 are given in Table 11.7. Compliance with the standards set has been made possible by the use of automobile emission control catalysts.[29] These were first used in the US during 1975 and in Europe from 1993. As a result of continuous improvement to design and manufacture, the catalysts have been able to conform with the increasing severity of the regulations.

11.2.1   Automobile Emission Control

Measures to control the level of carbon monoxide and hydrocarbons in automobile exhaust emissions began during the 1960s with modifications to the fuel management system. At the same time, base metal oxidation catalysts were tested and were found to be easily poisoned by other impurities. From 1975–77 precious metal catalysts containing platinum and palladium,[30] together with better fuel management, were found to be the essential to remove carbon monoxide and hydrocarbons from the exhaust. These early catalysts became more successful as the levels of lead additives in gasoline were first decreased and then com-

TABLE 11.7. Exhaust Emission Limits for US and European Union.

US Federal (g/mile)	HC	CO	NOX
Pre-control	15	90	6
1968-69	275 ppm	1.5%	–
1970	4.1	34	4
1975	1.5	15	3.1
1977	1.5	15	2.0
1981	0.41 (0.39)[a]	3.4	1.0
1994	0.25[a]	3.4	0.4
2003	0.125[a]	1.7	0.2
EU (g/km)	HC	CO	NOX
1993	–	2.72	0.97[b]
1996	0.5	2.2	–
2000 proposed	0.2	2.3	0.15
2005 indicative	0.1	1.0	0.08

[a]Non methane hydrocarbons. Californian regulations have generally been stricter than in the rest of the US and limits have been introduced on transitional low emission vehicles (TLEV-1994); low emission vehicles (LEV-1997); ultra-low emission vehicles (ULEV-1997) and zero emission vehicles (ZEV-1998).
[b]Includes hydrocarbons.

pletely removed. There was no NOX regulation during that period but, where necessary, some NOX reduction was made possible by recycling exhaust gases.

More stringent engine control and efficient catalysts were demanded from 1977 when legislation calling for NOX removal was also introduced. Nitrogen oxides were difficult to remove in conjunction with the other pollutants, particularly in the presence of oxygen. This led to strict control of the air/fuel ratio (A/F) in the engine by the use of oxygen sensors in the exhaust and the introduction of efficient *three-way* catalysts incorporating rhodium.[31]

In simple terms, the *two-way* catalyst is solely an oxidation catalyst, which is used to convert carbon monoxide and residual hydrocarbons to carbon dioxide. The *three-way* catalyst also provides a capability for reduction, which is required for the removal of nitrogen oxides.

The air/fuel ratio, based on the mass of air and fuel used in an engine, has a significant effect on the proportion of the undesirable impurities in exhaust gas as shown in Table 11.8. At the stoichiometric air/fuel ratio of about 14.7 there is just enough air for complete hydrocarbon combustion. With less air, that is with excess fuel, there will be incomplete combustion and the mixture is said to be *rich*. Under rich conditions, the exhaust will contain more reducing impurities such as carbon monoxide and hydrocarbons. With an excess of air, or a '*lean*' mixture, the exhaust will contain more oxidizing impurities such as oxygen and less nitrogen oxides and carbon monoxide. Exhaust gas compositions are related to the lambda (λ) value which is defined as the ratio of the actual and stoichiometric value of the A/F ratio. This variable ratio does not require knowledge of the gasoline composition.

TABLE 11.8. Variation of Exhaust Gas Composition with A/F Ratio.

Air/Fuel	λ-value	CO vol%	HC ppm	NOX ppm	O_2 vol%
13.2	0.9 (rich)	~3	~1000	~1600	~0.8
14.7	1.0 (stoichiometric)	~1.3	~400	~2000	~1.0
16.2	1.1 (lean)	~0.2	~250	~1500	~2.2

Notes: Usually about 3 mol H_2 /1 mol CO.
SO_2 content in ppm is (1.1837) x (mg of S/liter fuel) / (A/F ratio).
Engine performance is optimum at λ~0.9. λ = air:fuel / 14.7

Many chemical reactions can take place in the engine effluent to remove undesirable impurities and some of these are shown in Table 11.9. Reactions A-C can all take place in a catalytic converter although, if possible, reactions C-D should be avoided.

The basic catalytic systems that have been used to control emissions are:

- From 1977, closed loops with an oxygen sensor have been used to control the air/fuel ratio and to optimize conversion of carbon monoxide, hydrocarbons and NOX with a *three-way* catalyst system. Before the development of fuel management systems the lower conversions needed to meet legislation could be achieved with a variable air/fuel ratio.
- Two catalytic converters may be used with the engine operating in a rich mode. The first catalyst reduces nitrogen oxides in an oxygen deficient exhaust. Oxygen can then be added to the exhaust to remove carbon monoxide and hydrocarbons with an oxidation catalyst in the second container.
- If extra oxygen is added before treating a rich exhaust mixture in a catalytic converter only carbon monoxide and hydrocarbons are removed.
- Lean operating engines may operate with an air/fuel ratio as high as 26 (λ ~ 1.8) which produces low nitrogen oxide concentrations in the effluent. This demands the use of higher activity, oxidation catalysts because of the lower exhaust gas temperature.

Not all of these applications can meet the current strict emission legislation.

The autocatalyst operates under conditions that are constantly changing. Cold starts may be required several times a day and even on a motorway, automobiles are regularly accelerating and braking. This will influence the air/fuel ratio and engine temperature. These factors all have an effect on catalyst performance. A typical operating cycle is, therefore:

- Following ignition, the temperature increases from ambient to 600°C. Catalyst becomes active, or *lights off*, in the temperature range 250°–300°C.
- Conversion then increases rapidly up to ~600°C and conversion reaches 95–100%. Temperatures reach 900°C under full load.

TABLE 11.9. Catalytic Removal of Exhaust Gas Impurities.

Reactant	Reaction
A: Oxygen	$C_mH_n + (m + 0.25n)\,O_2 \rightarrow m\,CO_2 + 0.5n\,H_2O$
	$CO + 0.5\,O_2 \rightarrow CO_2$
	$H_2 + 0.5\,O_2 \rightarrow H_2O$
B: Nitrogen oxides	$2\,CO + 2\,NO \rightarrow 2\,CO_2 + N_2$
	$C_mH_n + 2(m + 0.25n)\,NO \rightarrow (m + 0.25n)\,N_2 + 0.5n\,H_2O + m\,CO_2$
	$2\,H_2 + 2\,NO \rightarrow N_2 + 2\,H_2O$
C: Rich gas	$CO + H_2O \leftrightarrow CO_2 + H_2 \; (HTS)$
	$C_mH_n + 2m\,H_2O \leftrightarrow m\,CO_2 + (2m + 0.5n)\,H_2$
D: Other reactions	$2\,SO_2 + O_2 \leftrightarrow 2\,SO_3$
	$SO_2 + 3\,H_2 \leftrightarrow H_2S + 2\,H_2O$
	$2\,NO + O_2 \rightarrow 2\,NO_2$
	$2\,NO + 5\,H_2 \rightarrow 2\,NH_3 + 2\,H_2O$
	$2\,NO + CO \rightarrow N_2O + CO_2$

- Air/fuel ratio is usually stoichiometric at about 14.7.
- Under fuel rich conditions NOX conversion is enhanced while under fuel lean conditions there is more carbon monoxide/hydrocarbon conversion.

There is normally some oscillation around the stoichiometric air/fuel ratio that is adjusted by the feedback system to control air addition to the engine. During acceleration, the conditions within the engine are richer and when decelerating conditions are leaner. Operation at the stoichiometric level provides sufficient carbon monoxide and hydrocarbons to reduce the NOX concentration.

11.2.2 Automobile Emission Control Catalysts

The choice of catalyst used is critical in removing the three major impurities and it must be operated under conditions that are effective for the oxidation of carbon monoxide and hydrocarbons and the reduction of the nitrogen oxides. This demanding operation was necessary from about 1980 when more comprehensive effluent control legislation was introduced in the US. By 1980, the original bead catalysts were being replaced by the more efficient extruded ceramic honeycombs or monoliths, which were impregnated with the active catalyst. Honeycombs have the advantage of a reduction in pressure drop through the catalyst bed, improved gas distribution, and a reduction in the size and weight of the equipment used. Metal monoliths were also developed, although these were not as widely used as ceramics. Platinum group metals are the most efficient and active catalysts available and are more resistant to poison than base metals.

11.2.2.1   *Bead Catalysts*

Bead catalysts were based on typical supports such as γ, δ or θ alumina with surface areas of about 100 m^2/g. Alumina was easily impregnated with the platinum and palladium and the resulting catalysts were stable at temperatures up to 1000–1100°C before the phase change to α alumina took place. Alumina can be stabilized to a limited extent by the addition of other refractory oxides but will still sinter at high temperature.

A disadvantage of the bead catalysts, when used in small volume reactors, is the need to select an appropriate size to compromise between an acceptable pressure drop through the bed and a satisfactory rate of gas mixing. A complex design of the catalyst container is also required, to provide for a uniform gas flow and to avoid disintegration of the catalyst by physical shock as the vehicle passes over uneven ground. They were, however, easily produced and withstood the significant thermal shock when used in an automobile exhaust. Under steady conditions, beads were stable for long periods.

Platinum group metals are best adsorbed onto the surface of alumina particles during impregnation, which limits the amount of metal used and minimizes costs.

Typical bead properties were in the ranges:

- Diameter   0.3–0.4 cm
- Pore volume   0.5–1.0 cm^3/g
- Bulk density   0.43– 0.67 kg/liter

11.2.2.2   *Monolith Catalysts*

The production of honeycomb monolith catalysts is relatively more complicated than the formation of bead catalysts. It requires several stages:

- Extrusion of the cordierite monolith followed by firing at a high temperature.
- Application of an alumina washcoat to the honeycomb surface again followed by firing at an appropriate temperature.
- Impregnation of the platinum group metals, together with other metal oxide promoters, onto the wash coat.

The most common support is a cylinder of cordierite ($2MgO.2Al_2O_3.5SiO_2$), extruded as a honeycomb, and covered with a washcoat of γ-, θ- and δ- alumina with a surface area of about 100 m^2g^{-1}. The components all have a similar thermal coefficient of expansion, and hence the composite is most able to withstand the heating and cooling cycles without disintegrating during use. Cordierite also has a bulk porosity in the range 20–40 %vol, mainly as macropores, so that good wash coat adhesion is achieved and the overall bulk density is low. Typical cor-

TABLE 11.10. Properties of Cordierite Monoliths.

Property	Values
Cell density	$400/in^2$ ($62/cm^2$)
Wall thickness (mm)	0.15
Channel size (mm)	0.10
Channel length (cm)	5–15
Open area (%)	75
Specific surface area (m^2/g)	2.8
Bulk density (kg/liter)	0.41
Maximum operating temperature (°C)	1200–1300

dierite honeycombs contain 400–600 channels per square inch of the face area and typical properties are given in Table 11.10.[32]

Metallic monoliths are very light and made up of alternating flat and corrugated foil plates about 0.05 mm thick in a cylindrical container. They present a large open area for gas to pass through the channels and have a high thermal conductivity. Metals are, however, non-porous and special alloys, containing

iron, chromium, aluminum (*Fecralite*), must be pre-treated to provide a suitable oxide layer which bonds with the alumina washcoat.

Ceramic or metal monoliths are packed into a stainless steel metal can with a ceramic or wire mesh mat as packing which acts as protection against mechanical vibration as well as providing insulation.

11.2.2.3 *Washcoat Composition*

The surface of the honeycomb is covered by the alumina washcoat, which can be stabilized against sintering at high temperature by the addition of more refractory materials such as barium oxide, calcium oxide, magnesium oxide or lanthanum oxide. A very thin surface layer of the washcoat, up to about 10-30μm, is applied although the thickness increases to about 150μm at the corners of the square channels.

The surface area of the washcoat can be varied within a range of 50–250 m^2/g of coating, depending on the platinum group metal loading and the stability required. The washcoat is applied as aqueous slurry that is then dried and calcined to provide the appropriate crystalline form and surface area. Some oxides are added as stabilisers, to prevent phase changes at the higher temperatures experienced in the exhaust, while others may also be added to improve the performance of the catalyst, as shown below:

- Cerium oxide acts as an oxygen *sink* which can absorb oxygen during lean operation to assist in oxidation when it is later released during periods of rich operation. Ceria will also stabilize the distribution of platinum during operation and promote the water gas shift reaction to produce hydrogen that improves the performance of the catalyst under rich conditions.

- Barium oxide has often been included in the washcoat because it can also absorb oxygen under lean conditions. The barium peroxide that formed also provides oxygen as it decomposes during rich operation.
- Zirconia is added to stabilize the ceria and to ensure than oxygen remains available from the surface oxides even if the ceria surface has been sintered at the high operating temperature of about 850°–900°C. Zirconia also plays an essential role in the *three-way* catalyst formulation as the support for rhodium.

Ceria and zirconia have been shown by electron microprobe spectroscopy to combine the preparation of the catalyst, forming a very thermally stable phase, and platinum group metals deposit preferentially on the alumina.[33] The washcoat has often been applied and fired with more alumina before the platinum group metals are impregnated. Alternatively, the washcoat and metals can be applied at the same time. In either case, the conditions for the deposition can be adjusted to provide a variable surface layer of alumina and the oxides, which absorb trace poisons and protect the active metals.

11.2.2.4   *Platinum Group Metal Catalysts*

When automobile emission regulations were first introduced, only platinum and palladium catalysts were used for the oxidation of carbon monoxide and residual hydrocarbons. Both metals were active oxidation catalysts although palladium was more temperature resistant than platinum but was more readily poisoned by sulfur. Rhodium became an important part of the *three-way* autocatalysts because it had high activity for the reduction of nitric oxide. Three-way catalysts were used in all cars from 1981 because of US Federal regulations although they had been introduced by California for 1978 model cars.

Although platinum and palladium have some activity for the reduction of NOX under stoichiometric conditions, the presence of oxygen in exhaust gas inhibits the conversion. Nitric oxide is, however, strongly adsorbed on the rhodium (III) surface to form nitrosyl groups.[34] These are reduced by adsorbed carbon monoxide and the nitrogen produced is desorbed at temperatures between 200°–300°C. Rhodium is less catalytically active than either platinum or palladium for the oxidation of carbon monoxide, because the adsorption of carbon monoxide by rhodium is inhibited by high concentrations of nitrogen atoms. Rhodium also has a relatively high water gas shift activity, does not sinter at high temperature and resists sulfur poisoning.

Some modifications to the operation of the car engine were made to resolve early difficulties in the optimisation of the removal of all three pollutants from the exhaust emission. Different combinations of the three catalysts were used in various locations behind the engine. The exact composition of the operating catalysts is not available in the literature, as this type of information is kept confidential within the business, but typical ranges are as follows:[32]

- Oxidation catalyst: platinum/palladium ratio of ~5:2 with 1.5 g of metals per liter of monolith volume.
- Three-way catalyst: platinum/rhodium ratio of 5:1 to 20:1 with 0.9–2.2 g of metal per liter of monolith volume.
- Three-way catalyst (containing palladium): metal contents of about 0.9–3.1 g Pt, 0–3.1 g Pd, and 0.15–0.5 g Rh per monolith.
- Palladium light-off catalyst: with 1.8–10.6 g of metal per liter of monolith volume.
- Three-way catalyst (all or most of the platinum replaced by palladium): platinum/palladium/rhodium ratio of 0–1:8–16:1 with 2–5.5 g of metal per liter of honeycomb volume.

The switch from platinum to palladium has been driven by the better availability and lower price of palladium, together with a higher level of combustion of the hydrocarbons. The improved performance of palladium has arisen because the amount of sulfur and lead impurities in the gasoline has decreased.

11.2.2.5   *Catalyst Poisons*

The performance of the catalyst is adversely affected by the presence of residual sulfur in the gasoline, and by tetraethyl lead, which used to be added to improve the octane rating of the gasoline. By the late 1990s, the use of lead additives had almost completely been discontinued in many parts of the world. Furthermore, the amount of sulfur in gasoline was also limited by environmental legislation. The sulfur components of gasoline, which were oxidised to sulfur dioxide in the engine and subsequently reduced to hydrogen sulfide in the exhaust, adversely affected the performance especially of palladium catalysts. Oil additives such as zinc dialkyl-dithiophosphates also have a poisoning effect on the catalyst due to the presence of zinc and the oxides of phosphorus in the exhaust gas. The total exposure of the catalyst to poisons[35] during a typical lifetime of 100,000 miles is shown in Table 11.11.

The addition of nickel oxide to a catalyst washcoat can minimize the formation of hydrogen sulfide during lean operation by absorbing some of the sulfur dioxide from exhaust gas. Nickel oxide can, therefore, store the sulfur dioxide as sulfate under reducing conditions and then release the sulfur dioxide under oxidizing conditions.

TABLE 11.11: Poisons in Catalysts Used Over 100,000 Miles.

Poison	Approximate poison content of fuel (g/kg catalyst)
Sulfur oxides from gasoline	6–20
Phosphates	0.08–0.2
Zinc (as dialkyl dithiophosphate)	0.32–0.48
Lead (as tetraethyl)	Depends on gasoline; ideally zero

11.2.3 Platinum Metal Group Availability

The platinum group metals are usually found as sulfides, arsenides or as the native metal, usually in conjunction with base metals. The concentration is almost always too low to justify mining for the precious metals alone, and the worldwide availability of the precious metal component tends to be determined by the demand for the other metal. For example, platinum is most commonly associated with nickel and copper sulfide deposits, and it is the extraction of the base metals from their ores that provides an economic route to the precious metal. In a typical operation in the US, 10 lbs copper can be extracted from a ton of ore, but the content of palladium is 0.000029 tr.oz and the platinum content is only 0.0000029 tr.oz. Significant quantities are only available in South Africa, Canada, Russia and the United States.

Spent honeycomb exhaust catalysts contain a relatively high concentration of the precious metal, and these can be recycled. It is certain that they will become an important and economic source of the metals as the use of automobiles continues to increase. About 15% platinum and 5% rhodium were recycled in 1990. The predicted demand for the metals to be used in autocatalysts, compared with the potentially available supplies, is shown in Table 11.12.[35]

11.2.4   Catalyst Operation

Three way auto-catalysts containing a high proportion of palladium were eventually preferred, because platinum was more expensive, potentially in short supply, and palladium is more active than platinum for hydrocarbon oxidation. Early palladium catalysts were easily deactivated by the sulfur and phosphorous impurities in the gasoline but new engines operating at higher temperature limited the adverse effect of these poisons. It is now common practice to replace the platinum with palladium completely in some catalysts although the loading of palladium is usually higher.

Different catalysts or combinations of catalysts can be used as outlined in the basic principles:

TABLE 11.12. Demand for Platinum Group Metals.

Year	Platinum		Palladium		Rhodium	
	World supply (tonne)	Catalyst use % total	World supply (tonne)	Catalyst use % total	World supply (tonne)	Catalyst use % total
1983	77	31	78	11	6	11
1988	107	34	104	7	10	85
1990	130	39	135	9	10	75
1993	144	39	131	10	12	85
1994	143	43	136	~10	12	~90
2000	137	24	224	70	–	–

- Palladium catalysts are very effective for the conversion of carbon monoxide and hydrocarbon residues, but the conversion of NO_x is low.
- Catalysts based on platinum and rhodium give good NO_x conversion, but an increased metal loading is required to give a satisfactory oxidation of the carbon monoxide and hydrocarbon components.
- Catalysts based on all three metals give an overall satisfactory performance, but conversion of NO_x is even higher when a binary platinum/rhodium catalyst is followed by a palladium catalyst.

Detailed information is never released by the catalyst supplier, as it is very difficult to guarantee adequate protection of proprietary information. Metal loadings vary depending on the engine size and the required performance. The catalyst monolith in a 1.8 liter car would have a volume of about 1.25 liters with 400 channels per square inch, or 60 channels per square centimeter. Up to 300 grams of an alumina washcoat containing 30% ceria and about 1% precious metals would be used, corresponding to about 0.1wt% of metal in the monolith. Some monoliths contain about 600 channels per square inch or 90 channels per square centimeter.

In common with all catalysts, there is a temperature, often referred to as the "light-off" temperature, below which the reaction does not proceed. As the temperature is increased above the light-off temperature, the reaction then proceeds with increasing rate. In the case of automobile exhaust catalysts, the temperature of the exhaust and the catalyst is too low for a period of about 30 seconds for the oxidation of exhaust hydrocarbon residues to take place. However, once the temperature reaches about 300°C, the catalyst becomes very active and the required reactions then take place. The composition of a typical exhaust gas under different conditions is shown in Table 11.13.[35]

Various modifications have been considered to decrease the time between starting the engine and reducing emissions, as shown below:

TABLE 11.13. Automobile Emissions at Increasing Engine Temperature.

	Cold	Warm	Normal
Temperature (°C)	0–250	250–300	500–900
Carbon monoxide (g/mile)	20–10	10–0.2	<1.0
Hydrocarbons (g/mile)	10 – <1	<1 – <0.1	<0.1
Nitrogen oxides (g/mile)	4 – <1	<1 – 0.1	0.1–0.2
Oxygen (%vol)	>1.0	<1.0	<0.5

- Move the catalyst closer to the engine. There is not much space to install a small catalyst bed in modern automobiles, and there are additional problems associated with the selection of a catalyst with sufficient thermal stability to operate at temperatures close to 1000°C. Catalysts derived from palladium may have the required thermal properties, but this approach is not favoured by the industry.
- Indirect electrical heating. This would certainly decrease the time before the catalyst became active, but the electrical load would be excessive, and the lifetime of the heater is likely to be too short to offer a viable solution.
- Use of a hydrocarbon adsorption trap. This approach has the potential to store unburnt hydrocarbons until the catalyst is sufficiently hot to enable oxidation of the hydrocarbon.

Compounds such as zeolites, perhaps containing a suitable metal promoter, can adsorb exhaust hydrocarbons that consist largely of olefin and aromatic compounds. The pore structure of many zeolites is broken down by the removal of alumina from the crystal lattice, particularly in the presence of steam at high temperatures. A practical solution to this problem would be to add a low alumina zeolite to the washcoat, prior to impregnation onto the monolith, and then to add an oxidation catalyst derived from palladium. Much of the hydrocarbon would be contained by the zeolite at low temperature and then oxidized over the palladium catalyst when the temperature became sufficiently high, both to desorb from the zeolite and to light off the catalyst.

A more direct way to avoid hydrocarbon escape during the heat up period has been the development of a very short monolith with about 400 channels per square centimeter (2500 channels per square inch) followed by a conventional monolith. The high gas velocity in the small channels allows faster heat and mass transfer so that the catalyst heats more rapidly to reaction temperature. Light-off is reported to take place within fifteen seconds of starting the engine.[35]

The use of a distillation device heated by the engine has also been suggested as a way of collecting some of the very lowest boiling fractions of gasoline. The low boiling fuel can be stored until the next cold start. The low-boiling components of gasoline are oxidised over the catalyst at a much lower temperature than the other higher-boiling components, and this approach[36] can lead to a reduction of up to 50% of the normal emissions during a cold-start.

11.2.5 Nitrogen Oxide Removal in Lean-Burn Engines

Lean-burn engines are more fuel-efficient when there is an excess of oxygen in the combustion gas and emit less carbon dioxide per mile travelled. Lean conditions lead to lower levels of NO_x in the exhaust gas, but the introduction of future legislation on the NO_x content of the exhaust will require that further reduction will have to be considered. NO_x removal in the presence of excess oxygen is more difficult than removal under stoichiometric conditions because normally the NO_x is reduced by some of the unburnt carbon monoxide and hydrocarbons in the exhaust.

A partial solution to the NO_X removal problem with lean-burn engines may be the use of a platinum/rhodium catalyst combined with a barium oxide *trap* that can absorb nitrogen dioxide as nitrate. The mechanism of the reaction is:

- nitric oxide is oxidized to nitrogen dioxide on a platinum site;
- the nitrogen dioxide forms barium nitrate on the trap before it can leave the catalytic converter;
- alternating rich operation for a few seconds every minute or so, releases the nitrogen dioxide which can then be reduced by carbon monoxide or hydrocarbon at a rhodium site.

The cycle then repeats during subsequent lean/rich operation.

The main problem with the barium oxide approach is that sulfur dioxide competes with NO_x for the basic sites, and is converted irreversibly to barium sulphate, which is quite inert. Thus, the active barium sites are quickly saturated, and the removal of NO_x from the emission is severely restricted. The sulfur content of the gasoline would need to be much lower for this approach to provide a long-term solution to the NO_x problem. There may also be problems associated with the thermal stability of the barium oxide traps during the many redox cycles, which the catalyst would be expected to experience over the lifetime of an autocatalyst.

Other procedures have been suggested for the treatment of the exhaust from lean-burn engines. One possibility is the absorption of NO_x on a suitable zeolite, followed by desorption and recycle of the NO_x back to the engine, where it would be reduced by some of the fuel in the combustion chambers. The problem with this approach is that most zeolites suffer from dealumination in the presence of steam at high temperature. An alternative approach could be the direct reduction of NO_x with hydrocarbons using a copper/ZSM-5 zeolite catalyst, but this has not yet been feasible, because the catalyst is deactivated at temperatures above 450°C. Tin oxide, supported on alumina, is also active for the reduction of NO_x by olefins in lean-burn exhaust gases. The olefin intermediate formed on the tin oxide surface can react with nitrogen oxide on an adjacent alumina site. Although such catalysts have not yet been used in autocatalysts, it is a further example of how autocatalysts, which already contain several components, must now be designed for specific duties.[37]

The use of lean-burn engines will therefore be limited until a viable solution to the NOX removal problem has been developed.

11.2.6 Diesel Engines

The efficiency of gasoline engines operating under lean-burn conditions has led to a resurgence of interest in diesel engines. The operation of the engine has been improved considerably by the use of better fuel injection systems, the use of turbo-chargers, and the recirculation of exhaust gases. These changes have led to a much lower concentration of pollutants in the exhaust, but there is still a need to reduce the concentrations of carbon monoxide, hydrocarbons and nitrogen oxides even further. The main problems have been caused by the presence of oxygen and sulfur dioxide in the exhaust gas from the diesel engine. It has not been possible to remove NO_x in the presence of oxygen to a satisfactory extent, and the sulfur dioxide is converted to sulphate on the washcoat, also contaminating any soot formed by partial combustion of the diesel fuel.

Since the 1980s, the main effort has been on the use of filters to remove up to 90% of the soot particles, which themselves contain a soluble organic fraction and some sulphuric acid. There is no regular regeneration of the filters, but as the filter takes up the soot particles, there is a gradual increase in pressure drop through the filter. This leads to an increase in the temperature of the filter, which in turn causes the soot to burn. High temperatures can be generated, and these can lead to damage to the filter.

Catalysts to remove carbon monoxide and hydrocarbons from exhaust gas have been used in Europe since 1991 with limited application to trucks in the US. Platinum or palladium have been used on ceramic monoliths with a special washcoat which minimizes sulfate formation. The main characteristics required by the catalysts are as follows:

- High carbon monoxide and hydrocarbon oxidation activity at a low engine temperature.
- Low sulfur dioxide oxidation activity and low reactivity with sulfur trioxide to form sulfate.
- Thermal stability at maximum engine temperature.

The concentration of the active metal usually lies in the range 0.35–1.76 g liter^{-1}. Smaller volumes of catalyst are required for diesel engines than for gasoline engines, and this results in a higher space velocity. Palladium catalysts have a lower activity for the oxidation of sulfur dioxide than platinum catalysts. Nevertheless, a lower metal content is used at the present time to avoid excessive oxidation of sulfur dioxide to the trioxide.

The activity of the catalyst is adversely affected by the deposition of soot particles. It has been shown that a filter that contains a platinum catalyst is active for the oxidation of nitric oxide to nitrogen dioxide at 195°C, when using low-sulfur fuel. The nitrogen dioxide then reacts with the soot at the same temperature with the formation of carbon dioxide and nitrogen, so that the filter is regenerated continuously, and the levels of NO_x are lowered.[38]

11.3   VOLATILE ORGANIC COMPOUNDS

The oxidation of trace amounts of volatile organic compounds (VOCs) by catalytic processes was used as early as the 1940s.[39] At that time, the main use of the procedure was to recover energy or remove unpleasant odors from waste gas streams. The US Clean Air Act (CAA) of 1970 led to a greater interest in the recovery and removal of VOC as oil prices increased and environmental standards became more stringent. As a result of the Clean Air Act, from 1990 an amendment required that a list of areas which did not comply with the National Ambient Air Quality Ozone Standard of 0.12 ppm over a one hour period was introduced.

A federal register was then issued for improvements to be made within a period of 3–20 years.[40] This meant that VOC removal from a wide range of process effluents became necessary. Of the 189 toxic compounds identified by the 1990 amendments, up to 154 can be removed by oxidation processes and more than 20,000 facilities in the US were reporting data to the EPA by 1994.

The main classification of VOCs include:

- Aliphatic and aromatic hydrocarbons;
- Organic oxygen and nitrogen compounds;
- Organic chlorine compounds.

A list of some of the main activities which could use catalytic oxidation processes, together with some of the VOCs emitted, is given in Table 11.14.[40]

Both thermal and catalytic oxidation processes have now been used for nearly thirty years and almost 100% removal efficiency can be achieved. Catalytic processes are usually felt to be more efficient and economical for the following reasons:

- lower operating temperature required;
- significantly lower fuel consumption;
- lower carbon dioxide emissions;
- much lower residence time and therefore a smaller reactor;
- lower capital and operating costs.

Oxidation catalysts can last for up to twenty years and are easily regenerated, if necessary, to remove coke formed by incomplete combustion or particles entering the bed with the gas stream.

TABLE 11.14. Sources of Volatile Organic Compounds.

Activity	Volatile Organic Compound
Chemical	Phthalic/maleic anhydrides
	Purified terephthalic acid
	Formaldehyde/methanol
	Ethylene oxide
	Cumene/acetone
	Polyethylene/polypropylene
	Acrylonitrile/acrylic acid
	Acetates/alcohols
Coating	Alcohols, ketones, aromatics, ethers
	Cyclohexanol, cellusolve
Bakeries	Ethanol
	Oils, fats, greases
Printing	Ink compounds, alcohols, glycols
Electronics	Ketones, cellusolve
Commercial	Coffee fumes
	Heavy oils, odours
Refining	Gasoline, volatile hydrocarbons

11.3.1 VOC Removal Processes

Early procedures to control VOC emissions included physical adsorption in beds of carbon which could be regenerated with steam or hot air to recover the organic impurity. Thermal combustion took place at temperatures in the range 800°–900°C. Catalytic oxidation procedures, however, had the advantage of operating at high conversion at much lower temperatures and higher space velocities. This meant that temperature control was easier and that the smaller reactor led to a reduction in the capital cost.[40]

Aliphatic compounds can be removed at temperatures between 200°–300°C while aromatic compounds oxidize at slightly higher temperature in the range 250°–400°C. An important benefit of catalytic oxidation is that the reaction tends to take place in the temperature range 200–350°C, compared with temperatures ranging between 600°C and 980°C for thermal oxidation. This has the advantage that any carbon monoxide that might be formed in the thermal process would easily be removed at very low temperature in the catalytic process. The temperature of the catalytic process is also too low for the formation of nitric oxide to occur by direct combination of the elements. Chlorinated compounds can also be oxidised at temperatures up to 450°C. A comparison of the differences between the operating temperatures for the oxidation of some VOCs in catalytic and thermal processes is shown in Table 11.15.[40,41]

TABLE 11.15. Catalytic and Thermal Oxidation Temperatures.

Organic Compound	Catalytic Temperature (^{0}C)	Thermal Temperature (^{0}C)
Benzene	200–300	800
Toluene	250–300	900
Xylene	250–300	–
Styrene	200–250	–
Methanol	200–250	–
Ethanol	200–250	–
Butanol	300–350	960
Acetone	200–250	–
Methyl ethyl ketone	300–350	980
Formaldehyde	150–200	–
Ethyl acetate	250–300	750
Carbon tetrachloride	300–350	780
Chlorobenzene	300–350	–
Carbon monoxide	<200	600
Acrylonitrile	250–300	–
Phthalic anhydride	250–300	–

When the concentration of hydrocarbons in the feed gas is greater than about 0.2%, it may be necessary to carry out the oxidation in tube-cooled converters, to provide adequate control over the temperature of the reaction.

Simple catalytic combustion units operate by preheating the contaminated gas with oxygen to the appropriate operating temperature before it passes directly to the catalyst bed. The VOCs should be less than 1% volume and the heat of combustion can be recovered by heat exchange with cold feed gas. It is usually recommended that the concentration of VOCs is less than 25% of the lower explosive limit.[40]

In a typical operation, 0.3–0.4% of mixed aromatic and oxygenated hydrocarbons are almost completely removed at a space velocity of 40,000 hrs^{-1} and an inlet temperature of 280°C. The increase in temperature is about 100°C per 0.1% hydrocarbon. When the concentration of hydrocarbons in the feed gas is greater than about 0.2%, it may be necessary to carry out the oxidation in tube-cooled converters, to provide adequate control over the temperature of the reaction. Some examples of operation with VOC oxidation catalysts are given in Table 11.16.

TABLE 11.16. Selected Operating Experiences with VOC Catalysts.

VOC	Catalyst Life	% Conversion Efficiency		
	Years	Initial	Pre-regen^n	Post-regen^n
Alcohols	>10	>95	70–90	>95
Cumene/acetone	> 5	>95	–	–
Phthalic anhydride	>17	95–98	70	92
Methanol/formaldehyde	>10	95–100	–	–
Toluene/xylene	> 5	>95	70	93

Operation at up to at least 100,000 Nm^3/hour.

11.3.2 VOC Oxidation Catalysts

The most active catalysts are derived from supported platinum group metals, or copper chromite when the VOC contains chlorine. The catalyst supports are produced as high surface area gamma-alumina spheres, extrudates or cylindrical pellets that can be used in dust free conditions. Alternatively, when treating large volumes of effluent gas, honeycomb monoliths made from cordierite or metal sheets are coated with a surface washcoat of the same alumina are particularly useful. The supports are selected to be stable at the operating temperature of the reaction.

The active component when treating hydrocarbons is usually high activity platinum or palladium, at a concentration of about 0.3-0.5wt%. In the presence of poisons or when oxidizing chlorinated hydrocarbons, catalysts such as copper chromite promoted with barium oxide, may be used although a higher temperature is required and the concentration of chlorine compound should be less than one volume percent. Details are given in Table 11.17.[42]

Other supports, such as ion exchange resins impregnated with chromia or titania can be used for the oxidation of halogenated hydrocarbons at temperatures around 250°C. Properly operated, precious metal catalysts can have lives exceeding fifteen years even if regenerated at intervals to remove carbon or dust deposits. Base metal catalysts can only operate for shorter periods of a few years before being replaced and can be deactivated at temperatures between 500°–700°C.

TABLE 11.17. Volatile Organic Compound Oxidation Catalysts.

Catalyst		
Platinum	Impregnated	0.3–0.5wt%
	Support	Alumina shapes or cordierite/metal monolith with alumina washcoat.
	Regeneration	Thermal at 400°–500°C or physical removal of dust.
	Life	Longer than 15 years in good conditions.
	Operation	250°–350°C
	Space velocity	40,000–100,000 h⁻¹ typical
	Conversion	Almost 100%
Copper chromite	Precipitated	
	Promoter	Barium oxide
	Life	~2 years
	Operation	Up to 450°C

REFERENCES

1. Nebel and Wright, *Environmental Science* (7th edition), Prentice Hall, NJ, 2000.
2. L R Raber, *Chem. Eng. News* (April 14 1997) 12.
3. EPA Toxic Release Inventory: reported *C & EN*, (April 3 1995) 4; *C & EN*, (July 15 1996) 29.
4. Worldwide Refining Survey, *Oil Gas Journal*, (Dec 18 2000).
5. Haldor Topsøe, *SNOX Process Leaflet*, 8.94; *Topsøe Topics* (July 1992).
6. Garten, Dalla Betta, Schlatter, *Catalytic Combustion*, Ed. by Ertl, p. 1668; Millard & Bowman, *Prog. Energy Combustion Science*, **15** (1989) 287.
7. Stambler et al, *Gas Turbine World*, (1993) 32.
8. Power Generation, *Hydrocarbon Processing*, (Sept 2000) 27.
9. Inui et al., *Catalysis Today*, **10** (1991) 1.
10. Spitznagel et al., *Environmental Catalysis*, Ed. by Armor, *ACS Symposium Series 552*, 250th National Meeting, Denver, Colorado, 1994, p. 172.
11. J N Armor, *Appl Catal* **1** (1992) 221; *Chem. Mat.* **6** (1994) 730; H. Bosch and F. J. Janssen, *Catal. Today* 2 (1987) 369; F. J. Janssen, *Catal Today* **16** (1993) 155.
12. *Norton Chemical Processes* (St. Gobain NorPro. Chem. Corpn.) Literature (1991); L. B. Sand, *Chem. Ind.* (1968) 71.
13. Heck et al., *Environmental Catalysis*, Ed. by Armor, ACS Symposium Series 552, 250th National Meeting, Denver, Colorado, 1994, p. 215.
14. Lowe and Ellison, *Environmental Catalysis*, Ed. by Armor, ACS Symposium Series 552, 250th National Meeting, Denver, Colorado, 1994, p. 190.
15. Janssen, *Environmental Catalysts – Stationary Sources*, in *Handbook of Heterogeneous Catalysis* Vol. 4, Ed. by G. Ertl, H. Knozinger and J. Weitkamp, VCH Weinheim, 1997, p. 1633.
16. G. Bond, *J. Cat*, **116** (1989) 531.
17. Haldor Topsøe, *WSA Process Leaflet*, 1.94.
18. Snyder et al., 85th Annual Meeting & Exhibition, Kansas City, Missouri (June 21-26, 1992) 2.
19. KTI/Shell DeNOX, *Gas Turbine World*, (July/Aug 1997).
20. J C Solt and J C Schlatter, Xonon Catalytic Process, *Oil Gas Journal*, (April 6, 1998) 76.
21. Furuya et al., *Proc* 2nd Int Workshop on Catalytic Combustion, Tokyo, Ed. by H. Arad, 1994, p. 162.
22. Rollbuhler, Paper A1AA-91-2463, 27th Jet Propulsion Conf., Sacramento, California, 1991.
23. M Machida, H Eguchi and H Arai, Garton, Dalla Betta, Schlatter, *Catalytic Combustion*, p. 1668; Avai, *Bull. Chem. Soc. Japan* **61** (1988) 3659.
24. Cohn, Steele, Anderson, USP 2975025 (1961).
25. CRI/KTI, Shell DeNOX System Literature (1999).

26. M Iwamoto in *Future Opportunities in Catalytic and Separation Technology*, Ed. by M. Misono, Y. Moro-oka and S. Krinura, Elsevier, Amsterdam, 1990, p. 121; *Environmental Catalysis*, Ed. by J. N. Armor, ACS Symposium Series 552,205[th] National Meeting, Denver, Colorado, 1994.
27. Manahan and Stanley, *General & Applied Chemistry,* 2nd Ed., William Grant Press, Boston, 1982, p. 381.
28. Haagen-Smit and Fox, *Ind. Eng. Chem.,* **48** (1956) 1484.
29. Acres and Cooper, *Plat. Metals Rev.,* **16**(3) (1972) 74; M. V. Twigg (Ibid), **43**(4) (1999) 168.
30. Summers et al., *Ind. Eng. Chem. Research & Development,* **11** (1972) 2; Barnes, *Advances in Chemistry Series 143,* Washington DC (1975).
31. Acres British Pat. 1390182 (1971); US Pat. 3951860 (1976).
32. M. V. Twigg, personal communication; E. S. J. Lox and B. T. Engler, in *Handbook of Heterogeneous Catalysis* 4, 1559, Ed by G. Ertl, Knozinger and Weitkamp, VCH Weinheim (1997).
33. M Jacoby, *C and EN* (Jan 25, 1999) 36.
34. J. M. Thomas and W. J. Thomas, *Principles and Practice of Heterogeneous Catalysis,* VCH Publishers Inc, NY and Weinheim, Germany, 1997, p. 578.
35. E S J Lox and B T Engler, Environmental Catalysis - Mobile Sources, p 1559 in *Handbook of Heterogeneous Catalysis* **4**, eds G Ertl, H Knozinger and J Weitkamp, VCH Weinheim (1997).
36. University of Texas, SW Research Inst., *New Scientist,* (Feb 3, 2001) 19.
37. H C Kung et al., *J. Catal.* **181** (1999) 1.
38. Hawker, *Platinum Metals Review,* **39** (1995) 2; Cook and Roth, *Plat. Metal Rev.* **39** (1995) 178; US Pat. 4902478.
39. Spivey, *Ind. Eng. Chem. Research,* **26** (1987) 2165; Jennings et al., *Catalytic Incineration for Control of VOC Emissions,* Noyes, N.Y., 1985.
40. *Catalytic Control of VOC Emissions,* Manufacturers of Emissions Control Association (MECA), 1707 L Street NW, Washington DC, Suite 5701, 1992.
41. Johnson Matthey and Haldor Topsøe catalyst literature.
42. *C & EN,* (Sept 7 1992) 34; Drago et al., *Environmental Catalysis*, Ed. by Armor, ACS Symposium Series 552, 250th National Meeting, Denver, Colorado, 1994, p. 341; Gervasini et al., *Environmental Catalysis*, Ed. by Armor ed., ACS Symposium Series 552, 250th National Meeting, Denver, Colorado, 1994, p. 353; Haldor Topsøe, *Catalytic Combustion Technology for Air Purification,* (10/96) and *Regenox Process* (9/90).

INDEX

A

Acetaldehyde
catalytica process, 303
Hoechst process, 303
palladium/copper catalysts,
303
Wacker process, 303
Acetic acid
acetaldehyde oxidation, 301
catalysts, 285, 294
methanol carbonylation, 302
operating conditions, 301
Acetone
Hock and Lang process, 266
isopropanol, 265
Acetylene, selective hydrogenation,
102–114
Acid catalysts
acid resins, 265
acid sites, 221
phosphoric acid/silica, 267
silica alumina, 267
zeolites, 267
Acid sites. *See* Fluid catalytic
cracking (FCC); Zeolite catalysts
Acrolein/acrylonitrile
acrylonitrile, 156–157
bismuth phosphomolybdate,
157, 158
catalyst life
structure, 158, 161
fluid bed process, 157
multi component catalyst, 157
propylene oxidation, 157
reaction mechanism, 159–161
Sohio process, 157–159, 161

Acrylic acid, mixed oxide catalysts,
161–162
Acrylonitrile, 4, 155, 157–160, 264,
466, 467
Active sites, 6, 7, 12, 48, 159–161,
189–190, 221, 225, 251, 252,
315, 320, 325–327, 339, 341,
344, 412, 418, 431, 432, 447,
448
Activity testing, 15–18
Adkins catalyst, 12
Air/fuel ratio
basic catalyst systems, 454
catalyst cycles, 453
exhaust gas composition, 453, 454
impurity removal, 454
photochemical smog, 452
precious metal catalysts, 452
three way catalysts, 453, 454
λ value, 453
Alkylation
liquid acid processes, 219
operating conditions, 220
production in USA, 218
reaction mechanism, 219–220
solid acid processes, 221
Alumina, 3, 45, 78, 142, 170, 215,
270, 313, 360, 397, 456
Ammonia oxidation
catalyst operation, 128–130
Kuhlmann discovery, 3, 120
modern process, 51
non platinum catalysts, 120
Ostwald process, 121
platinum catalysts, 120
platinum recovery, 131

L. Lloyd, *Handbook of Industrial Catalysts*, Fundamental and Applied Catalysis,
DOI 10.1007/978-0-387-49962-8, © Springer Science+Business Media, LLC 2011